COMPACT MODELS FOR INTEGRATED CIRCUIT DESIGN

COMPACT MODELS FOR INTEGRATED CIRCUIT DESIGN

CONVENTIONAL TRANSISTORS AND BEYOND

SAMAR K. SAHA

CRC Press
Taylor & Francis Group

CRC Press
Taylor & Francis Group
6000 Broken Sound Parkway NW, Suite 300
Boca Raton, FL 33487-2742

First issued in paperback 2017

ISBN-13: 978-1-4822-4066-5 (hbk)
ISBN-13: 978-1-138-82740-0 (pbk)

Visit the Taylor & Francis Web site at
http://www.taylorandfrancis.com

and the CRC Press Web site at
http://www.crcpress.com

In loving memory of my parents,

Mahamaya and Phani Bhusan Saha

Contents

Preface

Silicon integrated circuits (ICs) have ushered in an unprecedented revolution in many areas of today's society, including communications, medicine, military, security, and entertainment. This dramatic impact of ICs on society is due to the continuous miniaturization of metal-oxide-semiconductor (MOS) field-effect-transistor (FET) devices toward their ultimate dimensions of approximately 5 nm, thereby providing low-cost, high-density, fast, and low-power ICs. Our ability to fabricate billions of individual components on a silicon chip of a few centimeters squared has enabled the information age. However, with increase in the device densities in ICs, the complexities of IC design have increased significantly. Designing such complex IC chips is virtually impossible without computer-aided design (CAD) tools that help predict circuit behavior prior to manufacturing. However, the accuracy of CAD for ICs depends on the accuracy of the models, referred to as "compact models," of the active and passive elements used in the circuit. These compact models for circuit CAD have been the basic requirement for the analysis and design of ICs and are playing an ever-increasing role as the mainstream MOSFETs approach their fundamental scaling limit. Therefore, for efficient IC design using nanoscale devices, a detailed understanding of compact models for circuit CAD is crucial.

A large number of research articles as well as books are available on modeling nanoscale devices. Most of the published works on compact models for IC design CAD are extended user manuals of any industry standard compact MOS model and some are a collection of articles from contributed authors. Thus, the available books do not provide adequate background knowledge of compact models for beginners in industry as well as classroom teachers. In addition, the available titles on compact models do not deal with the major issue of process variability, which severely impacts device and circuit performance in advanced technologies and requires statistical compact models. Again, though the CMOS technology continues to be the pervasive technology of ICs, bipolar-junction transistors (BJTs) are an important element of IC chips. However, most of the compact modeling books do not discuss BJTs or BJT modeling for circuit CAD. Thus, a new treatise on compact modeling is crucial to address current modeling issues and understand new models for emerging devices.

With over 25 years in the field of semiconductor processes, device, and circuit CAD in industry and over 10 years in the teaching of compact modeling courses in academia, I felt the need for a comprehensive book that presents MOSFET, BJT, and statistical models and methodologies for IC design CAD. This book fulfills that need. Starting from basic semiconductor physics, this book presents advanced industry standard models for BJTs, MOSFETs,

FinFETs, and TFETs along with statistical MOS models. Thus, this book is useful to beginners as well as experts in the field of microelectronics devices and design engineering.

This book is intended for the senior undergraduate and graduate courses in electrical and electronics engineering programs and researchers and practitioners working in the area of electron devices. However, the presentation of the materials is such that even an undergraduate student not familiar with semiconductor physics can understand the basic concepts of compact modeling. A limited number of exercise problems are included at the end of each chapter, a feature that would help use of this book as a text for teaching at the senior undergraduate and graduate level courses in academia.

Chapter 1 provides an overview of compact transistor and interconnection models, a brief history of compact MOSFET models, and the motivation for compact models for very-large-scale-integrated (VLSI) circuit CAD. Chapter 2 reviews of basic semiconductor physics and *pn*-junction operations.

Chapter 3 presents MOS capacitor systems and the basic theory of two terminal devices. This chapter provides the background for developing four terminal MOSFET compact models for VLSI circuit CAD.

Chapter 4 describes the basic theory of long channel MOSFETs, including the Pao-Sah model, the charge-sheet model, and earlier generations of compact models. Chapter 5 provides detailed mathematical steps to derive the industry standard Berkeley Short Channel Insulated-Gate MOSFET version 4 (BSIM4) compact model. Chapter 5 also presents the parasitic models associated with MOSFET devices, including source/drain diode compact models. Chapter 6 presents the dynamic behavior and compact MOSFET intrinsic capacitance model. Chapter 7 describes the compact MOSFET modeling techniques for noise and radio-frequency circuit CAD.

Chapter 8 is dedicated to compact models for process variability analysis. This chapter describes the sources of variability, circuit model for process variability, and formulation of statistical models for variability-aware VLSI circuit design. This chapter also presents the techniques for mitigating the risk of process variability in advanced nanoscale VLSI circuits by novel device and process architectures.

Chapter 9 describes the basic theory and compact model for multi-gate transistors FinFETs and UTB-SOI MOSFETs, along with model parameter extraction procedures. Chapter 10 introduces compact models beyond CMOS devices including TFET.

Chapter 11 presents BJT compact models. Similar to Chapters 4 and 5, in Chapter 11, the industry standard BJT models have been derived from basic semiconductor theory and first generation models for easy understanding by beginners while retaining the rigor for the experts in the field.

Chapter 12 includes examples of compact model libraries for industry standard circuit simulation tools, calling the model in the circuit simulation net list (input file), and circuit simulation techniques to use the generated models.

An extensive set of references is provided at the end of this book to help the readers identify the evolution and development of compact models for VLSI circuit design and analysis.

Samar K. Saha
Santa Clara University, California

Author

Samar K. Saha received his PhD in physics from Gauhati University, Guwahati, India, and an MS degree in engineering management from Stanford University, Stanford, California. Currently, he is an adjunct professor in the electrical engineering department at Santa Clara University, Santa Clara, California, and a technical advisor at Ultrasolar Technology, Santa Clara, California. Since 1984, he has worked at various positions for National Semiconductor, LSI Logic, Texas Instruments, Philips Semiconductors, Silicon Storage Technology, Synopsys, DSM Solutions, Silterra USA, and SuVolta. He has also worked as a faculty member in the electrical engineering departments at Southern Illinois University at Carbondale, Illinois; Auburn University, Auburn, Alabama; the University of Nevada at Las Vegas, Nevada; and the University of Colorado at Colorado Springs, Colorado. He has authored more than 100 research papers, 1 book chapter on technology CAD (TCAD), and holds 10 U.S. patents. His research interests include nanoscale device and process architecture, TCAD, compact modeling, devices for renewable energy, and TCAD and R&D management.

Dr. Saha is the 2016–2017 president of the IEEE Electron Devices Society (EDS). He is a fellow of the Institution of Engineering and Technology, London, UK, and a distinguished lecturer of IEEE EDS. He has served as the vice president of EDS Publications; an elected member of the EDS Board of Governors; editor-in-chief of IEEE *QuestEDS*; chair of EDS George Smith and Paul Rappaport awards; editor of the Region-5&6 *EDS Newsletter*, chair of the EDS Compact Modeling Technical Committee, chair of the EDS North America West Subcommittee for Regions/Chapters; a member of the IEEE Conference Publications Committee; a member of the IEEE TAB Periodicals Committee; and the treasurer, vice chair, and chair of the Santa Clara Valley EDS chapter.

Dr. Saha has served as the head guest editor for the *IEEE Transactions on Electron Devices* (T-ED) special issues on *advanced compact models and 45-nm modeling challenges* and *compact interconnect models for giga scale integration*, and as a guest editor for the T-ED special issue on *advanced modeling of power devices and their applications*. He has also served as a member of the editorial board of the *World Journal of Condensed Matter Physics*, published by Scientific Research Publishing.

1

Introduction to Compact Models

1.1 Compact Models for Circuit Simulation

Compact models of a circuit element are simple mathematical descriptions of the behavior of that circuit element, which are used for computer-aided design (CAD) and analysis of integrated circuits (ICs). Compact models describe the device characteristics of a manufacturing technology by a set of physics-based analytical expressions with technology-dependent device model parameters that are solved by a circuit simulator for circuit analysis during IC design. *Compact modeling* refers to the art of generating compact models of an IC process technology by extracting elemental model parameters for accurate prediction of the behavior of the circuit elements of that technology in circuit simulation. In reality, the complete compact models include the modeling of each circuit element along with its parasitic components that run robustly for realistic assessment of the representative IC technology in circuit CAD [1,2].

Compact models of the circuit elements of an IC manufacturing technology have been the major part of electronic design automation (EDA) tools for circuit CAD since the invention of ICs in the year 1958 [3] and are playing an increasingly important role in the nanometer-scale system-on-chip design era. Today, compact models are the most important part of the process design kit [4,5], which is the interface between circuit designers and device technology. As the mainstream complementary metal-oxide-semiconductor (CMOS) technology is scaled down to the nanometer regime, a truly physical and predictive compact model for circuit CAD that covers geometry, bias, temperature, DC, AC, radio frequency (RF), and noise characteristics has become a major challenge for model developers and circuit designers [1]. A good compact model has to accurately capture all real-device effects and simultaneously produce them in a form suitable for maintaining high computational efficiency.

In the microelectronics industry, compact modeling includes (1) compact device models of the active devices such as bipolar junction transistors (BJTs) and metal-oxide-semiconductor field-effect transistors (MOSFETs) along with the parasitic elements of the active devices; and (2) compact interconnect models of the resistors, capacitors, and inductors of the metallization layers connecting the active devices in the ICs.

1.1.1 Compact Device Models

Compact device models describe the terminal behavior of a device in terms of the current-voltage (*I–V*), capacitance-voltage (*C–V*), and the carrier transport processes within the device. Figure 1.1 shows the basic features of a typical compact device model of a representative IC technology. As shown in Figure 1.1, a compact model is made of a *core model* along with the various models to account for the effects of the geometry and physical phenomena in the device. For a metal-oxide-semiconductor (MOS) transistor, the core model describes *I–V* and *C–V* behavior of an ideal large MOSFET device [4] of a target technology. The core model represents about 20% of the model code in terms of both execution time and the number of lines in the code. The rest of the model code comprises multiple models that describe the numerous real-device effects that are responsible for the accuracy of the compact

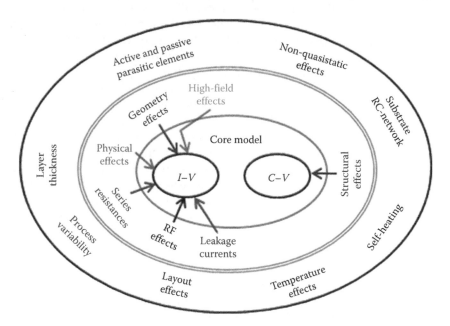

FIGURE 1.1
A typical composition of compact models of an IC technology: the core model includes the basic *I–V* and *C–V* behavior of a large geometry device in the inner circle; the core model is accompanied by the models for physical phenomena within the device and geometry and structural effects as shown in the middle circle; the final compact model with the geometrical and physical effects includes the external phenomena such as ambient temperature, layout effects, process variability, and NQS effects as shown in the outer circle of the model.

model. For MOSFET devices, device phenomena accompanying the core model include short-channel effects (SCEs), output conductance, quantum mechanical effects (QMEs), nonuniform doping effects, gate leakage current, band-to-band tunneling, noise, non-quasistatic (NQS) effect, intrinsic input resistance, and strain effect [4,6].

The compact model for circuit CAD is the bridge between the circuit design and processing groups and is a module of the extended technology CAD (TCAD) environment [7]. In the extended TCAD environment, the compact model plays an important role in developing next generation IC fabrication technology and assesses the manufacturability of IC fabrication processes by reverse modeling [7,8].

1.1.2 Compact Interconnect Models

Today's very-large-scale-integrated (VLSI) circuits consist of MOSFET devices and their interconnections, referred to as *interconnects*. In a typical VLSI chip, the active area is about 10% whereas the physical area is occupied by interconnect and isolation regions 6–10 times the active device area [9]. For this reason, the role of the interconnect is becoming increasingly important as the feature size is scaled down to decananometer regimes and the device density is increased on the chip. As VLSI technology shrinks below 22-nm geometries with Cu/low-k interconnections, parasitics due to interconnections are becoming a limiting factor in determining circuit performance. Therefore, accurate modeling of interconnect parasitic resistance (R), capacitance (C), and inductance (L) is essential in determining various on-chip interconnect-related issues, such as delay, cross talk, energy losses in R due to the current (I) flow or IR drop, and power dissipation. Accurate compact interconnect models are crucial for the design and optimization of advanced VLSI circuits for 22-nm CMOS technology and beyond. In addition, with the emergence of technologies such as carbon nanotubes and graphene nanoribbons, compact interconnection models that are suitable for these technologies are crucial for advanced circuit design. Currently available interconnect models, which are based on field solvers, are inadequate for accurate and meaningful analyses of today's chips, which house millions of devices. Interconnect models can accurately simulate on-chip global interconnections and speed-power optimization for advanced interconnect technologies. Modeling of these interconnect properties is thus important and must be included by the designer when checking circuit performance in circuit CAD. Though interconnect models are an essential part of optimizing VLSI circuit performance, interconnect modeling is outside the scope of this book; interested readers may refer Saha et al. [10] for recent development of interconnect models. In this treatise, compact modeling of field-effect transistors (FETs) and their parasitic components that are used in the mainstream VLSI circuit design are described.

1.2 Brief History of Compact Device Modeling

Since the 1960s, compact models for circuit CAD have continuously evolved [6]. After the invention of the bipolar transistor in 1947 [11,12], complete circuits including both active and passive devices were realized on monolithic silicon substrates by late 1950s. Computer simulation evolved as a practical way to predict circuit performance including nonlinearities because digital computers were capable of complex circuit analysis based on a network or matrix formulation. The 1950s and 1960s were dominated by BJT technology; the Ebers–Moll (EM) model has been the major large-signal compact model for bipolar transistors since its formulation in 1954 [13]. It is based directly on device physics and covers all operating regimes, that is, active, saturation, and cut-off operations of BJTs. However, various approximations limit the accuracy of the model. To overcome the limitation of the EM model, Gummel and Poon reported a BJT model based on *integrated charge control relations*, in 1970 [14]. The Gummel–Poon (GP) model offers a very clear and standardized description of existing physical effects in BJTs. Due to its simple yet physical model formulation, GP model remains the most popular BJT model till date. By the early 1970s, the circuit simulator had become a useful tool, essentially replacing the breadboarding of prototypes. The circuit CAD tool, *Simulation Program with Integrated Circuit Emphasis* (SPICE) from the University of California, Berkeley, became a widely used tool among the circuit design community [15]. Thus, with the introduction of SPICE, the compact model has become essential for circuit CAD. Meanwhile, the IC industry had reached an important juncture in its development. While the 1950s and 1960s were dominated by BJT technology, the 1970s saw MOS technology begin to overtake BJT technology in terms of functional complexity and level of integration. Thus, from simple basic compact MOSFET models, sophisticated models for FETs started to emerge. Today's sophisticated compact models for MOSFETs [4,16–20] evolved from models first developed 30 to 50 years ago [13,14,21–24]. A large number of developers have contributed to the evolution of compact modeling. In this section, we present only a brief history of the major development in the compact MOSFET modeling activities.

1.2.1 Early History of Compact MOSFET Modeling

In the early 1960s, MOSFET devices were introduced in fabricating ICs [25]. In order to understand the behavior of these emerging MOSFET devices, research effort on the development of semi-analytical models using simple device structures and simplified device physics started in the 1960s [21,26]. In 1964, Ihantola and Moll reported the design theory of MOSFET devices and developed the drain current (I_{ds}) equation to account for the varying bulk charge effect in the devices [21]. In the same year, Sah reported a simple theory of the MOSFET devices using valid approximations and simple assumptions and derived I_{ds} equations for circuit analysis [26]. In these models, the device is considered to be turned on above a certain applied input voltage, referred to as the *threshold voltage* (V_{th}), and turned off at the input bias below V_{th}. This approach is known as *threshold voltage–based* or, V_{th}-*based compact modeling*.

With the great potential of MOSFET devices in ICs during 1960s, a detailed understanding of MOSFET device physics became critical. In 1966, Pao and Sah [22] reported an I_{ds} equation to describe MOSFET device characteristics under varying biasing conditions in terms of a physical parameter called the *surface potential* (ϕ_s), where ϕ_s describes the mode of operation of MOSFET devices under the applied biasing conditions. It is to be noted that V_{th} is defined at a particular value of ϕ_s above which the device starts conducting whereas ϕ_s defines the entire range of operation of MOSFETs from *off-state* to *on-state*, depending on the applied biasing conditions. The value of ϕ_s is calculated, iteratively, from an implicit expression derived from Poisson's equation and Gauss's law. This I_{ds} model is a double integral equation, commonly known as the Pao-Sah model, that can only be solved numerically. Inherently, it takes into account both the drift and diffusion components of I_{ds}, and is valid in all regions of device operation: from the subthreshold (below V_{th}) to strong inversion region (above V_{th}). This method is now known as *surface potential*–based or, ϕ_s-based compact modeling. Sah's ϕ_s-based modeling requires iterations and integration and is computationally demanding for circuit CAD. Thus, the Pao-Sah model is inefficient for circuit CAD due to its complexities involving integration and iterations to get I_{ds} at each value of applied voltage. Thus, the search for simplified models for circuit CAD began in the late 1960s.

In the late 1970s, SPICE emerged as an essential circuit CAD tool to perform accurate and efficient design and analysis of ICs under the EDA environment [27]. In order to use SPICE, accurate and efficient compact models are required to describe the behavior of the devices used in the circuits. Thus, the explicit development of MOSFET compact models for circuit CAD started with the widespread usage of SPICE and continues today as the mainstream MOSFET devices rapidly approach their fundamental scaling limit near the 10-nm regime [1,28–33].

The first approach used in developing I_{ds} model is to circumvent the iterative computation of ϕ_s from the implicit relation [22] using V_{th} as the boundary between the off-state or weakly conducting state, referred to as the *weak inversion region*, and on-state, called the *strong inversion region*, of MOSFET devices, that is, use V_{th}-based compact modeling. This approach results in two current equations, one for the weak inversion and the other for strong inversion [25,34]. In V_{th}-based modeling, a linear approximation is made between ϕ_s and the applied input voltage to eliminate ϕ_s and relate the input voltage to the output current I_{ds}. This approach results in a simple *I–V* equation in the parabolic form and was first used for circuit simulation in 1968 [34]. This is the first known compact MOSFET model for circuit CAD and is referred to as the *Schichmann and Hodges* model. This model is implemented in SPICE as the MOS Level 1 model and is developed based on a number of simplifying assumptions and device physics appropriate for uniformly doped long-channel MOSFET devices. In addition, in MOS Level 1 model, the value of I_{ds} is zero below V_{th}, increases linearly above V_{th}, and remains constant above a drain saturation voltage (V_{dsat}). The MOS Level 1 model, though inaccurate, is widely used for hand calculation of I–V data and preliminary circuit simulation because of its simplicity and ease of use.

In order to account for the shortcomings of MOS Level 1 model such as small geometry effects, Ihantola and Moll [21] modified the device equation to use in SPICE as the MOS Level 2 model. The basic approach is to begin with the Level 1 model, and add equations and parameters to include the small geometry effects as corrections to the basic model. Unlike the Level 1 model, it is assumed that the depletion charge varies along the length of the channel; this results in a complex but more accurate expression for I_{ds} in SPICE Level 2 MOS model [35]. However, it is still not accurate for devices with submicron geometries.

In 1974, MOSFET scaling rule was established [36], and the MOSFET device and technology continued to evolve. As a result, the MOS device physics became complex, circuit density increased, and the device models were continually updated to account for emerging physics in scaled MOSFETs. The result is the evolution of MOS compact models. In 1978, Brews [23] reported a simplified model based on charge sheet approximation of the inversion charge density (Q_i) along with depletion approximation. With justified assumptions of Q_i, the total I_{ds} is shown to be the sum of the drift (I_{ds1}) and diffusion components (I_{ds2}). The values of ϕ_s at the source end (ϕ_{s0}) and drain end (ϕ_{sL}) of the devices required to calculate I_{ds} are obtained numerically by solving the implicit equation for ϕ_s along the channel at each applied biasing condition. In weak inversion, where ϕ_{s0} is almost equal to ϕ_{sL}, even a small error in the values of ϕ_{s0} and ϕ_{sL} can lead to a large error in the current I_{ds2}, which depends on the value of ($\phi_{sL}-\phi_{s0}$) [23]. Therefore, an accurate solution is required for the surface potential, particularly for weak inversion current calculations. There are several iterative schemes developed to solve the implicit equation for ϕ_s [37]. However, the available iterative schemes to solve this equation were relatively slow and did not include all regions of device operation while noniterative approximations did not extend to the accumulation region and were not sufficiently accurate, especially for computing the transcapacitances. Besides, the early ϕ_s-based models [23,37] consist of complex and lengthy expressions for currents, charges, and noise [38]. Thus, due to the complexity of the ϕ_s expression along with the lack of efficient techniques to compute ϕ_s, these models [23,37] were computationally challenging for circuit simulation in the early days of EDA environment. Therefore, search for different approaches continued to simplify the model for efficient solution of the model equations for circuit CAD in EDA environment.

In 1981, Level 3 MOS model was introduced for circuit CAD using SPICE2 [39]. Level 3 MOS model introduced many empirical parameters to model SCEs. However, the accuracy and scalability of the model for simulation of a wide range of channel length and width using one set of model parameters are not entirely satisfactory to the circuit designers. The short channel and narrow width effects are not modeled accurately in the MOS Level 1, 2, and 3 models and high field effects are not considered properly because of the limited understanding of the physics of small geometry devices at the time these models were developed. Thus, to keep parity with the continuous scaling down of MOSFETs, global effort continued for the development of accurate and efficient compact models for circuit CAD.

1.2.2 Recent History of Compact MOSFET Modeling

As the CMOS technology became the pervasive technology of ICs in 1970s, the complexities of MOSFET devices continued to increase. As a result, compact models based on simplified device physics became inadequate to analyze scaled geometry MOSFETs. The efforts for accurate and computationally efficient models continued using different approaches. The major modeling techniques used can be described as *threshold voltage-based, surface potential–based,* and *charge-based* as described in Sections 1.2.2.1 through 1.2.2.3.

1.2.2.1 Threshold Voltage–Based Compact MOSFET Modeling

The major development of V_{th}-based compact MOS model is the development of *Berkeley Short Channel IGFET Model,* commonly known as BSIM, in the year 1987 [24]. It incorporated some improved understanding of the SCEs and worked well for devices with channel length of 1 μm and above. However, it also introduced several empirical fitting parameters just to enhance the scalability of the model. Even then, the model scalability was not totally satisfactory. Also, circuit designers did not like the use of many fitting parameters, which do not have any physical meaning.

In order to address the shortcomings of the first generation of BSIM or BSIM1, BSIM2 was introduced in 1990 [40]. BSIM2 improved upon BSIM1 in several aspects such as model continuity, output conductance, and subthreshold current [40]. However, the model still could not use one set of parameters for wide range of device sizes. Users typically need to generate a few or many sets of model parameters, each covering a limited range of device geometries in order to obtain good accuracy over the full range of devices used in circuits. This makes the parameter extraction difficult. Also, it is difficult to use these parameters to perform statistical modeling or extrapolation of the model parameters from the present technology to a future one.

In early 1990s, the proprietary compact model, HSPICE Level 28, was released from Meta-Software to address the shortcomings of BSIM1 [41,42]; where 'H' in HSPICE abbreviates the initial of the family name, "Hailey' of the developers of the industrial SPICE circuit CAD and founders of the company Meta-Software. The widespread use of Meta-Software's circuit CAD tool, HSPICE, served as the vehicle for the Level 28 model, helping Level 28 to become the most widely used MOSFET model in the semiconductor industry. The HSPICE Level 28 model is based on BSIM, without many of BSIM's intrinsic shortcomings; it also has accurate capabilities for modeling both the analog and digital circuits in contrast to BSIM that has been mainly developed for modeling digital circuits.

In 1994, BSIM3 [43] was developed to account for the shortcomings of BSIM2. The device theory has been developed over a number of years [43–46]. The model explicitly takes into account the effects of many device sizes and process variables for good model scalability and predictability. The short channel and narrow width effects as well as high field effects are well modeled. The first released version of BSIM3, BSIM3v2 [43], offered better model accuracy and scalability than the previous BSIM models but it still suffers from discontinuity problems such as negative conductance and glitches in the g_m/I_{ds} versus V_g plot at the boundary between weak inversion and strong inversion; where g_m is the device transconductance. In the meantime, the need for a good open MOSFET model had been widely recognized by the semiconductor companies. To eliminate all the kinks and glitches in BSIM3v2, BSIM3v3 with a single-equation approach along with the enhanced modeling of small size and other physical effects [44–47] was developed. The BSIM3v3.0 model has been extensively verified and selected as the first industry standard compact MOSFET model in 1996 by Compact Modeling Council (CMC) [48]. The convergence performance of BSIM3v3.0 was enhanced in BSIM3v3.1 [45]. Version BSIM3v3.2 [47] introduced a new charge/capacitance model that accounts for the QM effect, and improves V_{th} model, substrate current model, NQS model, and others and was released in 1998 and 2005 [49–51].

During 1990s, Philips Laboratories started developing MOS Model 9 [52,53] and released the model in 1994 [54], making it widely available in mainstream circuit CAD tools. The basic features of MOS 9 include very clean and simple model equations, use of well-behaved hyperbolic expressions as smoothing functions for good behavior in circuit simulation, and less number of model parameters. The smoothing functions in MOS 9 serve continuous and smooth equations across the various transition points (such as V_{dsat}) of MOSFET operation and allow the realization of a single-model equation (e.g., I_{ds} equation) valid in all regions of device operation. Finally, MOS 9 includes some of the features of HSPICE Level 28, thus accommodating proper model binning. Unlike BSIM3, MOS 9 retains the existing approach in describing the geometry dependence of the model characteristics. While the basic method of the existing modeling know-how is used, the method is extensively modified to improve the circuit simulation results.

In the meanwhile, BSIM has been continuously updated and extended to accurately model the physical effects observed in sub-100 nm regime. In 2000, BSIM4, version BSIM4.1.0, was released [55]. BSIM4 offers several improvements over BSIM3, including the traditional *I–V* modeling of intrinsic transistor, the transistor's noise modeling, and the incorporation of extrinsic parasitics. Some of the salient features of BSIM4 are an accurate model of the intrinsic input resistance for RF, high-frequency analog and high-speed digital applications, flexible substrate resistance network for RF modeling, an accurate channel thermal noise model along with a noise partition model for the induced gate noise, an NQS model consistent with the gate resistance-based RF model, an accurate gate direct tunneling model, a geometry-dependent parasitics model for various source-drain connections and multifinger devices, improved model for steep vertical retrograde doping profiles, better model for halo-implanted devices in V_{th}, bulk charge effect model, and output resistance, asymmetrical and bias-dependent source-drain resistance, QM charge-layer model for both *I–V* and *C–V*, gate-induced drain/source leakage (GIDL/GISL) current model, and improved unified $1/f$ noise model [55–57].

1.2.2.2 Surface Potential–Based Compact MOSFET Modeling

In the surface potential–based modeling approach [23,37], ϕ_s is solved at the two ends of the MOS channel. The terminal charges, currents, and derivatives are then calculated from ϕ_s. During 1980s, a considerable progress has been made to solve ϕ_s efficiently from the implicit ϕ_s equation. In 1985, Bagheri and Tsividis reported an efficient algorithm [58] to solve these implicit ϕ_s equations using Schroder series method [59,60], which is based on Taylor series expansion of the inverse function, provided a good initial guess such as the *zero-order* relationship [61] is used. It is reported that at most only two iterations are required to achieve an excellent estimation of ϕ_{s0} or ϕ_{sL} in all operating regions.

In 1994, Arora et al. reported an efficient ϕ_s-based MOSFET model referred to as the "PCIM" for in-house circuit simulation of Digital Equipment Corporation's (DEC) Alpha chip [62]. Based on the source-side-only surface potential proposed by Park [63], Rios et al. in 1995 reported a model that is shown to be practical and efficient and used it in DEC's Alpha chip design from 1996, featuring automatic and physical transitions between partially and fully depleted modes of Silicon-on-Insulator (SOI) operations [64,65]. The source-side-only solution was used to offer a good compromise between the accuracy and simplicity, and the solution speed required for practical applications. This approach was shown to avoid solving for ϕ_s on the drain side, while providing a simple and self-consistent treatment of carrier velocity saturation. In addition, the appropriate treatment of the body charge linearization and the effective drain bias was used to maintain source-drain symmetry. The solution method preserves source-drain symmetry and produces the correct drain current behavior near drain voltage, $V_{ds} = 0$. It was reported that in the source-side-only approach, simple, explicit, and self-consistent V_{dsat}

solutions are possible by equating the saturation drain current to the model drain current equation, at $V_{ds} = V_{dsat}$. The velocity–field relation requires special treatment to be able to include the effect of longitudinal field-dependent mobility in the integration of the continuity equation. A good approximation was proposed by Arora et al. [62]. The small geometry effect and different physical effects including QM and polysilicon depletion effects are implemented in the CAD-oriented analytical MOSFET model [61]. QMEs on the inversion charge density can be handled in a physical manner by a bandgap-widening approach [65].

The development of ϕ_s-based *Hiroshima University STARC IGFET Model*, referred to as the HiSIM, has been started in the early 1990s based on the drift-diffusion concept and proved its feasibility for real applications [66–68]. Since 1993, the model has been successfully applied in the development of dynamic random-access memory (DRAM), subthreshold region ICs, and IC-card products at Siemens. In HiSIM, the surface potentials are obtained by solving the Poisson's equation iteratively both at the source side and at the drain side with an accuracy of 10 pV, and simulation speed is comparable to industry standard V_{th}-based models [66]. The reported accuracy is absolutely necessary for maintaining sufficient accurate solutions for transcapacitance values and achieving stable circuit simulation [69]. The salient features of HiSIM include accurate modeling of small geometry effects, polydepletion effects, and QM effects in MOSFETs. This is accomplished by modifying the generalized expression for ϕ_s to include a shift in V_{th} due to the above-referred physical effects. The HiSIM modeling approach automatically preserves scalability of model parameters, and thus, one model parameter set for all device dimensions is used. Since a complete ϕ_s-based model automatically preserves the overall model consistency through ϕ_s, the number of model parameters can be drastically reduced in comparison to the conventional V_{th}-based models [68]. This parameter reduction comes without any loss in the reproduction accuracy of measurement data (e.g., *I–V* characteristics). Moreover, it has been reported that the nonlinear phenomena such as harmonic distortions are accurately calculated automatically [69]. All higher-order phenomena observed such as noise have been shown to be determined by the potential gradient along the channel [69], which again highlights the strength of the concept of ϕ_s-based modeling. Investigations of the high-frequency small-signal behavior with HiSIM concluded that the NQS effect is not as strong as previously believed [70,71]. Three members of the HiSIM family have been selected as the industry standards by CMC [48]. HiSIM-HV (1st standard version released in January 2009) is the high-voltage MOS device model standard, HiSIM2 (1st standard version released in April 2011) is the second-generation MOSFET model standard, and HiSIM-SOI (1st standard version released in July 2012) is the surface-potential SOI-MOSFET model standard.

At Philips Semiconductors, the development of MOS model 11 or MM11 started in 1994, primarily aimed at simple and accurate digital, analog, and RF modeling [72] of advanced ICs using analytical solution of surface potential. The implicit ϕ_s equation is modified to include polysilicon depletion effect by including a potential across the depletion layer due to polysilicon depletion and an empirical parameter to account for SCEs. In order to obtain efficient expressions for model outputs, several approximations were made, mainly based on the linearization of the inversion charge as a function of ϕ_s. In MM11, a linearization is performed around the average of source and drain potentials given by $\overline{\phi_s} = 1/2(\phi_{s0} + \phi_{sL})$ [72]. This linearization technique was shown to yield simpler and accurate expressions for ϕ_s keeping model symmetry with respect to source-drain interchange. This linearization approach offers an easy implementation of well-known physical phenomena such as thermal noise [73], induced gate noise [73], and gate leakage [74] in ϕ_s-based models.

In MM11, an accurate description of mobility effects and conductance effects has been added with a special emphasis on distortion modeling. For an accurate description of distortion, MM11 model is shown to accurately describe the drain current and its higher-order derivatives (up to at least the 3rd order). Thus, MM11 models reported contain improved expressions for mobility reduction [75], velocity saturation, and various conductance effects [76]. The distortion modeling of MM11 has been rigorously tested on various MOSFET technologies [77], and is shown to offer an accurate description of modern CMOS technologies. MM11 model is shown to preserve the source-drain interchange symmetry in model expressions [75,78] and thus eliminates the discontinuities in the high-order derivatives of channel current at $V_{ds} = 0$ [79]. MM11 incorporates an accurate description of all-important physical effects, such as polydepletion [80], the effect of pocket implants [81], gate tunneling current [66,80], bias-dependent overlap capacitances [80,82], GIDL, and noise [68,83] and therefore offers an accurate description of advanced MOSFETs in circuit operation.

In the early 1990s, the development of ϕ_s-based model, called *SP model*, started at the Pennsylvania State University by the research group led by Gildenblat. The modeling algorithm has been developed over the years [84–90]. In SP, SCE is modeled using the reported [91] bias and geometry-dependent lateral gradient factor while the geometry-dependent technique was used in HiSIM [68]. To overcome the inherent complexities of ϕ_s-based compact model, especially the expressions for the intrinsic charges [38,92,93], various approximations were developed based, primarily, on the linearization of the inversion charge as a function of ϕ_s. It is observed that this linearization technique [79] is a critical step to preserving the Gummel symmetry test and to avoid difficulties in the simulation of passive mixers and related circuits [94]. The symmetric linearization method developed in SP [85,87,93] preserves the Gummel symmetry and produces expressions for both the drain current and the terminal charges that are as simple as those in V_{th}-based or Q_i-based models and are numerically indistinguishable from the original charge-sheet model equations [85,94].

It has been reported that the symmetric linearization approach is not particularly sensitive to the details of the velocity saturation model, which enabled the merger of the best features of the SP and MM11 models to create PSP model. In addition to charge linearization relative to the source causing violation of the Gummel symmetry test, the singular nature of the popular velocity saturation model [79,94] is a critical problem. The problem can be solved using different techniques such as adopting a V_{ds}-dependent critical field [38,62,72]. When combined with the symmetric linearization method, this technique automatically solves the singularity issue [85,94]. Some of the specific features of SP include its unique symmetric linearization method, completely noniterative formulation, nonregional description from accumulation to strong inversion, inclusion of all relevant short-channel and thin-oxide effects, bias-dependent effective doping to deal with *halo* effects, physical description of the overlap regions and of the *inner-fringing* effects, and the comprehensive and accurate NQS model based on the spline collocation method [93]. The latter has been recently extended to include the accumulation region [92] and the small-geometry effects [95]. Finally, it has been reported [96,97] that when combined with the general one-flux theory of the nonabsorbing barrier, SP model is capable of reproducing the quasi-ballistic effects using the one-flux method [98].

The new ϕ_s–based PSP model is obtained by merging and developing the best features of SP (developed at the Pennsylvania State University) and MM11 (developed at Philips) models. The first version of the compact MOS model PSP, Level 100, has been released to the public domain in April 2005. In December 2005, CMC elected PSP as the new industrial standard model for compact MOSFET modeling [48].

1.2.2.3 Charge-Based Compact MOSFET Modeling

During the late 1980s, the charge-based compact models emerged as a viable alternative to widely used V_{th}-based compact models due the increasing complexities of V_{th}-based modeling for scaled MOSFET devices and computationally demanding solution techniques for ϕ_s-based modeling. In 1987, Maher and Mead reported a drain current expression in terms of the inversion charge density (Q_i) at the source and the drain ends [99]. Subsequently, a unified charge control model (UCCM) relating charge densities in terms of terminal voltages was reported in the early 1990s [100,101]. In 1995, Cunha et al. derived expressions for the total charges and small signal parameters as a function of the source and drain channel charge densities [102]. In 2001, Gummel et al. derived a charge equation and reported a charge-based model, referred to as USIM [103]. In 2003, He et al. reported an alternative derivation of charge [104] using gradual channel [26] and charge-sheet [23] approximations and linearization of the bulk and inversion charges with respect to the surface potential at a fixed gate bias. Since there is no Q_i in the accumulation region, different approaches used include an equation for the accumulation charge similar to that for Q_i or accumulation surface potential.

In charge-based models, an implicit function is evaluated to find the charge density for each set of biasing voltages in SPICE iterations similar to ϕ_s calculation. Note that the current is an exponential function of ϕ_s whereas a linear or quadratic function of Q_i. Therefore, the accuracy of calculation of the Q_i is not as high as that of ϕ_s calculation. Some of the widely referred charge-based compact MOSFET models include ACM [102], EKV [16], and BSIM6 [4] as described below.

In 1995, Cunha et al. reported a charge-based compact model, called the *advanced compact MOSFET* or *ACM model* [102]. The basic formulation of the ACM model is based on the charge-sheet model [23], inversion charge versus current relationship [99], UCCM [100,101], and symmetrical MOSFET model [105]. Explicit expressions for the current, charges, transconductances, and the 16 capacitive coefficients are shown to be valid in the weak, moderate, and strong inversion regions. In 1997, the ACM model was implemented in a circuit simulator [106] and emerged out of the necessity of modeling MOS capacitor for analog design in digital CMOS technology. In order to model the weak nonlinearities of an MOS capacitor in the accumulation and moderate as well as strong inversion regimes, Behr et al. reported an improved capacitive model of the MOSFET gate in 1992 [107]. A link between the charge model by Cunha et al. [102] and the current-based model of Enz et al. [16] was established by Galup-Montoro et al. [105] and Cunha et al. [108]. The models for DC, AC, and NQS behaviors were developed [105,106]. In 1999, UCCM [100,101] was revisited [109,110] to enhance the basic ACM model [102].

The ACM model has been reported to have a hierarchical structure facilitating the inclusion of different physical phenomena into the model [111]. Because of its very simple expression for the derivative of the channel charge density, ACM has been reported to offer simple explicit expressions for all intrinsic capacitive coefficients even when SCEs are taken into account [111]. The parameters of the ACM can be easily extracted [108,110]. Recently, ACM has been reported to include unified $1/f$ noise and mismatch models [112,113].

In 1995, Enz et al. reported an analytical compact MOSFET model, referred to as the EKV model, by referencing all the terminal voltages to the substrate [16]. The primary objective of the EKV model was low-power analog IC CAD using an analytical model that is valid in all modes of device operation with accurate modeling of weak inversion regime [114,115]. The model uses the linearization of Q_i with respect to the channel voltage to derive I_{ds} based on the continuous g_m/I_{ds} characteristics. In 2003, a rigorous derivation of the charge-based EKV model along with the detailed technique of Q_i linearization was reported using the existing charge-based models [99,103,116,117].

The bulk voltage referencing makes the EKV model symmetric [118–120] and preserves the symmetry property with reference to effects such as velocity saturation and nonuniform doping in the longitudinal direction [121]. The EKV model uses normalized Q_i at the source and drain ends to determine all the important MOSFET variables including the current [118,122], the terminal charges [123], the transcapacitances [123–125], the admittances, the transadmittances, [125], and the thermal noise, including the induced-gate noise [126,127].

It is shown that in the charge-based EKV model, Q_i linearization offers a direct, simple relation between the surface potential ϕ_s and Q_i [118,122,128]. The EKV model has been evolved into a full featured scalable compact MOS model that includes all the major effects that have to be accounted for in deep submicron CMOS technologies [129–131]. The model has also been extended to double-gate device architectures using the EKV charge-based approach [132].

In 2003, He et al. reported the charge-based BSIM5 model that uses a single set of equations to calculate terminal charges throughout all the bias regions [104,133]. The BSIM5 Q_i equation is derived directly from the solution of Poisson's equation in terms of ϕ_s in contrast to the conventional charge-based models [16,102] to obtain the final explicit function relating Q_i with MOS terminal voltages. The core BSIM5 model is derived assuming gradual channel and constant quasi-Fermi level to the channel current, I_{ds} in terms of Q_i at the source and drain ends. The I_{ds} equation includes the diffusion and drift components in a very simplified form. The model is reported to offer symmetry, continuity, scalability, and computational efficiency with a minimal number of parameters. It can easily incorporate short-channel, nonuniform doping, and numerous other physical effects such as polydepletion, velocity saturation, and velocity overshoot to accurately model subtle details of the device behaviors including current saturation and QM effect. It is also reported that BSIM5 core model can be easily extended to model nonclassical devices such as ultrathin body SOI and multigate devices including FinFETs [134].

In late 2010, the BSIM group started the development of BSIM6 core model [4]. The basic objective of BSIM6 development is to solve the symmetry issue of BSIM4 while maintaining BSIM4's accuracy, speed, and user support. The core BSIM6 has been derived using the reported charge-based approach [99,128,131,133]. The main features of BSIM6 include: smooth and continuous behaviors of *I–V* and *C–V* and their derivatives; continuity around $V_{ds} = 0$ and symmetry issue; excellent scalability with geometry, bias, and temperature; robust and physical behavior; excellent analog and RF modeling capability; and maintaining BSIM4 user experience [135]. In May 2013, BSIM6 has been selected and released as the industry-standard compact model for the existing as well as advanced planar CMOS technology nodes [48]. The model has been coded in Verilog-A and implemented in major EDA environment [136].

1.3 Motivation for Compact Modeling

The major motivation for the use of compact model for circuit CAD in the semiconductor industry is the cost-effective and efficient design optimization of IC products [137] in EDA environment. The use of compact models in circuit CAD allows optimization of circuit performance for robust IC chip design. This optimization is a complex task due to the increasing complexities of the scaled MOSFET devices and technology. The continuous scaling of CMOS devices to sub-100 nm regime has resulted in higher device density, faster circuit speed, and lower power dissipation. Many new physical phenomena such as SCE and reverse SCE (RSCE), channel length modulation, drain-induced barrier lowering, remote surface roughness scattering, mobility degradation, impact ionization, band-to-band tunneling, velocity overshoot, self-heating, channel quantization, polysilicon depletion, RF behaviors, NQS effects, and discrete dopants become significant as the device dimension approaches its physical limit [51,55]. Thus, intuitive analysis of the performance of nanoscale VLSI circuits using first principle is no longer possible whereas trial-and-error experimentations using breadboarding prototype [27] to build and characterize advanced IC chips are time consuming and expensive. In addition, advanced VLSI circuits with scaled devices are susceptible to process variability, causing device and circuit performance variability [5]. As a result, the statistical analysis of circuits is critical to develop advanced VLSI chips. Therefore, the compact models are the desirable alternative for cost-effective and efficient design of robust VLSI circuits, analysis of statistical device performance, analysis of yield, and so on.

Again, by the introduction of the SPICE program from Berkeley in 1975, the circuit simulator became a useful design tool, essentially replacing the breadboarding of prototypes [27]. However, for accurate circuit analysis, compact device models are required. Thus, the widespread use of circuit simulation also motivated the early development and use of compact model for IC device analysis. For today's circuit design, the major motivations for compact modeling include:

1. Circumventing the inadequate conventional manual techniques for design and analysis of today's complex VLSI circuits consisting of billions of nanoscale devices
2. Designing an IC chip under the worst-case conditions so that manufacturing tolerances can be incorporated into the design, thus ensuring the target production yield of the chip
3. Performing statistical analysis to optimize circuits for process variability–induced circuit performance variability, and also ensuring the target production yield of the chip

4. Design-for-reliability, enabling designers to predict and optimize circuit performance

5. Improving design efficiency using compact models instead of measured data from billions of transistors with different dimensions operating under different voltages that are used in an IC chip

1.4 Compact Model Usage

Compact models are an integral part of circuit analysis in EDA environment. Typically, the analytical equations of the target compact model (e.g., BSIM4) are fitted to the device characteristics of an IC technology obtained under different biasing conditions and extract device model parameters. These model parameters are used to generate a technology-specific device model library. Similarly, compact model parameters for passive elements of a circuit are extracted from the respective model formulations. Thus, a model library includes compact models for active and passive elements describing the behavior of these elements in VLSI circuits. This model library is used as the input file along with the input circuit description, called the *netlist*, for circuit simulation using a circuit CAD tool [27] as shown in Figure 1.2. A netlist describes the detailed description of a circuit performance under the target biasing conditions.

Figure 1.2 shows a circuit netlist and compact model library as the input to circuit CAD and the output is the simulation results including circuit speed (delays), logic levels, circuit performance variability, and SRAM yield.

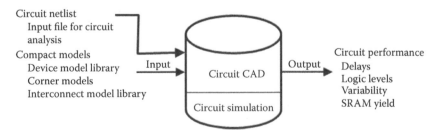

FIGURE 1.2
Usage of compact models in a circuit CAD: compact models describing the performance of circuit elements are used as the input to circuit CAD along with the description of the VLSI circuit for computer analysis of circuit performance; circuit CAD is a circuit simulation tool for computer analysis of VLSI circuits in EDA environment.

1.5 Compact Model Standardization

From the brief history of compact device models in Section 1.2, we find that a large number of compact MOSFET models have been developed over the past 40 years and are continued to date. Therefore, it is extremely difficult to generate, maintain, and support a large number of model libraries for a large number of process technologies for circuit CAD by device engineers of a manufacturing company. In order to improve the efficiency of compact modeling for circuit CAD, model developers and users have made a joint effort to establish a standard compact model for each IC device with robustness, accuracy, scalability, and computational efficiency to meet the needs of digital, analog, and mixed analog/digital designs. A standard model common to all or most semiconductor manufacturers and circuit CAD tools is desirable to facilitate intercompany collaborations.

With the objective of compact model standardization, an independent Compact Model Council, CMC was founded in 1996, consisting of many leading companies in the semiconductor industry. The charter of CMC is to promote the international, nonexclusive standardization of compact model formulations and the model interfaces. The CMC standardizes compact models for all major technologies to enhance the design efficiency, performs extensive model testing for model validation, and ensures robustness and accuracy of compact models for the latest technologies to shorten leading-edge design development cycle time. In 2013, CMC has become a part of an EDA standardization forum, Si2, to continue offering compact model standardization.

1.6 Summary

This chapter presents an overview of compact modeling for circuit CAD and the constituents of compact models to mathematically describe the real device effects. A brief history of compact MOSFET models for circuit simulation from the first Schimann-Hodges in 1970s to the recent surface potential–based and inversion charge–based models is presented. It is found that the early compact MOS models consist of physics-based analytical expressions to simulate the basic characteristics of devices in digital circuits. These models were continuously updated using empirical equations containing empirical fitting parameters to facilitate efficient circuit simulation. During 1980s physics-based compact MOS models with well-behaved mathematical smoothing functions were introduced, which describe the characteristics of scaled devices in all regions of circuit operation. With the increase in the complexities of MOS devices and technologies, compact MOS models based on surface potential and inversion charge started to emerge, especially to fulfil the increasing demands for analog and digital applications. These physics-based emerging models promise to simulate nano-MOSFET device characteristics in both digital and analog ICs. The accuracy, predictability, and longevity of these emerging models to meet the design challenges of MOS ICs down to 10 nm regimes are still to be seen. Finally, the motivation for compact modeling, the usage of compact models, and model standardization are briefly discussed.

Exercises

1.1 Double gate and multiple gate thin-body FETs like FinFETs and UTB-SOI FETs have emerged as the alternative devices to planar MOSFETs for advanced VLSI circuits. Write a brief history of the compact modeling of multiple gate FETs.

1.2 Write a brief history of compact modeling of the emerging devices including tunnel FETs as the potential alternative to next generation devices for VLSI circuits.

2

Review of Basic Device Physics

2.1 Introduction

Compact device models for circuit CAD (computer-aided design) requires detailed description of the transistor characteristics in the circuit environment under various biasing conditions. Transistor characteristics, however, depend on the material properties of the basic building blocks of each transistor along with its geometrical and structural information. IC (integrated circuit) transistors are fabricated on a semiconductor substrate such as silicon to achieve the desired device characteristics for the target circuit performance. These device characteristics are modulated by the transport of current-carrying fundamental constituents of matter referred to as the *electrons* and *holes*. Again, the electronic properties of semiconductors, primarily, depend on the transport of the majority carrier electrons or holes. The semiconductors with the majority carrier concentration as electrons are referred to as the n-type, whereas, the semiconductors with the majority carrier concentration as holes are referred to as the p-type. Thus, in order to understand the compact device models for circuit CAD, it is essential to understand the basic physics of the elemental n-type, and p-type semiconductors along with the transport properties of electrons and holes in building IC devices. Though a number of published titles are available on the subject, the objective of this chapter is to present a brief overview of the basic semiconductor theory along with the basics of n-type and p-type semiconductors in contact forming pn-junctions that are necessary to develop compact transistor models for circuit CAD. The review is brief and covers only those topics that have direct relevance to the field-effect transistor ICs. For more exclusive treatments, the readers are referred to textbooks on the subject [1–13].

2.2 Semiconductor Physics

Crystalline silicon is a widely used semiconductor-starting material in the fabrication of IC devices and chips. Thus, unless otherwise specified, in this book, the semiconductor physics is described with reference to silicon.

The silicon wafers used in the IC fabrication processes are cut parallel to either the <111> or <100> crystal planes. However, the <100> material is most commonly used due to the fact that, during IC fabrication processes, <100> wafers produce the lowest amount of charges at the silicon/silicon-dioxide (Si/SiO$_2$) interface and offer higher carrier mobility [14,15].

2.2.1 Energy Band Model

In a silicon crystal, each atom has four valence electrons and four nearest neighboring atoms. Each atom shares its valence electrons with its four neighbors in a paired configuration called *covalent bond*. It is predicted by quantum mechanics (QM) that the allowed energy levels of electrons in a solid is grouped into two bands, called the *valence band* (VB) and the *conduction band* (CB). These bands are separated by an energy range that the electrons in a solid cannot possess and is referred to as the *forbidden band* or *forbidden gap*. The VB is the highest energy band and its energy levels are mostly filled with electrons forming the covalent bonds. The CB is the next higher energy band with its energy levels nearly empty. The electrons that occupy the energy levels in the CB are called *free electrons* or *conduction electrons*.

Typically, the energy is a complex function of momentum in a three-dimensional space and there are many allowed energy levels for a large number of electrons in silicon, and therefore, the energy band diagram is also complex. For the simplicity of representation, only the edge levels of each of the allowed energy bands are shown in the energy band diagram in Figure 2.1. In Figure 2.1, E_c and E_v are the bottom edge of the CB and the top edge of the VB, respectively, and E_g is the bandgap energy separating E_c and E_v. And, at any ambient temperature $T(K)$, E_g is given by

$$E_g = E_c - E_v \tag{2.1}$$

When a valence electron is given sufficient energy ($\geq E_g$), it can break out of the chemical bonding state and excite into the CB to become a free electron leaving behind a vacancy, or *hole* in the VB. A hole is associated with a positive charge since a net positive charge is associated with the atom from which the electron broke away. Note that both the electron and hole are generated simultaneously from a single event. The electrons move freely in the CB and holes move freely in the VB. In silicon, the bandgap is small (~1.12 eV); therefore, even at room temperature a small fraction of the valence electrons are excited into the CB, generating electrons and holes. This allows limited conduction to take place from the motion of the electrons in the CB and holes in the VB. As shown in Figure 2.1, when an electron in the CB gains energy, it moves up to an energy $E > E_c$, while a hole in the VB gains energy, it moves down to an energy $E < E_v$. Thus, the energy of the electrons in the CB increases upward while the energy of the holes in the VB increases downward.

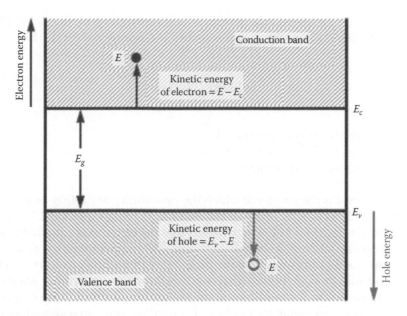

FIGURE 2.1
Energy band diagram of a semiconductor like silicon: E_c is the bottom edge of the CB and E_v is the top edge of the VB; the CB and VB are separated by an energy gap $E_g = E_c - E_v$.

The bandgap energy, E_g, for silicon at room temperature (300 K) is ~1.12 eV. As the temperature increases, the value of E_g for most semiconductors decreases due the increase in the crystal lattice spacing by thermal expansions. For silicon, the temperature coefficient of E_g at 300 K temperature is: $dE_g/dT \cong -2.73 \times 10^{-4} \text{eV/K}$ [16]. The temperature dependence of E_g for silicon can be modeled by using polynomial equations valid for different range of temperatures [16,17]. However, in circuit CAD tool SPICE (Simulation Program with Integrated Circuit Emphasis) [18], the temperature dependence of E_g is modeled by [19]

$$E_g(T) = 1.160 - \frac{7.02 \times 10^{-4} T^2}{1108 + T} \tag{2.2}$$

where:
  T is the temperature in Kelvin (K)
  $E_g(T)$ is in eV

2.2.2 Carrier Statistics

The electrical properties of a semiconductor are determined by the number of carriers available for conduction. This number is determined from the density of states and the probability that these states are occupied by carriers. The probability that an available state with energy E is occupied by an electron under a thermal equilibrium condition is given by the *Fermi–Dirac* probability density function $f(E)$, also called the *Fermi function* [1–11].

$$f(E) = \frac{1}{1+\exp\left[(E-E_f)/kT\right]} \qquad (2.3)$$

where:
   E_f is the *Fermi energy* or *Fermi level*
   $k = 1.38 \times 10^{-23}$ J K^{-1} is the Boltzmann constant
   T is the ambient temperature

The Fermi level is the energy at which the probability of finding an electron, at any $T > 0°$ K, is exactly one-half (Equation 2.3). From Equation 2.3, we find that when $E = E_f$, $f(E) = 1/2$, which means that the electron is equally likely to have an energy above E_f as below it. At absolute zero temperature ($T = 0°$ K): $f(E) = 1$ for $E < E_f$, indicating that the probability of finding an electron below E_f is unity and above E_f is zero (that is, $f(E) = 0$ for $E > E_f$). In other words, all energy levels below E_f are filled and all energy levels above E_f are empty. At finite temperatures, some states above E_f are filled and some states below E_f become empty. As T increases above absolute zero, the function $f(E)$ changes as shown in Figure 2.2. Thus, the probability that the energy levels above E_f are filled increases with temperature. It is important to note that the Fermi function or Fermi energy applies only under equilibrium conditions.

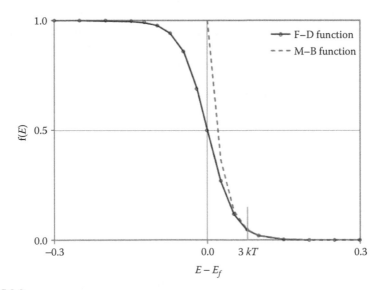

FIGURE 2.2
Fermi–Dirac (F–D) and Maxwell–Boltzmann (M–B) distribution functions in a semiconductor; the plots show that the F–D distribution can be approximated to M–B distribution at temperature, $T > 3$ kT.

Equation 2.3 describes the probability of an allowed energy state occupied by an electron with $E > E_f$. Then the probability of a state not occupied by an electron (with $E < E_f$) is given by

$$1 - f(E) = \frac{1}{1 + \exp\left[(E_f - E)/kT \right]} \tag{2.4}$$

Equation 2.4 is the probability function describing that a hole exists.

As shown in Figure 2.2, the probability distribution $f(E)$ makes a smooth transition from unity to zero as the energy increases across the Fermi level. The width of the transition is governed by the thermal energy kT. The value of thermal energy at room temperature is about 26 mV. Thus, for all energy at least several kT ($\sim 3\ kT$) above E_f, the function $f(E)$ in Equations 2.3 and 2.4 can be approximated by the simple expressions

$$f(E) \cong \exp\left(-\frac{E - E_f}{kT} \right) \quad \text{for } E > E_f \tag{2.5}$$

and

$$1 - f(E) \cong \exp\left(-\frac{E_f - E}{kT} \right) \quad \text{for } E < E_f \tag{2.6}$$

Equations 2.5 and 2.6 are identical to Maxwell Boltzmann density function for classical gas particles. For most device applications at room temperature, the function $f(E)$ given by Equation 2.5 is a good approximation as shown in Figure 2.2.

Fermi level can be considered to be the chemical potential for electrons and holes. Since the condition for any system in equilibrium is that the chemical potential must be constant throughout the system, it follows that the Femi level must be constant throughout a semiconductor in equilibrium.

2.2.3 Intrinsic Semiconductors

An *intrinsic semiconductor* is a perfect single crystal semiconductor with no impurities or lattice defects. In such materials, the VB is completely filled with electrons and the CB is completely empty. Therefore, in intrinsic semiconductors, there are no charge carriers at $0°$ K. At higher temperatures electron–hole pairs are generated as VB electrons are thermally excited across the bandgap to the CB. In intrinsic semiconductors, all the electrons in the CB are thermally excited from the VB. In other words, at a given temperature, the number of holes in the VB equals the number of electrons in the CB of an intrinsic semiconductor. Thus, if n and p are the concentrations of free electrons and holes, respectively, then

$$n = p = n_i \tag{2.7}$$

or,

$$np = n_i^2 \tag{2.8}$$

where:
   n_i is called the intrinsic carrier concentration and is the free electron (or hole) concentration in an intrinsic semiconductor

2.2.3.1 Intrinsic Carrier Concentration

From the effective densities of carriers and probability distribution function, we can derive the expression for the intrinsic carrier concentration in a semiconductor. Thus, from Equations 2.5 and 2.6, we can write the concentration of electrons in the CB as

$$n \cong N_c \exp\left(-\frac{E_c - E_f}{kT}\right) \tag{2.9}$$

and the concentration of holes in the VB as

$$p \cong N_v \exp\left(-\frac{E_f - E_v}{kT}\right) \tag{2.10}$$

where:
   N_c and N_v are the effective densities of states in the CB and VB, respectively

The expressions for N_c and N_v are derived from QM considerations [5]. Both N_c and N_v are proportional to $T^{3/2}$. For an intrinsic semiconductor, $n = p = n_i$ and E_f is called the intrinsic Fermi level, or the *intrinsic energy level*, E_i. Then (using $n = p = n_i$) we can write from Equations 2.9 and 2.10,

$$N_c \exp\left(-\frac{E_c - E_f}{kT}\right) = N_v \exp\left(-\frac{E_f - E_v}{kT}\right) \tag{2.11}$$

Now, solving Equation 2.11 for $E_f = E_i$, we get the expression for the *intrinsic energy* level as

$$E_i = E_f = \frac{E_c + E_v}{2} - \frac{kT}{2}\ln\left(\frac{N_c}{N_v}\right) \tag{2.12}$$

From Equation 2.12, it can be shown that the intrinsic Fermi level, E_i, is only about 7.3 meV below the mid-gap at $T = 300°\,\text{K}$. Since $kT << (E_c + E_v)$, Equation 2.12 can be simplified to

$$E_i = E_f \cong \frac{E_c + E_v}{2} \tag{2.13}$$

Thus, the intrinsic Fermi level in a semiconductor material is very close to the midpoint between the CB and the VB, and for all practical purposes, it can be assumed that E_i is in the middle of the energy gap. Thus, E_i is commonly referred to as the *mid-gap* energy level.

In order to derive an expression for the intrinsic carrier concentration as a function of T, we multiply Equations 2.9 and 2.10 to get

$$np = n_i^2(T) = N_c N_v \exp\left(-\frac{E_c - E_v}{kT}\right) = N_c N_v \exp\left[-\frac{E_g(T)}{kT}\right]$$

or $\tag{2.14}$

$$n_i(T) = CT^{3/2} \exp\left(-\frac{E_g(T)}{2kT}\right)$$

where:

C is a constant
E_g is the bandgap energy defined in Equation 2.1
k is the Boltzmann constant (8.62×10^{-5} eV K^{-1})
The term kT has the dimension of energy and is called *thermal energy* and is equal to 25.86 meV at $T = 300°$ K

Substituting the values for N_c and N_v [6], we can express Equation 2.14 as

$$n_i(T) = 3.9 \times 10^{16} T^{3/2} \exp\left(-\frac{E_g(T)}{2kT}\right) \tag{2.15}$$

If $E_g(T_{NOM})$ and $n_i(T_{NOM})$ are the values of E_g and n_i at the nominal or the reference temperature T_{NOM}, respectively, then we can show

$$n_i(T) = n_i(T_{NOM}) \cdot \left(\frac{T}{T_{NOM}}\right)^{3/2} \exp\left[-\frac{E_g(T)}{2kT} + \frac{E_g(T_{NOM})}{2kT_{NOM}}\right] \tag{2.16}$$

where $E_g(T)$ is given by Equation 2.2. The above expression is used in circuit CAD for calculation of n_i at any temperature T with $n_i = 1.45 \times 10^{10}$ cm^{-3} at $T = 300°$ K [6].

2.2.3.2 Effective Mass of Electrons and Holes

The electrons in the CB and holes in the VB move freely throughout the crystal like free particles, suffering only occasional scattering by impurities and defects present in the crystal. The free electrons experience Coulomb force

TABLE 2.1

Effective Mass Ratio for Silicon at 300 K (m_0 is the Free Electron Mass)

Carriers	Density of states effective mass (m_n^*/m_0)	Conductivity effective mass (m_n^*/m_0)
Electrons	1.08	0.26
Holes	0.81	0.386

due to the charged atomic cores of the host atoms in a regular lattice, giving rise to a periodic potential energy. The effect of the periodic potential of the crystal lattice on the motion of electrons in the CB and holes in the VB is represented by the effective masses of the electrons (m_n^*) and holes (m_p^*), respectively. In practice, there are several types of mass used for a given material and carrier type [1–11]. The effective mass required to calculate the carrier (electron and hole) concentration is called the *density of states effective mass*, whereas the mass required to calculate carrier mobility is called the *conductivity effective mass*. These effective masses depend on temperature. There is a large variation in the reported values of m_n^* and m_p^* [16]. The commonly used values for the effective mass for electrons and holes at room temperature are summarized in Table 2.1 [6].

2.2.4 Extrinsic Semiconductors

An *extrinsic semiconductor* is a semiconductor material with added elemental impurities called *dopants*. As we discussed in Section 2.2.3, the intrinsic semiconductor at room temperature has an extremely low number of free-carrier concentration, yielding very low conductivity. The added impurities introduce additional energy levels in the forbidden gap and can easily be ionized to add either electrons to the CB or holes to the VB, depending on the type of impurities and impurity levels.

Silicon is a column-IV element with four valence electrons per atom. There are two types of impurities in silicon that are electrically active: those from column V such as arsenic (As), phosphorous (P), and antimony (Sb); and those from column III such as boron (B). A column-V atom in a silicon lattice tends to have one extra electron loosely bound after forming covalent bonds with silicon atoms as shown in Figure 2.3a. In most cases, the thermal energy at room temperature is sufficient to ionize the impurity atom and free the extra electron to the CB. Such type of impurities (P, Sb, and As) are called *donor* atoms, since they donate an electron to the crystal lattice and become positively charged. Thus, the P, Sb, and As doped silicon is called *n*-type material that contains excess electrons and its electrical conductivity is dominated by electrons in the CB. On the other hand, a column-III impurity atom in a silicon lattice tends to be deficient of one electron when forming covalent bonds with other silicon atoms as shown in Figure 2.3b. Such an impurity (B) atom can also be ionized by accepting an electron from the VB, which leaves

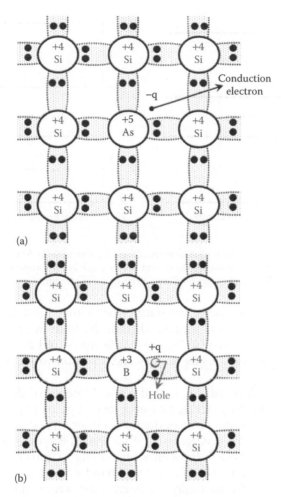

FIGURE 2.3
Extrinsic semiconductors forming covalent bonds: (a) an arsenic donor atom in silicon providing one electron for conduction in the CB and (b) a boron acceptor atom in silicon creating a
hole for conduction in the VB.

a freely moving hole that contributes to electrical conduction. These impurities (e.g., B) are called acceptors, since they accept electrons from the VB, and
the doped silicon is called *p*-type that contains excess holes.

Thus, we can see from Figure 2.3, the donor and acceptor atoms occupy
substitutional lattice sites and the extra electrons or holes are very loosely
bound, that is, can easily move to the CB or VB, respectively. In terms of
energy band diagrams, donors add allowed electron states in the bandgap
close to the CB edge as shown in Figure 2.4a whereas acceptors add allowed
states just above the VB edge as shown in Figure 2.4b. Figure 2.4 also shows
the positions of the Fermi level due to donors (Figure 2.4c) and acceptors
(Figure 2.4d). Donor levels contain positive charge when ionized (emptied).

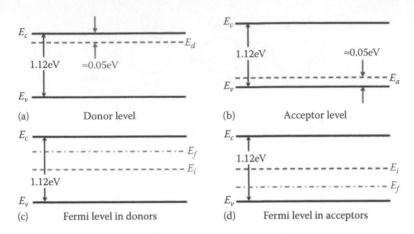

FIGURE 2.4
Energy band diagram representation in extrinsic semiconductors: (a) donor level E_d, (b) acceptor in silicon E_a, (c) intrinsic energy level and Fermi level in an *n*-type semiconductor, and (d) intrinsic energy level and Fermi level in a *p*-type semiconductor.

Acceptor levels contain negative charge when ionized (filled). A donor level E_d shown in Figure 2.4a is measured from the bottom of the CB whereas an acceptor level E_a shown in Figure 2.4b is measured from the top of the VB. The ionization energies for donors and acceptors are (E_c-E_d) and (E_a-E_v), respectively.

It is possible to dope silicon so that $p = n$. Material of this type is called *compensated* silicon. In practice, however, one type of impurity dominates over the other so that the semiconductor is either *n*-type or *p*-type. A semiconductor is said to be *nondegenerate* if the Fermi level lies in the bandgap more than a few kT ($\sim3\ kT$) from either band edge. Conversely, if the Fermi level is within a few kT ($\sim3\ kT$) of either band edge, the semiconductor is said to be *degenerate*. In the *nondegenerate* case, the carrier concentration obeys Maxwell-Boltzmann statistics given by Equations 2.5 and 2.6. However, for the degenerate case where the dopant concentration is in excess of approximately 10^{18} cm^{-3} (heavy doping), one must use Femi-Dirac distribution function given by Equations 2.3 and 2.4. Unless otherwise specified, we will assume the semiconductor to be nondegenerate.

2.2.4.1 Fermi Level in Extrinsic Semiconductor

In contrast to intrinsic semiconductor, the Fermi level in extrinsic semiconductor is not located at the mid-gap. The Fermi level in an *n*-type silicon moves up toward the CB, consistent with the increase in electron density described by Equation 2.9. On the other hand, the Fermi level in a *p*-type silicon moves toward the VB, consistent with the increase in hole density described by Equation 2.10. These cases are depicted in Figure 2.4c and d. The exact position of the Fermi level depends on both the ionization energy and concentration of dopants. For example for an *n*-type material with a donor impurity concentration N_d, the charge neutrality condition in silicon requires that

$$n = N_d^+ + p \tag{2.17}$$

where:
N_d^+ is the density of ionized donors

Using Equation 2.4 we can write

$$N_d^+ = N_d \left[1 - f(E_d)\right] = N_d \left\{ 1 - \frac{1}{1 + (1/2)\exp\left[(E_d - E_f)/kT\right]} \right\} \tag{2.18}$$

where:
$f(E_d)$ is the probability that a donor state is occupied by an electron in the normal state
E_d is the energy of the donor level

The factor 1/2 in the denominator of $f(E_d)$ arises from the spin degeneracy (up or down) of the available electronic states associated with an ionized level [20].

Substituting Equations 2.9 and 2.10 for *n* and *p*, respectively, and Equation 2.18 for N_d^+ in Equation 2.17, we get

$$N_c \exp\left(-\frac{E_c - E_f}{kT}\right) = \frac{N_d}{1 + 2\exp\left[-(E_d - E_f/kT)\right]} + N_v \exp\left(-\frac{E_f - E_v}{kT}\right) \tag{2.19}$$

Equation 2.19 can be solved for E_f. For an *n*-type semiconductor, $n \gg p$; therefore, the second term on the right hand side of Equation 2.19 can be neglected. Now, assuming $(E_d-E_f) \gg kT$, $\exp\left[-(E_d - E_f)/kT\right] \ll 1$. Therefore, from Equation 2.19 we get after simplification

$$E_c - E_f = kT \ln\left(\frac{N_c}{N_d}\right) \tag{2.20}$$

In this case, the Fermi level is at least a few kT below E_d and essentially all the donor levels are ionized, that is, $n = N_d^+ = N_d$ for an *n*-type semiconductor. Then from Equation 2.8, the hole density in an *n*-type semiconductor is given by

$$p = \frac{n_i^2}{N_d} \tag{2.21}$$

Similarly, for a p-type silicon with a shallow acceptor concentration N_a, the Fermi level is given by

$$E_f - E_v = kT \ln\left(\frac{N_v}{N_a}\right) \tag{2.22}$$

In this case, the hole density is $p = N_a^- = N_a$, and the electron density is

$$n = \frac{n_i^2}{N_a} \tag{2.23}$$

Instead of using Equations 2.20 and 2.22, we can express these in terms of E_f and E_i using Equations 2.9 and 2.10. From Equation 2.9, the intrinsic carrier concentration can be shown as

$$n_i \cong N_c \exp\left(-\frac{E_c - E_i}{kT}\right) \tag{2.24}$$

Or,

$$E_c = E_i + kT \ln\left(\frac{N_c}{n_i}\right) \tag{2.25}$$

Then substituting for E_c from Equation 2.25 into Equation 2.20, we get for an n-type silicon,

$$E_f - E_i = kT \ln\left(\frac{N_d}{n_i}\right) \tag{2.26}$$

Similarly, using Equation 2.10, we can express Equation 2.22 for a p-type silicon by

$$E_i - E_f = kT \ln\left(\frac{N_a}{n_i}\right) \tag{2.27}$$

Equations 2.26 and 2.27 are the measure of the Fermi level with reference to the mid-gap energy level for the n-type and p-type semiconductors, respectively.

2.2.4.2 Fermi Level in Degenerately Doped Semiconductor

For heavily doped silicon, the impurity concentration N_d or N_a can exceed the effective density of states N_c or N_v, so that $E_f \geq E_c$ and $E_f \leq E_v$ according to Equations 2.20 and 2.22. In other words, the Fermi level moves into the CB for $n+$ silicon, and into VB for the $p+$ silicon. In addition, when the impurity concentration is higher than 10^{18} cm^{-3}, the donor (or acceptor) levels broaden into bands. This results in an effective decrease in the ionization energy until finally the impurity band merges with the CB (or VB) and the ionization energy becomes zero. Under these circumstances, the silicon is said to be *degenerate*. Strictly speaking, Fermi statistics should be used for the calculation of electron concentration when $(E_c - E_f) \leq kT$ [20]. For practical purposes, it is a good approximation within a few kT to assume that the Fermi level of the degenerate $n+$ silicon is at the CB edge, and that the degenerate $p+$ silicon is at the VB edge.

2.2.5 Carrier Transport in Semiconductors

In thermal equilibrium, mobile (CB) electrons are in random thermal motion with an average velocity of thermal motion, $v_{th} \cong 1 \times 10^7$ cm sec^{-1} at 300° K. However, due to the random thermal motion of electrons, no net current flows through the material. On the other hand, in the presence of an electric field E, electrons move opposite to the direction of E. This process is called *electron drift* and causes a net current flow through the material. Also, if there is a carrier concentration gradient in the material, the carriers diffuse away from the region of higher concentration to the lower concentration, producing a net current flow in the semiconductor. Thus, the carrier transport or current flow in a semiconductor is the result of two different mechanisms: (1) the drift of carriers (electrons and holes), which is caused by the presence of an electric field and (2) the diffusion of carriers, which is caused by an electron or hole concentration gradient in the semiconductor. We will now consider factors involved in both phenomena.

2.2.5.1 Carrier Mobility and Drift Current

When an electric field is applied to a conducting medium containing free carriers, the carriers are accelerated in proportion to the force of the field. However, the accelerating carriers within a semiconductor will collide with various scattering centers including the atoms of the host lattice (lattice scattering), the impurity atoms (impurity scattering), and other carriers (carrier–carrier scattering). In the case of an electron, these different scattering mechanisms tend to redirect its momentum and in many cases tend to dissipate the energy gained from the electric field. Thus, under the influence of a uniform electric field, the process of energy gain from the field and energy loss due to the scattering balance each other and carriers attain a constant average velocity, called the *drift velocity* (v_d). At low electric fields, v_d is proportional to the electric field strength E and is given by

$$v_d = \mu E \qquad (2.28)$$

where:
  μ is the constant of proportionality and is called the mobility of the carriers
    in units of cm^2 V^{-1}sec^{-1}

The mobility is proportional to the time interval between collisions and inversely proportional to the effective mass of the carriers. The total mobility is determined by combining the mobilities for different scattering mechanisms such as mobility due to *lattice scattering* μ_L and mobility due to *ionized impurity scattering* μ_I. Assuming different scattering mechanisms are independent, we can write the expression for total mobility using Mathiessen's rule

$$\frac{1}{\mu} = \frac{1}{\mu_L} + \frac{1}{\mu_I} + \ldots \tag{2.29}$$

The measurement data show that the electron mobility (μ_n) in an n-type silicon is about three times the hole mobility (μ_p) in a p-type silicon since the effective mass of electrons in the CB is much lighter than that of holes in the VB.

Carrier mobility in bulk silicon is a function of the doping concentrations. Figure 2.5 shows plots of electron and hole mobilities in silicon as a function of doping concentration at room temperature. It is observed from the plots that at low impurity levels, the mobilities are mainly limited by carrier collisions with the silicon lattice or acoustic phonons. As the doping concentration increases beyond 1×10^{15} cm^{-3}, the mobilities decrease due to the increase in the collisions with the charged (ionized) impurity atoms through Coulomb interaction. At high temperatures, the mobility tends to be limited by lattice scattering and is proportional to $T^{-3/2}$, relatively insensitive to the doping concentration. At low temperatures, the mobility is higher; however, it strongly depends on doping concentration as it becomes more limited by

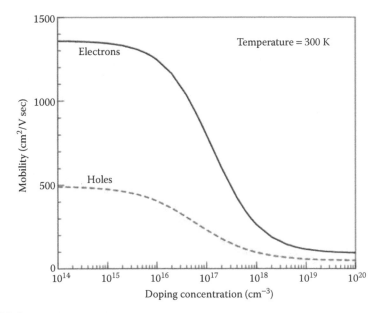

FIGURE 2.5
Electron and hole mobilities in bulk silicon at 300 K as a function of doping concentration.

impurity scattering. The detailed temperature dependence of mobility can be found in Arora and Arora et al. [17,21].

The carrier mobility discussed earlier is the *bulk mobility* applicable to conduction in the silicon substrate far away from the surface. In the channel region of MOSFET (metal-oxide-semiconductor field-effect transistor) devices, the current flow is governed by the *surface mobility*. The surface mobility is much lower than the bulk mobility due to additional scattering mechanism between the carriers and Si/SiO_2 interface in the presence of the high electric field normal to the channel as discussed in Section 5.3.1 of Chapter 5.

2.2.5.2 Electrical Resistivity

The drift of charge carriers under an applied electric field E results in a current, called the *drift current*. For a homogeneous n-type silicon, if there are n number of electrons per unit volume each carrying a charge q flow with a drift velocity v_d, then the *electron drift current* density is given by

$$J_{n,drift} = qnv_d = qn\mu_n E \tag{2.30}$$

where we have used Equation 2.28 for v_d; in Equation 2.30, $q = 1.6 \times 10^{-19}$ C is the electronic charge and μ_n is the electron mobility. From Ohm's law, the resistivity ρ of a conducting material is defined by E/J_n; therefore, from Equation 2.30, the resistivity ρ_n to electron current flow is given by

$$\rho_n = \frac{1}{qn\mu_n} \tag{2.31}$$

Similarly, for a p-type silicon, the hole drift current density, $J_{p,drift}$, and resistivity, ρ_p are given by

$$J_{p,drift} = qpv_d = qp\mu_p E \tag{2.32}$$

$$\rho_p = \frac{1}{qp\mu_p} \tag{2.33}$$

where:
μ_p is the hole mobility

If the silicon is doped with both donors and acceptors, then the total resistivity can be expressed as

$$\rho = \frac{1}{qn\mu_n + qp\mu_p} \tag{2.34}$$

Thus, the resistivity of a semiconductor depends on the electron and hole concentrations and their mobilities. Empirical resistivity versus impurity

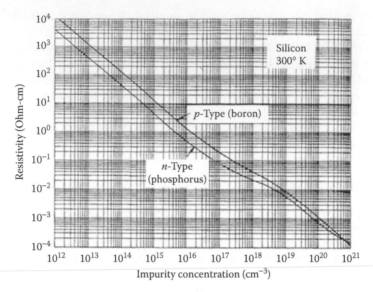

FIGURE 2.6
Impurity concentration versus resistivity for n-type and p-type silicon at 300° K. (After Sze 2007.)

concentrations plots are shown in Figure 2.6 for uniformly doped silicon at 300° K. The plot for n-type is lower than p-type doped silicon because electron mobility is higher than the hole mobility.

2.2.5.3 Sheet Resistance

The resistance of a uniform conductor of length L, width W, and thickness t is given by

$$R = \rho \frac{L}{tW} \tag{2.35}$$

where:
ρ is the resistivity of the conductor in ohm-centimeter

Typically, in an IC technology, the thickness t of a diffusion region is uniform and normally much less than both L and W of the region. Therefore, it is useful to define a new variable ρ_{sh}, called the *sheet resistance*, which has the dimension of Ohm (Ω) and is given by

$$\rho_{sh} = \frac{\rho}{t} \tag{2.36}$$

Then Equation 2.35 becomes

$$R = \rho_{sh} \frac{L}{W} \tag{2.37}$$

From Equation 2.37, it is found that when $L = W$, the diffused layer becomes a square with $R = \rho_{sh}$. Thus, the total resistance of a diffusion line is simply ρ_{sh} times the number of squares in the path of current and is expressed in units of Ω per square ($\Omega/\square$). The process parameters that determine the sheet resistance of a layer are the resistivity and thickness t of the layer. Since the resistivity is a function of carrier concentration and mobility, both of which are functions of temperature, ρ_{sh} is temperature dependent.

2.2.5.4 Velocity Saturation

The field versus velocity linear relationship, given by Equation 2.28 in Section 2.2.5.1, is valid only for low electric field ($<1 \times 10^4$ V cm^{-1}) and carriers are in equilibrium with the lattice. At higher electric fields, the average carrier energy increases and carriers lose their energy by optical-phonon emission nearly as fast as they gain it from the field. This causes a decrease in μ from its low field value as the field increases until finally the drift velocity reaches a limiting value v_{sat}, referred to as the *saturation velocity*. This phenomenon is called the *velocity saturation*. For silicon, a typical value of $v_{sat} = 1.07 \times 10^7$ cm sec^{-1} for electrons and occurs at an electric field of about 2×10^4 V cm^{-1}. The corresponding values for holes are $v_{sat} = 8.34 \times 10^6$ cm sec^{-1} and $E \cong 5.0 \times 10^4$ V cm^{-1}.

It is found that the measured value of drift velocity for electrons and holes in silicon is a function of the applied field E and can be approximated by the following expression

$$v_d = v_{sat} \frac{E/E_c}{\left[1+\left(E/E_c\right)^\beta\right]^{1/\beta}} \tag{2.38}$$

where:
E_c is the critical electric field at which carrier velocity saturates

The parameters v_{sat}, E_c, and β in Equation 2.38 are given in Table 2.2.

Figure 2.7 shows the simulated value of drift velocity for electrons and holes at 300°K in silicon as a function of the applied field E obtained by Equation 2.38. It is observed from Figure 2.7 that at low fields, the carrier

TABLE 2.2

Parameters for Field Dependence of Drift Velocity for Silicon at 300 K

Parameter	v_{sat} (cm sec^{-1})	E_c (V cm^{-1})	β
Electrons	1.07×10^7	6.91×10^3	1.11
Holes	8.34×10^6	1.45×10^4	2.637

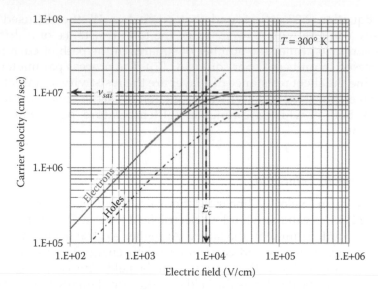

FIGURE 2.7
Drift velocities of electrons and holes in silicon at room temperature as a function of applied electric field showing velocity saturation at high electric fields.

velocity increases linearly with the electric field indicating constant mobility. When the field exceeds about 2×10^4 V cm^{-1}, carriers begin to lose energy by scattering with optical phonons and their velocity saturates. As the field exceeds 100 KV cm^{-1}, carriers gain more energy from the field than what they can lose by scattering. Consequently, their energy with respect to the bottom of the CB (for electrons) or top of the VB (for holes) begins to increase. The carriers are no longer at thermal equilibrium with the lattice. Since they acquire energy higher than the thermal energy (kT) they are called hot carriers.

It is these hot carriers that are responsible for reducing the mobility at high fields. For a more heavily doped material, the low-field mobility is lower because of the impurity scattering. However, v_{sat} remains the same, independent of impurity scattering. Also, v_{sat} is weakly dependent on temperature and decreases slightly as the temperature increases [17]. Figure 2.7 shows carrier velocity as a function of electric field. It is observed from the plots that the carrier velocity increases linearly at low electric field, then the increase in the carrier velocity slows down with the increase in electric field, and finally above a certain critical electric field the carrier velocity saturates.

2.2.5.5 Diffusion of Carriers

In addition to the drift of electrons under the influence of an electric field, the carriers also diffuse if the carrier concentration is not uniform within

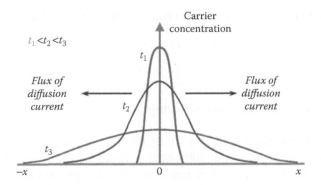

FIGURE 2.8
Diffusion of carriers from high concentration to low concentration due to concentration gradient over different time intervals $t_1 < t_2 < t_3$; t_1 is the initial time and the background concentration ≈ 0.

a semiconductor. This leads to an additional component of current in proportion to the concentration gradient and is called the *diffusion current*. Thus, the diffusion is a gradient driven motion and occurs from high-concentration regions toward low-concentration regions as shown in Figure 2.8.

The diffusion flux is given by Fix's first law,

$$F = -D\frac{dC}{dx} \tag{2.39}$$

where:
  F, D, and C are the flux of carriers, diffusion constant, and carrier density, respectively

The negative sign is due to the fact that the carriers flow from the higher concentration to lower concentration; that is, dC/dx is negative. If the carrier flow in a semiconductor material is electron, then the diffusion current flow due to the electron concentration gradient dn/dx is given by

$$J_{n,diff} = qD_n\frac{dn}{dx} \tag{2.40}$$

Similarly, the hole diffusion current due to hole concentration gradient dp/dx is given by

$$J_{p,diff} = -qD_p\frac{dp}{dx} \tag{2.41}$$

where:
  D_n and D_p are called the *diffusivity* or *diffusion constants* for electrons and holes in the material, respectively

and are related to the respective mobility by the relationship [6]

$$\frac{D_n}{\mu_n} = \frac{D_p}{\mu_p} = \frac{kT}{q} \equiv v_{kT} \tag{2.42}$$

where:
$v_{kT} \equiv kT/q$ is called the thermal voltage

Equation 2.42 is often referred to as the Einstein's relation. For lightly doped silicon (e.g., $N_d \cong 1 \times 10^{15}$ cm^{-3}) at room temperature, $D_n = 38$ cm^2 sec^{-1} and $D_p = 13$ cm^2 sec^{-1}. The negative sign in Equation 2.41 implies that the hole current flows in a direction opposite to the hole concentration gradient.

2.2.5.6 Nonuniformly Doped Semiconductors and Built-In Electric Field

Let us consider an n-type material with nonuniformly doped N_d donor atoms as shown in Figure 2.9. Considering complete ionization of donor atoms, we have $n = N_d^+ = N_d$.

Due to the concentration gradient, electrons diffuse from the high-concentration region to the low-concentration region. Then from Equation 2.39 the diffusion flux of electrons is given by

$$F_{n,diff} = -D_n \frac{dn(x)}{dx} \tag{2.43}$$

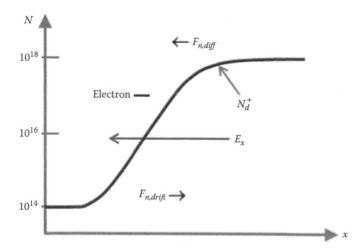

FIGURE 2.9
Drift and diffusion of carriers in a nonuniformly doped n-type semiconductor: $F_{n,diff}$ is the electron diffusion flux from the high concentration to low concentration; $F_{n,drift}$ is the drift flux of electrons due to the built-in electric field, E_x set up by the ionized donors and diffused electrons in the semiconductor.

where:

the subscript n represents the parameters for electrons

As the electrons move (diffuse) away, they leave behind positively charged donor ions (N_d^+), which try to pull electrons back causing drift flux of electrons from the low- to high-concentration region. This drift of electrons from low- to high-concentration regions sets up an electric field, E_x from the high-concentration to the low-concentration regions as shown in Figure 2.9. Then from Equation 2.30, the flux due to the drift of electrons is given by

$$F_{n,drift} = n(x)v_d = n\mu_n E_x \tag{2.44}$$

An equilibrium is established when *diffusion = drift*. Here $n(x)$ is the number of electrons in the diffusion flux at any point x in the distribution and $\neq N_d(x)$. Therefore, a *built-in* electric field is established that prevents diffusion of electrons. Then from Equations 2.43 and 2.44, we get the expression for the built-in electric field for electrons in an n-type nonuniformly doped substrate as

$$E_x = -\frac{D_n}{\mu_n}\frac{1}{n}\frac{dn(x)}{dx} = -v_{kT}\frac{1}{n}\frac{dn(x)}{dx} \tag{2.45}$$

Similarly, the built-in electric field for holes in a nonuniform p-type substrate is given by

$$E_x = \frac{D_p}{\mu_p}\frac{1}{p}\frac{dp(x)}{dx} = v_{kT}\frac{1}{p}\frac{dp(x)}{dx} \tag{2.46}$$

In Equations 2.45 and 2.46 we have used Einstein's relation given in Equation 2.42. This built-in electric field favors the transport of the minority carriers if created by an external source.

2.2.6 Generation–Recombination

In a semiconductor under thermal equilibrium, carriers possess an average thermal energy corresponding to the ambient temperature. This thermal energy excites some valence electrons to reach the CB. This upward transition of an electron from the VB to CB leaves behind a hole in the VB and an electron–hole pair is created. This process is called the *carrier generation* (G). On the other hand, when an electron makes a transition from the CB to the VB, an electron–hole pair is *annihilated*. This reverse process is called carrier *recombination* (R). Under thermal equilibrium, G = R so that the carrier concentration remains the same and the condition $pn = n_i^2$ is maintained. The thermal G–R process is shown in Figure 2.10.

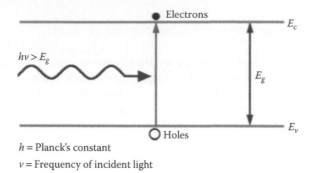

h = Planck's constant

v = Frequency of incident light

FIGURE 2.10
Band-to-band generation of electron–hole pairs under optical illumination of photon energy
hv, where h and v are the Planck's constant and the frequency of incident light, respectively.

The equilibrium condition of a semiconductor is disturbed by optically
or electrically introducing free carriers exceeding their thermal equilibrium
values resulting in $pn > n_i^2$ or by electrically removing carriers resulting in
$pn < n_i^2$. The process of introducing carriers in access of thermal equilibrium
values is called the *carrier injection* and the additional carriers are called the
excess carriers. In order to inject excess carriers optically, we shine light with
energy $E = hv > E_g$ on an intrinsic semiconductor so that the valence electrons
can be excited into the CB by the excess energy $\Delta E = (hv - E_g)$, where h and v
are Planck's constant and frequency of light, respectively. In this process, we
get optically generated excess electrons (n_L) and holes (p_L) in the semiconductor as shown in Figure 2.10. Therefore, the total nonequilibrium values of
carrier concentration is given by

$$\left. \begin{array}{l} n = n_i + n_L \\ p = n_i + p_L \end{array} \right\} \quad \text{Injection of carriers by light} \qquad (2.47)$$

2.2.6.1 Injection Level

From Equation 2.47, we observe that both n and p are greater than the intrinsic carrier concentration of the semiconductor, and therefore, $pn > n_i^2$ for
injection of carriers into the semiconductor. If the injected carrier density
is lower than the majority carrier density at equilibrium so that the latter
remains essentially unchanged while the minority carrier density is equal
to the excess carrier density, then the process is called the *low-level injection*.
If the injected carrier density is comparable to or exceeds the equilibrium
value of the majority carrier density, then it is called the *high-level* injection.
    To illustrate the injection levels, we consider an n-type *extrinsic* semicon-
ductor with $N_d = 10^{15}$ cm^{-3}. Then from Section 2.2.4.1, the equilibrium majority carrier electron concentration is given by $n_{no} = 1 \times 10^{15}$ cm^{-3}, whereas
from Equation 2.21, the minority carrier hole concentration is given by

$p_{no} = 1 \times 10^5$ cm^{-3}. Here, n_{no} and p_{no} define the equilibrium concentrations of electrons and holes, respectively, in an n-type material. Now, we shine light on the sample so that 1×10^{13} cm^{-3} electron–hole pairs are generated in the material. Then using Equation 2.47, the total number of electrons $n_n = n_{no} = 1 \times 10^{15}$ cm^{-3} and $p_n = 1 \times 10^{13}$ cm^{-3}. Thus, the majority carrier concentration n_n remains unchanged, whereas the minority carrier concentration p_n is increased significantly. This is an example of low-level injection. On the other hand, if 1×10^{17} cm^{-3} electron–hole pairs are generated by incident light, then from Equation 2.47, we get $n_n \cong 1 \times 10^{17}$ cm^{-3} and $p_n = 1 \times 10^{17}$ cm^{-3} changing both the electron and hole concentrations in the semiconductor, resulting in a high-level injection. *The mathematics for high-level injection are complex, and therefore, we will consider only low-level injection.*

2.2.6.2 Recombination Processes

The semiconductor material returns to equilibrium through recombination of injected minority carriers with the majority carriers in the case of carrier injection or through generation of electron–hole pairs in the case of extraction of carriers.

The electron–hole recombination process occurs by transition of electrons from the CB to the VB. In a direct bandgap semiconductor like GaAs where the minimum of the CB aligns with the maximum of the VB, an electron in the CB can give up its energy to move down to occupy the empty state (hole) in the VB without a change in the momentum as shown in Figure 2.11a. Since the momentum (k) must be conserved in any energy level transition, an electron in GaAs can easily make direct transition from E_c to E_v across E_g. This is called the *direct* or *band-to-band recombination*. When direct recombination happens, the energy given up by electron will be emitted as a photon, which makes it useful for light-emitting diodes.

If we generate excess carriers (Δn, Δp) at a rate G_L due to the incident light, then for low-level injection, we get $\Delta p = \Delta n = U\tau = G_L\tau$, where U is the net recombination rate and τ is the excess carrier lifetime. If p_o and n_o are the equilibrium concentrations of electrons and holes, respectively, and p and

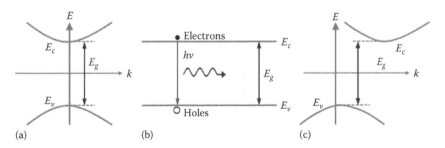

FIGURE 2.11
Bandgap in semiconductors: (a) direct bandgap, (b) band-to-band recombination in a direct bandgap semiconductor, and (c) indirect bandgap.

n are the respective total concentrations due to generation, then $\Delta p = p - p_o$ and $\Delta n = n - n_o$ and the net recombination rate due to direct recombination is given by

$$U = \frac{\Delta n}{\tau_n} = \frac{\Delta p}{\tau_p} \qquad (2.48)$$

where:

τ_n and τ_p are the excess carrier electron and hole lifetime, respectively

For band-to-band recombination, the excess carrier lifetime for an electron is equal to that of a hole since the single phenomenon annihilates an electron and a hole simultaneously.

For *indirect bandgap* semiconductors such as silicon and germanium (Figure 2.11c), the probability of direct recombination is very low. Physically, this means that the minimum energy gap between E_c and E_v does not occur at the same point in the momentum space as shown in Figure 2.11c. In this case, for an electron to reach the VB, it must experience a change of momentum as well as energy to satisfy the conservation principle. This can be achieved by recombination processes through intermediate trapping levels, called the *indirect recombination* as shown in Figure 2.12.

Impurities that form electronic states deep in the energy gap assist the recombination of electrons and holes in the indirect bandgap semiconductors. Here the word *deep* indicates that the states are far away from the band edges and near the center of the energy gap. These deep states are commonly referred to as *recombination centers* or *traps*. Such recombination centers are usually unintentional impurities, which are not necessarily ionized at room temperature. These deep level impurities have concentrations far below the concentration of donor or acceptor impurities, which have shallow energy levels. Gold (Au) is a deep level impurity intentionally used in silicon to increase the recombination rate. This recombination via deep level impurities or traps is often referred to as the *indirect recombination*. The process shown in Figure 2.12 consists of (1) an electron capture by an empty center, (2) electron emission from an occupied center, (3) hole capture by an occupied center, and (4) hole emission by an empty center.

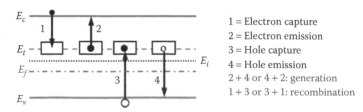

FIGURE 2.12
Generation and recombination in an indirect bandgap semiconductor; E_t is the trap level deep into the bandgap; 1, 2, 3, and 4 represent the generation and recombination processes.

Let us consider the following example where an impurity like Au is introduced that provides a *trapping level* or a set of allowed states at energy E_t. The trap level E_t is assumed to act like an acceptor (it can be neutral or negatively charged). Recombination is accomplished by trapping an electron and a hole. (The analysis can be easily extended to the case where the trap acts like a donor, that is, positively charged or neutral charge states.) The indirect recombination process was originally proposed by Shockley and Read [22] and independently suggested by Hall [23] and, therefore, is often referred to as the *Shockley–Read–Hall* (SRH) recombination. By considering the transition processes shown in Figure 2.12, Shockley, Read, and Hall showed that for low-level injection, the net recombination rate is given by

$$U = \frac{v_{th}\sigma N_t \left(pn - n_i^2\right)}{n + p + 2n_i \cosh\left[\left(E_t - E_i\right)/kT\right]} \tag{2.49}$$

where:
v_{th} is the carrier thermal velocity ($\approx 1 \times 10^7$ cm sec^{-1})
σ is the carrier capture cross section ($\approx 10^{-15}$ cm^2)
N_t is the density of trap centers
$v_{th}\sigma N_t$ is the *capture probability* or capture cross section

From Equation 2.49 we observe the following:

1. The "driving force" or the rate of recombination is proportional to $\left(pn - n_i^2\right)$, that is, the deviation from the equilibrium condition
2. $U = 0$ when $\left(np = n_i^2\right)$, that is, equilibrium condition
3. U is maximum when $E_t = E_i$, that is, trap levels near the mid-band are the most efficient recombination centers

Thus, for the simplicity of understanding, let us consider the case when $E_t = E_i$. Then from Equation 2.49, the net recombination rate is given by

$$U = \frac{v_{th}\sigma N_t \left(pn - n_i^2\right)}{n + p + 2n_i} \tag{2.50}$$

For an *n-type* semiconductor with low-level injection, $n \gg p + 2n_i$; denoting $p = p_n$ as the total excess minority carrier concentration and $\left(p_{no} = n_i^2/n\right)$ as the equilibrium minority carrier concentration, we get after simplification of Equation 2.50

$$U = v_{th}\sigma N_t \left(p_n - p_{no}\right) = \frac{\Delta p}{\tau_p} \tag{2.51}$$

where the minority carrier hole lifetime in an *n*-type semiconductor is given by

$$\tau_p = \frac{1}{v_{th}\sigma_p N_t} \tag{2.52}$$

In an *n*-type material, lots of electrons are available for capture. Therefore, Equation 2.51 shows that the minority carrier hole lifetime τ_p is the limiting factor in recombination process in an *n*-type material.

Similarly, for a *p-type* semiconductor, we can show from Equation 2.50 that the net recombination rate for electrons is given by

$$U = \frac{\Delta n}{\tau_n} \tag{2.53}$$

where

$$\tau_n = \frac{1}{v_{th}\sigma_n N_t} \tag{2.54}$$

is the minority carrier electron lifetime. Thus, for a *p*-type semiconductor the minority carrier electron lifetime is the limiting factor in the recombination process.

The other recombination process in silicon that does not depend on deep level impurities and that sets an upper limit on lifetime is *Auger recombination*. In this process, the electrons and holes recombine without trap levels and the released energy (of the order of energy gap) is transferred to another majority carrier (a hole in a *p*-type and electron in an *n*-type silicon). Usually, Auger recombination is important when the carrier concentration is very high ($>5 \times 10^{18}$ cm^{-3}) as a result of high doping or high-level injection.

2.2.7 Basic Semiconductor Equations

2.2.7.1 Poisson's Equation

Poisson's equation is a very general differential equation governing the operation of IC devices and is based on Maxwell's field equation that relates the charge density to the electric field potential. Conventionally, the electrostatic potential, ϕ in a semiconductor is defined in terms of the intrinsic Fermi level (E_i) such that

$$\phi = -\frac{E_i}{q} \tag{2.55}$$

The negative sign in Equation 2.55 is due to the fact that E_i is defined as the electron energy whereas ϕ is defined for a positive charge. The electric field E, which is defined as the electrostatic force per unit charge, is equal to the negative gradient of ϕ, such that

$$E = -\frac{d\phi}{dx} \tag{2.56}$$

Mathematically, Poisson's equation (for silicon) is stated as

$$\frac{dE}{dx} = \frac{\rho(x)}{K_{si}\varepsilon_0} \tag{2.57}$$

or, using Equation 2.56,

$$\frac{d^2\phi}{dx^2} = -\frac{\rho(x)}{K_{si}\varepsilon_0} \tag{2.58}$$

where
   $\rho(x)$ is the net charge density at any point x
   ε_0 ($=8.854 \times 10^{-14}$ F cm^{-1}) is the permittivity of free space
   K_{si} ($=11.8$) is the relative permittivity of silicon

If n and p are the free electron and hole concentrations, respectively, corresponding to N_d^+ and N_a^- ionized acceptor and donor concentrations, respectively, in silicon, we can express Equation 2.58 as

$$\frac{d^2\phi}{dx^2} = -\frac{dE}{dx} = -\frac{q}{K_{si}\varepsilon_0}\left\{\left[p(x)-n(x)\right]+\left[N_d^+(x)-N_a^-(x)\right]\right\} \tag{2.59}$$

Assuming complete ionization of dopants, $N_d^+ = N_d$ and $N_a^- = N_a$, we can write Poisson's equation as

$$\frac{d^2\phi}{dx^2} = -\frac{q}{K_{si}\varepsilon_0}\left\{\left[p(x)-n(x)\right]+\left[N_d(x)-N_a(x)\right]\right\} \tag{2.60}$$

Equation 2.60 is a one-dimensional (1D) equation and can easily be extended to three-dimensional (3D) space. 1D-Poisson equation is adequate for describing most of the basic device operations. However, for small geometry advanced devices 2D (two-dimensional) or 3D Poisson's equation must be used.

   Another form of Poisson's equation is Gauss's law, which is obtained by integrating Equation 2.57:

$$E = \frac{1}{K_{si}\varepsilon_0}\int \rho(x)dx = \frac{Q_s}{K_{si}\varepsilon_0} \tag{2.61}$$

It is to be noted that the semiconductor as a whole is charged neutral, that is, ρ must be zero. However, when the space charge neutrality does not apply, Poisson's equation must be used.

2.2.7.2 Carrier Concentration in Terms of Electrostatic Potential

In an *n*-type nondegenerate semiconductor the Fermi level E_f (or Fermi potential $\phi_f = -E_f/q$) lies above the intrinsic level E_i (or intrinsic potential $\phi_i = -E_i/q$) as shown in Figure 2.4c. Then from Equation 2.26 we can write

$$N_d = n_i \exp\left(\frac{E_f - E_i}{kT}\right) = n_i \exp\left[\frac{q}{kT}(\phi_i - \phi_f)\right] \tag{2.62}$$

while in a *p*-type semiconductor the Fermi level E_f (or Fermi potential ϕ_f) lies below the intrinsic level E_i (or intrinsic potential ϕ_i) as shown in Figure 2.4d, and from Equation 2.27 we can show

$$N_a = n_i \exp\left(\frac{E_i - E_f}{kT}\right) = n_i \exp\left[\frac{q}{kT}(\phi_f - \phi_i)\right] \tag{2.63}$$

At room temperature, the available thermal energy is sufficient to ionize nearly all acceptor and donor atoms due to their low ionization energies. Hence it is safe approximation to say that in a nondegenerate silicon at room temperature:

$$n \approx N_d \, (n-\text{type}) \tag{2.64}$$

$$p \approx N_a \, (p-\text{type}) \tag{2.65}$$

where:
  N_d is the concentration of donor atoms
  N_a is the concentration of acceptor atoms

In an *n*-type material, where $N_d \gg n_i$, electrons are majority carriers whose concentration is given by Equation 2.64, while the hole concentration p_n (representing concentration of p in an *n*-type material) from Equation 2.64 is given by

$$p_n \cong \frac{n_i^2}{N_d} \tag{2.66}$$

The hole concentration p_n is much smaller than n_n in an *n*-type semiconductor. Thus, holes are minority carriers in an *n*-type semiconductor. Similarly, in a *p*-type semiconductor where $N_a \gg n_i$, holes are the majority carriers given by Equation 2.65, while the minority carrier electron concentration is given by

$$n_p \cong \frac{n_i^2}{N_a} \tag{2.67}$$

Since $n_p \ll p$, electrons are minority carriers in a p-type semiconductor. Consequently, we often use the terminology of majority and minority carriers.

From Equation 2.62, we can write for an n-type semiconductor

$$\phi_i - \phi_f = \frac{kT}{q} \ln\left(\frac{N_d}{n_i}\right) = v_{kT} \ln\left(\frac{N_d}{n_i}\right) \equiv -\phi_B \qquad (2.68)$$

where:

$\phi_B \equiv (\phi_f - \phi_i)$ is called the *bulk potential* and is negative for n-type semiconductors

Similarly, from Equation 2.63, for p-type semiconductor, we can show

$$\phi_f - \phi_i = v_{kT} \ln\left(\frac{N_a}{n_i}\right) \equiv \phi_B \qquad (2.69)$$

Thus, we can write a generalized expression for bulk potential in semiconductors as

$$\phi_B = \left(\phi_i - \phi_f\right) = \pm v_{kT} \ln\left(\frac{N_b}{n_i}\right) \qquad (2.70)$$

where:

the "+" sign is for p-type semiconductors with $N_b = N_a$
the "−" sign is for n-type semiconductors with $N_b = N_d$

Note that the Fermi potential, ϕ_f, is not only a function of carrier concentration but also dependent on temperature through n_i. From Equation 2.70, we observe that since n_i increases with temperature according to Equation 2.15, the magnitude of ϕ_B decreases and as n_i approaches to N_b, ϕ_f approaches to ϕ_i. Thus, with an increase of temperature, the Fermi level approaches the mid-gap position, that is, the intrinsic Fermi level, showing thereby that the semiconductor becomes intrinsic at high temperature. Thus, the doped or extrinsic silicon will become intrinsic if the temperature is high enough. The temperature at which this happens depends upon the dopant concentration. When the material becomes intrinsic, the device can no longer function, and therefore, the intrinsic region is avoided in device operation.

The temperature coefficient of ϕ_f can be obtained by differentiating Equation 2.70 giving

$$\frac{d\phi_f}{dT} = \frac{1}{T}\left[\phi_f - \left(\frac{E_g}{2} + \frac{3}{2}v_{kT}\right)\right] \qquad (2.71)$$

Equation 2.71 gives $d\phi_f/dT \sim 1$ mV K^{-1}. If we use Equation 2.15 for n_i, then ϕ_f with reference to $\phi_i = 0$ at any temperature T can be written in terms of T_{NOM} as

$$\phi_f(T) = \phi_f(T_{NOM}) \cdot \left(\frac{T}{T_{NOM}}\right) - v_{kT} \left\{ \frac{3}{2} \ln\left(\frac{T}{T_{NOM}}\right) + \left[-\frac{E_g(T)}{2kT} + \frac{E_g(T_{NOM})}{2kT_{NOM}} \right] \right\} \quad (2.72)$$

Equation 2.72 is used in circuit CAD tools for modeling the temperature dependence of ϕ_f.

2.2.7.3 Quasi-Fermi Level

Under thermal equilibrium conditions, the electron and hole concentrations are given by Equations 2.62 and 2.63 (using $n = N_d$ and $p = N_a$), respectively, maintaining the condition $pn = n_i^2$. However, when carriers are injected into the semiconductor or extracted out from the semiconductor, the equilibrium condition is disturbed. In nonequilibrium conditions: (1) injection, $np > n_i^2$ or (2) extraction, $np < n_i^2$, we cannot use Equations 2.62 and 2.63. And, the carrier densities can no longer be described by a constant Fermi level through the system. Here, we define *quasi-Fermi* levels such that Equations 2.62 and 2.63 hold as given by

$$n = n_i \exp\left(\frac{E_{fn} - E_i}{kT}\right) = n_i \exp\left[\frac{q}{kT}(\phi_i - \phi_{fn})\right] \quad (2.73)$$

$$p = n_i \exp\left(\frac{E_i - E_{fp}}{kT}\right) = n_i \exp\left[\frac{q}{kT}(\phi_{fp} - \phi_i)\right] \quad (2.74)$$

where:
E_{fn} and E_{fp} are the electron and hole quasi-Fermi levels, respectively

It is to be noted that E_{fn} *and* E_{fp} *are the mathematical tools; their values are chosen so that the accurate carrier concentrations are given in the nonequilibrium situations.* In general, $E_{fn} \neq E_{fp}$.

From Equations 2.73 and 2.74, we can show

$$pn = n_i^2 \exp\left(\frac{E_{fn} - E_{fp}}{kT}\right) \quad (2.75)$$

In equilibrium condition, $E_{fn} = E_{fp} = E_f$ and $\phi_{fn} = \phi_{fp}$ so that Equations 2.73 and 2.74 become same as Equations 2.62 and 2.63 for $n = N_d$ and $p = N_a$, respectively. And, Equation 2.75 becomes $pn = n_i^2$.

2.2.7.4 Transport Equations

In Section 2.2.5.5, we have shown that the electron diffusion current density $J_{n,diff}$ due to concentration gradient in a semiconductor is given by Equation 2.40. On the other hand, the electron current density due to drift of electrons by an applied electric field described in Section 2.2.5.2 is given by Equation 2.30. Thus, when an electric field is present in addition to a concentration gradient, both the drift and diffusion current will flow through the semiconductor. The total electron current density J_n at any point x is then simply the sum of the diffusion and drift currents, that is, J_n $(=J_{n,drift} + J_{n,diff})$. Therefore, the total electron current in a semiconductor is given by

$$J_n = qn\mu_n E + qD_n \frac{dn}{dx} \qquad (2.76)$$

Similarly, the total hole current density J_p $(=J_{p,drift} + J_{p,diff})$ is given by

$$J_p = qp\mu_p E - qD_p \frac{dp}{dx} \qquad (2.77)$$

so that the total current density $J = J_n + J_p$. The current Equations 2.76 and 2.77 are often referred to as the transport equations.

Under thermal equilibrium no current flows inside the semiconductor and therefore, $J_n = J_p = 0$. However, under nonequilibrium conditions J_n and J_p can be written in terms of quasi-Fermi potentials ϕ_n and ϕ_p for electric field, E, in Equations 2.76 and 2.77, respectively, to get

$$J_n = -qn\mu_n \frac{d\phi_n}{dx}$$
$$\qquad (2.78)$$
$$J_p = -qp\mu_p \frac{d\phi_p}{dx}$$

2.2.7.5 Continuity Equations

When carriers diffuse through a certain volume of semiconductor, the current density leaving the volume may be smaller or larger depending upon the recombination or generation taking place inside the volume. Let us consider a small length Δx of a semiconductor as shown in Figure 2.13 with cross-sectional area A in the yz plane.

From Figure 2.13, the hole current density entering the volume $A.\Delta x$ is $J_p(x)$ whereas the density leaving is $J_p(x + \Delta x)$. From the conservation of charge, the rate change of hole concentration in the volume is the sum of (1) net holes flowing out of the volume and (2) net recombination rate. That is,

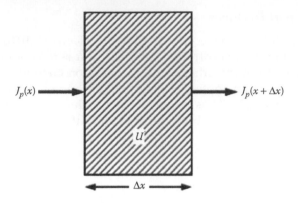

FIGURE 2.13
Current continuity in a semiconductor: $J_p(x)$ is the hole currents flowing into an elemental length Δx of the semiconductor and $J_p(x + \Delta x)$ is the net current flowing out after carrier generation–recombination processes inside the element; U is the net recombination rate.

$$-\frac{\partial p}{\partial t}\Delta x = \left[\frac{1}{q}J_p(x+\Delta x) - \frac{1}{q}J_p(x)\right] + \left(G_p - R_p\right)\Delta x \qquad (2.79)$$

The negative sign is due to the decrease of holes due to recombination; and G_p and R_p are the generation and recombination rate of holes in the volume, respectively. Then from Equation 2.79, we can show

$$-\frac{\partial p}{\partial t} = \frac{1}{q}\frac{\partial J_p}{\partial x} + \left(G_p - R_p\right) \qquad (2.80)$$

Similarly, for electrons we can show

$$-\frac{\partial n}{\partial t} = -\frac{1}{q}\frac{\partial J_n}{\partial x} + \left(G_n - R_n\right) \qquad (2.81)$$

where:
    R_n and G_n are the recombination and generation rate of electrons, respectively

Equations 2.80 and 2.81 are called the *continuity equations* for holes and electrons, respectively, and describe the time-dependent relationship between current density, recombination and generation rates, and space. They are used for solving transient phenomena and diffusion with recombination–generation of carriers.

Equations 2.60, 2.78, 2.80, and 2.81 constitute a complete set of 1D equations to describe carrier, current, and field distributions in a semiconductor; however, they can easily be extended to 3D space. Given appropriate boundary conditions, we can solve them for any arbitrary device structure. Generally, we will be able to simplify them based on physical approximations.

2.3 Theory of *n*-Type and *p*-Type Semiconductors in Contact

We have discussed the basic theory of intrinsic, *n*-type, and *p*-type semiconductors in Section 2.2. In this section, we will discuss the underlying physics of a semiconductor substrate when one region is *n*-type and the immediate adjacent region is *p*-type, forming a junction called the *pn-junction* or *pn-junction diode* or simply *diode*. In reality, a silicon *pn*-junction is formed by counter doping a local region of a larger region of doped silicon as shown in Figure 2.14. The *pn* junctions form the basis for all advanced semiconductor devices. Therefore, understanding their operation is basic to the understanding of most advanced IC devices.

2.3.1 Basic Features of *pn*-Junctions

A silicon *pn*-junction structure is an alternating type of *p*-type and *n*-type doped silicon layers. The *pn*-junctions can be fabricated in a variety of techniques on a silicon substrate using photo mask → Implant › Drive-in. A typical final impurity profile along the active region can be simplified as an *erfc* or *Gaussian* as shown in Figure 2.14b and c.

As shown in Figure 2.14a, the basic structure includes an *n*-region doped on a *p*-type substrate. The vertical cross section of the intrinsic or active *pn*-junction is shown in Figure 2.14a by a vertical cutline *A*. The 1D-doping

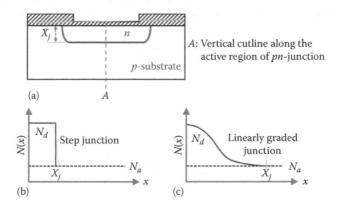

FIGURE 2.14
A typical *pn*-junction: (a) 2D cross section showing the cutline along the depth of the structure to obtain 1D doping profiles, (b) 1D-doping profile of an abrupt junction, and (c) 1D-doping profile of a graded junction.

profile along the cutline of the active device is shown in Figure 2.14b and c. The metallurgical junction depth X_j is indicated as the point where the net impurity concentrations of donors and acceptors are equal. For compact modeling, the actual impurity profile is approximated by a *step* or *abrupt* (high–low) shallow junctions, Figure 2.14b or a *linearly graded* (deep) junctions, Figure 2.14c, so that a tractable circuit model can be developed. A step doping profile is characterized by constant *p*-type dopant concentration N_a that changes with position in a stepwise fashion to a constant *n*-type dopant concentration N_d.

From the 1D impurity profiles in Figure 2.14b and c, we find that there is a large carrier concentration gradient at the junction resulting in carrier diffusion. Holes from the *p*-side diffuse into the *n*-side, leaving behind negatively charged acceptor ions $\left(N_a^-\right)$ and electrons from the *n*-side diffuse into the *p*-side leaving behind positively charged donor ions $\left(N_d^+\right)$. Consequently, a space charge region is formed (negative charge on the *p*-side and positive charge on the *n*-side), creating thereby an electric field *E*, and, hence, a potential difference as shown in Figure 2.15. The direction of the field (*n*-region to *p*-region) is such that it opposes further diffusion of carriers so that, in thermal equilibrium, the net flow of carriers is zero; that is, an electric field is set up, which tends to pull *electrons* and *holes* back to the original positions. The internal potential difference between the two sides of the junction is called the *built-in potential* or *barrier height*, ϕ_{bi}. The space charge region on two sides of the metallurgical junction is often called the *depletion region*, because the region is depleted of the free carriers.

Figure 2.16a shows the energy-band diagram of a *p*-type silicon and *n*-type silicon physically separated from each other. As discussed in Section 2.2.4, the Fermi level for an *n*-type silicon lies close to its CB, and for a *p*-type silicon lies close to its VB. Also, as we will show later, the Fermi level of a semiconductor is flat, that is, spatially constant, when there is no current flow in it. Therefore, as the *p*-type region and the *n*-type region are brought

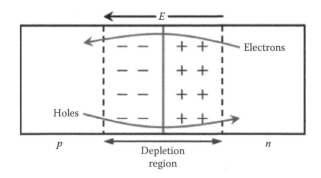

FIGURE 2.15
Formation of built-in electric field due to the space charges left behind by mobile carriers after diffusion from the high- to the low-concentration region on either side of the junction.

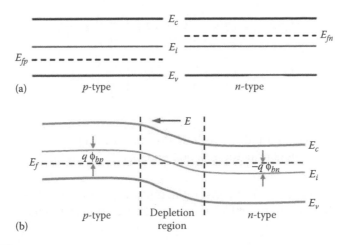

FIGURE 2.16
Energy band diagram of a *pn*-junction at equilibrium: (a) isolated *n*- and *p*-regions and (b) *p-n* regions are in contact to form a *pn*-junction.

together to form a *pn*-junction, the Fermi level must remain flat across the entire structure if there is no current flow in and across the junction. This causes the energy band bending, as shown in Figure 2.16b. The potential difference between the corresponding energy bands on the *p*- and *n*-sides is called the *built-in potential*, ϕ_{bi}, of the *pn*-junction as shown in Figure 2.16b.

2.3.2 Built-In Potential

In *pn*-junctions at equilibrium, the diffusion of carriers is balanced by the drift of carriers by the *built-in* electric field. To facilitate the description of both the *n*-side and the *p*-side of a *pn*-junction simultaneously, when necessary for clarity, we will distinguish the parameters on the *n*-side from the corresponding ones on the *p*-side by adding a subscript *n* to the symbols associated with the parameters on the *n*-side, and subscript *p* to the symbols associated with the parameters on the *p*-side. For example, E_{fp} and E_{fn} denote the Fermi level, respectively, on the *p*-side and *n*-side. Similarly, n_n and p_n denote the electron concentration and hole concentration, respectively, on the *n*-side, and n_p and p_p denote the electron concentration and hole concentration, respectively, on the *p*-side. Thus, n_n and p_p specify the majority carrier concentrations, while n_p and p_n specify the minority carrier concentrations.

Consider the *n*-side of a *pn*-junction at thermal equilibrium. If the *n*-side is nondegenerately doped to a concentration of N_d, then the separation between its Fermi level, which is flat across the junction, and its intrinsic Fermi level is given by Equations 2.62 and 2.63:

$$E_{fn} - E_{in} = kT \ln\left(\frac{N_d}{n_i}\right) = kT \ln\left(\frac{n_{no}}{n_i}\right) \equiv -q\phi_{bn}$$

$$E_{ip} - E_{fp} = kT \ln\left(\frac{N_a}{n_i}\right) = kT \ln\left(\frac{p_{po}}{n_i}\right) \equiv q\phi_{bp}$$

(2.82)

where:
   n_{no} and p_{po} represent the equilibrium concentrations in the *n*-type and
   p-type semiconductors, respectively

Since at equilibrium, E_f is a constant across the *pn*-junction, that is, $E_{fp} = E_{fn}$,
therefore, the built-in potential across the *pn*-junction is given by

$$q\phi_{bi} = E_{ip} - E_{in} = kT \ln\left(\frac{n_{no}p_{po}}{n_i^2}\right)$$

(2.83)

From *pn*-product equation, $n_{no}p_{no} = n_i^2 = n_{po}p_{po}$, therefore, Equation 2.83 can
also be written as

$$\phi_{bi} = \phi_{bp} - \phi_{bn} = v_{kT} \ln\left(\frac{N_a N_d}{n_i^2}\right) = v_{kT} \ln\left(\frac{n_{no}p_{po}}{n_i^2}\right)$$

(2.84)

$$\phi_{bi} = \phi_{bp} - \phi_{bn} = v_{kT} \ln\left(\frac{n_{no}}{n_{po}}\right) = v_{kT} \ln\left(\frac{p_{po}}{p_{no}}\right)$$

(2.85)

Thus, ϕ_{bi} given by Equation 2.84 or 2.85 exists across a *pn*-junction without an
applied bias at thermal equilibrium to counteract diffusion. The typical value
of ϕ_{bi} is in between 0.5 and 0.9 V for silicon junctions and is strongly dependent on temperature due to dependence on n_i. And, ϕ_{bi} across a *pn*-junction
increases as N_d or N_a increases.

2.3.3 Step Junctions

The analysis of *pn*-junction is much simpler if the junction is assumed to
be abrupt, that is, the doping impurities are assumed to change abruptly
from *p*-type on one side to *n*-type on the other side of the junction. The
abrupt junction approximation is reasonable for modern VLSI (very-large-
scale-integrated) devices, where the use of ion implantation for doping
the junctions, followed by low thermal cycle diffusion and/or annealing,
resulting in junctions that are fairly abrupt. Besides, the abrupt-junction
approximation often leads to closed-form solutions for easier understanding of device physics.

2.3.3.1 *Junction Potential and Electric Field*

The analysis of an abrupt junction becomes even simpler in the depletion approximation in which the *pn*-junction is approximated by three regions as illustrated in Figure 2.17. Both the bulk *p*-region, that is, the region with $x < -x_p$, and the bulk *n*-region, that is, the region with $x > x_n$, are assumed to be charge neutral, while the transition region, that is, the region with $-x_p < x < x_n$, is assumed to be depleted of mobile electrons and holes. The width W_d of the depletion region can be obtained by solving Poisson's equation 2.60 as repeated below:

$$\frac{d^2\phi}{dx^2} = -\frac{q}{K_{si}\varepsilon_0}\left\{\left[p(x)-n(x)\right]+\left[N_d(x)-N_a(x)\right]\right\} \qquad (2.86)$$

Let us assume that the free carrier concentrations n and p are negligibly small compared to the fixed ionized impurities $N_a^- \cong N_a$ and $N_d^+ \cong N_d$ over the entire region defined by the depletion width bounded by $-x_p$ and x_n, that is, $N_d \gg n_n$ or p_n and $N_a \gg p_p$ or n_p as shown in Figure 2.17. This assumption is often referred to as the *depletion approximation*. It is often used during the development of analytical device models.

For the simplicity of modeling, we will assume that all the donors and acceptors within the depletion region are ionized, and that the junction is abrupt and not compensated; that is, there are no donor impurities on the *p*-side and no acceptor impurities on the *n*-side. With these assumptions, Equation 2.86 becomes

$$\frac{d^2\phi}{dx^2} = \frac{qN_a(x)}{K_{si}\varepsilon_0} \quad \text{for} -x_p < x < 0 \qquad (2.87)$$

and,

$$\frac{d^2\phi}{dx^2} = -\frac{qN_d(x)}{K_{si}\varepsilon_0} \quad \text{for } 0 < x < x_n \qquad (2.88)$$

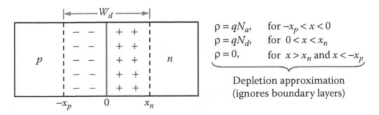

FIGURE 2.17
The *pn*-junction charge condition under *depletion approximation* in three different regions: the equilibrium depletion region is bounded by $-x_p$ and x_n on the *p*-region and *n*-regions, respectively; the depletion region is assumed to be free of mobile carriers with $\rho = 0$.

Integrating Equation 2.87 from $x = -x_p$ to at any point $x < 0$ and Equation 2.88 from $x > 0$ to $x = x_n$ using the boundary condition $d\phi/dx = 0$ at $x = -x_p$ and $x = x_n$, we get the electric field distribution in the depletion region. Thus, assuming a step pn-junction so that N_a and N_d are uniform in p- and n-regions, respectively, and depletion approximation the electric field, $E(x)$ distribution within the depletion region can be shown as

$$E(x) = -\frac{qN_a}{K_{si}\varepsilon_0}(x_p - x) \quad \text{for} - x_p < x < 0 \tag{2.89}$$

$$E(x) = -\frac{qN_d}{K_{si}\varepsilon_0}(x_n - x) \quad \text{for} \, 0 < x < x_n \tag{2.90}$$

Since the electric field must be continuous at $x = 0$, we get from Equations 2.89 and 2.90 the maximum electric field E_{max} as

$$E_{max} = -\frac{qN_a}{K_{si}\varepsilon_0}x_p = -\frac{qN_d}{K_{si}\varepsilon_0}x_n \tag{2.91}$$

or

$$qN_a x_p = qN_d x_n \tag{2.92}$$

which gives the distribution of charge on either side of the junction and shows that the negative charge on the p-side exactly equals the positive charge on the n-side. Equation 2.92 also shows that the width of the depletion region on each side of the junction varies inversely with the dopant concentration; the higher the doping concentration, the narrower the depletion region. Equations 2.89 and 2.90 also show that E varies linearly between 0 and E_{max} as shown in Figure 2.17.

Let ϕ_m is the total potential drop across the pn-junction; that is, $\phi_m = [\phi(x_n) - \phi(x_p)]$. Then the total potential drop can be obtained by integrating Equations 2.89 and 2.90 from $x = -x_p$ to $x = x_n$. Now, we can get:

$$\phi_m = \int_{-x_p}^{x_n} d\phi(x) = -\int_{-x_p}^{x_n} E(x)dx$$

$$= \frac{E_{max}(x_n + x_p)}{2} = \frac{E_{max}}{2}W_d \tag{2.93}$$

where:
$W_d = (x_n + x_p)$ is the total width of the depletion layer

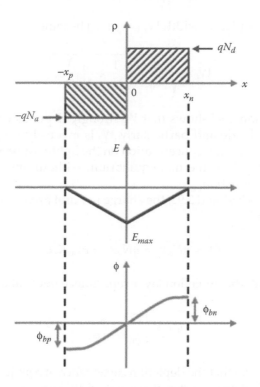

FIGURE 2.18

Depletion approximation of a *pn*-junction: the equilibrium distribution of charge, ρ; electric field, *E*; and electrostatic potential, φ within the depletion region.

It can be seen from Equation 2.93 that ϕ_m is equal to the area under the $E(x)$ versus x plot, that is, Figure 2.18. Eliminating E_{max} from Equations 2.91 and 2.93, we can show that

$$W_d = \sqrt{\frac{2\varepsilon_0 K_{si}(N_a + N_d)}{qN_aN_d}\phi_m}$$
(2.94)

In order to derive expressions for x_p and x_n, we integrate Equations 2.89 and 2.90 once again. Remembering that $E = -d\phi/dx$, and the potential difference between the p and n sides is ϕ_{bi}, it can be shown that

$$x_p = \sqrt{\frac{2K_{si}\varepsilon_0}{q}\frac{N_d}{N_a(N_a + N_d)}\phi_{bi}}$$
(2.95)

And,

$$x_n = \sqrt{\frac{2K_{si}\varepsilon_0}{q}\frac{N_a}{N_d(N_a + N_d)}\phi_{bi}}$$
(2.96)

So that the total depletion width W_d ($=x_p + x_n$) becomes

$$W_d = \sqrt{\frac{2K_{si}\varepsilon_0}{q}\left(\frac{1}{N_a}+\frac{1}{N_d}\right)\phi_{bi}}$$

(2.97)

Note that Equation 2.97 shows that W_d strongly depends on the doping on the lightly doped side and particularly W_d is inversely proportional to the square root of the doping concentration on the lightly doped side. The value of W_d given above is at thermal equilibrium without any external voltage applied to the *pn*-junction.

From Equations 2.91 and 2.92, the charge per unit area on either side of the depletion region is

$$Q_d = qN_dx_p = qN_dx_n = E_{max}K_{si}\varepsilon_0$$

(2.98)

We can show that, the depletion layer capacitance per unit area is given by

$$C_d = \frac{d|Q_d|}{d\phi_m} = \frac{K_{si}\varepsilon_0}{W_d}$$

(2.99)

Equation 2.99 shows that the depletion capacitance of a *pn*-junction is equivalent to a parallel-plate capacitor of separation W_d and dielectric constant K_{si}. Physically, this is due to the fact that only the mobile charge at the edges of the depletion layer, but not the space charge within the depletion region, responds to changes of the applied voltage.

2.3.4 *pn*-Junctions under External Bias

An externally applied voltage, V_d across a *pn*-junction has the effect of shifting the Fermi level of the bulk neutral *n*-region relative to that of the bulk neutral *p*-region. That is, the total potential drop is the sum of the built-in potential and the externally applied potential:

$$\phi_m = \phi_{bi} \pm V_d$$

(2.100)

where:
"+" sign is for the case where the junction is reverse biased and $\phi_m > \phi_{bi}$ the "–" sign is for the case where the junction is forward biased and $\phi_m < \phi_{bi}$

Thus, when the *pn*-junction is in a nonequilibrium condition, with voltage V_d applied to it, then, as stated earlier, the potential barrier height becomes $(\phi_{bi}-V_d)$, so that the depletion width as a function of voltage becomes

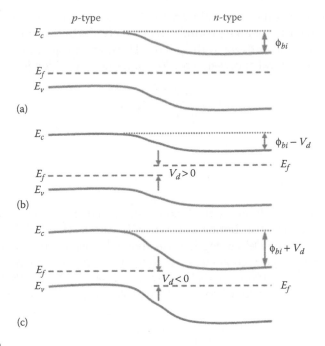

p-type   n-type

(a)

(b)

(c)

FIGURE 2.19
The *pn*-junction in equilibrium and under external bias: (a) equilibrium, (b) forward bias, and (c) reverse bias.

$$W_d = \sqrt{\frac{2K_{si}\varepsilon_0}{q}\left(\frac{1}{N_a}+\frac{1}{N_d}\right)\cdot\left(\phi_{bi}-V_d\right)} \qquad (2.101)$$

This shows that a forward bias V_d ($\equiv V_f$) will result in a decrease in the depletion width due to the decrease in the barrier height, while a reverse bias $-V_d$ ($\equiv V_r$) will result in an increase in the depletion width due to a higher barrier height as shown in Figure 2.19.

Using Equation 2.95 for x_p or 2.96 for x_n in Equation 2.91, the maximum electric field E_{max} in the depletion region becomes

$$E_{max} = \sqrt{\frac{2q}{K_{si}\varepsilon_0}\frac{N_a N_d}{(N_a+N_d)}\left(\phi_{bi}-V_d\right)} \qquad (2.102)$$

Equation 2.102 shows that the higher the reverse voltage (e.g., $-V_d$), the higher is the electric field across the *pn*-junction.

2.3.4.1 One-Sided Step Junctions

If the impurity concentration on one side of a *pn*-junction is much higher than the other side, the junction is called a *one-sided step junction*. In this case, the depletion region extends almost totally into the lighter doped side. For example, in the case of an $n+p$ junction ($N_d \gg N_a$ and $x_n \ll x_p$), the depletion width W_d is almost entirely in the p-side. Thus, from Equation 2.101, we can show that the general expression for W_d for a one-sided step junction is

$$W_d = \sqrt{\frac{2K_{si}\varepsilon_0}{qN_b} \cdot \left(\phi_{bi} \pm V_d\right)} \tag{2.103}$$

where:
$N_b = N_a$ for $n+p$ junction
$N_b = N_d$ for $p+n$ junction

A more accurate result for the depletion width can be obtained by considering the majority carrier distribution tails or spillover (electrons in the n-side and holes in the p-side by Debye length, L_d) as shown by dashed lines in Figure 2.20. Each contributes a correction factor v_{kT} to ϕ_{bi}. Thus, the depletion width is still given by Equation 2.103 except that ϕ_{bi} is replaced by ($\phi_{bi} - 2v_{kT}$) so that, using this more accurate expression, W_d for a one-sided step junction becomes

$$W_d = \sqrt{\frac{2K_{si}\varepsilon_0}{qN_b} \cdot \left(\phi_{bi} - 2v_{kT} \pm V_d\right)} \tag{2.104}$$

However, Equation 2.103 is accurate to within about 3% for the biases normally encountered in the VLSI circuits.

2.3.5 *pn*-Junction Equations

In considering *I–V* characteristics of a *pn*-junction, it is much more convenient to work with the quasi-Fermi potentials, instead of the intrinsic potential.

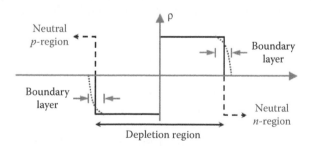

$\rho \cong 0$ outside depletion region; $\rho \cong |N_a - N_d|$ within depletion region; boundary layer spread $\approx 3L_d$.

FIGURE 2.20
Majority carrier spillover (broken lines) outside the depletion region forming a boundary layer of about $3L_d$ at the boundary of the neutral bulk region; L_d is the Debye length defining the abruptness of the junction.

The quasi-Fermi potentials and the current densities for doped semiconductors given by Equations 2.73, 2.74, and 2.78 can be expressed as

$$J_n = -qn\mu_n \frac{d\phi_n}{dx}$$

$$J_p = -qp\mu_p \frac{d\phi_p}{dx}$$

(2.105)

where:

$$\phi_n = \phi_i - v_{kT} \ln\left(\frac{n}{n_i}\right)$$

$$\phi_p = \phi_i + v_{kT} \ln\left(\frac{p}{n_i}\right)$$

(2.106)

where:
ϕ_n and ϕ_p are the quasi-Fermi potentials for electrons and holes, respectively

2.3.5.1 Relationship between Minority Carrier Density and Junction Voltage

Under forward bias V_d, the barrier to majority carrier flow is reduced. And, electrons are injected from *n*-region to *p*-region and holes are injected from *p*-region to *n*-region. The electrons going from *n*-region to *p*-region become minority carriers in the *p*-region. Similarly, *holes* going from *p*-region to *n*-region become minority carriers in the *n*-region. Therefore, the minority carrier behavior is of fundamental importance to understand the behavior of a *pn*-junction. The minority carriers injected across the barrier will tend to recombine if given sufficient time. They will also tend to diffuse away from the region of the junction.

In order to calculate diode current in thermal equilibrium, let us consider n_{no} and p_{po} are the equilibrium majority carrier concentrations in the neutral *n*- and *p*-regions, respectively; and n_{po} and p_{no} are the equilibrium minority carrier electron and hole concentrations in the neutral *p*- and *n*-regions, respectively, as shown in Figure 2.21. Then from carrier statistics discussed in Section 2.2.7.2, we have in the *neutral n*-region

$$n_{no} \cong N_d; \quad p_{no} \cong \frac{n_i^2}{N_d}$$

(2.107)

and, in the *neutral p*-region

$$p_{po} \cong N_a; \quad n_{po} \cong \frac{n_i^2}{N_a}$$

(2.108)

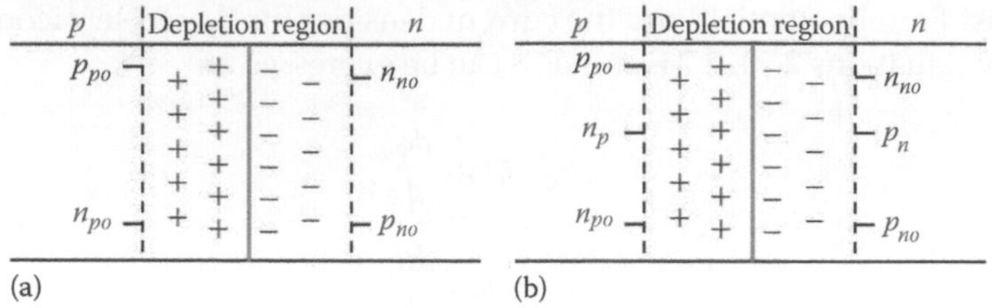

FIGURE 2.21
Carrier concentrations at the edge of depletion region: (a) *pn*-junction at equilibrium where p_{po} and n_{no} are the equilibrium majority carrier hole and electron concentrations in the *p*-type and *n*-type regions, respectively, whereas n_{po} and p_{no} are the equilibrium minority carrier electron and hole concentrations in the *p*-type and *n*-type regions, respectively and (b) *pn*-junction after minority carrier n_p and p_n injection in the bulk *p*-region and *n*-region, respectively.

From Equation 2.85, the equilibrium carrier concentrations in a *pn*-junction are given by the expressions

$$\phi_{bi} = \begin{cases} v_{kT} \ln\left(\dfrac{n_{no}}{n_{po}} \right) \\[2em] v_{kT} \ln\left(\dfrac{p_{po}}{p_{no}} \right) \end{cases} \tag{2.109}$$

Therefore, from Equation 2.109, we can write for a *pn*-junction at equilibrium

$$n_{no} = n_{po} \exp\left(\frac{\phi_{bi}}{v_{kT}} \right)$$
$$p_{po} = p_{no} \exp\left(\frac{\phi_{bi}}{v_{kT}} \right) \tag{2.110}$$

Now, under the applied bias V_d, we replace ϕ_{bi} by $(\phi_{bi} \pm V_d)$; therefore, from Equation of 2.110, the nonequilibrium carrier concentrations are given by

$$n_n = n_p \exp\left(\frac{\phi_{bi} - V_d}{v_{kT}} \right)$$
$$p_p = p_n \exp\left(\frac{\phi_{bi} - V_d}{v_{kT}} \right) \tag{2.111}$$

where:
  n_p is the nonequilibrium minority electron concentration at the edge of the depletion region in the neutral *p*-region
  p_n is the nonequilibrium hole concentration at the edge of the depletion region in the neutral *n*-region as shown in Figure 2.21b

Let us further assume *low-level injection*, that is, the injected carrier densities are lower than the background concentrations, so that $n_n = n_{n0}$ and $p_p = p_{p0}$. Then from Equations 2.110 and 2.111, we get

$$n_p = n_{po} \exp\left(\frac{V_d}{v_{kT}}\right)$$

$$p_n = p_{no} \exp\left(\frac{V_d}{v_{kT}}\right)$$

(2.112)

In Equation 2.112 n_p and p_n are the injected minority carrier concentrations at the edge of the depletion region in the p- and n-regions, respectively. The expressions in Equation 2.112 define the minority carrier densities at the *edge* of the space charge region under an applied bias and are the most important boundary conditions governing a pn-junction. They relate the minority carrier concentrations at the boundaries of the depletion layer to their thermal equilibrium values and to the applied voltage across the junction. They apply to both a forward-biased ($V_d > 0$) junction resulting in $n_p \gg n_{po}$ at $x = -x_p$ and $p_n \gg p_{no}$ at $x = x_n$, and to a reverse-biased ($V_d < 0$) junction resulting in $n_p \ll n_{po}$ at $x = -x_p$ and $p_n \ll p_{no}$ at $x = x_n$. Expressions in Equations 2.112 can be expressed as

$$n_p = \frac{n_i^2}{p_{po}} \exp\left(\frac{V_d}{v_{kT}}\right)$$

$$p_n = \frac{n_i^2}{n_{no}} \exp\left(\frac{V_d}{v_{kT}}\right)$$

(2.113)

Again, for low-level injection in the p-region, $p_{po} = p$ and $n_p = n$; similarly, in the n-region, $n_{no} = n$ and $p_n = p$; therefore, we get from Equation 2.112 or Equation 2.113

$$pn = n_i^2 \exp\left(\frac{V_d}{v_{kT}}\right)$$

(2.114)

Equation 2.114 defines the pn-product of carriers at the depletion edge under the applied voltage V_d as shown in Figure 2.21. Thus, the applied bias in a pn-junction sets up the following processes as shown in Figure 2.22:

- The injected carriers in the n- and p-regions momentarily set up an electric field (from n to p)
- This field draws in majority carriers in each region

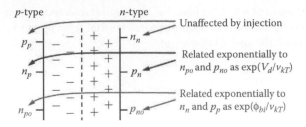

FIGURE 2.22
Carriers in a *pn*-junction under applied bias showing the corresponding dependence on built-in potential and applied bias.

- These majority carriers neutralize the injected carriers and reestablish the charge neutrality
- While this process is going on, the injected minority carriers diffuse into the *n*- and *p*-regions; that is, recombination process takes place over some distance

The distribution of carriers in the *n*-region of the *pn*-junction is shown in Figure 2.23. The majority carrier concentration shown by broken line remains unchanged whereas the minority carrier concentration decays exponentially and approaches to the equilibrium concentration in each side of the junction.

The injected excess carriers set up a momentary electric field, E, in the regions of excess carrier concentration. Then the current due to this drift electric field in the *n*-region is $I_{drift} = q\mu_n nE$ for majority carrier *electrons* and $I_{drift} = q\mu_p pE$ for minority carrier *holes*. Since $n \gg p$, the hole drift current is negligible in the *n*-region. Similarly, electron drift current is negligible in the neutral *p*-region. The minority carriers move primarily by diffusion while the majority carriers are pulled to the junction by drift. Since the injected

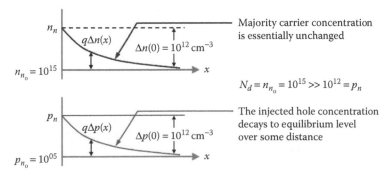

FIGURE 2.23
The carrier profile in the *n*-region of a *pn*-junction with applied bias; the majority carrier electron concentration, n_{no} is 1×10^{15} cm^{-3} and injected carrier concentration is 1×10^{12} cm^{-3} describing low-level injection.

minority carriers control the current flow in a *pn*-junction, the current flow in *pn*-junctions can be considered as the diffusion current only. *Thus, we see that the minority carriers really control the behavior of pn-junctions.*

2.3.6 *pn*-Junctions *I–V* Characteristics

We discussed in Section 2.3.2 that the drift component of the current caused by the electric field in the depletion region is exactly balanced out by the diffusion component of the current caused by the electron and hole concentration gradient across the junction, resulting in zero current flow in the *pn*-junction device. When an external voltage is applied, this current component balance is upset, and current will flow in the diode. If carriers are generated by light or some other external means, thermal equilibrium is disturbed, and current can also flow in a *pn*-junction. Here, the current flow in a *pn*-junction as a result of an external applied voltage is described.

Let us consider a forward-biased *pn*-junction. Electrons are injected from the *n*-side into the *p*-side, and holes are injected from the *p*-side to *n*-side. If the generation and recombination in the depletion region are negligible, then the hole current leaving *p*-side is the same as the hole current entering the *n*-side. Similarly, the electron current leaving the *n*-side is equal to the electron current entering the *p*-side. To determine the total current flowing in the *pn*-junction, we need to determine either hole current entering the *p*-side or electron current entering to *n*-side of the *pn*-junction.

The starting point for describing *I–V* characteristics of a *pn*-junction is the continuity equations. From Equation 2.81, the electron continuity equation is given by

$$-\frac{\partial n}{\partial t} = -\frac{1}{q}\frac{\partial J_n}{\partial x} + \left(G_n - R_n\right) \tag{2.115}$$

where:
R_n and G_n are the electron recombination and generation rates, respectively

Equation 2.115 can be rewritten as

$$\frac{\partial n}{\partial t} = \frac{1}{q}\frac{\partial J_n}{\partial x} - \frac{n-n_o}{\tau_n} \tag{2.116}$$

where τ_n is the electron lifetime defined in terms of the excess electron concentration n over the thermal equilibrium value n_o in Equations 2.48 and 2.53 and is given by

$$\tau_n \equiv \frac{n-n_o}{R_n - G_n} \tag{2.117}$$

Now, substituting Equation 2.76 for J_n in Equation 2.116, we get

$$\frac{\partial n}{\partial t} = n\mu_n \frac{\partial E}{\partial x} + \mu_n E \frac{\partial n}{\partial x} + D_n \frac{\partial^2 n}{\partial x^2} - \frac{n - n_0}{\tau_n} \tag{2.118}$$

Equation 2.118 is the general equation that is solved under appropriate boundary conditions to derive an expression for electron current flow across a *pn*-junction under an applied bias.

In order to calculate the diode current, we assume that the injected minority carriers move away from the depletion region by diffusion only—*diffusion approximation*. We calculate the diode current under the following assumptions:

1. The step junction profile is applicable
2. The depletion approximation is valid
3. Low-level injection is maintained in the bulk
4. No generation–recombination takes place in the depletion region
5. There is no voltage drop in the bulk region so that V_d is sustained entirely across the depletion region
6. The width of the bulk *p*- and *n*-regions outside the depletion region is much longer than the minority carrier diffusion length for holes and electrons L_p and L_n, respectively (long-base diode)

With the above simplifying assumptions, the current through a *pn*-junction can be shown to be

$$I_d = I_s \left[\exp\left(\frac{V_d}{v_{kT}}\right) - 1 \right] \tag{2.119}$$

where I_s is called the reverse saturation current and is given by

$$I_s = \begin{cases} qA_d n_i^2 \left[\dfrac{D_p}{N_d L_p} + \dfrac{D_n}{N_a L_n} \right]; & W_n > L_p \quad \text{and} \quad W_p > L_n \\[4mm] qA_d n_i^2 \left[\dfrac{D_p}{N_d W_n} + \dfrac{D_n}{N_a W_p} \right]; & W_n < L_p \quad \text{and} \quad W_p < L_n \end{cases} \tag{2.120}$$

where:
  A_d is the active area of the *pn*-junction
  W_n and W_p are the width of the neutral *n*- and *p*-regions, respectively
  D_n and D_p are the minority carrier electron and hole diffusion constants, respectively
  L_n and L_p are the minority carrier electron and hole diffusion lengths, respectively

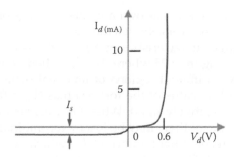

FIGURE 2.24
Current voltage characteristics of a typical *pn*-junction; I_s is the reverse saturation current; an applied voltage of about 0.6 V is required to overcome the built-in voltage and device conduction.

Actual diodes may represent intermediate cases, that is, $W_n > L_p$ and $W_p < L_n$ and vice versa. In either case, the lightly doped side of the junction largely determines the diode current I_d in Equation 2.119. Figure 2.24 shows a typical *I–V* characteristics of a *pn*-junction.

2.3.6.1 Temperature Dependence of pn-Junction Leakage Current

From Equation 2.120 we see that the temperature dependence of the electron and hole diffusion currents is dominated by the temperature dependence of the parameter n_i^2, which is proportional to $\exp(-E_g/kT)$ as shown in Equation 2.14, where E_g is the bandgap energy. Then substituting for $n_i(T)$ from Equation 2.14 in Equation 2.120, we can show the temperature dependence of I_s with reference to T_{NOM} as

$$I_s(T) = I_s(T_{NOM})\left(\frac{T}{T_{NOM}}\right)^3 \exp\left[\frac{E_g(T_{NOM})}{kT_{NOM}} - \frac{E_g(T)}{kT}\right]$$

$$= I_s(T_{NOM})\left(\frac{T}{T_{NOM}}\right)^{XTI} \exp\left[\frac{E_g(T_{NOM})}{kT_{NOM}} - \frac{E_g(T)}{kT}\right]$$

(2.121)

where exponent 3 is replaced by the parameter *XTI*. In advance *pn*-junction model for circuit CAD, two parameters *XTI* and *NJ*, called *temperature exponent coefficient fitting* parameters, are used to express Equation 2.121 as

$$I_s(T) = I_s(T_{NOM})\cdot\exp\left\{\frac{\left[E_g(T_{NOM})/kT_{NOM}\right] - \left[E_g(T)/kT\right] + XTI\ln(T/T_{NOM})}{NJ}\right\}$$ (2.122)

2.3.6.2 Limitations of pn-Junction Current Equation

The ideal *pn*-junction current Equation 2.119 accurately describes the device characteristics of *pn*-junctions over a certain range of applied voltage. However,

Equation 2.119 becomes inaccurate over a significant range of device operations both in the forward- and reverse-biased modes.

The current voltage characteristics of a *forward-biased* silicon *pn*-junction diode are shown in Figure 2.25 where the ideal diode current is shown by the broken line. Two different regions of nonideal behavior are shown in this plot. At a very low value of the forward bias ($V_d < 0.3$ V), the injected carrier densities are relatively small. When these carriers move through the depletion region, some of them may be lost by recombination in this region, thereby forming a recombination current I_{rec}, which is added to the ideal diode diffusion current. The result is a larger total current than that predicted by the ideal diode Equation 2.119, particularly in the low current level, and violates assumption 4. Thus, I_{rec} dominates in the silicon diode at very low current levels and negligibly small at higher current levels.

In deriving Equation 2.119, we have assumed that all the minority carriers cross the depletion region. In practice, some recombine through trapping centers. Then, using the SRH theory of generation and recombination, it can be shown that the space-charge recombination current I_{rec} is

$$I_{rec} = \frac{qA_d n_i W}{2\tau_{rec}} \exp\left(\frac{V_d}{2v_{kT}}\right) \tag{2.123}$$

In Equation 2.123, τ_{rec} is the lifetime associated with the recombination of excess carriers in the depletion region. τ_{rec} is analogous to, but usually greater than, τ_n and τ_p for the neutral regions and is generally approximately equal

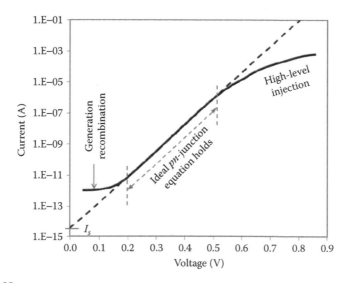

FIGURE 2.25
Forward characteristics of a real *pn*-junction: plot shows the deviation of ideal current equation at the low- and high-current levels due to generation–recombination and high-level injections, respectively.

to $2\sqrt{\tau_p \tau_n}$. Thus, the total diode saturation current, I_s, is the sum of Equations 2.120 and 2.123. In general, until V_d reaches a value of about 0.4 V, the neutral region diffusion current will be less than I_{rec}.

At high current levels, the injected minority carrier density is comparable to the majority carrier concentration (high-level injection), and therefore, assumption 3 is invalid. For high-level injection, majority carrier concentration increases significantly above its equilibrium value, giving rise to an electric field. Thus, in such cases both drift and diffusion components must be considered. The presence of the electric field results in a voltage drop across this region and thus reduces the applied voltage across the junction, resulting in a lower current than expected. It can be shown that under high-level injection the diode current I_d is

$$I_d = \frac{qA_d n_i D_p}{W} \exp\left(\frac{V_d}{2v_{kT}}\right) \quad \text{(high-level injection)} \tag{2.124}$$

which indicates that high-level current depends on $1/2v_{kT}$ rather than on $1/v_{kT}$ as shown in Figure 2.25. Thus, depending on the magnitude of the applied forward voltage, the current through a *pn*-junction can be represented by an empirical expression

$$I_d = I_s \left[\exp\left(\frac{V_d}{n_E v_{kT}}\right) - 1 \right] \tag{2.125}$$

where n_E is called the *ideality factor* and is a measure of the deviation of the real and the ideal *I–V* plots. When recombination current dominates or when there is high-level injection $n_E = 2$ and when diffusion current dominates $n_E = 1$.

In the case of a *reverse-biased pn*-junction, Figure 2.26 shows the current through the *pn*-junction where I_s is the current due to an ideal *pn*-junction (Equation 2.119). Clearly, the current in a real *pn*-junction does not saturate at $-I_s$ as predicted by Equation 2.119. This is because when the *pn*-junction is reverse biased, generation of electron–hole pairs in the depletion region takes place, which was neglected in the ideal *pn*-junction equation. In fact, the generation current dominates because carrier concentrations are smaller than their thermal equilibrium values. Again, using SRH theory, it can be shown that the generation current I_{gen} is

$$I_{gen} = \frac{qA_d n_i W_d}{2\tau_{gen}} \tag{2.126}$$

where:

τ_{gen} is the generation lifetime of the carriers in the depletion region and is approximately equal to $2\tau_p$ if we assume $\tau_p = \tau_n$

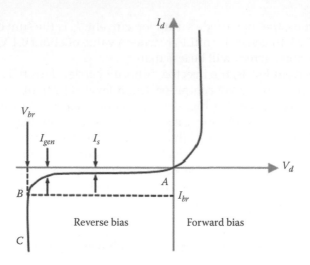

FIGURE 2.26
Reverse characteristics of a real *pn*-junction; V_{br} and I_{br} are the breakdown voltage and current, respectively; I_s is the ideal reverse saturation current; and I_{gen} is the generation current in the depletion region.

Note that while I_s is proportional to n_i^2, I_{gen} is proportional to n_i only. Thus, I_{gen} will dominate when n_i is small as is the case at room and low temperatures. Further, since the space charge width W_d increases as the square root of the reverse bias (Equation 2.103), the generation current increases with reverse bias voltage as shown in Figure 2.26. Thus, taking into account I_{gen}, the total reverse current I_r becomes $I_r \equiv -I_d = -(I_s + I_{gen})$. This value of I_r agrees well with the measured value of reverse current and also it provides proper voltage dependence of the reverse current in properly constructed silicon planar *pn*-junctions.

In real *pn*-junctions there is a third component of leakage current, called the surface leakage current I_{sl}. This current can be treated as a special case of I_{gen} modeled at the surface where a high concentration of dislocations at the oxide-silicon interface, often referred to as fast surface states, provides additional generation centers over those present in the bulk. It is very much process dependent and is responsible for large variation in the leakage current. Both process-induced and electrically induced defects at the surface generally increase the generation rate by an order of magnitude compared with the bulk recombination–generation rate. In that case I_{sl} dominates over the other components of I_r and is thus responsible for higher leakage current for a *pn*-junction compared to that predicted by the sum of I_{gen} and I_s. Leakage current is highly temperature dependent due to the presence of n_i term. Also, note that the generation limited leakage current is proportional to n_i while diffusion limited leakage current is proportional to n_i^2.

2.3.6.3 Bulk Resistance

At high current levels, bulk resistance and the metal–silicon contact resistance can produce a significant voltage drop (assumption 5), resulting in a smaller voltage across the junction and thus a lower current. Usually, the bulk resistance and contact resistance are combined into one resistor called series resistance r_s (Figure 2.27). Thus, if V_d is the applied voltage to the diode terminals and V_d' is the voltage across the diode junction, resulting in the current I_d as shown in Figure 2.24, we have

$$V_d = V_d' + r_s I_d \qquad (2.127)$$

Under the ideal conditions when $r_s = 0$, $V_d = V_d'$, that is related to I_d by Equations 2.119 or 2.125. Thus, in the presence of the series resistance, I–V expression of a *pn*-junction becomes

$$I_d = I_s \left[\exp\left(\frac{V_d - I_d r_s}{n_E v_{kT}} \right) - 1 \right] \qquad (2.128)$$

Rearranging this equation yields

$$V_d = n_E v_{kT} \ln\left(1 + \frac{I_d}{I_s} \right) + I_d r_s \qquad (2.129)$$

Clearly, when I_d is large, the terminal voltage V_d will increase linearly with I_d because $I_d r_s$ increases faster than the logarithmic term.

2.3.6.4 Junction Breakdown Voltage

From Equation 2.126, we observe that the *reverse* (or leakage) current of a *pn*-junction depends on W_d, and from Equation 2.101 we observe that W_d depends on the reverse bias $V_d = V_r$. Also, we notice from Equation 2.102 that the electric field in the depletion region increases with the increase of V_r. When the field reaches a certain critical field E_c corresponding to the reverse voltage $V_r = V_{br}$ called the *breakdown voltage*, a slight increase of reverse

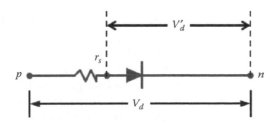

FIGURE 2.27
Diode model at high-level current: r_s is the diode resistance due to contact and the neutral bulk regions.

voltage causes a very large increase of current as shown in Figure 2.26 (region BC). This condition is often called the *breakdown condition* and is a most important consideration in device design. The breakdown occurs because carriers, while moving through the depletion region, acquire sufficient energy to create new electron–hole pairs through impact ionization [24,25]. The newly generated electron–hole pairs can also acquire sufficient energy from the field to create additional electron–hole pairs. Since the electrons and holes travel in opposite directions, the carriers can multiply a few times in the depletion region before they reach the electrodes. This multiplicative process results in an avalanche effect. The resulting breakdown voltage, V_{br}, is called the avalanche breakdown voltage and can be obtained using Equation 2.102.

$$E_{max} = \sqrt{\frac{2q}{K_{si}\varepsilon_0} \frac{N_a N_d}{(N_a + N_d)} (\phi_{bi} + V_r)} \qquad (2.130)$$

At the breakdown condition, $E_{max} = E_c$ and $V_r = V_{br}$; since $V_{br} \gg \phi_{bi}$, we can safely neglect ϕ_{bi} in Equation 2.130 to obtain the expression for breakdown voltage for a *pn*-junction

$$V_{br} = \frac{K_{si}\varepsilon_0 E_c^2}{2q} \left(\frac{1}{N_a} + \frac{1}{N_d} \right) \qquad (2.131)$$

Equation 2.131 shows that any increase in the doping, either of *n*- or *p*-region, results in a decrease in the breakdown voltage V_{br}. Further, it shows that V_{br} is controlled by the concentration N_b of the lightly doped region and is proportional to $1/N_b$. In a *pn*-junction, V_{br} generally varies as $N^{-2/3}$ [13]. For moderately doped silicon (1×10^{14} to 1×10^{16} cm^{-3}), the value of the critical field is $E_c \sim 4 \times 10^5$ V cm^{-1} and for a first approximation V_{br} is independent of doping [26].

If the *pn*-junction is heavily doped (concentration $>1 \times 10^{18}$ cm^{-3}) on both sides, the depletion layer is very narrow. Carriers cannot gain enough energy within the depletion region so that avalanche breakdown is not possible. However, in the depletion region, the electric field is high; E_{max} can be close to 1×10^6 V cm^{-1}. In such a heavily doped *p+ n+* junction under reverse bias, electrons at the VB of the *p+* side *tunnel* through the forbidden gap into the CB of the *n+* side. This tunneling process can be approximated by a particle penetrating a triangular potential barrier, with a height higher than its energy by the semiconductor bandgap E_g. This *tunneling* process contributes to the current resulting in breakdown of the junction. This mechanism of breakdown is called the *Zener breakdown*. In the source-drain *pn*-junction of a MOSFET, the avalanche breakdown dominates [27,28].

2.3.7 *pn*-Junction Dynamic Behavior

Besides electrostatic behavior, *pn*-junctions are often subject to varying voltages. In such dynamic operations, charges in the *pn*-junction vary, resulting in an additional current not predicted by the DC current (Equation 2.119). There are two types of stored charge in a *pn*-junction: (1) the charge Q_{dep} due to the depletion or space-charge region on each side of the junction and (2) the charge Q_{dif} due to minority carrier injection. Remember that it is these injected (excess) mobile carriers that generate current I_d and also represent a stored charge Q_{dif} in a *pn*-junction. The latter is given by the area between the curve representing p_n (or n_p) and the steady state level p_{no} (or n_{po}) as shown in Figure 2.23. These two types of stored charges result in two types of capacitances: the junction capacitance C_j due to Q_{dep} and the diffusion capacitance due to Q_{dif} as discussed in Sections 2.3.7.1 and 2.3.7.2, respectively.

2.3.7.1 *Junction Capacitance*

In a *pn*-junction, a small change in the applied voltage causes an incremental change in the depletion region charge Q_{dep} due to the corresponding change in the depletion width. If the applied voltage is returned to its original value, carriers flow in such a direction that the previous increment of charge is neutralized. The response of the *pn*-junction to the incremental voltage thus results in a generation of an effective capacitance C_j referred to as the *transition capacitance, junction capacitance*, or *depletion layer capacitance*. Recalling the definition of capacitance per unit area in terms of an incremental charge dQ_{dep} per unit area induced by an applied voltage dV_d, we have

$$C_j = \frac{dQ_{dep}}{dV_d} \qquad (2.132)$$

Considering, $Q_{dep} = qN_a x_p = qN_d x_n$ from Equation 2.92, we can show

$$C_j = qN_a \frac{dx_p}{dV_d} = qN_d \frac{dx_n}{dV_d} \qquad (2.133)$$

Then using Equation 2.95 or 2.96, the *pn*-junction capacitance per unit area can be shown as

$$C_j = \sqrt{\frac{qK_{si}\varepsilon_0}{2(\phi_{bi} - V_d)} \left(\frac{N_a N_d}{N_a + N_d} \right)} \qquad (2.134)$$

Equation 2.134 is the expression for the diode capacitance for a step profile in terms of the physical parameters of the device. Remember that Equation 2.134 is valid for $V_d < \phi_{bi}$, that is, for reverse bias only. Comparing Equations 2.134 and 2.101, it is easy to see that

$$C_j = \frac{K_{si}\varepsilon_0}{W_d} \qquad (2.135)$$

Equation 2.135 states that the junction capacitance is equivalent to that of a parallel plate capacitor with silicon as the dielectric and separated by a distance W_d, the depletion width. Though the derivation of Equation 2.134 is based on a step profile, it can be shown that *the relationship is valid for any arbitrary doping profile.*

It should be pointed out that although the *pn*-junction capacitance can be calculated using the parallel plate capacitor formula, there are differences between the two types of capacitors. While true parallel plate capacitance is independent of applied voltage, *pn*-junction capacitance given by Equation 2.134 becomes voltage dependent through W_d. Therefore, the total charge in a *pn*-junction cannot be obtained by simply multiplying the capacitance by the applied voltage, although a small variation in the charge can still be obtained by multiplying a small variation in the voltage by the instantaneous capacitance value. Another difference is that, in a *pn*-junction, the dipoles in the transition region have their positive charge in the *n*-side depletion region and negative charge in the *p*-side depletion region, while in a parallel plate capacitor the separation between the charges in the dipoles is much less and the dipoles are distributed homogenously throughout the dielectric.

For a one-sided step junction, for example, $n+p$ diode with $N_d \gg N_a$, Equation 2.134 becomes

$$C_j = \sqrt{\frac{qK_{si}\varepsilon_0 N_a}{2(\phi_{bi} - V_d)}} \qquad (2.136)$$

For the circuit CAD, it is more convenient to express capacitance in terms of model parameters. If C_{j0} is the junction capacitance at equilibrium, that is, at $V_d = 0$, then from Equation 2.134 we get

$$C_{j0} = \sqrt{\frac{qK_{si}\varepsilon_0}{2\phi_{bi}}\left(\frac{N_a N_d}{N_a + N_d}\right)} \qquad (2.137)$$

Then using Equation 2.137 in Equation 2.134, the junction capacitance for a *pn*-junction is given by

$$C_j = \frac{C_{j0}}{\sqrt{1-(V_d/\phi_{bi})}} \qquad (2.138)$$

In IC *pn*-junctions, the doping profile is neither abrupt nor linearly graded as assumed in the derivation for C_j, and therefore, to calculate the capacitance for real devices, we replace the one-half power in Equation 2.138 by m_j, called the junction *grading coefficient*, resulting in the following generalized equation for C_j as

$$C_j = \frac{C_{j0}}{\left[1-\left(V_d/\phi_{bi}\right)\right]^{m_j}}$$ (2.139)

For IC *pn*-junctions, m_j ranges between 0.2 and 0.6. Figure 2.28 shows a plot of the junction capacitance C_j as a function of junction voltage V_d. Note that the capacitance C_j decreases as the reverse-biased $|V_d|$ increases (V_d is negative). When the diode is forward biased (V_d is positive), the capacitance C_j increases and becomes infinite at $V_d = \phi_{bi}$ as shown in Figure 2.28 (Curve 1). This is because Equation 2.139 no longer applies due to the depletion approximation becoming invalid. A more exact analysis of the C_j as a function of the behavior of the forward bias V_d is shown by Curve 2. However, in SPICE a straight line is used instead of Curve 2 in Figure 2.28. In this case, we define a parameter F_c, $0 < F_c < 1$, such that when the *pn*-junction is forward biased and $V_d \geq F_c\phi_{bi}$, the following equation for C_j is used. By Taylor series expansion of $\left[1-\left(V_d/\phi_{bi}\right)\right]^{-m_j}$ at $V_d = F_c.\phi_{bi}$, we can show

$$\left(1-\frac{V_d}{\phi_{bi}}\right)^{-m_j} = \left(1-FC\right)^{-\left(m_j+1\right)}\left[m_j\left(\frac{V_d}{\phi_{bi}}\right)+1-FC\left(m_j+1\right)\right]$$ (2.140)

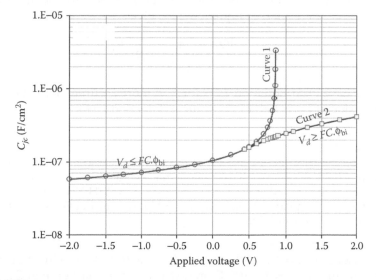

FIGURE 2.28

Junction capacitance of a typical *pn*-junction obtained by using the expressions in Equation 2.141; curve 1 represents Equation 2.138 for $V_d < \phi_{bi}$ and curve 2 is obtained by analytical expression to ensure convergence in circuit simulation during forward biasing a *pn*-junction.

Then we can show

$$
C_j = \begin{cases}
\dfrac{C_{j0}}{\left(1-\dfrac{V_d}{\phi_{bi}}\right)^{m_j}}; & V_d \le FC\phi_{bi} \\[4ex]
\dfrac{C_{j0}}{(1-FC)^{m_j+1}}\left[\dfrac{m_j}{\phi_{bi}}V_d+1-FC(m_j+1)\right]; & V_d \ge FC\phi_{bi}
\end{cases}
\tag{2.141}
$$

2.3.7.2 Diffusion Capacitance

The diffusion capacitance C_{dif} is associated with the rearrangement of the excess minority carriers in response to an incremental change in the applied forward voltage. The variation in the stored charge Q_{dif} associated with the excess minority carrier injection in the bulk region under forward bias, is modeled by the capacitance C_{dif}. The capacitance C_{dif} is called the diffusion capacitance, because the minority carriers move across the bulk region by diffusion; since Q_{dif} is proportional to the current I_d, for an $n+\,p$ junction we can write

$$
Q_{dif} = \frac{1}{A_d}\tau_p I_d
\tag{2.142}
$$

For a short base diode, τ_p is replaced by τ_t, the transit time of the *pn*-junction. For the case of a long base diode the transit time is the excess minority carrier lifetime. Differentiating Equation 2.142 gives

$$
C_{dif} = \frac{dQ_{dif}}{dV_d} = \frac{\tau_p I_s}{A_d \upsilon_{kT}}\exp\left(\frac{V_d}{\upsilon_{kT}}\right)
\tag{2.143}
$$

where we have used Equation 2.119 for I_d. A more accurate derivation shows that the value of C_{dif} is half of the value in Equation 2.143.

> **EXAMPLE:**
> Let us compare the magnitude of the two capacitances for a forward bias of 0.3 V; assume we have an $n+\,p$ diode with $N_a = 1 \times 10^{15}$ cm^{-3} and $N_d = 1 \times 10^{19}$ cm^{-3}; then Equation 2.84 gives $\phi_{bi} = 0.814$ V. For a forward bias of 0.3 V, Equation 2.101 gives $W_d = 8.15 \times 10^{-5}$ cm and Equation 2.134 gives $C_j = 1.27 \times 10^{-8}$ F cm^{-2}.

Again, assuming $\tau_t = 1 \times 10^{-7}$ sec, and $I_s = 4 \times 10^{-12}$ A for a junction area of 20×20 μm^2 gives $C_{dif} = 4 \times 10^{-7}$ F cm^{-2}, which is much larger than C_j.

It should be noted that under forward bias, C_{dif} increases much faster with increasing $V_d\,(=V_f)$, due to the exponential dependence on V_d, as compared to C_j. However, under reverse bias, C_j decreases much more slowly with increasing $V_d (=-V_r)$, as compared to C_{dif}. Therefore, C_j is the dominant capacitance for reverse bias and small for forward bias ($V_d < \phi_{bi}/2$), while diffusion capacitance C_{dif} is dominant for forward bias ($V_d > \phi_{bi}/2$).

2.3.7.3 Small Signal Conductance

In the model discussed in Section 2.3.7.2, referred to as the large-signal model, we did not place any restriction on the allowed voltage variation. However, in some circuit situations, voltage variations are sufficiently small so that the resulting small current variations can be expressed using linear relationships. This is the so called small signal behavior of a *pn*-junction. An example of linear relations are the capacitances C_j and C_{dif} in Equations 2.141 and 2.143, respectively, as they represent an overall nonlinear charge storage effect in terms of linear circuit elements (capacitors), although we did not label them as such.

For small variations about the operating point, which is set by the DC condition, the nonlinear junction current can be linearized so that the incremental diode current is proportional to the incremental applied bias. This linear relationship is used to calculate the small signal conductance g_d

$$g_d = \frac{dI_d}{dV_d} \tag{2.144}$$

Using (2.119) for I_d, we have

$$g_d = \frac{I_s}{v_{kT}} \exp\left(\frac{V_d}{v_{kT}}\right) = \frac{1}{v_{kT}}(I_d + I_s) \tag{2.145}$$

Thus, Equation 2.145 clearly shows that g_d is proportional to the slope of the DC characteristics at the operating point. When the diode is forward biased, I_d is much larger than I_s and therefore, g_d is proportional to I_d. However, when the diode is reverse biased, $I_d = -I_s$ and therefore, from Equation 2.145, g_d becomes zero. But in real diodes, $g_d \neq 0$ in the reverse bias condition due to the fact that the generation current I_{gen} (Equation 2.126) is dominant conduction mechanism.

2.3.8 Diode Equivalent Circuit for Circuit CAD

The small signal equivalent circuit of a *pn*-junction is shown in Figure 2.29. In Figure 2.29, r_s represents the series resistance due to ohmic drop across the neutral *n*- and *p*-regions; C_j is junction capacitance; C_d is the diffusion capacitance due to the minority carrier diffusion through the neutral regions; and g_d is the small signal conductance of the *pn*-junctions.

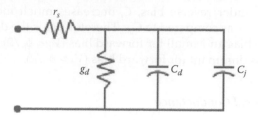

FIGURE 2.29
An equivalent circuit for a *pn*-junction showing the relevant circuit elements: r_s is the series resistance of the neutral *n*- and *p*-regions; C_j is junction capacitance; C_d is the diffusion capacitance; and g_d is the small-signal conductance.

2.4 Summary

This chapter presented a brief overview of the basic semiconductor physics and basic theory of extrinsic semiconductors forming *pn*-junctions. First of all, the basic properties of intrinsic semiconductor materials including bond and band structures, intrinsic carrier concentration, and energy levels are discussed. Then the behavior of extrinsic semiconductors, carrier statistics of electrons and holes, carrier transport, and transport equations are discussed. After the discussion of *p*-type and *n*-type semiconductors, the basic properties of *n*- and *p*-type semiconductors forming *pn*-junctions are described. Then the basic theory of *pn*-junctions, current transport, and dynamic characteristics are discussed. Finally, a basic equivalent circuit model of *pn*-junction for circuit CAD is presented.

Physical Constants

Constants	Symbol	Magnitude	Units
Electronic charge	q	1.602×10^{-19}	C
Free-electron mass	m	9.11×10^{-28}	g
Boltzmann's constant	k	1.38×10^{-23}	J K^{-1}
		8.62×10^{-5}	eV K^{-1}
Planck's constant	h	6.25×10^{-34}	J s
Permittivity of free space	ε_0	8.854×10^{-14}	F cm^{-1}
Thermal voltage at 300 K	$kT/q = v_{kT}$	0.02586	V
Thermal energy at 300 K	kT	0.02586	eV

Exercises

2.1 Experimental results show that the bandgap energy (E_g) in silicon decreases with temperature (T). The E_g versus T behavior is modeled by an empirical relation in circuit CAD given by

$$E_g(T) = 1.160 - \frac{7.02 \times 10^{-4} T^2}{1108 + T} \text{ (eV)} \qquad \text{(E2.1)}$$

Here, T is in Kelvin.

a. Compute and plot E_g for $0 \le T \le 600$ K.

b. From the plot extract $E_g(T = 300$ K$)$.

c. E_g versus T is also modeled by polynomial equations given below:

$$E_g(T) = 1.206 - 2.73 \times 10^{-4} T \text{ (eV)} \qquad \text{(E2.2)}$$

and

$$E_g(T) = 1.16 - 3 \times 10^{-4} T \text{ (eV)} \qquad \text{(E2.3)}$$

Calculate $E_g(T)$ using the polynomial equations and plot E_g versus T characteristics on the same graph in part (a) (superimpose). From the plots, show the range of temperature at which the polynomial equations are valid. Extract the values of $E_g(T = 300° K)$ from the polynomial equations and compare with that in part (a).

2.2 Use Equation E2.1 to compute and plot n_i versus T for $0 \le T \le 600$ K from the following equation:

$$n_i(T) = 3.9 \times 10^{16} T^{3/2} \exp\left[-\frac{E_g(T)}{2kT}\right] \qquad \text{(E2.4)}$$

From the plot, extract n_i at $T = 300°$ K and compare your results with that obtained for silicon.

2.3 A p-type semiconductor is doped with $N_a = 1 \times 10^{16}$ cm^{-3} and has the minority carrier lifetime $= 10$ μsec.

a. Calculate the steady state electron and hole concentrations under light that creates 10^{18} cm^{-3} sec^{-1} electron–hole pairs.

b. Calculate and sketch the position of equilibrium Fermi level E_f relative to E_i.

c. Calculate and sketch the position of quasi-Fermi levels E_{fn} and E_{fp} relative to E_i.

d. Compare the position of equilibrium Fermi level in part (b) with that of the steady state quasi-Fermi levels under the light in part (c). What are the similarities and differences? Explain.

e. Calculate and compare the *pn*-products under the equilibrium and nonequilibrium conditions at room temperature.

2.4 Consider an abrupt *n+p*-junction with $N_d = 10^{20}$ cm^{-3}, $N_a = 1 \times 10^{16}$ cm^{-3}, and area $= 20 \times 20$ μm^2:

a. Calculate the built-in potential (ϕ_{bi}) and zero-bias capacitance (C_{j0}).

b. Calculate the junction capacitance for an applied bias $V = -5$ V.

2.5 An IC resistor is shown in Figure E2.1. The doping concentrations for the *n*- and *p*-type regions are $N_d = 2.5 \times 10^{16}$ cm^{-3} and $N_a = 2.5 \times 10^{15}$ cm^{-3}, respectively. The junction depth $X_j = 0.4$ μm, the width of the *n*-type region $W = 2.5$ μm, and its length is $L = 20$ μm. The contact regions are each $3W \times 3W$ in area as shown in Figure E2.1.

a. Calculate the depletion width into the *n*- and *p*-sides of the *pn*-junction at $V_d = 0$.

b. Calculate the sheet resistance of the *n*-type region. Assume that the depletion region does not contribute to resistivity.

c. Calculate and sketch the position of quasi-Fermi levels E_{fn} and E_{fp} relative to E_i.

d. Calculate the maximum electric field at the *pn*-junction.

e. Assuming the DC voltage $V_d = 0$, calculate the depletion capacitance C_d in fF between the *n*-region and the *p*-type substrate.

f. Compute and plot C_j–V characteristics for applied bias range −2.0 V to ϕ_{bi} of the *pn*-junction for the doping gradient factor $m = 0.3, 0.4,$ and 0.5. Explain your results.

g. Use series expansion to show that the expression in Equation 2.141 is valid for $V_d \geq FC.\phi_{bi}$.

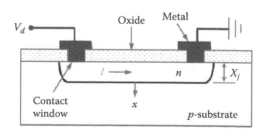

FIGURE E2.1
pn-junction capacitance modeling.

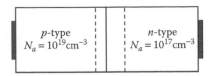

FIGURE E2.2
pn-junction *I–V* characteristics.

h. Compute and plot C_j–V characteristics for $-1.2\ V \leq V_d \leq 1.2\ V$ using Equation 2.141 for $V_d \leq FC.\phi_{bi}$ and $V_d \geq FC.\phi_{bi}$. Consider $m = 0.36$. State any assumptions you make including the fitting parameter FC. Explain your results.

2.6 In the derivation of the forward *I–V* characteristic of a *pn*-junction, we assumed *quasi-equilibrium*; that is, we assumed that we could simply subtract V_d as a small perturbation on the equilibrium situation. We will examine the validity of this assumption in this problem. Consider the diode shown in Figure E2.2 (the contacts are remote).

a. Assuming $D_n = 25\ cm^2\ sec^{-1}$, $D_p = 10\ cm^2\ sec^{-1}$, and $L_n = L_p = 10\ \mu m$, calculate the current that flows across the junction at an applied forward bias of 0.4 V.

b. With $V_d = 0$, electrons and holes will flow across the junction due to drift and diffusion, such that the currents due to drift and diffusion exactly cancel each other ($I = 0$). Estimate the hole diffusion current that would flow if there were no electric field to stop it.

c. What do your answer in part (a) and (b) tell you about the validity of our quasi-equilibrium assumption.

3

Metal-Oxide-Semiconductor System

3.1 Introduction

The metal-oxide-semiconductor (MOS) structure, commonly referred to as the MOS capacitor, is a two-terminal device with one electrode connected to the metal and the other electrode connected to the semiconductor, forming a voltage-dependent capacitor. The acronym MOS is used even if the top electrode is not a metal and the insulator is not an oxide. An MOS capacitor is a very useful device both for evaluating the MOS integrated circuit (IC)–fabrication process and for predicting the MOS transistor performance. Therefore, MOS capacitors are included in the test chip for IC process and device characterization.

The MOS capacitor systems have been the subject of numerous investigations and the detailed description of the early development can be found in the literature [1]. The major objective of this chapter is to build the foundation for the development of MOS transistor theory and models that will be used in Chapters 4, 5, and 9. In order to achieve our objective, we first discuss the behavior of an MOS capacitor system and then develop the charge-voltage (Q–V) and capacitance-voltage (C–V) relationships, which will be used later in the development of MOS transistor model.

3.2 MOS Capacitor at Equilibrium

In order to describe the basic performance of MOS capacitor system, let us consider the two-dimensional (2D) cross section of an ideal MOS capacitor shown in Figure 3.1. The structure includes a p- or n-type semiconductor substrate such as silicon, a dielectric layer such as silicon dioxide (SiO_2), a metal or polysilicon gate, a gate electrode (G), and a body (back or bulk) electrode (B) for operating the MOS capacitor system at the intended applied bias V_g and V_b. Typically, the SiO_2 layer is thermally grown on silicon substrate with a typical thickness between 10 and 100 nm. The gate

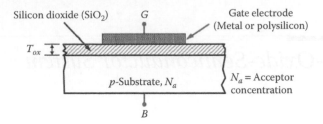

FIGURE 3.1
2D cross section of an ideal MOS capacitor structure fabricated on a uniformly doped p-type substrate with doping concentration, N_a; here G and B denote the gate and body electrodes for applied biases to the gate and substrate, respectively.

metal or degenerately doped polysilicon or a combination of polysilicon and silicide (e.g., $TiSi_2$, $CoSi_2$) is formed on the top of the gate dielectric by masking, photolithography, and annealing processes. The body electrode is obtained by deposited metal to achieve an ohmic contact. If the substrate conducts sufficiently to support the displacement currents, the structure in Figure 3.1 forms a parallel-plate capacitor with G as one electrode, B as the second electrode, and SiO_2 as the dielectric. This is referred to as the MOS capacitor system. This system is in thermal equilibrium with applied DC bias, and if the change in voltage is sufficiently slow, it is approximated to be a constant. Thus, from the parallel-plate capacitance formulation, we can write the oxide capacitance (C_{ox}) per unit area between the metal and silicon surface as:

$$C_{ox} = \frac{\varepsilon_0 K_{ox}}{T_{ox}} \tag{3.1}$$

where:
   ε_0 is the permittivity of free space or vacuum
   K_{ox} is the dielectric constant of oxide
   T_{ox} is the gate oxide thickness

In order to study the behavior of MOS capacitor system, let us consider the metal, oxide, and semiconductor (p-type silicon) as three separate materials, that is, materials before brought into contact. The energy band diagram of each material is shown separately in Figure 3.2. In Figure 3.2, E_0 denotes a convenient reference potential energy level, which is the vacuum or free electron energy level. In reality, E_0 is the level at which the Coulombic potential of an isolated positive charge becomes zero. It is to be noted that the reported value of bandgap energy for SiO_2 layer is in the range of 8.0–9.0 eV [1–3]. In Figure 3.2, we have used 8.0 eV as the bandgap energy for SiO_2 to discuss the behavior of MOS capacitor. The other characteristic parameters of the three materials in Figure 3.2 are defined in the next subsection.

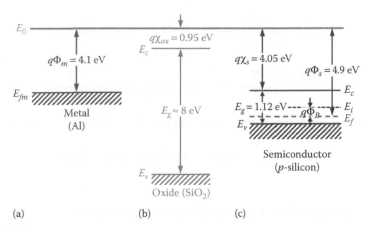

FIGURE 3.2
The energy band diagram of three separate materials that form an MOS capacitor system: (a) aluminum, (b) thermally grown SiO_2, and (c) p-type silicon substrate with $N_a = 1 \times 10^{15}$ cm^{-3}; here, E_0 = vacuum energy level (reference energy), E_c = bottom edge of conduction band, E_v = top-edge of valence band, E_f = Fermi level, E_g = forbidden energy (energy gap), E_i = Intrinsic energy level, $E_{fm} = E_c$ = Fermi level in metal; Φ_m = metal work function, χ_s = electron affinity in silicon, χ_{ox} = electron affinity in oxide, Φ_s = semiconductor work function, and q = electronic charge.

3.2.1 Work Function

Figure 3.2 shows the energy band diagrams of the metal, oxide, and semiconductor materials relative to vacuum level, E_0. In Figure 3.2a, Φ_m is defined as the metal work function in units of volts or ($q\Phi_m$) in units of energy. Φ_m is the energy required to take an electron across the surface energy barrier of metal at the Fermi level E_{fm} to E_0. However, for a metal, the Fermi level E_{fm} is at E_c. Thus, Φ_m is the energy difference between E_0 and E_{fm}, that is $\left(q\Phi_m = E_0 - E_{fm}\right)$. For pure metals without impurities and contamination, the value of Φ_m depends only on the charge distribution of the atomic core or the type of atom involved. For aluminum metal shown in Figure 3.2a, the value of $\Phi_m = 4.10$ V.

In semiconductors and insulators, the height of the surface energy barrier is defined by electron affinity, χ_s and χ_{ox} as shown in Figure 3.2c and b, respectively. As shown in Figure 3.2b and c, χ is the energy difference between the vacuum level E_0 and the bottom of the conduction band edge E_c at the surface, and for a semiconductor material, $q\chi_s = \left(E_0 - E_c\right)$. And, χ defines the basic property of a material independent of the presence of impurities or imperfections and only varies from one atomic type to another or is changed by alloy composition. Unlike metals, the Fermi level, E_f, is not a constant in semiconductors and depends on the doping concentration of impurities. Since the work function is the energy required to take an electron from E_f to E_0, the electron affinity χ_s is used to define the work function Φ_s in semiconductors. Thus, for a p-type semiconductor, the work function is given by

$$q\Phi_{sp} = q\chi_s + \frac{E_g}{2} + q\phi_{Bp} \quad (p\text{-type semiconductor}) \tag{3.2}$$

where:

E_g is the bandgap energy

ϕ_{Bp} is the bulk or Fermi potential for a p-type semiconductor

Similarly, the work function for an n-type semiconductor is given by

$$q\Phi_{sn} = q\chi_s + \frac{E_g}{2} - q\phi_{Bn} \quad (n\text{-type semiconductor}) \tag{3.3}$$

where:

ϕ_{Bn} is the bulk or Fermi potential for an n-type semiconductor

If the doping concentration for both the n-type and p-type semiconductors is the same, then $|\phi_{Bp}| = |\phi_{Bn}| \equiv \phi_B$ and is given by Equation 2.70 as

$$\phi_B = v_{kT} \ln\left(\frac{N_b}{n_i}\right) \tag{3.4}$$

where:

$v_{kT} \left(= kT/q\right)$ is the *thermal voltage* at the ambient temperature T

n_i is the intrinsic carrier concentration

In v_{kT}, the parameters k and q represent the Boltzmann constant and electronic charge, respectively. In order to calculate the value of the semiconductor work function, Φ_s, the magnitude of ϕ_B is calculated from Equation 3.4 as shown in the following example.

Let us consider a p-type silicon with $N_b = N_a = 1 \times 10^{15}$ cm^{-3} at room temperature 300 K so that $v_{kT} \cong 0.0259$ V. Then using $n_i = 1.45 \times 10^{10}$ cm^{-3}, we can show from Equation 3.4 that the value of $\phi_B \cong 0.29$ V. Now, considering $q\chi_s = 4.05$ eV and $E_g = 1.12$ eV for silicon, we get from Equation 3.2, $q\Phi_{sp} \equiv q\Phi_s \cong 4.90$ eV. For aluminum, $q\Phi_m = 4.1$ eV; therefore, for a p-type silicon, $\Phi_m < \Phi_s$, that is, the energy required to free an electron from the p-type silicon is higher than the energy required to free an electron from aluminum.

In order to calculate Φ_s for polysilicon gate, it is assumed that the polysilicon is degenerately doped so that the Fermi energy lies at the band edges, that is, E_f is at E_c for an n-type polysilicon and E_f is at E_v for a p-type polysilicon. For nanoscale CMOS (complementary metal-oxide-semiconductor) technology, work function engineering is used to achieve the target value of metal gate work function [4]. The work functions of commonly used gate material for IC technology are shown in Table 3.1 [5,6].

Now, let us consider the energy bands of three materials shown in Figure 3.2a–c are brought in contact to form an MOS capacitor system. It can be shown that when different materials are in contact with each other, the work

TABLE 3.1

Work Function of Different Materials Used as Gate Materials

Material	Work Function (eV)
Al	4.10
Au	5.27
MoSi$_2$	4.73
TiSi$_2$	3.95
n-type degenerately doped polysilicon	4.05
p-type degenerately doped polysilicon	5.17

function between the two ends of the composite system of materials depends only on the first and the last materials [7]. Thus, for an MOS system, the work function difference between the metal and the semiconductor defines the behavior of the system. The work function difference between two materials in contact can be visualized as the contact potential between them. For an MOS capacitor system, the work function difference between the metal and semiconductor ($\Phi_m - \Phi_s$) causes distortion in the band structure of the system as shown in Figure 3.3a. This is because when *three* materials are in contact, E_f is constant at equilibrium and E_0 is continuous; *holes* flow from *p*-type semiconductor to metal and *electrons* flow from metal to *p*-type semiconductor on contact until a potential is built up to counterbalance the difference in work function. However, the currents through SiO$_2$ are very small. Thus, there is a variation in electrostatic potential from one region to another, causing band

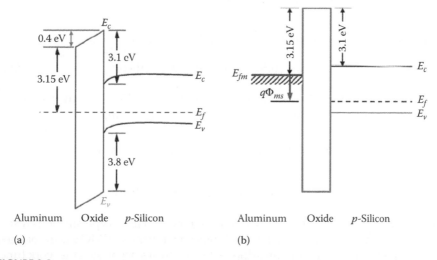

(a)   (b)

FIGURE 3.3
MOS capacitor system at applied gate voltage, $V_g = 0$, showing (a) band bending at the surface due to Φ_{ms} between aluminum metal and *p*-type semiconductor and (b) flat band condition for structure shown in (a); oxide is assumed to be free of any charges.

bending in the oxide and silicon. Since metal is an equipotential region, there is no band bending in metal.

For p-type silicon and aluminum metal MOS capacitor (Al-SiO$_2$-Si) system, $\Phi_m < \Phi_s$, therefore, energy bands bend downward in the oxide and silicon near the surface as shown in Figure 3.3a. And, there is an abrupt transition in E_c and E_v levels at the material interfaces. The metal and semiconductor work function difference (Φ_m–Φ_s) causes a potential drop in oxide and near the silicon surface due to band bending. A typical potential drop in oxide is about 0.4 V. This potential drop depends on the doping level in silicon and can be supported since no current flows through oxide. The values shown in Figure 3.3a for the band bending in the oxide and silicon are obtained by assuming that the oxide is an ideal insulator without any charges. We can compensate for this band bending by applying an external voltage $V_{fb} = (\Phi_m$–$\Phi_s)$, which caused the band bending in the first place. V_{fb} is referred to as the *flat band voltage* and the band structure for an MOS capacitor at flat band condition is shown in Figure 3.3b.

Thus, the condition for flat band voltage at the Si/SiO$_2$ interface is given by

$$V_{fb} = \Phi_m - \Phi_s \equiv \Phi_{ms} \tag{3.5}$$

where:

Φ_{ms} is the work function difference between the gate electrode and bulk silicon (in units of volts)

Then for an Al-SiO$_2$-pSi system,

$$q\Phi_{ms} = q\Phi_m - q\Phi_{sp} = q\Phi_m - \left(q\chi_s + \frac{E_g}{2} + q\phi_B \right) \tag{3.6}$$

Considering the values shown in Figure 3.2a–c, we get

$$\Phi_{ms} = -(0.51 + \phi_B) \quad \text{for } p\text{-type silicon} \tag{3.7}$$

Since $\phi_B \cong 0.29\,\text{V}$ for substrate concentration $N_b = 1 \times 10^{15}\,\text{cm}^{-3}$; therefore, Φ_{ms} is a negative number. Similarly, for an Al-SiO$_2$-nSi system,

$$q\Phi_{ms} = q\Phi_m - q\Phi_{sn} = q\Phi_m - \left(q\chi_s + \frac{E_g}{2} - q\phi_B \right) \tag{3.8}$$

$$\Phi_{ms} = -(0.51 - \phi_B) \quad \text{for } n\text{-type silicon} \tag{3.9}$$

Equation 3.9 shows that Φ_{ms} for MOS capacitors with an n-type silicon is also a negative number for $N_b < 1 \times 10^{18}\,\text{cm}^{-3}$. Since in advanced CMOS technologies, the channel is undoped or lightly doped, Φ_{ms} is always negative. This work function difference causes band bending when the materials are brought in contact.

For degenerately doped polysilicon gate electrode, the band structure for an MOS capacitor system is shown in Figure 3.4.

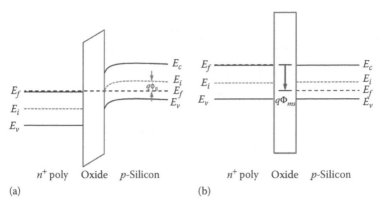

FIGURE 3.4
MOS capacitor system with degenerately doped $n+$ polysilicon gate electrode and p-type silicon (a) band bending at the surface due work function difference, Φ_{ms}, (b) flat band condition; oxide is assumed to be free of any charges.

Thus, with reference to Figure 3.4, Φ_{ms} for an $n+$ polysilicon gate on a p-type substrate MOS capacitor system is

$$q\Phi_{ms} = q\chi_s - \left(q\chi_s + \frac{E_g}{2} + q\phi_B\right) = -(0.56 + \phi_B), \quad \text{for } p\text{-type silicon} \quad (3.10)$$

Similarly, it can be shown that Φ_{ms} for $p+$ polysilicon gate and n-type substrate MOS capacitor system is

$$q\Phi_{ms} = \left(q\chi_s + E_g\right) - \left(q\chi_s + \frac{E_g}{2} - q\phi_B\right) = (0.56 + \phi_B), \quad \text{for } n\text{-type silicon} \quad (3.11)$$

Equation 3.10 shows that even for an $n+$ polysilicon gate with p-type silicon MOS capacitor system, Φ_{ms} is still negative. On the other hand, Equation 3.11 shows that for a $p+$ polysilicon gate with n-type substrate, Φ_{ms} is a positive quantity. The value of Φ_{ms} for polysilicon gate is found to be dependent on polysilicon doping concentration and grain structure [8,9].

3.2.2 Oxide Charges

During oxide growth process or subsequent IC fabrication processing steps, some impurities or defects are inadvertently incorporated into the oxide. As a result, the oxide is contaminated with various types of charges and traps. Typically, four different types of charge have been identified in thermally grown oxide on a silicon surface as shown in Figure 3.5 [10]. These charges are (1) interface-trapped charge Q_{it}, (2) fixed-oxide charge Q_f, (3) oxide-trapped charge Q_{ot}, and (4) mobile ionic charge Q_m. All of these charges are dependent on IC fabrication processing steps. The detailed description of the origin and techniques of measurements of different oxide charges are available in the literature [1,11]. In the following subsection, the basic properties of these charges are described.

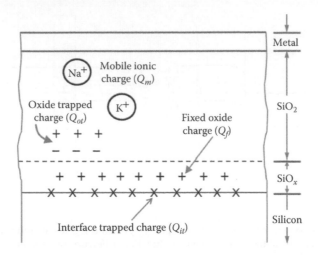

FIGURE 3.5
Types and location of the charges associated with thermally grown SiO_2 on silicon.

3.2.2.1 Interface-Trapped Charge

The interface-trapped charge density, Q_{it}, also referred to as the *surface states, fast states*, or *interface states*, exists at the Si/SiO_2 interface as shown in Figure 3.5. It is caused by defects at that interface which gives rise to charge *traps* or *electronic energy levels* with energy states (E_s) in the silicon bandgap that can capture or emit mobile carriers. These electronic states are due to lattice mismatch at the interface, dangling bonds, the adsorption of foreign impurity atoms at the silicon surface, and other defects caused by radiation or any bond-breaking process. Q_{it} is the most important type of charge because of its wide-ranging and degrading effect on device characteristics. Under the equilibrium condition, the occupancy of the interface states or traps depends on the position of the Fermi level.

Typically, the interface trap levels with density, D_{it} (traps cm^{-2} eV^{-1}), are distributed over energies within the silicon energy gap [1–3,5]. D_{it} varies significantly from process to process and is dependent on crystal orientation. In thermally grown SiO_2 on silicon, the most of the interface-trapped charge is neutralized by low temperature ($\leq 500°C$) hydrogen annealing. D_{it} correlates with the density of available bonds at the surface. Therefore, in <100> orientation with lower density of silicon atoms (available bonds) at the surface, D_{it} is about an order of magnitude lower than that in <111> oriented silicon with higher available bonds at the surface. The value of D_{it} at mid-gap for <100> oriented silicon in modern MOS VLSI (very-large-scale-integrated) process can be as low as 5×10^9 cm^{-2} eV^{-1}. *Higher values of* D_{it} *cause instabilities in the MOS transistor behavior.*

3.2.2.2 Fixed-Oxide Charge

The fixed charge density, Q_f, is the immobile charge always present and located within 1 nm transition layer of nonstoichiometric silicon oxide

(SiO$_x$) at the boundary between the silicon and SiO$_x$ layer as shown in Figure 3.5. Generally, Q_f is positive and appears to arise from incomplete silicon-to-silicon bonds and depends on the oxidation ambient, temperature and annealing conditions, and silicon orientation. Since the density of atoms at the surface of a silicon crystal depends on the crystal orientation, Q_f is higher in <111> silicon than in <100> wafers. However, it is independent of the doping type and concentration in the silicon, oxide thickness, and oxidation time. Q_f can be minimized by annealing the oxide in an inert ambient, such as Argon at a temperature in excess of 900°C. A typical value of Q_f for a carefully treated Si/SiO$_2$ system is about 1×10^{10} cm^{-2} for the <100> surface. Because of the low values of Q_{it} and Q_f, the <100> orientation is preferred for silicon MOSFETs (metal-oxide-semiconductor field-effect transistors).

3.2.2.3 Oxide-Trapped Charge

The oxide-trapped charge density, Q_{ot}, is associated with defects in SiO$_2$. Q_{ot} is located in traps distributed throughout the oxide layer. The oxide traps are usually electrically neutral and are charged by introducing electrons and holes into the oxide through ionizing radiation such as implanted ions, X-rays, and electron beams. The magnitude of Q_{ot} depends on the amount of radiation dose and energy and the field across the oxide during irradiation. Like Q_{it}, these charges could be positive (trapped holes) or negative (trapped electrons). Q_{ot} resembles Q_f in that its magnitude is not a function of silicon surface potential and there is no capacitance associated with it.

3.2.2.4 Mobile Ionic Charge

The mobile ionic charge density, Q_m, is due to sodium (Na$^+$) or other alkali ions that get into the oxide during cleaning, processing, and handling of MOS devices. These ions move very slowly within the oxide; their transport depends strongly on the applied electric field (~1 MV cm^{-1}) and temperature (30°C–400°C). Positive voltages push the ions toward the Si/SiO$_2$ interface while the negative voltages draw them toward the gate. A current is observed in the external circuit during ion drift. The drift of ions changes the centroid of charge within the oxide layer, resulting *in a shift of the flat band voltage of MOS capacitor system and may cause an unexpected device failure*. Different approaches are used to reduce mobile ion contamination in gate oxide and mitigate the risk of mobile ionic induced device failure [1,5].

The earlier described oxide charges cause an additional band bending at the silicon surface of an MOS capacitor system and shift the value of V_{fb} caused by Φ_{ms} as described in the following section.

3.2.3 Flat Band Voltage

In order to determine the total shift in the flat band voltage (ΔV_{fb}) by various oxide charges, let us consider $\rho(x)$ as the charge density per unit volume within the oxide. Then from Gauss's law (Equation 2.61), we can show

$$\Delta V_{fb} = -\frac{1}{K_{ox}\varepsilon_0}\int_0^{T_{ox}} x\rho(x)dx = -\frac{1}{C_{ox}}\int_0^{T_{ox}} \frac{x}{T_{ox}}\rho(x)dx \qquad (3.12)$$

where $\rho(x)$ includes the charge densities due to Q_{it}, Q_f, Q_{ot}, and Q_m. Q_f and Q_{it} are located at or near the Si/SiO$_2$ interface (i.e., $x = T_{ox}$) whereas Q_{ot} and Q_m are distributed throughout the oxide. Therefore, we only integrate Q_{ot} and Q_m that are distributed throughout the oxide to get

$$\Delta V_{fb} = -\frac{Q_{it}+Q_f}{C_{ox}} - \frac{1}{C_{ox}}\int_0^{T_{ox}} \frac{x}{T_{ox}}\left[Q_{ot}(x)+Q_m(x)\right]dx \qquad (3.13)$$

In compact modeling for circuit simulation, Equation 3.13 is expressed as

$$\Delta V_{fb} = -\frac{Q_o}{C_{ox}}(V) \qquad (3.14)$$

where:
  Q_o is the *equivalent interface charge* located at the Si/SiO$_2$ interface and causes the same effect as that of the actual charges of unknown distribution

Q_o is always positive for both p- and n-type substrates. ΔV_{fb} is the gate voltage that is needed to cause Q_o to be imaged in the gate electrode so that none is induced in the silicon. However, when gate "floats" or the gate electrode is absent, the oxide charges will seek all their image charges in the silicon.

In Figure 3.3a, we have shown the band bending of an MOS capacitor system due to work function difference between the metal and semiconductor. The corresponding flat band voltage is given by Equation 3.5. Now, the shift in work function due to band bending by oxide charges is given by Equations 3.13 and 3.14. Thus, combining Equations 3.5 and 3.14, the total V_{fb} due to Φ_{ms} and Q_o is given by

$$V_{fb} = \Phi_{ms} - \frac{Q_o}{C_{ox}} \qquad (3.15)$$

Typically, Q_o/C_{ox} is much smaller than Φ_{ms} in Equation 3.15. Therefore, for an MOS capacitor with p-substrate and $n+$ polysilicon gate, V_{fb} is a negative number since Φ_{ms} is negative from Equation 3.10. On the other hand, for MOS capacitor with n-substrate and $p+$ polysilicon gate, V_{fb} is positive since Φ_{ms} is positive from Equation 3.11.

3.2.4 Effect of Band Bending on the Semiconductor Surface

Let us now consider an Al-SiO$_2$-Si MOS capacitor system on a p-type substrate to discuss the effect of band bending at the silicon surface on the surface behavior of MOS capacitors. We know that the concentration of holes in a p-type substrate is given by (Equation 2.63),

$$p = n_i \exp\left(\frac{E_i - E_f}{kT}\right) \tag{3.16}$$

The band structure of the system is shown in Figure 3.6. It is seen from Figure 3.6 that as the bands bend downward, the energy difference $(E_i - E_f)$ gradually decreases as we approach the silicon surface at $x = 0$ from the bulk at $(x = \infty)$. Then from Equation 3.16, the decrease in $(E_i - E_f)$ results in a decrease in the hole concentration p. This implies that the holes are depleted at the surface, giving rise to a space charge region. On the other hand, if the bands bend upward, as in the case of an MOS capacitor system with $(\Phi_m > \Phi_s)$, the value of $(E_i - E_f)$ increases at the surface, resulting in an increase in the hole concentration (accumulation) at the surface. Thus, even without an applied external voltage to an MOS capacitor, the carrier concentration at the surface differs from that in the bulk due to Φ_{ms} and Q_o. This change in the concentration sets up an electric field at the surface and hence a voltage difference between the silicon surface and bulk. This voltage difference is referred to as the *surface potential* ϕ_s and represents the electrostatic potential at the surface measured from the bulk intrinsic level E_i. Thus, ϕ_s is the difference between $E_i(x = 0)$ at the surface and $E_i(x = \infty)$ at a point deep into the substrate. As shown in Figure 3.6, ϕ_s is a measure of the amount of total band bending at the silicon surface. And, at a depth x into the surface, the potential is given by $\phi(x)$.

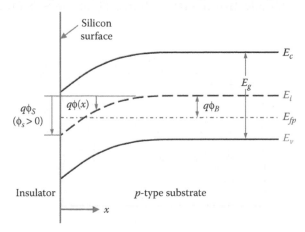

FIGURE 3.6

MOS capacitor system: Band bending showing the surface potential ϕ_s at the surface of a p-type silicon; here x is the distance from the insulator/substrate interface into the substrate with $x = 0$ at the surface.

The band bending described earlier can be compensated by applying an external gate voltage given by Equation 3.15. The condition to achieve the flat bands at the surface is called the *flat band condition* and the corresponding gate voltage required to achieve the flat band condition is called the *flat band voltage, V_{fb}*. Thus, *V_{fb} is the applied gate voltage to have zero surface potential with flat energy bands over the entire semiconductor surface*. The flat band condition is often used as a reference state along with V_{fb} as a reference voltage and, thus, can be considered as an important figure of merit for an MOS capacitor system.

3.3 MOS Capacitor under Applied Bias

In the previous section, we described the behavior of an MOS capacitor system without the application of any external bias. Now, let us discuss the behavior of the system under the applied gate bias V_g as shown in Figure 3.7. The applied V_g is shared between the voltage across the oxide V_{ox}, surface potential ϕ_s, and the work function Φ_{ms} between the metal and the semiconductor to achieve flat band condition. Thus,

$$V_g = V_{ox} + \phi_s + \Phi_{ms}$$
$$= V_{ox} + \phi_s + V_{fb}$$

(3.17)

With reference to charges, an MOS capacitor consists of three different charges under the applied V_g such as: (1) gate charge Q_g due to the applied V_g to the gate, (2) effective interface charge Q_o at the Si/SiO$_2$ interface for

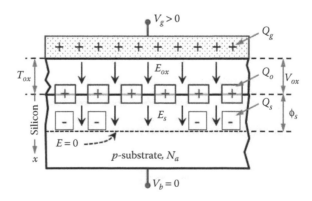

FIGURE 3.7
An MOS capacitor system under the applied gate bias V_g showing various charges, electric fields, and potentials. E_{ox} is the electric field in oxide; E_s is the electric field in substrate.

nonideal insulator as discussed in Section 3.2.2, and (3) the induced charge Q_s in the silicon underneath the gate oxide. Then from the charge neutrality condition we get

$$Q_g + Q_o + Q_s = 0 \qquad (3.18)$$

If the applied voltage V_g is positive, then the electric field E_s is directed into the silicon surface at the interface and will induce a charge Q_s in the silicon. The density of the induced charge Q_s per unit area can be calculated by applying Gauss's law at the Si/SiO$_2$ interface. Thus, Q_s per unit area is given by

$$Q_s = -\varepsilon_0 K_{si} E_s \qquad (3.19)$$

where:
K_{si} is the permittivity of silicon
ε_0 is the permittivity of vacuum

Similarly, applying Gauss's law at the metal-oxide interface gives

$$Q_g = \varepsilon_0 K_{ox} E_{ox} \equiv V_{ox} C_{ox} \qquad (3.20)$$

where:
$E_{ox} = V_{ox}/T_{ox}$ is the electric field in the oxide

The field E_{ox} and E_s are related by Equation 3.18. For an ideal oxide, $Q_o = 0$, and we have from Equation 3.18, $Q_g = -Q_s$; then from Equations 3.19 and 3.20, we get

$$\varepsilon_0 K_{si} E_s = \varepsilon_0 K_{ox} E_{ox}$$

$$\text{or} \qquad (3.21)$$

$$E_s = \frac{K_{ox} E_{ox}}{K_{si}}$$

Now, substituting for E_s from Equation 3.21 in Equation 3.19 we get

$$Q_s = -\varepsilon_0 K_{si} \left(\frac{K_{ox} E_{ox}}{K_{si}} \right) = -\varepsilon_0 K_{ox} E_{ox} = -V_{ox} C_{ox}$$

$$(3.22)$$

$$\therefore V_{ox} = -\frac{Q_s}{C_{ox}}$$

Now, substituting for V_{ox} from Equation 3.22 in Equation 3.17, we get

$$V_g = V_{fb} + \phi_s - \frac{Q_s}{C_{ox}} \qquad (3.23)$$

Equation 3.23 relates the applied bias V_g and the surface potential ϕ_s. At the flat band condition, $\phi_s = 0$ and $Q_s = 0$; therefore, from Equation 3.23, $V_g = V_{fb}$. Within the range $0 > V_g > 0$, different surface conditions result in an MOS capacitor system as discussed in Sections 3.3.1 through 3.3.3.

3.3.1 Accumulation

To continue our discussion on Al/SiO$_2$/p-silicon MOS capacitor system, let us apply a negative gate voltage V_g with body grounded such that $V_g < V_{fb}$. The negative voltage at the gate creates an upward electric field E_{ox} from the substrate to metal as shown in Figure 3.8. Since the applied negative voltage depresses the electrostatic potential of the metal relative to the substrate, electron energies are raised in the metal relative to the substrate. As a result, the Fermi level E_{fm} for the metal moves up above its equilibrium position by qV_g. Since Φ_m and Φ_s do not change with V_g, moving E_{fm} up in energy relative to E_f causes the oxide conduction band to bend upward, consistent with the direction of the field E_{ox} causing gradient in the energy bands [2,12].

With reference to charge, the negative voltage at the gate results in a negative charge ($Q_g < 0$) on the gate. This in turn induces an equal amount of positive charge Q_s at the silicon surface. This amount of positive charge in the p-type silicon means excess hole concentration is created at the surface as shown in Figure 3.8. These holes are accumulated at the surface and known as the *accumulation* charges. We know from Equation 3.16 that, as the hole concentration increases at the surface, $(E_i–E_f)$ increases, resulting in the bands bending upward as shown in Figure 3.9. Thus, in accumulation for p-type silicon we have

$$\text{Accumulation} \begin{cases} V_g < V_{fb} \\ \phi_s < 0 \\ Q_s > 0 \end{cases} \tag{3.24}$$

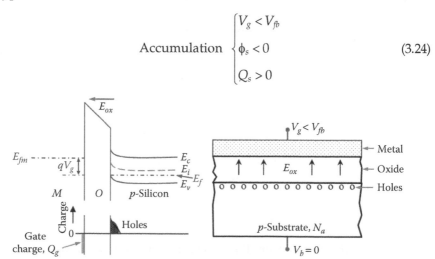

FIGURE 3.8

Effect of applied voltage, $V_g < V_{fb}$ on a p-type MOS capacitor system: the applied negative bias $V_g < V_{fb}$ causes hole accumulation at the silicon surface.

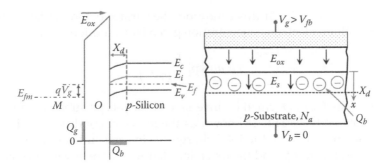

FIGURE 3.9
Effect of applied voltage, $V_g > V_{fb}$ on a p-type MOS capacitor system: the applied positive bias $V_{go} = (V_g - V_{fb})$ depletes the holes from the silicon surface. Q_b is the depletion or bulk charge; Q_g is the gate charge; X_d is the depletion region width.

This bias condition (Equation 3.24) is useful in the characterization of MOS capacitor system.

3.3.2 Depletion

Now, let us apply a positive gate voltage $V_g > V_{fb}$ with body grounded. This positive V_g will create a downward electric field E_{ox} from the gate into the substrate as shown in Figure 3.9. A positive gate voltage raises the potential of the gate, lowering the Fermi level E_{fm} by qV_g. Moving E_{fm} down in energy relative to E_f causes band bending downward in the oxide conduction band in accordance to the direction of E_{ox}.

Again, with reference to charge, a positive voltage at the gate deposits positive charge ($Q_g > 0$) on it. Due to $V_g > 0$, the holes are repulsed away from the silicon surface, leaving behind negatively charged acceptor ions. Thus, a positive charge on the gate induces a negative charge Q_s at the surface due to the depletion of holes creating a depletion region of width X_d. This is known as the *depletion* condition. Since the hole concentration decreases at the surface, then from Equation 3.16, $(E_i - E_f)$ must decrease. As a result, E_i slowly approaches to E_f thereby bending the bands downward near the surface as shown in Figure 3.9. Thus, the depletion condition is given by

$$\text{Depletion} \begin{cases} V_g > V_{fb} \\ \phi_s > 0 \\ Q_s < 0 \end{cases} \tag{3.25}$$

3.3.3 Inversion

If we further increase the positive gate voltage, the downward band bending will further increase. At a sufficiently large $V_g \gg V_{fb}$, the band bending may pull down the mid-gap energy level E_i below the constant E_f at the silicon surface, that is, $E_f > E_i$. At this condition, the surface behaves like an n-type material with an electron concentration given by (Equation 2.62)

$$n = n_i \exp\left(\frac{E_f - E_i}{kT}\right) \tag{3.26}$$

Thus, the n-surface is formed by inversion of the p-type substrate due to the applied gate voltage. This is known as the *inversion* condition as shown in Figure 3.10. In inversion, the total charge, Q_s, in the semiconductor consists of depletion charge, Q_b, and the inversion charge, Q_i. The inversion condition for MOS capacitor with p-type substrate is defined by

$$\text{Inversion} \quad \begin{cases} V_g \gg V_{fb} \\ \phi_s > 0 \\ Q_s < 0 \end{cases} \tag{3.27}$$

Under the applied $V_g \gg V_{fb}$, the p-type surface is inverted as soon as E_i is pulled below E_f. However, for small $(E_f{-}E_i)$, the electron concentration remains very small and the inversion is *weak*. This is referred to as the *weak inversion regime*. If we increase V_g such that $(E_f{-}E_i)$ at the surface equals $(E_i{-}E_f)$ at the p-type bulk, the concentration of electrons at the surface will be equal to that of holes in the bulk. This is called the *strong inversion* regime. On further increase of V_g, the electron concentration will exceed the concentration of the holes in the inversion region. Under the inversion condition, the depth of the inversion region (X_{inv}) into the substrate can be defined at $E_f = E_i$ and is about 3 nm [2].

Now, let us discuss how the inversion layer is formed in the substrate. At the onset of inversion, the minority carrier electrons in the p-type substrate of

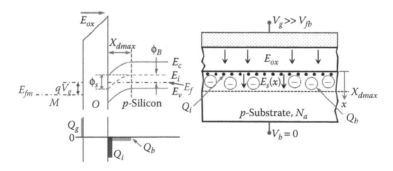

FIGURE 3.10
Effect of an applied voltage, $V_g \gg V_{fb}$ on a p-type MOS capacitor system: a large positive bias $V_g \gg V_{fb}$ causes inversion of the p-type surface forming an n-type layer. The gate charge is compensated by the depletion charge Q_b and the inversion charge Q_i in the semiconductor. Q_b is the depletion or bulk charge; Q_g is the gate charge; Q_i is the inversion charge; X_{dmax} is the maximum depletion width.

an MOS capacitor system originate from thermally generated electron–hole pairs within the depletion region. The rate of thermal generation depends upon the minority carrier lifetime, which is of the order of microseconds. It is found that the time required to form an inversion layer at the surface is about 0.2 sec [3]. Thus, the *formation of the inversion layer is a relatively slow process compared to the time required for the holes (majority carriers) to flow from or to the silicon surface*, which is of the order of picoseconds. Once the inversion layer is formed, it shields the underneath depletion layer, thus limiting the maximum width, X_{dmax}, of the depletion layer.

So far, we have presented a qualitative overview of the basic operation of an MOS capacitor system. In the following section, we will develop MOS capacitor theory that can be extended to develop the operational theory of MOSFET devices in Chapters 4, 5, and 9.

3.4 MOS Capacitor Theory

Now, let us derive the relation between the surface potential (ϕ_s), electric field (E_s), and charge (Q_s) by solving Poisson's equation for potential (ϕ) near the surface region of the silicon substrate of an MOS capacitor system. The Poisson's equation (Equation 2.58) is given by

$$\frac{d^2\phi}{dx^2} = -\frac{1}{K_{si}\varepsilon_0}\rho(x) \tag{3.28}$$

where:
  $\rho(x)$ is the charge density at any point x along the depth of the substrate and is given by

$$\rho(x) = q\left[p(x) - n(x) + N_d^+(x) - N_a^-(x)\right] \tag{3.29}$$

where:
  $p(x)$ is the hole concentration
  $n(x)$ is the electron concentration
  $N_d^+(x)$ is the ionized donor concentration in the semiconductor substrate
  $N_a^-(x)$ is the ionized acceptor concentration in the semiconductor substrate

Thus, combining Equations 3.28 and 3.29 we get

$$\frac{d^2\phi}{dx^2} = -\frac{q}{K_{si}\varepsilon_0}\left[p(x) - n(x) + N_d^+(x) - N_a^-(x)\right] \tag{3.30}$$

Before solving Equation 3.30 for $\phi(x)$ at any point x of the surface, let us review the relevant semiconductor equations in the following subsection.

3.4.1 Formulation of Poisson's Equation in Terms of Band-Bending Potential

In order to solve Poisson's Equation 3.30 for $\phi(x)$ at any point x near the surface of an MOS capacitor system, we express carrier density $\rho(x)$ in terms of potentials ϕ_i, ϕ_f, and $\phi(x)$. In an n-type semiconductor with doping concentration, N_d, the majority carrier *electron* concentration n is given by (Equations 2.62 and 2.64)

$$n \cong \begin{cases} n_i \exp\left(\dfrac{E_f - E_i}{kT}\right) = n_i \exp\left[\dfrac{q\left(\phi_i - \phi_f\right)}{kT}\right] \\ \\ N_d^+ \end{cases} \tag{3.31}$$

In Equation 3.31, $\phi_f = -E_f/q$ is the Fermi potential and $\phi_i = -E_i/q$ is the intrinsic potential; then the minority carrier concentration p_n in an n-type semiconductor is given by (Equation 2.66)

$$p_n \cong \frac{n_i^2}{N_d^+} \tag{3.32}$$

Similarly, the majority carrier concentration in a p-type semiconductor with doping concentration N_a is given by (Equations 2.63 and 2.65)

$$p \cong \begin{cases} n_i \exp\left(\dfrac{E_i - E_f}{kT}\right) = n_i \exp\left[\dfrac{q(\phi_f - \phi_i)}{kT}\right] \\ \\ N_a^- \end{cases} \tag{3.33}$$

And, the minority carrier concentration n_p in a p-type semiconductor is given by (Equation 2.67)

$$n_p \cong \frac{n_i^2}{N_a^-} \tag{3.34}$$

In order to develop a generalized expression for both n-type and p-type substrates, we define, N_b as the substrate concentration. Then from Equations 3.31 and 3.33, we can show that the bulk (Fermi) potential is

$$\phi_B = \left|\phi_f - \phi_i\right| = v_{kT} \ln\left(\frac{N_b}{n_i}\right) \tag{3.35}$$

In Equation 3.35, N_b represents the donor-type doping concentration for an n-type substrate and acceptor-type doping concentration for a p-type substrate; v_{kT} is the thermal voltage.

Now, in order to express $\rho(x)$ in terms of band bending $\phi(x)$ at any point x near the surface of a semiconductor, we consider the band structure of an $Al/SiO_2/p$-silicon MOS capacitor system as shown in Figure 3.11.

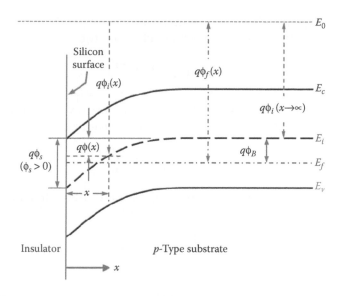

FIGURE 3.11

Equilibrium band structure of a *p*-type MOS capacitor system showing band bending in the substrate; here, $x = 0$ at the Si/SiO$_2$ interface and increases with the depth into the substrate; ϕ_s is the surface potential representing the total band bending at the surface.

From Figure 3.11, the amount of band bending at any point x near the silicon surface with reference to E_0 is given by

$$\phi(x) = \phi_i(x) - \phi_i(x \to \infty) \tag{3.36}$$

where $\phi(x = 0) = \phi_s$ = surface potential; $\phi_i(x \to \infty) = \phi_i$ is the intrinsic potential. Also, from Figure 3.11, the bulk-potential $\phi_B = (\phi_f - \phi_i)$. It is seen from Figure 3.11 that $\phi_i(x) > \phi_i$ when bands bend downward at the interface, resulting in $\phi(x) > 0$. Thus, it is clear that $\phi(x) > 0$ when bands bend downward in the *depletion* and *inversion* conditions and $\phi(x) < 0$ when bands bend upward in the *accumulation* condition.

Now, let us express $p(x)$, $n(x)$, $N_d^+(x)$, and $N_a^-(x)$ in Poisson's Equation 3.30 in terms of potential $\phi(x)$ and solve for ϕ_s and E_s at the surface of the MOS system. For a *p*-type silicon substrate, we can express the majority carrier concentration given in Equation 3.33 at any point x by substituting for $\phi_i(x)$ from Equation 3.36 so that

$$p(x) = n_i \exp\left\{\frac{\left[\phi_f - \phi_i(x)\right]}{v_{kT}}\right\} = n_i \exp\left\{\frac{\phi_f - \left[\phi_i(x \to \infty) + \phi(x)\right]}{v_{kT}}\right\} \tag{3.37}$$

Since $\left[\phi_f - \phi_i(x \to \infty)\right] = (\phi_f - \phi_i) = \phi_B$; then we can express (3.37) as

$$p(x) = n_i \exp\left[\frac{(\phi_f - \phi_i) - \phi(x)}{v_{kT}}\right] = n_i \exp\left(\frac{\phi_f - \phi_i}{v_{kT}}\right) \exp\left(-\frac{\phi(x)}{v_{kT}}\right) \tag{3.38}$$

Now, using Equation 3.33 in Equation 3.38, we get for the majority carrier concentration at any point x in a p-type substrate from

$$p(x) = N_a \exp\left(-\frac{\phi(x)}{v_{kT}}\right) \tag{3.39}$$

Then from Equation 3.34, the minority carrier electron concentration at any point x near the surface of a p-type substrate is given by

$$n(x) \cong \frac{n_i^2}{p(x)} = \frac{n_i^2}{N_a}\exp\left(\frac{\phi(x)}{v_{kT}}\right) \tag{3.40}$$

Again, from Equations 3.34 and 3.35, we can show that for a p-type substrate

$$\frac{n_i^2}{N_a} = \begin{cases} n_i \exp\left(-\dfrac{\phi_B}{v_{kT}}\right) \\[4mm] N_a \exp\left(-\dfrac{2\phi_B}{v_{kT}}\right) \end{cases} \tag{3.41}$$

Then the minority carrier concentration given by Equation 3.40 can also be written as

$$n(x) = \begin{cases} n_i \exp\left[\dfrac{\phi(x)-\phi_B}{v_{kT}}\right] \\[4mm] N_a \exp\left[\dfrac{\phi(x)-2\phi_B}{v_{kT}}\right] \end{cases} \tag{3.42}$$

Substituting the expressions for $p(x)$ and $n(x)$ from Equations 3.39 and 3.40, respectively, in Equation 3.29 we get

$$\rho(x) = q\left\{ N_a e^{-[\phi(x)/v_{kT}]} - \frac{n_i^2}{N_a}e^{\phi(x)/v_{kT}} + N_d^+(x) - N_a^-(x) \right\} \tag{3.43}$$

Again, assuming complete ionization of acceptor atoms, for a p-type substrate we get from Equations 3.34 and 3.35, $p \cong N_a^- \equiv N_a$ and $n \cong N_d^+ = n_i^2/N_a$. Therefore, for a uniformly doped p- substrate, we can write

$$N_d^+(x) - N_a^-(x) = \frac{n_i^2}{N_a} - N_a \tag{3.44}$$

Then combining Equations 3.43 and 3.44, the charge density in the substrate (assuming complete ionization of dopant atoms in silicon) is given by

$$\rho(x) = q\left[N_a\left(e^{-[\phi(x)/v_{kT}]} - 1\right) - \frac{n_i^2}{N_a}\left(e^{[\phi(x)/v_{kT}]} - 1\right) \right] \tag{3.45}$$

It is to be noted that the first term in Equation 3.45 represents the charge density in the p-type substrate due to the majority carrier concentration and the second term represents the charge density due to minority carrier concentration. Now, substituting Equation 3.45 in Equation 3.30, we get Poisson's equation in terms of band-bending potential $\phi(x)$ at a depth x near the surface of a p-type substrate as

$$\frac{d^2\phi(x)}{dx^2} = -\frac{q}{\varepsilon_0 K_{si}}\left[N_a\left(e^{-[\phi(x)/v_{kT}]}-1\right) - \frac{n_i^2}{N_a}\left(e^{[\phi(x)/v_{kT}]}-1\right)\right] \quad (3.46)$$

We will solve Equation 3.46 for $\phi(x)$ to obtain MOS capacitor behavior under different operating conditions. Again, we notice from Equation 3.46 that the first term inside the square bracket is due to the majority carrier charge density whereas the second term is due to minority carriers in a p-type semiconductor substrate.

3.4.2 Electrostatic Potentials and Charge Distribution

In order to solve Equation 3.46 for potential distribution in silicon, we use the mathematical identity

$$\frac{d}{dx}\left(\frac{d\phi}{dx}\right)^2 = 2\frac{d\phi}{dx}\frac{d^2\phi}{dx^2} \quad (3.47)$$

Then multiplying both sides of Equation 3.46 by $2[d\phi(x)/dx]$, we get

$$2\frac{d\phi(x)}{dx}\frac{d^2\phi(x)}{dx^2} = -\frac{2q}{\varepsilon_0 K_{si}}\left[N_a\left(e^{-[\phi(x)/v_{kT}]}-1\right) - \frac{n_i^2}{N_a}\left(e^{[\phi(x)/v_{kT}]}-1\right)\right]\frac{d\phi(x)}{dx} \quad (3.48)$$

Now, using Equation 3.47 in the left-hand side of Equation 3.48, we can show that

$$\frac{d}{dx}\left(\frac{d\phi(x)}{dx}\right)^2 = -\frac{2q}{\varepsilon_0 K_{si}}\left[N_a\left(e^{-[\phi(x)/v_{kT}]}-1\right) - \frac{n_i^2}{N_a}\left(e^{[\phi(x)/v_{kT}]}-1\right)\right]\frac{d\phi(x)}{dx} \quad (3.49)$$

We integrate Equation 3.49 from the bulk $(\phi(x)=0, d\phi(x)/dx=0)$ toward the surface at any point x $(\phi(x), d\phi(x)/dx)$ near the surface shown in Figure 3.11 so that

$$\int_0^{d\phi(x)/dx} d\left(\frac{d\phi(x)}{dx}\right)^2 = -\frac{2q}{\varepsilon_0 K_{si}}\int_0^{\phi(x)}\left[N_a\left(e^{-[\phi(x)/v_{kT}]}-1\right) - \frac{n_i^2}{N_a}\left(e^{[\phi(x)/v_{kT}]}-1\right)\right]d\phi(x) \quad (3.50)$$

We know that the electric field at any point x near the silicon surface is given by $E(x) = -[d\phi(x)/dx]$; therefore, after integration and simplification of Equation 3.50, we can show

$$\left(\frac{d\phi(x)}{dx}\right)^2 = \frac{2qN_a v_{kT}}{\varepsilon_0 K_{si}}\left[\left(e^{-[\phi(x)/v_{kT}]} + \frac{\phi(x)}{v_{kT}} - 1\right) + \frac{n_i^2}{N_a^2}\left(e^{[\phi(x)/v_{kT}]} - \frac{\phi(x)}{v_{kT}} - 1\right)\right] \quad (3.51)$$

$$= E_s^2(x)$$

In silicon substrate at the Si/SiO$_2$ interface, $x = 0$, $\phi(x) = \phi_s$, and $E(x) = E_s$; then from Gauss's law, the total charge per unit area induced in silicon (equal and opposite to the charge on the metal gate) is given by $Q_s = -\varepsilon_0 K_{si} E_s$. Then from Equation 3.51, we can show that the charge per unit area at the surface of the substrate is given by

$$Q_s = \pm\sqrt{2qK_{si}\varepsilon_0 N_a v_{kT}}\left[\left(e^{-(\phi_s/v_{kT})} + \frac{\phi_s}{v_{kT}} - 1\right) + \frac{n_i^2}{N_a^2}\left(e^{(\phi_s/v_{kT})} - \frac{\phi_s}{v_{kT}} - 1\right)\right]^{1/2} \quad (3.52)$$

The expression (Equation 3.52) is valid for all regions of MOS capacitor operations: accumulation, depletion, and inversion. The positive sign indicates the induced charge is positive for accumulation and negative sign represents that for depletion and inversion in a *p*-type substrate. Equation 3.52 can also be expressed as

$$Q_s = \pm\frac{\sqrt{2}v_{kT}K_{si}\varepsilon_0}{L_d}\left[\left(e^{-(\phi_s/v_{kT})} + \frac{\phi_s}{v_{kT}} - 1\right) + \frac{n_{po}}{p_{po}}\left(e^{(\phi_s/v_{kT})} - \frac{\phi_s}{v_{kT}} - 1\right)\right]^{1/2} \quad (3.53)$$

where n_i^2/N_a^2 is expressed in terms of the equilibrium majority and minority carrier concentrations p_{po} and n_{po} in a *p*-type substrate, respectively, such that $n_i^2/N_a^2 = (n_i^2/N_a).(1/N_a) = n_{po}/p_{po}$ and is related to ϕ_B by Equation 3.41, whereas L_d is the Debye length defined by

$$L_d = \sqrt{\frac{\varepsilon_0 K_{si} kT}{q^2 N_a}} \quad (3.54)$$

Again, in Equations 3.52 and 3.53, the first term within parenthesis is the majority carrier charge in *p*-type substrate whereas the second term is due to the minority carrier electrons.

The variation of the induced charge Q_s as a function of ϕ_s using Equation 3.52 for *p*-type substrate is illustrated in Figure 3.12, which clearly shows all three regimes of MOS capacitor operation.

From Figure 3.12, the different regimes of MOS capacitor operation are easily identified. Let us use Equation 3.52 to analyze different regions of MOS capacitor operation.

1. When $\phi_s < 0$, the MOS structure is in the accumulation mode and the inversion carrier term is negligible. Then the dominant term in Equation 3.52 is $\exp(-\phi_s/v_{kT})$. Thus, Q_s varies with ϕ_s as

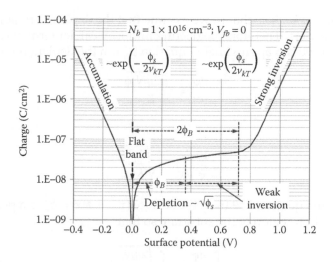

FIGURE 3.12

Variation of the induced charge density Q_s in a p-type silicon as a function of surface potential ϕ_s in all regimes of MOS capacitor operation; plot is obtained by Equation 3.52 using $N_a = 1 \times 10^{16}$ cm^{-3}, and $V_{fb} = 0$.

$$Q_s \approx \exp\left(-\frac{\phi_s}{2v_{kT}}\right) \text{ (accumulation)} \tag{3.55}$$

2. When $\phi_s > 0$ such that $0 < \phi_s < 2\phi_B$, the MOS capacitor structure is in the depletion and weak inversion regime. In this case, the term that dominates in Equation 3.52 is $\sqrt{\phi_s}$ and therefore Q_s varies with ϕ_s as

$$Q_s \approx \sqrt{\phi_s} \text{ (depletion and weak inversion)} \tag{3.56}$$

3. When $\phi_s > 2\phi_B$, the MOS capacitor structure is in the strong inversion regime, and to a first approximation, the majority carrier term in Equation 3.52 can be neglected. Then Q_s varies with ϕ_s as

$$Q_s \approx \exp\left(\frac{\phi_s}{2v_{kT}}\right) \text{ (strong inversion)} \tag{3.57}$$

The accumulation, depletion, and inversion conditions described by Equations 3.55, 3.56, and 3.57, respectively, are for p-type substrates. For n-type substrates, these conditions will be reversed.

3.4.2.1 MOS Capacitor at Depletion: Depletion Approximation

In the depletion region, $0 < \phi_s < 2\phi_B$ and the induced charge Q_s within the space charge region near the surface of the substrate is obtained by solving Poisson's equation. This induced charge in the depletion region is the depletion or *bulk charge* denoted by Q_b. For the simplicity of calculation, we use *depletion approximation*; that is, we assume that the depletion region is free of minority carriers. Therefore, neglecting minority carrier term and recognizing $\phi(x) > 0$, we can approximate Equation 3.51 in the depletion region as

$$\frac{d\phi}{dx} \cong -\sqrt{\frac{2qN_a\phi}{K_{si}\varepsilon_0}} \tag{3.58}$$

or

$$\frac{1}{\sqrt{\phi}}\frac{d\phi}{dx} = -\sqrt{\frac{2qN_a}{K_{si}\varepsilon_0}} \tag{3.59}$$

We integrate Equation 3.59 from the surface ($x = 0$, $\phi(x) = \phi_s$) to a point (x, $\phi(x)$) in the depletion region of the substrate so that

$$\int_{\phi_s}^{\phi(x)} \frac{d\phi}{\sqrt{\phi}} = -\int_0^x \sqrt{\frac{2qN_a}{K_{si}\varepsilon_0}}dx \tag{3.60}$$

Integrating Equation 3.60 and after simplification, we can show

$$\phi(x) = \phi_s\left(1 - \sqrt{\frac{qN_a}{2K_{si}\varepsilon_0\phi_s}}x\right)^2 \tag{3.61}$$

If we define

$$X_d = \sqrt{\frac{2K_{si}\varepsilon_0\phi_s}{qN_a}} \tag{3.62}$$

Then we can express Equation 3.61 as

$$\phi(x) = \phi_s\left(1 - \frac{x}{X_d}\right)^2 \tag{3.63}$$

Equation 3.63 is parabolic with the vertex at $\phi(x) = 0$, $x = X_d$. Thus, X_d is the distance to which band bending extends and is the width of the depletion region as shown in Figure 3.13. Then using Equation 3.62 for X_d, the depletion layer (bulk) charge density is given by

$$Q_b = -qN_aX_d = -\sqrt{2qK_{si}\varepsilon_0N_a\phi_s} \tag{3.64}$$

Equation 3.64 also easily follows from Equation 3.52, using depletion approximation. In an MOS capacitor system, the onset of strong inversion is defined by

$$\phi_s = 2\phi_B = 2v_{kT}\ln\left(\frac{N_a}{n_i}\right) \tag{3.65}$$

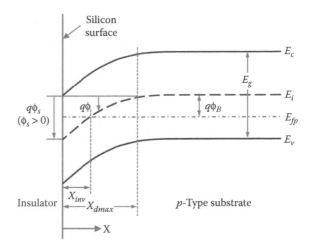

FIGURE 3.13
Variation of band-bending potential $\phi(x)$ along the depth of a p-type substrate showing the maximum width, X_{dmax}, of the depletion region at strong inversion and the width of the inversion layer, X_{inv}.

where ϕ_B is given by Equation 3.35. At strong inversion, X_d reaches a maximum, X_{dmax}, when $\phi_s = 2\phi_B$. This is because at strong inversion, the inversion layer shields the depletion charge so that the surface below can no longer respond to the applied V_g. Therefore, from Equations 3.62 and 3.65, the maximum width of the depletion layer is given by

$$X_{dmax} = \sqrt{\frac{4K_{si}\varepsilon_0 v_{kT}}{qN_a} \ln\left(\frac{N_a}{n_i}\right)} \tag{3.66}$$

3.4.2.2 MOS Capacitor at Inversion

In Section 3.3.3 we discussed that a sufficiently high gate voltage can cause enough band bending to pull the mid-gap energy E_i below the constant Fermi level E_f, that is, $E_i > E_f$. Under this condition, the surface of the p-type semiconductor is inverted and behaves like an n-type material with an electron concentration, n, given by Equation 3.42. In the inversion region $\phi_B < \phi_s < 2\phi_B$, the inversion layer charge Q_i can be calculated by considering the electron concentration (second term) from the general solution of Poisson's equation in Equation 3.52. In Equation 3.52, we observe that for $\phi_s > 0$, $\exp(-\phi_s/v_{kT})$ is negligibly small, the term "–1" is negligibly small since $\exp(\phi_s/v_{kT}) \gg -1$ in strong inversion, and the term $(-\phi_s/v_{kT})$ is negligibly small in weak inversion. Therefore, from Equation 3.52, the induced charge in the semiconductor under the inversion condition is given by

$$Q_s \cong -\sqrt{2qK_{si}\varepsilon_0 N_a v_{kT}} \left[\frac{\phi_s}{v_{kT}} + \frac{n_i^2}{N_a^2} e^{(\phi_s/v_{kT})} \right]^{1/2} \tag{3.67}$$

Using the relation $n_i^2 / N_a^2 = \exp(-2\phi_B/v_{kT})$ from Equation 3.41 in Equation 3.67, we get for the total induced charge density in the semiconductor as

$$Q_s = -\sqrt{2qK_{si}\varepsilon_0 N_a v_{kT}} \left[\frac{\phi_s}{v_{kT}} + e^{(\phi_s - 2\phi_B)/v_{kT}} \right]^{1/2} \tag{3.68}$$

Note that the induced charge represented by Equation 3.68 is the sum of the inversion charge Q_i and the depletion charge Q_b, that is,

$$Q_s = Q_i + Q_b \tag{3.69}$$

Using the expressions for $Q_b = -\sqrt{2qK_{si}\varepsilon_0 N_a \phi_s}$ from Equation 3.64 and Q_s from Equation 3.68 in Equation 3.69, we get the inversion charge per unit area as

$$Q_i = -\sqrt{2qK_{si}\varepsilon_0 N_a} \left[\left(\phi_s + v_{kT} e^{(\phi_s - 2\phi_B)/v_{kT}} \right)^{1/2} - \sqrt{\phi_s} \right] \tag{3.70}$$

Equation 3.70 shows the relation between the inversion charge density Q_i and surface potential ϕ_s for an MOS capacitor system. Figure 3.14 shows the dependence of Q_i, Q_b, and Q_s on ϕ_s. It is observed from Figure 3.14 that Q_b does not vary significantly. On the other hand, Q_i and Q_s clearly show two distinct regions of operation depending on the value of ϕ_s. These regions become more apparent on $\log(Q_i)$ versus ϕ_s plot as shown in Figure 3.15. These regions are (1) weak inversion for lower values of ϕ_s and (2) strong inversion at higher values of ϕ_s. Classically, the condition separating the

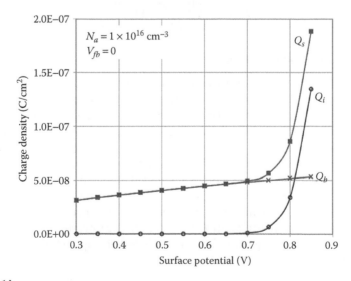

FIGURE 3.14

Variation of Q_b, Q_s, and Q_i as a function of ϕ_s obtained by Equations 3.64, 3.68, and 3.70, respectively, for an MOS capacitor system on a p-type substrate with $N_a = 1 \times 10^{16}$ cm^{-3}.

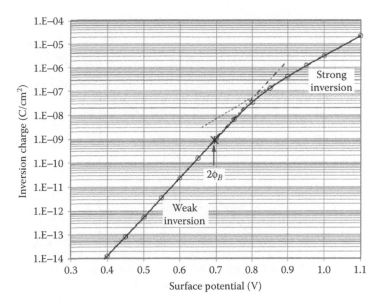

FIGURE 3.15
Variation of Q_i as function of ϕ_s for an MOS capacitor with p-type silicon substrate obtained by Equation 3.70; the plot shows the weak and strong inversion regions; here $N_a = 1 \times 10^{16}$ cm^{-3} and $V_{fb} = 0$.

weak and strong inversion regions is defined by $\phi_s = 2\phi_B$; that is, the inversion carrier concentration becomes equal to that of the majority carrier at the surface.

In order to distinguish the transition region between the weak and strong inversions, the inversion regime is divided into three regions. The third region that lies between the weak and strong inversion is called the moderate inversion region, which lies between $2\phi_B$ and $(2\phi_B + 6v_{kT})$ as shown in Figure 3.16. According to this convention, the region beyond $(2\phi_B + 6v_{kT})$ is the strong inversion region [7].

From the general expression of Q_i Equation 3.70, we can derive the regional expressions for Q_i at weak and strong inversions.

Weak inversion sets in when the band bending at the surface exceeds ϕ_B and extends to $2\phi_B$, that is, $\phi_B < \phi_s < 2\phi_B$. Within this region, the inversion charge Q_i is small compared to the depletion layer charge Q_b, that is,

$$|Q_i| \ll |Q_b| \quad (\text{weak inversion}) \tag{3.71}$$

For a small ϕ_s, Equation 3.70 can be simplified to obtain Q_i at weak inversion as

$$Q_i = -\sqrt{2qK_{si}\varepsilon_0 N_a v_{kT}}\, e^{(\phi_s - 2\phi_B)/2v_{kT}} \quad (\text{weak inversion}) \tag{3.72}$$

Equation 3.72 implies that in the weak inversion regime Q_i is essentially an exponential function of the surface potential ϕ_s as shown in Figure 3.15.

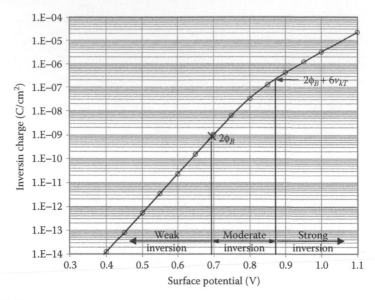

FIGURE 3.16
Variation of Q_i as function of ϕ_s for an MOS capacitor with p-type silicon substrate showing the weak, moderate, and strong inversion regions; Q_i is obtained by Equation 3.70; here $N_b = 1 \times 10^{16}$ cm^{-3} and $V_{fb} = 0$.

Strong inversion is defined when the band bending at the surface is such that $\phi_s \gg 2\phi_B$ so that the inversion charge Q_i is large compared to the depletion region charge Q_b, that is,

$$|Q_i| \gg |Q_b| \quad (\text{strong inversion}) \tag{3.73}$$

In this case, the exponential term in Equation 3.70 is large compared to ϕ_s and $\phi_s \gg 2\phi_B$. Thus, at strong inversion we get

$$Q_i \cong -\sqrt{2qK_{si}\varepsilon_0 N_a v_{kT}}\, e^{(\phi_s/2v_{kT})} \quad (\text{strong inversion}) \tag{3.74}$$

Thus, the inversion charge is an exponential function of the surface potential. Therefore, *a small increment of the surface potential induces a large change in the inversion layer charge.*

Let us now find out the relation between the gate voltage V_g and surface potential ϕ_s. From Equation 3.23 we get

$$V_g = V_{fb} + \phi_s - \frac{Q_s}{C_{ox}} \tag{3.75}$$

Now, substituting for Q_s from Equation 3.68 in Equation 3.23, we get

$$V_g = V_{fb} + \phi_s + \frac{\sqrt{2qK_{si}\varepsilon_0 N_a v_{kT}}}{C_{ox}}\left[\frac{\phi_s}{v_{kT}} + e^{(\phi_s - 2\phi_B)/v_{kT}}\right]^{1/2} \tag{3.76}$$

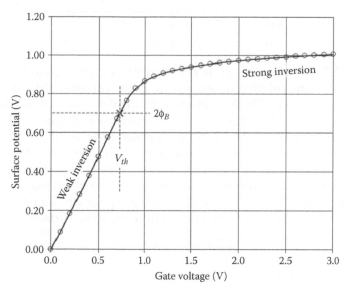

FIGURE 3.17

Surface potential versus gate voltage for a typical MOS capacitor system with p-type substrate obtained by numerical solution of Equation 3.76; the condition for strong inversion, at $\phi_s = 2\phi_B$, is also shown on the plot; the parameters used to compute surface potential are: $T_{ox} = 1.5$ nm, $N_a = 1 \times 10^{16}$ cm^{-3}, and $V_{fb} = 0$.

Equation 3.76 is an implicit relation in ϕ_s and must be solved numerically. Figure 3.17 shows the results of ϕ_s versus V_g characteristics obtained by numerical simulation. At low V_g ($>V_{fb}$), ϕ_s and X_d increase reasonably rapidly with V_g. This regime corresponds to the depletion and weak inversion regions of device operation. At larger gate biases, ϕ_s is almost constant and is pinned. This pinning occurs at the onset of strong inversion and the classical condition for pinning is $\phi_s = 2\phi_B$. This condition is referred to as the condition for *threshold* and the corresponding gate voltage is called the *threshold voltage*, V_{th}. Thus, at the onset of inversion, $\phi_s = 2\phi_B$ and $V_g = V_{th}$; then from (3.76), we get

$$V_{th} = V_{fb} + 2\phi_B + \frac{\sqrt{2qK_{si}\varepsilon_0 N_a (2\phi_B)}}{C_{ox}}$$

$$= V_{fb} + 2\phi_B + \gamma\sqrt{2\phi_B}$$

(3.77)

where:

$\gamma \equiv \sqrt{2qK_{si}\varepsilon_0 N_a} / C_{ox}$ is called the *body effect coefficient* and is dependent on the substrate doping and gate oxide thickness.

V_{th} is one of the most important parameters for MOSFET devices and will be discussed in Chapter 4

Beyond the strong inversion, the concentration of the inversion charge $n(x)$ becomes significant. Therefore, from Equation 3.51 we get

$$\frac{d\phi(x)}{dx} = -\sqrt{\frac{2qN_a v_{kT}}{K_{si}\varepsilon_0}\left[\frac{\phi(x)}{v_{kT}} + \frac{n_i^2}{N_a^2}e^{\phi(x)/v_{kT}}\right]} \tag{3.78}$$

Equation 3.78 must be solved numerically with boundary condition, $\phi(x) = \phi_s$ at $x = 0$. From the solution of $\phi(x)$, the inversion carriers, $n(x)$, can be calculated from Equation 3.42. The numerically calculated $n(x)$ versus depth plot is shown in Figure 3.18. It is seen that the inversion charge distribution is extremely close to the surface with an inversion layer width < 5 nm.

From the previous mathematical formulation, let us find a simple analytical expression for *inversion layer thickness*. We have shown earlier that the general expression for inversion carrier charge is given by $Q_i = -\sqrt{2qK_{si}\varepsilon_0 N_a v_{kT}}\,e^{(\phi_s - 2\phi_B)/2v_{kT}}$. And, from the expression (Equation 3.42), we can show that the minority carrier concentration at the surface $x = 0$ is

$$\sqrt{n(0)} = \sqrt{N_a}\,e^{(\phi_s - 2\phi_B)/2v_{kT}} \tag{3.79}$$

Thus, combining Equations 3.72 and 3.79, we get

$$Q_i = -\sqrt{2qK_{si}\varepsilon_0 v_{kT}n(0)} \tag{3.80}$$

Again, if the inversion layer thickness is X_{inv}, then $Q_i = qn(0)X_{inv}$; then from Equation 3.80 we can show that the classical inversion charge thickness is given by

$$X_{inv} = \frac{Q_i}{qn(0)} = \frac{2K_{si}\varepsilon_0 v_{kT}}{Q_i} \tag{3.81}$$

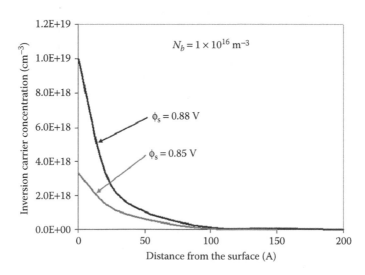

FIGURE 3.18
Calculated minority carrier electron distribution in a *p*-type silicon substrate of an MOS capacitor system for different ϕ_s.

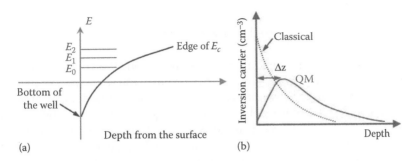

FIGURE 3.19

Inversion layer quantization: (a) minority carrier electron in a potential well of an MOS capacitor system on a *p*-type silicon substrate; the potential well is bounded by potential barrier at the Si/SiO$_2$ interface and conduction band bending due to high V_g to achieve $\phi_s \geq 2\phi_B$; (b) typical minority carrier electron concentration in silicon surface as a function of silicon depth for classical and QM model; Δz is the shift in the centroid of inversion charge due to quantization.

It is also observed from Equation 3.79 that at the onset of strong inversion defined by $\phi_s = 2\phi_B$, the inversion layer concentration at the surface becomes equal to the majority carrier concentration in the surface, that is, $n(0) = N_b$.

Generally, inversion carriers must be treated quantum-mechanically as a 2D gas molecules. According to quantum mechanical (QM) model, the inversion layer carriers occupy discrete energy bands as shown in Figure 3.19a and the peak distribution is about 1–3 nm away from the surface as shown in Figure 3.19b. Thus, near the silicon surface, the inversion layer charges are confined to a potential well bounded by (1) oxide barrier height at the Si/SiO$_2$ interface and (2) bend silicon conduction band at the surface due to sufficiently high gate voltage V_g as shown in Figure 3.19a.

Due to QM confinement of inversion layer electrons in the *p*-type silicon surface, the *electron* energy levels are grouped in discrete sub-bands of energy, E_j, where $j = 0, 1, 2, \ldots$ quantized states as shown in Figure 3.19a. Each E_j corresponds to a quantized level for electron motion in the normal direction. The net result of QM effect is that the inversion layer density peaks below the SiO$_2$/Si interface with about zero value at the surface contrary to the classical inversion carrier distribution as shown in Figure 3.19b. Therefore, for accurate computation of inversion carrier distribution at the silicon surface, we have to solve both Schrödinger and Poisson equations self-consistently with boundary conditions: $\phi(x) = 0$ for $x < 0$ in the oxide; and $\phi(x) = 0$ @ $x = \infty$ deep into the silicon substrate.

As observed from Figure 3.19b, the silicon surface is depleted of mobile carriers due to inversion layer quantization. This depletion region in silicon can be considered as an insulating layer of silicon increasing the effective gate oxide thickness. This increase in the effective gate thickness is given by

$$\Delta T_{ox} = \frac{\varepsilon_{ox}}{\varepsilon_{si}} \Delta z \qquad (3.82)$$

where:

Δz is the shift in the centroid of inversion charge

Since the peak of the inversion charge is away from the surface due to QM effect, a higher V_g overdrive is required to produce the same level of inversion charge density predicted by classical theory. In other words, QM effect can be considered to reduce the net inversion charge density. Thus, the inversion layer quantization can be modeled as *bandgap widening* due to an increase in the effective bandgap energy, E_g, by an amount ΔE_g [13]. Then from Equation 2.14 relating E_g and the intrinsic carrier concentration, we can show that the intrinsic carrier concentration n_i^{QM} due to QM effect is given by

$$n_i^{QM} = n_i^{CL} \exp\left(-\frac{\Delta E_g}{2kT} \right) \tag{3.83}$$

where:

$\Delta E_g = \left(E_g^{QM} - E_g^{CL} \right)$ is the increase in the apparent value of E_g due to QM effect; here, E_g^{QM} is the energy gap due to QM effect and E_g^{CL} and n_i^{CL} are the energy gap and intrinsic carrier concentration, respectively, without the QM effect denoting the classical expression

Equation 3.83 shows that the inversion layer quantization decreases the intrinsic concentration compared to the classical value. We know from Equation 3.72, Q_i is proportional to n_i through the term $\exp(-2\phi_B/v_{KT})$. Thus, Q_i decreases due to QM effect. This decrease in Q_i due to QM effect has severe consequences on MOS transistor device performance as we will discuss in Chapter 9.

3.5 Capacitance of MOS Structure

In the previous section, we developed the mathematical foundation of MOS capacitor system relating the charge and potential under different gate biasing conditions. In this section, we will discuss the basic characteristics of MOS capacitor system under an applied bias. We know that the capacitance of any system is the ratio of the variation in charge due to the corresponding variation in the small signal voltage. Thus, the total capacitance (C) of an MOS structure in equilibrium is given by

$$C = \frac{d\left(-Q_s\right)}{dV_g} \tag{3.84}$$

From Equation 3.17, the applied bias of an MOS system is $V_g = V_{fb} + V_{ox} + \phi_s$; since V_{fb} is a constant, we can write

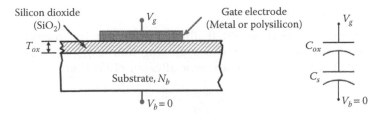

FIGURE 3.20
The individual component of the total MOS capacitor system; C_{ox} and C_s are the capacitance per unit area of the gate oxide capacitance and substrate capacitance, respectively. Nb is the substrate concentration.

$$dV_g = dV_{ox} + d\phi_s \tag{3.85}$$

From Equations 3.84 and 3.85, we can show that

$$\frac{dV_g}{d(-Q_s)} = \frac{dV_{ox}}{d(-Q_s)} + \frac{d\phi_s}{d(-Q_s)}$$

$$\therefore \quad \frac{1}{C} = \frac{1}{C_{ox}} + \frac{1}{C_s} \tag{3.86}$$

Thus, the total capacitance of an MOS structure equals the oxide capacitance C_{ox} and the substrate capacitance C_s connected in series as shown in Figure 3.20. Here, C_s is the capacitance per unit area of the space charge region in the silicon.

Equation 3.86 along with Equation 3.53 for Q_s is used to calculate C–V characteristics for the target range of operation of MOS capacitor systems. In order to generate C–V plot, first of all, we calculate the general expression for the space charge region capacitance C_s in the semiconductor from the total charge Q_s given by Equation 3.53. Then we select an appropriate value of ϕ_s for each mode of operation of an MOS capacitor system to obtain the corresponding C–V characteristics.

3.5.1 Low Frequency C–V Characteristics

In order to obtain the low frequency (LF) C–V characteristics of MOS capacitors, we can show from Equation 3.53

$$C_s = \frac{d(-Q_s)}{d\phi_s} = -\frac{\sqrt{2}K_{si}\varepsilon_0}{L_d} \frac{\left[1 - e^{-(\phi_s/v_{kT})} + \left(n_{p0}/p_{p0}\right)\left(e^{\phi_s/v_{kT}} - 1\right)\right]}{2\left[\left(e^{-(\phi_s/v_{kT})} + \left(\phi_s/v_{kT}\right) - 1\right) + \left(n_{p0}/p_{p0}\right)\left(e^{(\phi_s/v_{kT})} - \left(\phi_s/v_{kT}\right) - 1\right)\right]^{1/2}} \tag{3.87}$$

Equation 3.87 is the general expression for C_s in an MOS capacitor system that we will apply to analyze C–V characteristics in the different operational regions of the system discussed in Section 3.3.

3.5.1.1 Accumulation

In strong accumulation, $\phi_s \ll 0$; then considering the majority carrier term only in Equation 3.87 and recognizing that $\exp(-\phi_s/v_{kT}) \gg 1$ and $\exp(-\phi_s/v_{kT}) \gg (\phi_s/v_{kT})$, we can show after simplification

$$C_s = \frac{K_{si}\varepsilon_0}{\sqrt{2}L_d} e^{-(\phi_s/2v_{kT})} \tag{3.88}$$

Since $\phi_s \ll 0$, for large ϕ_s, C_s becomes very large; therefore, from Equation 3.86, we get

$$\frac{1}{C} = \frac{1}{C_{ox}} + \frac{1}{C_s} \cong \frac{1}{C_{ox}} \tag{3.89}$$

Thus, in accumulation region

$$C \cong C_{ox}\,(\text{accumulation}) \tag{3.90}$$

3.5.1.2 Flat Band

At flat band condition, $\phi_s = 0$, the inversion charge term (i.e., term containing n_{po}/p_{po}) in Equation 3.87 can be neglected to get

$$C_s = \frac{d(-Q_s)}{d\phi_s} = -\frac{\sqrt{2}K_{si}\varepsilon_0}{L_d} \frac{\left[1 - e^{-(\phi_s/v_t)}\right]}{2\left[\left(e^{-(\phi_s/v_t)} + (\phi_s/v_t) - 1\right)\right]^{1/2}} \tag{3.91}$$

Since the direct substitution of $\phi_s = 0$ in Equation 3.91 results in indeterminate, we simplify Equation 3.91 by series expansion using $e^x = 1 + (x/1!) + (x^2/2!) + (x^3/3!) + \cdots, -\infty < x < \infty$, where $x \equiv -\phi_s/v_{kT}$, then

$$C_s \cong -\frac{\sqrt{2}K_{si}\varepsilon_0}{L_d} \frac{\left(1 - \left\{1 - (\phi_s/v_{kT}) + (1/2)\left[-(\phi_s/v_{kT})\right]^2 + (1/6)\left[-(\phi_s/v_{kT})\right]^3 + \cdots\right\}\right)}{2\left(\left\{1 - (\phi_s/v_{kT}) + (1/2)\left[-(\phi_s/v_{kT})\right]^2 + (1/6)\left[-(\phi_s/v_{kT})\right]^3 + \cdots\right\} + (\phi_s/v_{kT}) - 1\right)^{1/2}}$$

$$= -\frac{\sqrt{2}K_{si}\varepsilon_0}{L_d} \frac{\left[-(\phi_s/v_{kT}) + (1/2)(\phi_s/v_{kT})^2 - (1/6)(\phi_s/v_{kT})^3 + \cdots\right]}{2\left[(1/2)(\phi_s/v_{kT})^2 - (1/6)(\phi_s/v_{kT})^3 + \cdots\right]^{1/2}}$$

$$= -\frac{\sqrt{2}K_{si}\varepsilon_0}{L_d} \frac{\left[-(\phi_s/v_{kT})\right]\left[1 - (1/2)(\phi_s/v_{kT}) - (1/6)(\phi_s/v_{kT})^2 + \cdots\right]}{\sqrt{2}\,(\phi_s/v_{kT})\left[1 - (1/6)(\phi_s/v_{kT}) + \cdots\right]^{1/2}}$$

$$= \frac{K_{si}\varepsilon_0}{L_d} \frac{\left[1 - (1/2)(\phi_s/v_{kT}) - (1/6)(\phi_s/v_{kT})^2 + \cdots\right]}{\left[1 - (1/6)(\phi_s/v_{kT}) + \cdots\right]^{1/2}}$$

Now we substitute for $\phi_s = 0$ to get

$$C_s(\text{flat band}) \cong \frac{K_{si}\varepsilon_0}{L_d} \frac{\left[1-(1/2)(\phi_s/v_{kT})-(1/6)(\phi_s/v_{kT})^2 + \cdots\right]}{\left[1-(1/6)(\phi_s/v_{kT})+\cdots\right]^{1/2}}$$

$$= \frac{K_{si}\varepsilon_0}{L_d} \tag{3.92}$$

Then from Equation 3.86, the total capacitance of an MOS structure at flat band condition is given by

$$C \equiv C_{fb} = \left(\frac{1}{C_{ox}} + \frac{L_d}{K_{si}\varepsilon_0}\right)^{-1} \quad (\text{flat band}) \tag{3.93}$$

Since $(L_d/K_{si}\varepsilon_0)$ is a finite number, Equation 3.93 shows that C_{fb} is somewhat less than C_{ox}.

3.5.1.3 Depletion

In the depletion regime $(0 < \phi_s < 2\phi_B)$, the general expression for C_d is given by Equation 3.87. However, we can derive an approximate expression from depletion approximation discussed in Section 3.4.2.1. We know that at the depletion condition $Q_s \cong Q_b$, then substituting for $Q_s = Q_b$ in Equation 3.23, we get

$$V_g = V_{fb} + \phi_s - \frac{Q_b}{C_{ox}} \tag{3.94}$$

Again, from Equation 3.64 we get:

$$Q_b = -\sqrt{2qK_{si}\varepsilon_0 N_b \phi_s}$$

or

$$\phi_s = \frac{Q_b^2}{2qK_{si}\varepsilon_0 N_b} \tag{3.95}$$

Then substituting for ϕ_s from Equation 3.95 to Equation 3.94, we can show after simplification

$$V_g - V_{fb} = \frac{Q_b^2}{2qK_{si}\varepsilon_0 N_b} - \frac{Q_b}{C_{ox}}$$

or

$$Q_b^2 - \left(\frac{2qK_{si}\varepsilon_0 N_b}{C_{ox}}\right)Q_b - 2qK_{si}\varepsilon_0 N_b\left(V_g - V_{fb}\right) = 0 \tag{3.96}$$

Equation 3.96 is a quadratic equation in Q_b with solution given by

$$Q_b = \frac{qK_{si}\varepsilon_0 N_b}{C_{ox}} \pm \sqrt{\left(\frac{qK_{si}\varepsilon_0 N_b}{C_{ox}}\right)^2 + 2qK_{si}\varepsilon_0 N_b \left(V_g - V_{fb}\right)} \qquad (3.97)$$

Then from Equation 3.97 we can show that the depletion capacitance is given by

$$C = \frac{dQ_b}{dV_g} = \frac{C_{ox}}{\sqrt{1 + \left[2C_{ox}^2 \left(V_g - V_{fb}\right)/qK_{si}\varepsilon_0 N_b\right]}} \quad \text{(depletion)} \qquad (3.98)$$

From Equation 3.98 we observe that the depletion capacitance C decreases with the increase in V_g. It is clear from Equation 3.98 that for a given voltage $(V_g - V_{fb})$, the capacitance in the depletion region will be higher for higher N_b as well as lower C_{ox} or thicker T_{ox}. It is also seen from Equation 3.98 that at $V_g = V_{fb}$, $C = C_{ox}$; this is because Equation 3.98 is derived assuming depletion approximation; that is, the transition between the accumulation and depletion region is abrupt.

3.5.1.4 Inversion

In strong inversion, $\phi_s \gg 0$; considering only the minority carrier term in Equation 3.87 and recognizing that $\exp(\phi_s / v_{kT}) \gg 1$ and $\exp(\phi_s / v_{kT}) \gg (\phi_s / v_{kT})$, we can show after simplification

$$C_s \cong -\frac{\sqrt{2}K_{si}\varepsilon_0}{L_d} \frac{\left(n_{p0}/p_{p0}\right)e^{\phi_s/v_{kT}}}{2\sqrt{n_{p0}/p_{p0}}\,e^{\phi_s/2v_{kT}}} = -\frac{K_{si}\varepsilon_0}{\sqrt{2}L_d} \frac{n_i}{N_b} e^{\phi_s/2v_{kT}} \qquad (3.99)$$

where we have used $N_b = p_{po} = N_a$ and $n_{po} = n_i^2 / N_a$. In Equation 3.99, the negative sign indicates that the charge has changed sign. Since $\phi_s \gg 0$, for large value of ϕ_s, C_s becomes very large. Therefore, from Equation 3.86, the total capacitance of an MOS capacitor system at strong inversion is given by

$$\frac{1}{C} = \frac{1}{C_{ox}} + \frac{1}{C_s} \approx \frac{1}{C_{ox}} \qquad (3.100)$$

In a MOS capacitor system, the inversion layer is formed by thermally generated minority carriers (electrons for p-type substrate). The concentration of minority carriers can change only as fast as carriers can be generated within the depletion region near the surface. As a result, the MOS capacitance at inversion is a function of the frequency of the AC signal used to measure the capacitance. If the AC signal is sufficiently low (typically, 10 Hz), the inversion layer charge can respond to the AC bias and the DC sweeping voltage, generating LF C–V plot. In this condition, Equation 3.99 is valid and therefore

$$C \cong C_{ox} \quad \text{(inversion at LF signal)} \qquad (3.101)$$

However, if the frequency of the AC voltage signal is too high (typically, above 1×10^5 Hz), the inversion charge cannot respond to the changes in AC voltage. As a result, the measured C–V curve, called the high frequency (HF) C–V plot is significantly different from the LF C–V plot. Also, we derived Equation 3.87 assuming that all charges in the depletion region of an MOS capacitor follow the variation of ϕ_s. Thus, Equation 3.87 is valid for LF C–V curve. On the other hand, the HF capacitance of an MOS system can be obtained by considering only the depletion charge, Q_b, at the surface and the effective depth, X_d, of the depletion region at inversion condition using parallel plate capacitance formula, $C = K_{si}\varepsilon_0 / X_d$.

In the accumulation and depletion regions of an MOS capacitor system, the minority carrier charge is negligibly small compared to the bulk charge and the total charge at the surface is primarily due to the majority carriers. As a result, MOS capacitance in these regions is independent of frequency at all range of operational frequency.

3.5.2 Intermediate and High Frequency C–V Characteristics

We discussed in the previous section that there are plenty of majority carriers in the substrate that can respond to the AC signal. However, the minority carriers are scarce and have to originate by diffusion from the bulk substrate, by generation in the depletion region, or by external sources (e.g., $n+$ diffusion region in MOS transistors). Thus, the inversion charge cannot respond to the applied AC signal higher than 100 Hz. Therefore, at any HF-applied signal to V_g, the inversion charge Q_i is assumed to be a constant, and therefore $Q_b = Q_s$. Thus, at any HF-applied signal to V_g, only Q_b will vary with the signal around its maximum value Q_{bmax}, and X_d will vary with the signal around its maximum value X_{dmax}. Thus, the HF capacitance is given by:

$$\frac{1}{C} = \frac{1}{C_{ox}} + \frac{X_{dmax}}{K_{si}\varepsilon_0} \equiv \frac{1}{C_{min}} \quad (\text{inversion at HF}) \qquad (3.102)$$

Now, substituting for X_{dmax} from Equation 3.66 to Equation 3.100, we can show

$$\frac{1}{C_{min}} = \frac{1}{C_{ox}} + \sqrt{\frac{4v_{kT}}{qK_{si}\varepsilon_0 N_b} \ln\left(\frac{N_b}{n_i}\right)} \quad (\text{inversion at HF}) \qquad (3.103)$$

3.5.3 Deep Depletion C–V Characteristics

We discussed in the previous section that the inversion layer is formed by thermal generation of carriers in the substrate of an MOS capacitor. However, if an MOS capacitor is swept from the accumulation to the inversion regions at a relatively fast rate (e.g., 10 V sec^{-1} and higher) so that there is not enough time for the thermal generation of the minority carriers, the capacitance will continue to fall with increasing magnitude of V_g along the depletion curve.

This is a nonequilibrium condition under which the depletion width continues to widen beyond its maximum value, X_{dmax}, in order to balance the increased gate charge and C does not reach a minimum. This expansion of the depletion region deep into the substrate is referred to as the *deep depletion*. From Equation 3.98, the capacitance in the deep depletion mode is given by

$$C = \frac{K_{si}\varepsilon_0}{X_d} = \frac{C_{ox}}{\sqrt{1 + \left[2C_{ox}^2\left(V_g - V_{fb}\right)/qK_{si}\varepsilon_0 N_b\right]}} \quad \text{(deep depletion)} \quad (3.104)$$

The capacitance in the deep depletion is obtained when the rate of DC voltage sweep is high, independent of the frequency of the AC signal voltage (HF) and no inversion charge can form. *The easiest way to obtain deep depletion is to sweep the DC voltage by either applying a voltage step or using a fast voltage ramp on the gate.*

Thus, from the previous mathematical analysis, we find that, depending on the frequency of the AC signal and measurement conditions, three types of C–V plots are obtained as shown in Figure 3.21. *It should be pointed out that the frequency dependence of capacitance in inversion is true only for MOS capacitor*

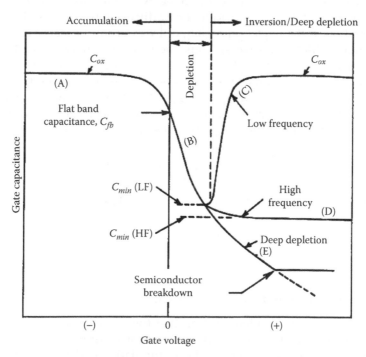

FIGURE 3.21

C–V characteristics of an ideal MOS capacitor system from accumulation to inversion regimes: regions A and B represent the accumulation and depletion, respectively; regions C and D represent the LF and HF inversion capacitances, respectively; and plot E shows deep depletion.

and not for MOS transistors. In the case of MOS transistors, the source and drain diffusions can supply minority carriers to the inversion layer almost instantaneously.

3.5.4 Deviation from Ideal C–V Curves

The ideal MOS capacitance plots shown in Figure 3.21 are obtained by assuming that the gate oxide is a perfect insulator free of charges ($Q_0 = 0$) and $\Phi_{ms} = 0$, so that $V_{fb} = 0$. However, due to the nonideal nature of the MOS structures, experimental LF and HF C–V plots deviate from the ideal behavior by one or more of the following parameters: (1) nonzero Φ_{ms}, (2) interface traps, (3) mobile ions in the oxide, (4) fixed charge, and (5) nonuniform substrate doping. The detailed description of the nonideal behavior of MOS capacitor system is available in the literature [2].

3.5.5 Polysilicon Depletion Effect on C–V Curves

In our discussions so far we assumed that the polysilicon gate is degenerately doped with concentration in excess of 5×10^{19} cm^{-3}. However, if the gate is nondegenerately doped, it can no longer be treated as an equipotential area like metal gate. In this case, the capacitance given by Equation 3.86 for a MOS capacitor system must be modified to include the capacitance C_{poly} due to polysilicon depletion at the polysilicon-gate/gate-dielectric interface. For polysilicon gate MOS capacitor, the capacitance, C, of the system is a series combination of: (1) polysilicon depletion capacitance, C_{poly}; (2) oxide capacitance, C_{ox}; and (3) substrate capacitance, C_s, as shown in Figure 3.22b. Then the resulting gate capacitance at strong inversion is given by [14]

$$\frac{1}{C} = \frac{1}{C_{poly}} + \frac{1}{C_{ox}} + \frac{1}{C_s} \tag{3.105}$$

Typical MOS C–V characteristics due to polysilicon gate depletion effect is shown in Figure 3.23. It is observed from Figure 3.23 that as V_g increases in the inversion regime, C_{poly} decreases due to the increase in the depletion width, $X_{poly-depletion}$ of polysilicon gate causing a decrease in the total capacitance C and thereby C/C_{ox}. Therefore, LF C–V plots show a local maximum at a certain V_g.

The C–V behavior shown in Figure 3.23 is attributed to the deviation from the nondegenerate doping of the polysilicon gate [14–19]. The result of the nondegenerate polysilicon doping is that the LF capacitance in inversion ($C_{g,inv}$) is much smaller than that of accumulation, and $C_{g,inv}$ decreases slightly with gate bias. However, at gate bias larger than a certain voltage $C_{g,inv}$ recovers to C_{ox} rather abruptly [14–19].

The decrease in capacitance due to polysilicon depletion effect can be expressed as an increase in the effective gate oxide thickness. The increase in

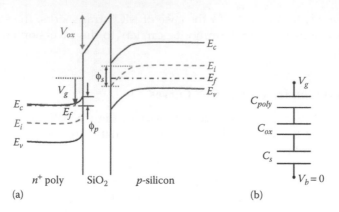

(a)   (b)

FIGURE 3.22
Polysilicon depletion effect in a polysilicon/SiO_2/p-type silicon substrate MOS capacitance system: plot shows (a) the band bending at strong inversion along with (b) the components of the associated capacitance with the structure.

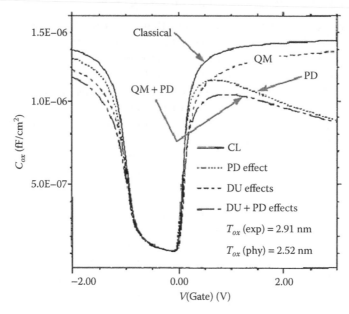

FIGURE 3.23
Low frequency $C–V$ characteristics of polysilicon gate MOS capacitor system showing the impact of QM and poly-depletion (PD) effects on MOS capacitance system. (Data from S. Saha et al., *Mater. Res. Soc. Symp. Proc.*, 567, 275–282, 1999.)

effective SiO_2 gate oxide thickness can be shown to be ΔT_{ox}(poly-depletion) $=$ $(K_{ox}/(K_{si})(X_{d,PD})$, where $X_{d,PD}$ is the width of the depletion region in the polysilicon at the polysilicon/SiO_2 interface. The total increase in the physical gate oxide thickness is the sum of the effective SiO_2 thickness due to QM effect expressed in Equation 3.82 and polysilicon depletion effect and is given by

$$\Delta T_{ox} = \frac{K_{ox}}{K_{si}} \left(\Delta z + X_{d,PD} \right) \tag{3.106}$$

where:

Δz and $X_{d,PD}$ are the centroid of inversion charge and width of the polysilicon depletion region, respectively

The increase in the effective gate oxide thickness due to polysilicon depletion and QM effects is about 0.5–0.7 nm and depends on the gate voltage and polysilicon doping density [18–19].

3.6 Summary

This chapter presented the basic structure and operation of an MOS capacitor system to build the foundation for developing MOS transistor compact models. We have discussed the basic MOS structure by considering the energy band model of metal, oxide, and semiconductors. The basic operation of an MOS capacitor system is discussed at equilibrium and under biasing conditions. The important parameter of the MOS structure is the flat band voltage V_{fb}. The significance of V_{fb} and the work function difference between the metal and semiconductor for MOS operation is clearly discussed using the energy band diagram. Analytical model of MOS capacitor system is developed to discuss the accumulation, depletion, and inversion mode operations of MOS capacitor structures. Finally, the analytical expressions to understand the C–V characteristics of MOS structures at different operation regimes are also presented.

Exercises

3.1 Consider an MOS capacitor system on a uniformly doped p-type substrate with doping concentration $N_a = 1 \times 10^{17}$ cm^{-3}. Calculate the flat band voltage. Assume that Si/SiO$_2$ interface charge is negligibly small. *Clearly state any assumptions you make.*

3.2 Consider an MOS capacitor system on a uniformly doped p-type substrate with doping concentration $N_a = 1 \times 10^{17}$ cm^{-3} operating in the accumulation region.

a. Sketch the band diagram into the substrate and clearly explain and label all relevant parameters such as:

   i. Surface potential

   ii. Fermi potential

   iii. Energy levels with reference to Si/SiO$_2$ interface.

b. Write one-dimensional Poisson equation that you will solve to obtain the surface potential in the accumulation region.

c. Derive an expression for the accumulation charge.

d. Derive an expression for surface potential.

Clearly state any assumption you make.

3.3 In Equation 3.40, we find that the inversion carrier density, $n(x)$ in a p-type substrate increases exponentially as $exp(\phi(x)/v_{kT})$ as x approaches to the surface at $x = 0$. In other words, we can consider, $n(x)$ decreases away from the surface as $exp(-\phi(x)/v_{kT})$ and approaches to a minimum value at $x = X_{inv}$, defined as the inversion layer thickness. Thus, all the minority carrier electrons are confined in a region bounded by the depth X_{inv}, where the intrinsic band energy, E_i intersects the Fermi level, E_f. Here, $\phi(x)$ is the potential at any point x from the surface and v_{kT} is the thermal voltage at the ambient temperature T. If an MOS capacitor system on a uniformly doped p-type substrate with doping concentration $N_a = 1 \times 10^{17}$ cm^{-3} operating in the inversion region,

a. Calculate the maximum width of the depletion layer into the silicon (x-direction).

b. Considering the inversion carrier concentration, $n_{inv} \propto exp(-\phi/v_{kT})$ along x-direction into the silicon and $n_{inv} \sim 0$ @ $x = X_{inv}$ whereas $n_{inv} = n_{surf}$ and $\phi = \phi_s$ @ $x = 0$:

   i. Calculate the potential drop $\Delta\phi$ @ $x = X_{inv}$.

   ii. Now assume that the device is in *weak inversion* ($\phi_B < \phi_s < 2\phi_B$) so that $Q_b \gg Q_{inv}$, then use Gauss's law and $\Delta\phi$ expression from part (b)(i) to show:

$$X_{inv} = v_{kT}\sqrt{\frac{K_{si}\varepsilon_0}{2qN_a\phi_s}}$$

where:

   ϕ is the potential at any point in silicon

   K_{si} is the dielectric constant of silicon

   ε_0 is the permittivity of free space

   ϕ_s is the surface potential

c. Calculate the thickness of the inversion layer, X_{inv} in silicon.

d. Sketch the band diagram into the substrate and clearly explain and label the following parameters with reference to Si/SiO$_2$ interface:

   i. Surface potential, ϕ_s

   ii. Bulk potential, ϕ_B

    iii. Energy levels $(E_c, E_i, E_f, E_v, E_g)$

    iv. Width of the depletion layer, X_d

    v. Width of the inversion layer, X_{inv}

Clearly state any assumptions you make.

3.4 Use Equation 3.52 to calculate and plot the total charge (Q_s) in semiconductor as a function of surface potential ϕ_s of an MOS capacitor system for $-0.4\ V \le \phi_s \le 1.4\ V$. Label all the operating regions of the MOS capacitor system. Assume that the MOS capacitor is fabricated on a uniformly doped p-type substrate with doping concentration $N_a = 1 \times 10^{17}\ cm^{-3}$. *Clearly state any assumptions you make.*

3.5 Consider an MOS capacitor system with uniformly doped p-type substrate with doping concentration, N_a, to show the following:

    a. $n_i^2 / N_a^2 = \exp(-2\phi_B / v_{kT})$, where n_i, ϕ_B, and v_{kT} are the intrinsic carrier concentration, bulk potential, and thermal voltage, respectively.

    b. Complete the mathematical steps to express Equations 3.52 to 3.53 in terms of Debye length L_d given by Equation 3.54.

3.6 Consider a double gate MOS capacitor system shown in Figure E3.1 fabricated on a lightly doped, $N_a = 5 \times 10^{15}\ cm^{-3}$, p-type substrate with $t_{si} = 30\ nm$, $L_g = 45\ nm$, and $T_{ox} = 1.5\ nm$ as shown in Figure E3.1, and gates connected together. *Clearly state any assumptions you make* to answer the following questions.

    a. Calculate the total width of the depletion region in silicon at strong inversion, if $t_{si} \to \infty$.

    b. Calculate the total width of the inversion layer at strong inversion. Assume that the inversion layer thickness by a single gate is given by

$$X_{inv} = \frac{kT}{q} \sqrt{\frac{K_s \varepsilon_0}{2qN_a\phi_s}}$$

    where:

        k is the Boltzmann constant

        T is the temperature

        q is the electronic charge

        K_{si} is the dielectric constant in silicon

        ε_0 is the permittivity of free space

        N_a is the channel doping concentration

        ϕ_s is the surface potential

    c. Use the values of the parameters in parts (a) and (b) to sketch the band diagram into silicon from $x = 0$ at the top gate Si/SiO$_2$

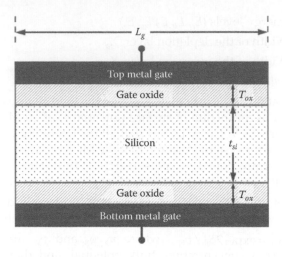

FIGURE E3.1
An ideal double gate MOS capacitance structure.

interface to $x = t_{si} = 30$ nm at the bottom gate Si/SiO$_2$ interface. Clearly define and label all relevant data points such as

   i. Surface potential

   ii. Fermi potential

  iii. Depletion width

  iv. Inversion layer thickness

   v. Energy levels.

  d. Calculate and plot the potential into the silicon from the top gate to the bottom gate at strong inversion for the top gate only. Repeat the calculation for the bottom gate only and plot on the same X–Y graph. Show the overall potential distribution from the top gate to the bottom gate when both gates are considered. Clearly define and label all relevant data points. Explain.

3.7 FinFETs with intrinsic or very lightly doped channel doping shown in Figure E3.2 are the most realistic double gate MOSFETs. Consider a simple FinFET capacitor fabricated on a lightly doped p-type substrate with $N_a = 3 \times 10^{15}$ cm^{-3}, $T_{fin} = 20$ nm, $H_{fin} = 50$ nm, $L_g = 50$ nm, $T_{ox} = 15$ nm, and gates connected together, as shown in Figure E3.2. (Ignore the effect of Source/Drain on the MOS capacitor system).

  a. Calculate the depletion width of silicon at strong inversion.

  b. Calculate the width of the inversion layer at strong inversion. Assume that the inversion layer thickness for a single gate is given by

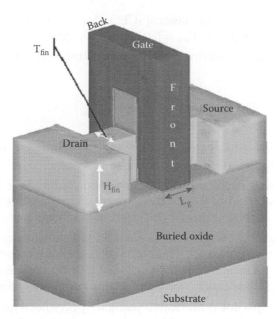

FIGURE E3.2
FinFET MOS device structure.

$$X_{inv} = \frac{kT}{q}\sqrt{\frac{K_s\varepsilon_0}{2qN_a\phi_s}}$$

c. Use the values of the parameters in parts (a) and (b) to sketch the band diagram into the silicon fin from $x = 0$ at the front-gate Si/SiO$_2$ interface to $x = T_{fin}$ at the back-gate Si/SiO$_2$ interface. Clearly define and label all relevant data points such as:

   i. Surface potential

   ii. Fermi potential

   iii. Depletion width

   iv. Inversion layer thickness

   v. Energy levels.

d. Calculate and plot the potential into the silicon fin from the front gate to the back gate at weak inversion ($\phi_s = 1.5\phi_B$) and strong inversion on the same X–Y graph. Clearly define and label all relevant data points. Explain.

3.8 A p-type MOS capacitor with $N_a = 1 \times 10^{18}$ cm^{-3} and $T_{ox} = 3$ nm is fabricated to characterize the van Dort's analytical bandgap widening QM model. Due to the inversion layer quantization, the increase in the effective bandgap $\Delta E_g = E_g^{QM} - E_g^{CL} = 104$ mV. Here, E_g^{QM} and E_g^{CL}

represent the QM and classical (CL) values of bandgap (E_g), respectively. Assume $Q_o = 0$, $V_{sub} = 0$, and $n+$ poly gate.

a. Show that the intrinsic carrier concentration due to QM effect is given by

$$n_i^{QM} = n_i^{CL} \exp\left(-\frac{\Delta E_g}{2kT} \right)$$

b. Calculate the value of n_i^{QM} at the ambient temperature $T = 300°K$.

c. If the shift in the centroid of inversion charge, $\Delta z \cong 1$ nm due to the QM effect in the p-substrate, estimate the value of $T_{ox}(\text{eff})$ at strong inversion due to QM effect of the metal gate capacitor.

d. Estimate the ratio of C/C_{ox} at strong inversion for metal gate capacitor in part (c). Here, C = bias-dependent capacitance of MOS system at strong inversion and C_{ox} = gate oxide capacitance. What is your conclusion on (C/C_{ox}) at strong inversion?

e. If the capacitor is fabricated using $n+$ polysilicon gate with doping concentration 1×10^{19} cm^{-3}, calculate the value of $T_{ox}(\text{eff})$ at strong inversion. Assume that the depletion width of silicon and polysilicon is given by the same expression.

f. Estimate the ratio of C/C_{ox} at strong inversion for polysilicon gate capacitor in part (e). Here, C = bias-dependent capacitance of MOS system at strong inversion and C_{ox} = gate oxide capacitance. What is your conclusion on (C/C_{ox}) at strong inversion?

3.9 An MOS capacitor is built with the structure shown in Figure E3.3. Both the n-type and p-type silicon regions are uniformly doped with concentration 1×10^{17} cm^{-3} and the areas of both the capacitors over the n-type and p-type regions are the same and $T_{ox} = 200$ nm. The threshold voltages for the n- and p-substrates are -1 V and $+1$ V, respectively. Sketch the shape of the HF (1 MHz) C–V curve that you would expect to measure for this structure. Calculate the maximum and minimum values of the composite capacitor per unit area. Label as many points as possible and explain.

3.10 Consider an MOS capacitor system on a uniformly doped p-type silicon substrate with doping concentration $N_a = 1 \times 10^{16}$ cm^{-3} at room temperature. Assume $V_{fb} = 0$ and wherever necessary $T_{ox} = 20$ nm.

a. Calculate and plot the inversion, bulk, and total charge Q_i, Q_b, and Q_s, respectively, in the substrate of the capacitor on the same plot. Clearly label all relevant parameters and explain.

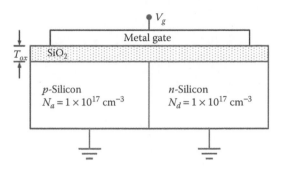

FIGURE E3.3
An MOS capacitor system.

b. Calculate and plot $\log(Q_i)$ versus surface potential ϕ_s for $\phi_s = 0.3$ to 1.1. Clearly, label different operating regimes of the MOS capacitor system and explain.

c. Calculate and plot ϕ_s versus gate voltage V_{gs} for $V_{gs} = 0$ to 3.0 V. Clearly label different operating regimes of the MOS capacitor system and explain.

3.11 Consider an MOS capacitor system on a uniformly doped p-type silicon substrate. In Section 3.4, MB probability distribution function is used to solve Poisson's equation in obtaining the surface charge Q_s in the semiconductor of the MOS capacitor system as a function of surface potential. Use the same procedure as in Section 3.4 and FD probability distribution function to obtain an expression similar to Equation 3.52 for Q_s as a function of surface potential in the semiconductor of the same MOS-capacitor system. Plot Q_s versus surface potential and compare with the plot in Figure 3.12.

4

Large Geometry MOSFET Compact Models

4.1 Introduction

In the two-terminal MOS (metal-oxide-semiconductor) capacitor system in Chapter 3, we have discussed that an inversion condition can be reached by a certain applied gate bias to form a thin layer of minority carrier concentration (e.g., electron) in the majority carrier (e.g., p-type) silicon surface at the silicon/SiO_2 interface. Under this inversion condition, the thermally generated minority carriers diffuse to the surface to form the inversion layer in the majority carrier substrate. However, it is difficult to sustain this minority carrier inversion layer in a majority carrier substrate from thermal generation and subsequent diffusion of these carriers to the surface without a steady source of carrier supply. Therefore, a heavily doped minority carrier region (e.g., $n+$ region in a p-type substrate), called the *source* (s), is added to the MOS structure as a terminal for the steady supply of minority carriers at the inversion condition. And, a complete MOSFET (metal-oxide-semiconductor field-effect transistor) structure is formed by adding one more terminal, called the *drain* (d), with a heavily doped region with the doping-type same as the source region. These source and drain terminals contact the two opposite ends of the inversion layer so that a potential difference can be applied across this layer and cause a current flow in the MOSFET structure. In this chapter, we will develop the basic mathematical models of this current flow from the source to drain of MOSFET devices, referred to as the *drain current model*.

Since the conception of MOSFETs in 1920s [1], there has been a continuous research and development effort on MOSFET device, technology, and modeling [2–5]. As stated in Chapter 1, the basic theory of MOSFETs has been developed in 1960s. In 1970s, complementary MOS (CMOS) technology with MOSFET devices became the pervasive technology of mainstream VLSI (very-large-scale-integrated) circuits. In the last five decades, there has been a relentless pursuit of developing MOSFET compact models that accurately simulate the experimental behavior of MOSFET devices in VLSI circuits. In this chapter, we will present the basic MOSFET drain current models for large geometry devices to lay the foundation for the understanding of advanced industry standard models for circuit computer-aided design (CAD). Our objective is to determine the drain current for any combination of DC voltages. First of all, we will present a brief overview of the basic MOSFET structure as used in VLSI technology, its features and behavior under operating biases, and the basic theory of MOSFET device operation and characteristics. One of the most important physical parameters of MOSFET device operation is the threshold voltage, V_{th}, defined as the gate voltage at which the device starts to turn on. In this chapter, we will develop the basic theory of V_{th} modeling for long channel devices. Throughout this chapter we will assume that the channel is sufficiently *long* and *wide*, so that the edge effects are negligibly small. Unless stated otherwise, we will also assume that the substrate is uniformly doped p-type silicon. We will introduce the relevant basic drain current models in a systematic way, deriving them from an important model and relating them to the source and to each other. Before, describing large geometry model, we first present a brief overview of MOSFET device architecture for better appreciation of MOSFET compact models.

4.2 Overview of MOSFET Devices

An ideal MOSFET device structure is shown in Figure 4.1 and a 2D (two-dimensional) cross section is shown in Figure 4.2. The structure includes a semiconductor substrate such as silicon on which a thin insulating layer such as SiO_2 of thickness T_{ox} is grown. A conducting layer (a metal or degenerately doped polycrystalline silicon) called gate electrode is deposited on the top of the gate oxide. Two heavily doped regions of depth X_j, called the source and drain, are formed in the substrate on either side of the gate. The source and drain regions overlap with the gate at its two ends. The source-to-drain regions are equivalent to two back-to-back *pn*-junctions. This region between the source and drain near the silicon surface is called the *channel region*. Thus, in essence, a MOSFET is essentially an MOS capacitor with two back-to-bask *pn*-junctions at the two ends of the gate. In advanced VLSI circuits, NMOS (p-type body with n+ source-drain) and PMOS (n-type body with p+ source-drain) are fabricated together using shallow trench isolation (STI) and is called the CMOS transistor. Thus, the STI shown in Figure 4.2 is used to isolate various devices fabricated on the same substrate. For device operation, a MOSFET is a four-terminal device with *gate g, source s, drain d*, and *substrate or body b*. The device is symmetrical and cannot be distinguished without the applied bias. The body terminal allows to modulating the inversion layer from the gate as well as body to offer more flexibility of devices at circuit operation.

As shown in Figure 4.1, a MOSFET device is characterized by channel length L, channel width W, gate oxide with thickness T_{ox}, substrate doping N_b, and source-drain with junction depth X_j. In advanced VLSI circuits, NMOS (p-type body with n+ source-drain) and PMOS (n-type body with p+ source-drain) are used together and is called the complementary MOS transistor.

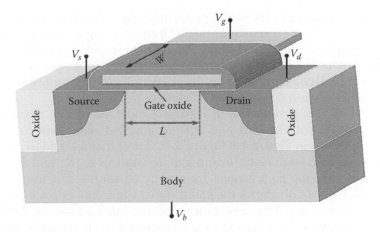

FIGURE 4.1
An ideal structure of a four-terminal MOSFET device; here V_g, V_s, V_d, and V_b are the gate, source, drain, and body terminals, respectively; and W and L are the channel width and channel length of the device, respectively.

4.2.1 Basic Features of MOSFET Devices

A 2D cross section of an advanced CMOSFET (CMOS field-effect transistor) structure along with its basic technology parameters is shown in Figure 4.2 [6]. It is observed from Figure 4.2 that the basic device engineering includes: (1) gate engineering to integrate dual-polysilicon (degenerately doped $n+$ and $p+$) gates or work function engineering for metal gate, (2) channel engineering with p-type and n-type well implants as well as threshold-voltage adjust implants

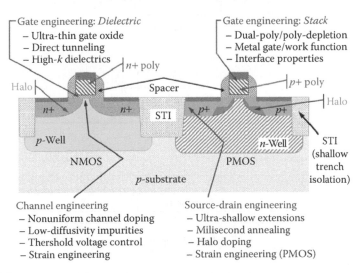

FIGURE 4.2
A typical 2D-cross section of an ideal advanced CMOS device showing major technology elements.

with impurities same as the well type, (3) halo implants with impurities same as the well type, and (4) source-drain (S/D) engineering to implant S/D with impurities opposite to the well-type. The channel engineering with V_{th} adjust implant produces vertically nonuniform channel doping profile and S/D engineering with halo implants produces laterally nonuniform channel doping profile in advanced bulk MOSFET devices [7–11].

A MOSFET structure shown in Figures 4.1 and 4.2 can be characterized by different circuit elements and current flow between source-drain terminals as shown in Figure 4.3. Figure 4.3a shows that a MOSFET structure includes parasitic source, drain, and gate resistances R_s, R_d, and R_g, respectively; and parasitic *pn*-junctions from body to source and from body to drain [12]. Figure 4.3b shows the small signal capacitances associated with a MOSFET structure. The capacitances include intrinsic source, drain, and body capacitances C_{GS}, C_{GD}, and C_{GB}, respectively, with reference to source and the extrinsic gate overlap capacitances C_{GSO}, C_{GDO}, and C_{GBO} with source, drain, and body, respectively, with reference to the gate and the junction capacitances C_{jS} and C_{jD} due to body to source and body to drain *pn*-junctions, respectively, as shown in Figure 4.3b [13].

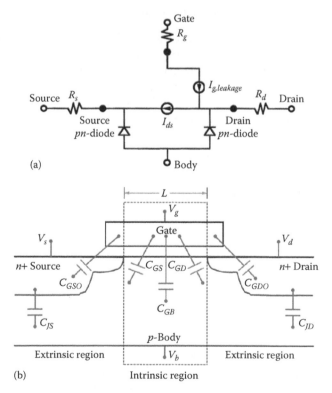

FIGURE 4.3
Basic features of MOSFETs: (a) basic circuit elements and current flows and (b) intrinsic and extrinsic parasitic capacitance elements.

4.2.2 MOSFET Device Operation

A MOSFET device has three modes of operation such as *accumulation*, *depletion*, and *inversion* similar to an MOS capacitor system. Therefore, the theory developed for an MOS capacitor system can be directly extended to MOSFETs by considering the channel potential due to the lateral electric field from the source to drain terminals of the structure shown in Figure 4.1.

In conventional MOSFET device operation, the source is used as the reference terminal with bias $V_s = 0$ and a drain voltage V_{ds} with reference to the source is applied to the drain so that the S/D *pn*-junctions are reverse biased. Under this biasing condition, the body or substrate current, $I_{bs} = 0$, and the gate current, $I_{gs} = 0$. The gate bias, V_{gs}, controls the surface carrier densities. A certain value of V_{gs}, referred to as the threshold voltage (V_{th}), is required to create the channel *inversion layer*. The parameter, V_{th} is determined by the properties of the structure. Thus, with reference to source potential,

- For $V_{gs} < V_{th}$, the MOSFET structure consists of two back-to-back *pn*-junctions and only leakage currents ($\sim I_o$ of S/D *pn*-junctions) flow from source to drain of the device, that is, $I_{ds} \sim 0$;
- For $V_{gs} > V_{th}$, an inversion layer exists, that is, a conducting channel exists from the drain to source of the device and a drain current I_{ds} will flow.

The body or bulk terminal allows modulating the inversion layer from the bottom by body bias, V_{bs}, as well as from the top by V_{gs} to offer more flexibility of devices at circuit operation. In normal MOSFET operation, V_{bs} is applied to reverse bias the source-drain *pn*-junctions.

4.3 MOSFET Threshold Voltage Model

All MOS capacitor equations derived in Chapter 3 are valid for large L and large W MOSFETs with proper consideration of the lateral electric field, E_y, due to the applied drain bias, V_{ds}, as shown in Figure 4.4.

Let us consider the source potential $V_s = 0$ as the reference voltage for MOSFETs. Due to the applied V_{ds}, the surface potential, ϕ_s, is a function of location, y, along the channel such that $\phi_s = \phi_s(y)$. Therefore, a channel potential, $V_{ch}(y)$, exists along the channel from the source to drain such that

$$V_{ch}(y) = \begin{cases} V_{sb}; & \text{at } y = 0 \\ V_{sb} + V_{ds}; & \text{at } y = L \end{cases} \tag{4.1}$$

Similar to an MOS capacitor, a MOSFET V_{th} model is obtained by solving Poisson's equation relating the charge density, ρ, to the electrostatic potential ϕ (or, electric field E) given by

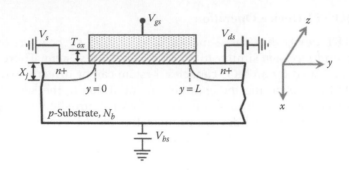

FIGURE 4.4
Schematic 2D cross section of an *n*-channel MOSFET showing the biasing conditions and the coordinate system; *x*, *y*, and *z* represent the distances along the depth, length, and width of the device, respectively.

$$\nabla^2 \phi = -\frac{\rho}{K_{si}\varepsilon_0} \tag{4.2}$$

Generally, a MOSFET is a 3D (three-dimensional) problem; however, for all practical purposes (except for very small W and L), we can treat the system as a 2D problem in the x and y directions only. We can further convert the 2D problem to 1D (one-dimensional) by a set of simplifying assumptions. First of all, we assume that the variation of the electrical field E_y in the y direction along the channel is much less than the corresponding variation of the electrical field E_x in the x direction down to the substrate. Then we have

$$\frac{\partial E_y}{\partial y} \ll \frac{\partial E_x}{\partial x}; \quad \therefore \frac{\partial^2 \phi}{\partial y^2} \ll \frac{\partial^2 \phi}{\partial x^2} \tag{4.3}$$

Equation 4.3 is referred to as the *gradual channel approximation* (GCA) [2]. Therefore, like an MOS capacitor, we solve for ϕ in the x direction along the depth of the channel only to obtain the total charge, Q_s, in the semiconductor. For MOSFETs in *inversion* ($\phi_B < \phi_s < 2\phi_B$), the total charge $Q_s = Q_s(y)$ due to the channel potential $V_{ch}(y)$ can be derived from the MOS capacitor theory (Equation 3.52). In Equation 3.52, we observe that for $\phi_s > 0$, $\exp(-\phi_s/v_{kT})$ is negligibly small, the term "−1" is negligibly small since $\exp(\phi_s/v_{kT}) \gg -1$ in strong inversion, and the term $(-\phi_s/v_{kT})$ is negligibly small in weak inversion. Therefore, from Equation 3.52, $Q_s(y)$ for MOSFETs at inversion can be shown as

$$Q_s(y) = -\sqrt{2K_{si}\varepsilon_0 q N_b} \left[\phi_s(y) + v_{kT} \frac{n_i^2}{N_b^2} \left(e^{(\phi_s(y) - V_{ch}(y))/v_{kT}} \right) \right]^{1/2}$$

$$= -\sqrt{2K_{si}\varepsilon_0 q N_b} \left[\phi_s(y) + v_{kT} \left(e^{(\phi_s(y) - 2\phi_B - V_{ch}(y))/v_{kT}} \right) \right]^{1/2} \tag{4.4}$$

$$= -\sqrt{2K_{si}\varepsilon_0 q N_b} f\left(\phi_s(y), \phi_B, V_{ch}(y) \right)$$

where we have used $n_i^2/N_b^2 = \exp(-2\phi_B/v_{kT})$ from Equation 3.41, and defined

$$f\big(\phi_s(y),\phi_B,V_{ch}(y)\big) \equiv \left[\phi_s(y) + v_{kT}\left(e^{(\phi_s(y)-2\phi_B-V_{ch}(y))/v_{kT}}\right)\right]^{1/2} \quad (4.5)$$

Now, from Equation 3.23, the gate voltage V_{gb} with reference to bulk can be represented as

$$V_{gb} = V_{fb} + \phi_s(y) - \frac{Q_s(y)}{C_{ox}} \quad (4.6)$$

Substituting for $Q_s(y)$ from Equation 4.4 to Equation 4.6, we can show

$$V_{gb} = V_{fb} + \phi_s(y) + \frac{\sqrt{2K_{si}\varepsilon_0 qN_b}}{C_{ox}}\left[\phi_s(y) + v_{kT}e^{((\phi_s(y)-2\phi_B-V_{ch}(y))/v_{kT})}\right]^{1/2} \quad (4.7)$$

Conventionally, *strong inversion* is defined at $\phi_s = 2\phi_B$. Therefore, assuming $V_{ds} \cong 0$, we get from Equation 4.1, $V_{ch}(y) \cong V_{sb}$. Then the surface potential, $\phi_s(y)$, at strong inversion due to V_{sb} is a constant along the channel and is given by

$$\phi_s(y) = \phi_s = 2\phi_B + V_{sb} \quad (4.8)$$

Thus, under the condition $V_{ds} \cong 0$, substituting for $\phi_s(y)$ from Equation 4.8 and $V_{ch}(y) = V_{sb}$ in Equation 4.4, we get:

$$Q_s(y) = -\sqrt{2K_{si}\varepsilon_0 qN_b}\left[(2\phi_B + V_{sb}) + v_{kT}\left(e^{((2\phi_B+V_{sb})-2\phi_B-V_{sb})/v_{kT}}\right)\right]^{1/2}$$
$$\cong -\sqrt{2K_{si}\varepsilon_0 qN_b(2\phi_B + V_{sb})} \quad (4.9)$$

Now, substituting for $Q_s(y)$ from Equation 4.9 to Equation 4.6, we get

$$V_{th} = V_{gb} = V_{fb} + 2\phi_B + \frac{\sqrt{2K_{si}\varepsilon_0 qN_b}}{C_{ox}}\sqrt{(2\phi_B + V_{sb})}$$
$$= V_{fb} + 2\phi_B + \gamma\sqrt{(2\phi_B + V_{sb})} \quad (4.10)$$

where the parameter γ strongly depends on channel doping concentration and called the *body factor* given by

$$\gamma = \frac{\sqrt{2K_{si}\varepsilon_0 qN_b}}{C_{ox}} \quad (4.11)$$

Thus, from Equation 4.10, the threshold voltage for long channel devices is given by

$$V_{th} = V_{fb} + 2\phi_B + \gamma\sqrt{(2\phi_B + V_{sb})} \quad (4.12)$$

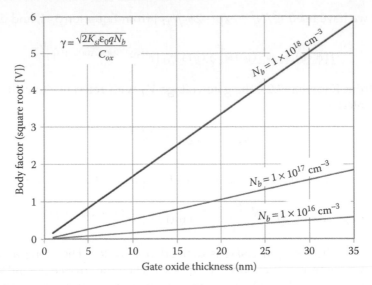

FIGURE 4.5
Effect of gate oxide thickness T_{ox} and channel concentration N_b on body coefficient of long channel MOSFETs.

where from Equation 3.15, $V_{fb} = \phi_{ms} - Q_o/C_{ox}$. Now, if the body bias $V_{bs} = V_{sb} = 0$, then from Equation 4.12, the threshold voltage without body bias is given by

$$V_{th0} = V_{fb} + 2\phi_B + \gamma\sqrt{2\phi_B} \qquad (4.13)$$

Combining Equations 4.12 and 4.13, we get the expression for threshold voltage at any biasing condition as

$$V_{th} = V_{th0} + \gamma\left(\sqrt{(2\phi_B + V_{sb})} - \sqrt{2\phi_B}\right) \qquad (4.14)$$

Equation 4.14 is the general expression for V_{th} of MOSFETs in inversion condition. The body effect parameter γ depends on the gate oxide thickness and channel doping concentration. Figure 4.5 shows the effect of T_{ox} and N_b on γ.

4.4 MOSFET Drain Current Model

In order to obtain the device characteristics of MOSFETs, let us consider an n-channel MOSFET (nMOSFET) device with uniformly doped substrate concentration N_b (cm^{-3}), the structure and dimensions of which are shown in Figure 4.6. For the sake of simplicity we will assume this to be a large

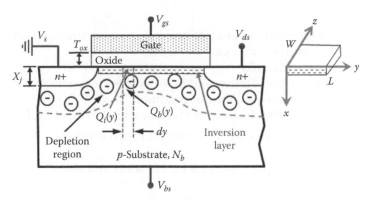

FIGURE 4.6
Schematic of an *n*MOSFET device showing different biases and reference direction; *x*, *y*, and *z* distance into the silicon, along the channel, and along the channel width of the device, respectively.

geometry device so that the short channel and narrow width effects can be neglected. We will develop a generalized large geometry MOSFET drain current model using several simplifying assumptions.

4.4.1 Drain Current Formulation

In general, the static and dynamic characteristics of a semiconductor device under the influence of external fields can be described by the following three sets of coupled differential equations

1. The Poisson's equation for electrostatic potential ϕ is described in Equation 4.2 and is given by

$$\nabla^2\phi = -\frac{\rho}{K_{si}\varepsilon_0} \tag{4.15}$$

   where:
   ρ is the charge density
   K_{si} is the dielectric constant of silicon
   ε_0 is the permittivity of free space

2. The *current density* equations for electron current density (J_n) and hole current density (J_p),

$$J_n = q\mu_n nE + qD_n\nabla n \quad \text{(electrons)}$$
$$J_p = q\mu_p pE - qD_p\nabla p \quad \text{(holes)} \tag{4.16}$$

   Equation 4.16 under nonequilibrium condition is represented by

$$J_n = -qn\mu_n\nabla\phi_n \quad \text{(electrons)}$$
$$J_p = -qp\mu_p\nabla\phi_p \quad \text{(holes)} \tag{4.17}$$

where:

n and p represent the electron and hole concentrations, respectively

q is the electronic charge

E is the electric field

ϕ_n and ϕ_p are the electron and hole quasi-Fermi potentials, respectively, in nonequilibrium condition

μ_n and μ_p are the electron and hole mobilities, respectively.

The total current density (J) flowing through the device is given by $J = J_n + J_p$.

3. The *current continuity* equations for electrons and holes

$$\frac{\partial n}{\partial t} = \frac{1}{q} \nabla \cdot J_n - R_n + G_n \text{ (electrons)}$$

$$\frac{\partial p}{\partial t} = -\frac{1}{q} \nabla \cdot J_n - R_p + G_p \text{ (holes)}$$

(4.18)

In Equation 4.18, G_n and G_p are the generation rates for electrons and holes, respectively, whereas R_n and R_p represent the recombination rates for electrons and holes, respectively.

As pointed out in Section 4.3, modeling of a MOSFET device is a 3D problem; however, for all practical purposes (except for small geometry devices), we can treat a MOSFET device as a 2D problem in the x and y directions only (Figure 4.6). Even as a 2D problem, the mathematical expressions are fairly complex and can only be solved exactly using numerical techniques used in 2D/3D device simulators including MEDICI [14], MINIMOS [15], Sentaurus Device [16], and ATLAS [17]. However, in order to obtain simplified analytical solutions for circuit CAD, we make a number of valid simplifying assumptions to develop compact device equations that accurately describe the behavior of semiconductor devices in circuit operation.

Now, we make a number of valid simplifying assumptions to develop a generalized expression for drain current, I_{ds}, of large geometry MOSFETs on a uniformly doped substrate as described below:

Assumption 1: We assume that the variation of the electric field E_y in the y direction (along the channel) is much less than the corresponding variation of the electric field E_x in the x direction into the substrate. Thus, as discussed in Section 4.3, here again, we assume GCA [2] so that we need to solve only 1D Poisson's equation described in Equation 2.58, which is given by

$$\frac{d^2\phi}{dx^2} = -\frac{\rho(x)}{K_{si}\varepsilon_0}$$

(4.19)

2D numerical analysis shows that the GCA is valid for most of the channel length except near the drain end of the channel region. Near the drain end of the channel, the longitudinal electric field E_y is comparable to the transverse electric field E_x even for long channel devices and GCA breaks down. In spite of its failure near the drain end, the GCA is used as it reduces the system to a 1D current flow problem. The fact that we have to solve only a 1D Poisson's equation means that the charge expressions developed in Chapter 3 for an MOS capacitor system could be used for an MOS transistor, with the modification that charge and potential will now be position dependent in the y direction.

Assumption 2: Assume that only minority carriers contribute to I_{ds}; for example, for an nMOSFET device, the hole current can be neglected. In nMOSFETs, the majority carrier holes are created by impact ionization and become important in describing the device characteristics in the avalanche or breakdown regime. However, in the normal operation range of MOSFET devices, the drain current does not include breakdown regime, and therefore, the assumption that the current in MOSFETs is due to the minority carriers is valid under the normal biasing conditions, for example, for nMOSFETs $V_{ds} \geq 0$ and $V_{bs} \leq 0$. Thus, the drain current model needs to consider only the minority carrier current density, J_n, for nMOSFET devices.

Assumption 3: Assume there are no generation and recombination of carriers, that is, for an nMOSFET device $R_n = 0 = G_n$. Then considering only the static characteristics of the device, the continuity equation 4.18 becomes

$$\nabla \cdot J_n = 0 \tag{4.20}$$

This implies that the total drain current I_{ds} is a constant at any point along the channel of the device.

Assumption 4: Assume that the current flows in the y direction along the channel only, that is, $d\phi_n/dx = 0$. Thus, the electron quasi-Fermi potential, ϕ_n, is a constant in the x direction. Then from Equation 4.17, the electron current density is given by

$$J_n(x,y) = -qn(x,y)\mu_n(x,y)\frac{\partial\phi_n}{\partial y} \tag{4.21}$$

Since the cross-sectional area of the channel in which the current flows is the channel width, W, times the channel length, L, integrating Equation 4.21 across the depth x and width z, we get I_{ds} at any point y in the channel as

$$I_{ds}(y) = -W \int_0^\infty \left[qn(x,y)\mu_n(x,y)\frac{\partial \phi_n}{\partial y} \right] dx = \text{constant} \qquad (4.22)$$

where μ_n in Equation 4.22 is the channel electron surface mobility for nMOSFETs, often referred to as the surface mobility μ_s in order to distinguish it from the bulk mobility deep into the substrate described in Section 2.2.5.1. In the rest of the discussion, we will replace μ_n by μ_s to emphasize that the inversion layer mobility we deal with for MOSFET devices is the surface mobility.

In MOSFET devices, the application of source and drain voltages relative to the substrate results in a lowering of the quasi-Fermi level F_n (or potential ϕ_n) at the source end of the device by an amount qV_{sb}, and the drain end of the device by an amount $q(V_{sb} + V_{ds})$, relative to equilibrium Fermi level E_f in the substrate. It is this difference in ϕ_n between the source and drain that drives the electrons down the channel. Now, the channel potential $V_{ch}(y)$ at any point y in the channel in Figure 4.7 is given by

$$V_{ch}(y) = \phi_n(y) - \phi_n\big|_{\text{source}} \qquad (4.23)$$

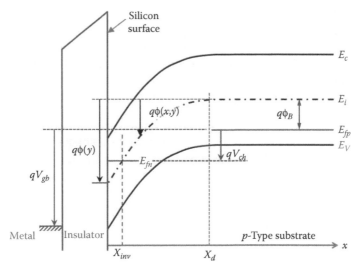

FIGURE 4.7
Energy band diagram of an nMOSFET device shown in Figure 4.6; E_c, E_v, and E_i represent the bottom of the conduction band, top of the valence band, and intrinsic band, respectively, of the p-type substrate; E_{fn} and E_{fp} are the quasi-Fermi level of electrons and holes, respectively; qV_{ch} is the channel potential due to the difference in $E_{fn} - E_{fp}$ caused by the difference in source and drain potentials.

At the source end of the channel, $V_{ch}(0) = 0$, and at the drain-end of the channel, $V_{ch}(L) = (V_{sb} + V_{ds})$. Thus, compared to the case of an MOS capacitor, the quasi-Fermi potential is lowered by an amount $V_{ch}(y)$ at the surface region of a MOSFET device. As a result, the surface electron concentration (n_s) is lowered by a factor $\exp(-V_{ch}(y)/v_{kT})$. Then following the derivation of minority carrier density expression (Equation 3.42) for an MOS capacitor, we can write the minority carrier surface electron concentration at any point y in a MOSFET device as

$$n(y) = N_b \exp\left[\frac{\phi(y) - 2\phi_B - V_{ch}(y)}{v_{kT}} \right] \tag{4.24}$$

where the parameters have their usual meanings as defined in Section 3.4.1. The minority carrier concentration changes due to the applied bias; however, the majority carrier hole concentration does not change with bias, and therefore, following MOS capacitor Equation 3.39, we can write for the majority carrier concentration in MOSFETs as $p = N_b \exp[(-\phi(y)/v_{kT}]$.

Then using Equation 4.23, Equation 4.22 can be written as:

$$I_{ds}(y) = -W \frac{dV_{ch}(y)}{dy} \int_0^\alpha qn(x,y)\mu_s(x,y)dx \tag{4.25}$$

Assumption 5: For the simplicity of long channel I_{ds} calculation, we assume μ_s = constant at some average gate and drain electric field; however, μ_s depends on both E_x and E_y as we will discuss in Chapter 5. With this assumption, we can write Equation 4.25 as:

$$I_{ds}(y) = -W\mu_s \frac{dV_{ch}(y)}{dy} \int_0^\alpha qn(x,y)dx \tag{4.26}$$

Now, we define Q_i as the mobile minority carrier charge density, that is

$$Q_i(y) = -q \int_0^\alpha n(x,y)dx \tag{4.27}$$

Using Equation 4.27 in Equation 4.26, we get the general expression for $I_{ds}(y)$ as

$$I_{ds}(y)dy = W\mu_s Q_i(y)dV_{ch}(y) \tag{4.28}$$

Again, assuming GCA is valid along the entire length of the channel, we get after integrating Equation 4.28 along the channel length from $y = 0$ to $y = L$

$$I_{ds} = \mu_s \left(\frac{W}{L}\right) \int\limits_{V_{sb}}^{V_{sb}+V_{ds}} Q_i(y) dV_{ch} \tag{4.29}$$

Equation 4.29 is the general expression for I_{ds} flowing through a MOSFET device. In order to calculate I_{ds}, we need to calculate the mobile inversion charge density $Q_i(y)$ in the channel region. A number of I_{ds} models have been developed depending on different approaches to compute $Q_i(y)$. We will discuss some of the early generation of compact models in the following section to appreciate the rigor of the advanced industry standard compact models.

4.4.2 Pao-Sah Model

In this model, $Q_i(y)$ is calculated numerically by integrating the electron concentration in the x direction. In order to evaluate $Q_i(y)$, let us change the variable of integration in Equation 4.27 from x to ϕ and integrate from $\phi(x = 0) = \phi_s$ to $\phi(x = \infty) = \phi_B$ so that

$$Q_i(y) = -q \int\limits_{0}^{\infty} n(x,y) dx = -q \int\limits_{\phi_s}^{\phi_B} n\big(\phi, V_{ch}(y)\big) \frac{dx}{d\phi} d\phi \tag{4.30}$$

where:

ϕ_s is the surface potential (at $x = 0$) and is position dependent due to the applied voltage between the source and drain

Since the inversion layer is formed when the minority carrier concentration exceeds the majority carrier concentration, that is, $\phi \geq \phi_B$, the upper limit of integration is ϕ_B where the inversion layer ends. In Equation 4.30, $(d\phi/dx)^{-1} = -1/E_x$, where E_x is the vertical electric field along the depth of the channel. In order to obtain $Q_i(y)$ from Equation 4.30, we need to determine the electron concentration along the channel $n\big(\phi, V_{ch}(y)\big)$ and the variation of potential representing the inverse of the vertical electric field $(d\phi/dx)^{-1} = -1/E_x$ along the depth of the channel.

Derivation of $n(\phi, V_{ch}(y))$: As pointed out earlier, we can use MOS capacitor equation with appropriate modification to include the channel potential $V_{ch}(y)$ to account for the applied drain bias in MOSFETs. Therefore, considering the channel potential, $V_{ch}(y)$, due to the applied V_{ds} in Equation 3.42, we can write the expression for the inversion carrier at a point y along the channel as

$$n(y) = N_b e^{(\phi(y) - 2\phi_B - V_{ch}(y))/v_{kT}} \equiv n\big(\phi, V_{ch}(y)\big) \tag{4.31}$$

Derivation of $(d\phi/dx)^{-1}$: From Gauss's law given in Equation 2.61, the total induced charge in the p-type semiconductor of an nMOSFET device is given by

$$Q_s = -\varepsilon_0 K_{si} E_s = \varepsilon_0 K_{si} \left(\frac{d\phi}{dx} \right)$$

$$\therefore \left(\frac{d\phi}{dx} \right) = \frac{Q_s}{\varepsilon_0 K_{si}}$$

(4.32)

where we have used $E_s = -(d\phi/dx)$ in Equation 4.32. Again, repeating Equation 4.4 for $Q_s(y)$ of an nMOSFET device, we get

$$Q_s(y) = -\sqrt{2K_{si}\varepsilon_0 q N_b} \, f\left(\phi_s(y), \phi_B, V_{ch}(y) \right)$$

(4.33)

Combining Equations 4.32 and 4.33, we get

$$\frac{d\phi}{dx} = -\sqrt{\frac{2qN_b}{K_{si}\varepsilon_0}} \, f\left(\phi_s, \phi_B, V_{ch}(y) \right)$$

(4.34)

Now, substituting for $n(\phi, V_{ch}(y))$ and $(d\phi/dx)^{-1}$ from Equations 4.31 and 4.34, respectively, into Equation 4.30, we can show

$$Q_i(y) = \sqrt{\frac{K_{si}\varepsilon_0 N_b q}{2}} \int_{\phi_s}^{\phi_B} \frac{e^{\left[\phi(y) - 2\phi_B - V_{ch}(y) \right]/v_{kT}}}{f\left(\phi_s, \phi_B, V_{ch}(y) \right)} d\phi$$

(4.35)

Equation 4.35 is the generalized expression for the inversion charge $Q_i(y)$ in a MOSFET device. Then substituting for $Q_i(y)$ from Equation 4.35 to Equation 4.29, we get the expression for the drain current as

$$I_{ds} = \mu_s \frac{W}{2L} C_{ox} \gamma \int_{V_{sb}}^{V_{sb}+V_{ds}} \int_{\phi_s}^{\phi_B} \frac{e^{\left[\phi(y) - 2\phi_B - V_{ch}(y) \right]/v_{kT}}}{f\left(\phi_s, \phi_B, V_{ch}(y) \right)} d\phi dV_{ch}$$

(4.36)

In Equation 4.36, we have used $\gamma = \sqrt{2qK_{si}\varepsilon_0 N_b}/C_{ox}$ and γ is defined in Equation 4.11 as the body effect coefficient. Equation 4.36 was fist derived by Pao and Sah [2] and is called the *Pao-Sah* or *double integral* model for MOSFET devices. Equation 4.36 can only be solved numerically using ϕ_s from Equation 4.7 given by

$$V_{gb} = V_{fb} + \phi_s(y) + \gamma \left[\phi_s(y) + v_{kT} e^{\left[(\phi_s(y) - 2\phi_B - V_{ch}(y)) / v_{kT} \right]} \right]^{1/2}$$

(4.37)

As we can see Equation 4.37 is an implicit equation in ϕ_s and must be solved for a given bias condition using an iterative procedure. The Pao-Sah model given by Equation 4.36 provides a unified description of both the drift and diffusion components of I_{ds} and is valid in all regions of a MOSFET device operation. However, due to long computation time to generate *I–V* characteristics by solving double numerical integration along with the iterative solution of ϕ_s at each bias point, the model is too complex and unsuitable for circuit CAD. Therefore, various simplifications of Pao-Sah model have been used to develop computationally efficient compact models suitable for circuit analysis [3–5,12–22]. Pao-Sah model is used to benchmark the accuracy of other simplified models.

4.4.3 Charge-Sheet Model

In order to derive an accurate and simplified I_{ds} model from the generalized Equation 4.29, let us assume that the inversion layer is a sheet of charge without any finite thickness. Then assuming that the depletion approximation is valid, we can show from Equation 3.64 that the induced depletion charge in terms of the body factor γ is given by

$$Q_b(y) = -\gamma C_{ox}\sqrt{\phi_s(y)} \tag{4.38}$$

Again, from Equation 4.6, the expression for the total charge in the semiconductor is given by

$$Q_s(y) = -C_{ox}\left[V_{gb} - V_{fb} - \phi_s(y)\right] \tag{4.39}$$

We know that $Q_i(y) = Q_s(y) - Q_b(y)$; therefore, from Equations 4.38 and 4.39, the expression for the sheet of inversion charge with zero thickness is given by

$$Q_i(y) = -C_{ox}\left[V_{gb} - V_{fb} - \phi_s(y) - \gamma\sqrt{\phi_s(y)}\right] \tag{4.40}$$

Rearranging Equation 4.28, using Equations 4.38 and 4.40, Brews [5] showed that the total drain current can be expressed as

$$I_{ds}(y) = -\mu_s W\left[Q_i(y)\frac{d\phi_s}{dy} - v_{kT}\frac{dQ_i(y)}{dy}\right]$$

$$= I_{ds1}(y) + I_{ds2}(y) \tag{4.41}$$

where I_{ds1} and I_{ds2} are given by

$$I_{ds1}(y) = -\mu_s W Q_i(y)\frac{d\phi_s}{dy}$$

$$I_{ds2}(y) = \mu_s W v_{kT}\frac{dQ_i}{dy} \tag{4.42}$$

We know that under the lateral electric field $E(y)$ from the source to drain along the channel, the electrons move with a drift velocity v_d and the drain current due to drift of electrons is given by

$$I_{ds}(\text{drift}) = W\left(J_{\text{drift}}\right) = W\left(nqv_d\right) = WQ_i\mu_s E(y) = -WQ_i\mu_s\frac{d\phi_s}{dy} = I_{ds1} \tag{4.43}$$

where n is the inversion layer electron concentration for nMOSFETs, q is the electronic charge so that $Q_i = nq$ and $v_d = \mu_s E(y) = -\mu_s (d\phi_s/dy)$.

Again, if the electron transport is due to the concentration gradient (dn/dy) along the channel, then from Fick's first law of diffusion (Equation 2.39), the electron diffusion current along the channel is given by

$$I_{ds}(\text{diffusion}) = W(J_{\text{diffusion}}) = W\left(qD_n\frac{dn}{dy}\right) = qW\left(v_{kT}\mu_s\right)\frac{dn}{dy} = W\mu_s v_{kT}\frac{dQ_i}{dy} = I_{ds2} \quad (4.44)$$

where we have used Einstein's relation $D_n/\mu_s = kT/q = v_{kT}$. Thus, we find that the total drain current in a MOSFET device is the sum of the drift and diffusion components I_{ds1} and I_{ds2} as given by Equations 4.43 and 4.44, respectively. In general, I_{ds1} and I_{ds2} are coupled differential equations and cannot be integrated separately. However, for simplicity of compact device modeling, we solve each component separately under the appropriate boundary conditions and add them together to obtain the expression for the total drain current I_{ds}.

4.4.3.1 Drift Component of Drain Current

Substituting for Q_i from Equation 4.40 to Equation 4.43, we get for the drift component of the drain current as

$$I_{ds1}(y) - \mu_s WC_{ox}\left[V_{gh} - V_{fb} - \phi_s(y) - \gamma\sqrt{\phi_s(y)}\right]\frac{d\phi_s(y)}{dy} \quad (4.45)$$

In order to solve Equation 4.45, we use the boundary condition

$$\phi_s(y) = \begin{cases} \phi_{s0} & \text{at } y = 0 \\ \phi_{sL} & \text{at } y = L \end{cases} \quad (4.46)$$

where:

ϕ_{s0} and ϕ_{sL} represent the surface potential at the source end and at the drain end of the channel, respectively, as shown in Figure 4.8

Therefore, using the boundary condition from Equation 4.46, we get from Equation 4.45

$$\int_0^L I_{ds1}(y)dy = \mu_s WC_{ox}\int_{\phi_{s0}}^{\phi_{sL}}\left[V_{gb} - V_{fb} - \phi_s(y) - \gamma\sqrt{\phi_s(y)}\right]d\phi_s(y) \quad (4.47)$$

After integration and simplification, we get the drift component of the drain current in MOSFETs as

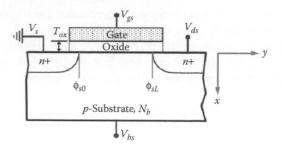

FIGURE 4.8
MOSFET device structure showing the boundary conditions to solve current equations for the drift and diffusion components of the drain currents; ϕ_{s0} and ϕ_{sL} are the surface potentials at the source end ($y = 0$) and drain end ($y = L$) of the channel, respectively.

$$I_{ds1} = \mu_s C_{ox} \frac{W}{L}\left[\left(V_{gb} - V_{fb}\right)\left(\phi_{sL} - \phi_{s0}\right) - \frac{1}{2}\left(\phi_{sL}^2 - \phi_{s0}^2\right) - \frac{2}{3}\gamma\left(\phi_{sL}^{3/2} - \phi_{s0}^{3/2}\right)\right] \quad (4.48)$$

where, ϕ_{s0} and ϕ_{sL} are the surface potentials as shown in Figure 4.8 and are computed iteratively for each bias point from the surface potential Equation 4.37.

4.4.3.2 Diffusion Component of Drain Current

From Equation 4.44, we get for the *diffusion* component of the drain current using the boundary condition from Equation 4.46 as

$$\int\limits_0^L I_{ds2}dy = \mu_s Wv_{kT}\int\limits_{\phi_{s0}}^{\phi_{sL}} dQ_i \quad (4.49)$$

Substituting for Q_i from Equation 4.40 in Equation 4.49, we can show

$$\int\limits_0^L I_{ds2}dy = \mu_s C_{ox}Wv_{kT}\int\limits_{\phi_{s0}}^{\phi_{sL}}\left(d\phi_s + \frac{\gamma}{2}\frac{d\phi_s}{\sqrt{\phi_s}}\right) \quad (4.50)$$

Therefore, after integration and simplification, we get for diffusion component of drain current as

$$I_{ds2} = \mu_s C_{ox} \frac{W}{L}v_{kT}\left[\left(\phi_{sL} - \phi_{s0}\right) + \gamma\left(\phi_{sL}^{1/2} - \phi_{s0}^{1/2}\right)\right] \quad (4.51)$$

In order to solve I_{ds1} and I_{ds2} from Equations 4.50 and 4.51, respectively, we obtain ϕ_{s0} at $y = 0$ at the source end and ϕ_{sL} at $y = L$ at the drain end of the MOSFET channel from Equation 4.37. The total current is obtained by adding Equations 4.48 and 4.51. The values of ϕ_{s0} and ϕ_{sL} required to calculate I_{ds} are obtained numerically by solving the implicit Equation 4.37 under the boundary conditions

$$\phi_s(y) = \begin{cases} \phi_{s0} & \text{at } y = 0 \\ \phi_{sL} & \text{at } y = L \end{cases}$$

and (4.52)

$$V_{ch}(y) = \begin{cases} V_{sb} & \text{at } y = 0 \\ V_{sb} + V_{ds} & \text{at } y = L \end{cases}$$

Using the boundary conditions (Equation 4.52) in Equation 4.37, we can show that the implicit equations for ϕ_{s0} and ϕ_{sL} are given by

$$\phi_{s0} = V_{gb} - V_{fb} - \gamma\sqrt{\phi_{s0} + v_{kT}e^{(\phi_{s0} - 2\phi_B - V_{sb})/v_{kT}}}$$ (4.53)

$$\phi_{sL} = V_{gb} - V_{fb} - \gamma\sqrt{\phi_{sL} + v_{kT}e^{[\phi_{sL} - 2\phi_B - (V_{sb} + V_{ds})]/v_{kT}}}$$ (4.54)

From Equations 4.48 and 4.51, we find that both the drift and diffusion components of I_{ds} depend on ($\phi_{sL} - \phi_{s0}$). In weak inversion, $\phi_{s0} \approx \phi_{sL}$, so that even small errors in the values of ϕ_{s0} and ϕ_{sL} can lead to a large error in I_{ds2}. Therefore, an accurate solution is required for the surface potential, particularly for weak inversion conditions. In reality, the accuracy of calculation for ϕ_s must be ~1×10^{-12} V. The implicit Equation 4.37 can be solved iteratively as well as by using Taylor series expansion [23] to obtain ϕ_{s0} and ψ_{sL} at each biasing condition.

Figure 4.9 shows the total drain current I_{ds} and its components I_{ds1} and I_{ds2} as function of V_{gb} at $V_{db} = 3$ V and $V_{sb} = 1$ V. Figure 4.9 shows that in strong inversion, $I_{ds} \approx I_{ds1}$, and therefore, the total current is mainly due to the drift of electrons due to V_{ds}. In weak inversion, $I_{ds} \approx I_{ds2}$, and the current is mainly due to diffusion of minority carriers from the source end to the drain. However, there is a region between the weak inversion and the strong inversion, called *moderate inversion*, where both the drift and diffusion components are important. The width of the moderate inversion in terms of voltage is several tenths of a volt [24,25]. It is shown that the lower limit of $\phi_s \equiv \phi_{mL}$ in the moderate inversion is ~$(2\phi_B - v_{kT})$, whereas the upper limit $\phi_s = \phi_{mU} \sim (2\phi_B + 6v_{kT})$. And, the corresponding values for V_{gb} are V_{gbL} and V_{gbU}, respectively, are obtained from Equation 4.37 by solving for $\phi_s = \phi_{mL}$ and ϕ_{mU}, respectively.

The comparison of $I_{ds} - V_{ds}$ characteristics shows that the Brews charge-sheet model predicts I_{ds} within 1% of that calculated using the Pao-Sah model under most operating conditions [13]. Although, the charge-sheet model is simpler compared to the Pao-Sah model, it still requires time-consuming iterations to calculate ϕ_{s0} and ϕ_{sL}. Therefore, it is computationally intensive. Hence, in spite of its advantages, this model has not been widely used in real circuit CAD until the development and release of the Hiroshima University STARC IGFET Model (HiSIM) [26] in 2006. HiSIM basic current equations are based on Brews charge-sheet model.

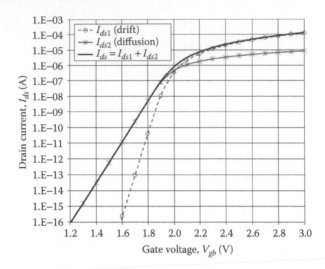

FIGURE 4.9
Drain current as a function of gate voltage obtained by charge sheet model; the plots are obtained for nMOSFET devices with $W/L = 1$ and $T_{ox} = 600$ nm for biasing condition $V_{fb} = -1$ V, $V_{sb} = 1$ V, and $V_{db} = 3$ V.

Though the Pao-Sah and charge-sheet models are complete and most accurately describe MOSFET device characteristics, they are complex and computationally inefficient for circuit CAD. Therefore, simplified analytical compact drain current models have been developed from the generalized drain current (Equation 4.29) based on additional approximations to circumvent solving the implicit Equation 4.37 for calculating ϕ_s. This is achieved by separately modeling each distinct regions of device operation with appropriate boundary conditions. The most commonly used boundary between the *weak* and *strong* inversion regions is the threshold voltage, V_{th}. Based on this approach, we will develop a current equation for *strong inversion* region and the other for *weak inversion* region of device operation to analyze each region independently. Note that the Pao-Sah and charge-sheet models model the entire range of device operation and have completely natural transitions between different regions. The regional models, also known as the *piece-wise* multisectional models, are most commonly used for circuit CAD because of their simplicity and computational efficiency. In the following section, we will develop the first-order piecewise model for large geometry devices and subsequent improvement of the basic models for improved accuracy. In Chapter 5, we will develop more accurate compact industry standard models for short channel VLSI devices for circuit CAD.

4.4.4 Regional Drain Current Model

Equation 4.29 represents the generalized expression for I_{ds} that is derived using five appropriate assumptions to include both the drift and diffusion components of current and is repeated here

$$I_{ds} = \mu_s \left(\frac{W}{L}\right) \int_{V_{sb}}^{V_{sb}+V_{ds}} Q_i(y)dV_{ch} \tag{4.55}$$

In order to derive simplified regional compact MOSFET models for circuit CAD, we will make further simplifying assumptions.

Assumption 6: Let us assume that the diffusion current is negligibly small so that the current flow along the channel in the device is only due to the drift of minority carriers by the applied drain voltage, V_{ds}. This is a fairly good assumption provided the device is in strong inversion; that is, the gate voltage is greater than the threshold voltage ($V_{gs} > V_{th}$ or $\phi_s > 2\phi_B$). If the diffusion current is neglected, then from Equation 4.16 we can write for an nMOSFET device

$$J_n(y) \cong qn(x,y)\mu_s E(y) = -qn(x,y)\mu_s(x,y)\frac{\partial\phi_s(y)}{\partial y} \tag{4.56}$$

where we can safely use $\left(\partial\phi_n/\partial y\right) = \left(\partial\phi_s/\partial y\right)$, that is, the gradients of quasi-Fermi potential and surface potential along the channel are the same in strong inversion. Therefore, for nMOSFET devices at strong inversion, we can write

$$\phi_s(y) = \phi_s(0) + V_{ch}(y) \tag{4.57}$$

where:
$\phi_s(0)$ is the surface potential at $y = 0$ (source end)

For the simplicity of calculation, it is more convenient to express the channel potential in terms of the source potential, V_{sb} and channel voltage $V(y)$ at any point y in the channel due to the applied drain voltage V_{ds} so that

$$V_{ch}(y) = V_{sb} + V(y) \tag{4.58}$$

where $V(y)$ now varies from 0 at $L = 0$ at the source end to V_{ds} at $y = L$ at the drain end of the channel. Then using Equation 4.58, at strong inversion ($\phi_s = 2\phi_B$), Equation 4.57 can be expressed as

$$\phi_s(y) = 2\phi_B + V_{sb} + V(y) \tag{4.59}$$

where $\phi_s(0) = 2\phi_B$ at strong inversion, and $V(y)$ varies from 0 at the source end to V_{ds} at the drain end. Now, substituting Equation 4.59 in 4.56, we get

$$J_n(y) = qn(x,y)\mu_s E(y) = -qn(x,y)\mu_s(x,y)\frac{dV}{dy} \tag{4.60}$$

And, therefore,

$$I_{ds}(y) = -W\frac{dV}{dy}\int_0^\infty qn(x,y)\mu_s(x,y)dx \qquad (4.61)$$

We know that the minority carrier charge density, Q_i, is given by

$$Q_i(y) = -q\int_0^\infty n(x,y)dx \qquad (4.62)$$

Then using Equation 4.62 in Equation 4.61, we get the simplified expression for $I_{ds}(y)$ as

$$I_{ds}(y)dy = W\mu_s Q_i(y)dV(y) \qquad (4.63)$$

Again we assume that GCA is valid along the entire length of the channel; then integrating Equation 4.63 along the channel length from $y = 0$ to $y = L$ we get

$$I_{ds} = \mu_s\left(\frac{W}{L}\right)\int_0^{V_{ds}} Q_i(y)dV \qquad (4.64)$$

Equation 4.64 is the simplified drain current equation for I_{ds} expression to develop compact MOSFET model for circuit CAD in the different regions of device operation. Thus, to calculate I_{ds} in a MOSFET device, we need to calculate Q_i. In the following section, we will derive simple and more useful expression for Q_i using charge balance equation given by Equation 4.40.

4.4.4.1 Core Model

In terms of source as the reference terminal, the expression for inversion charge $Q_i(y)$ in Equation 4.40 can be expressed as

$$Q_i(y) = Q_s(y) - Q_b(y) = -C_{ox}\left[V_{gs} + V_{sb} - V_{fb} - \phi_s(y)\right] - Q_b(y) \qquad (4.65)$$

Linear Region Operation: In order to develop the linear region I_{ds} model, we substitute for $\phi_s(y)$ from Equation 4.59 into Equation 4.65 to obtain inversion charge at strong inversion as

$$Q_i(y) = -C_{ox}\left[V_{gs} - V_{fb} - 2\phi_B - V(y)\right] - Q_b(y) \qquad (4.66)$$

And, substituting $\phi_s(y)$ from Equation 4.59 into Equation 4.38, we get for the depletion charge

$$Q_b(y) = -\gamma C_{ox}\sqrt{2\phi_B + V_{sb} + V(y)} \qquad (4.67)$$

Assumption 7: For a first-order model, we assume that Q_b is a constant along the length of the channel, independent of the applied drain voltage V_{ds} so that $\phi_s(y) = 2\phi_B + V_{sb}$ is a constant along the length of the channel. Therefore, Equation 4.67 can be approximated to

$$Q_b(y) \cong -\gamma C_{ox}\sqrt{2\phi_B + V_{sb}} \qquad (4.68)$$

Now, substituting for $Q_b(y)$ from Equation 4.68 into Equation 4.66, we get after simplification

$$Q_i(y) = -C_{ox}\left[V_{gs} - \left(V_{fb} + 2\phi_B + \gamma\sqrt{2\phi_B + V_{sb}}\right) - V(y)\right] \qquad (4.69)$$

In Equation 4.12, we have shown that $V_{th} = V_{fb} + 2\phi_B + \gamma\sqrt{(2\phi_B + V_{sb})}$; therefore, we can express Equation 4.69 as

$$Q_i(y) = -C_{ox}\left[V_{gs} - V_{th} - V(y)\right] \qquad (4.70)$$

Now, substituting for $Q_i(y)$ from Equation 4.70 in Equation 4.64, we get

$$I_{ds} = \mu_s C_{ox}\left(\frac{W}{L}\right)\int_0^{V_{ds}}\left[V_{gs} - V_{th} - V(y)\right]dV \qquad (4.71)$$

After integration of Equation 4.71, we get the first-order drain current model as

$$I_{ds} = \mu_s C_{ox}\left(\frac{W}{L}\right)\left[V_{gs} - V_{th} - \frac{V_{ds}}{2}\right]V_{ds}; \quad V_{gs} > V_{th} \qquad (4.72)$$

This current equation was derived by Sah [27] and later used by Shichman and Hodges [28] for modeling MOSFET devices in circuit simulation. This is known as the Simulation Program with Integrated Circuit Emphasis (SPICE) MOS Level 1 model. The factor $\mu_s C_{ox}$ is a model parameter and is referred to as the process transconductance, κ, so that

$$\kappa = \mu_s C_{ox} \qquad (4.73)$$

The parameter κ describes the effect of process variation in the drain current. Also, $\kappa(W/L)$ is called the gain factor of a MOSFET device and is defined as

$$\beta = \mu_s C_{ox}\left(\frac{W}{L}\right) \qquad (4.74)$$

For small $V_{ds} \leq (V_{gs} - V_{th}) \leq 0.1$ V, Equation 4.72 can be approximated using β from Equation 4.74 as

$$I_{ds} \cong \beta\left[V_{gs} - V_{th}\right]V_{ds}; \quad V_{gs} > V_{th} \qquad (4.75)$$

Equation 4.72 shows that I_{ds} varies linearly with V_{ds}. Consequently, this region of MOSFET device performance is called the *linear region* operation. From Equation 4.75, we get

$$\frac{V_{ds}}{I_{ds}} \cong \frac{1}{\beta\left[V_{gs}-V_{th}\right]} = \frac{1}{\beta V_{gst}} \equiv R_{ch} \tag{4.76}$$

where:

R_{ch} is called the channel resistance and is the effective resistance between the source and drain regions of MOSFET channel

Note that R_{ch} varies linearly with $(V_{gs} - V_{th}) \equiv V_{gst}$. V_{gst} is referred to as the *effective gate voltage* or *gate over drive voltage*. Thus, MOSFET devices are sometime referred to as the voltage-controlled variable resistors.

Figure 4.10 shows I_{ds} versus V_{ds} plots for different values of V_{gs} as calculated from Equation 4.72. It is seen from Figure 4.10 that for a given value of V_{gs}, the drain current I_{ds} initially increases with increasing V_{ds}, reaches a peak value and then begins to decrease with further increase in V_{ds}. This decrease in I_{ds} for higher values of V_{ds} is in contrast to the experimental observation, which shows saturation of I_{ds} at its peak value with further increase in V_{ds}. The discrepancy between the measured and computed value of I_{ds} by Equation 4.72 is due to the breakdown of GCA at high V_{ds} beyond

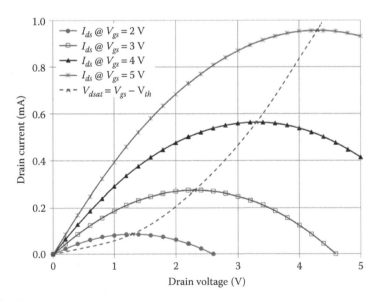

FIGURE 4.10
The current voltage characteristics of an nMOSFET device using Equation 4.72 with $T_{ox} = 20$ nm, $W/L = 1$, $V_{th} = 0.7$ V, and electron mobility $= 600$ cm^2 V^{-1} sec^{-1}; the dashed line shows the saturation drain voltage, V_{dsat} for each gate voltage.

the peak value of I_{ds}. Now, by differentiating Equation 4.72 with respect to V_{ds} we get

$$\frac{dI_{ds}}{dV_{ds}} = \mu_s C_{ox}\left(\frac{W}{L}\right)\left[V_{gs} - V_{th} - V_{ds}\right] \tag{4.77}$$

We know that at the point of inflexion, that is, at the peak location of $I_{ds} - V_{ds}$ plot, the slope of I_{ds} versus V_{ds} plot, $dI_{ds}/dV_{ds} = 0$. Therefore, equating Equation 4.77 to zero, we get

$$V_{ds} = V_{gs} - V_{th} \tag{4.78}$$

Equation 4.78 shows the condition at which I_{ds} peaks. Now, at this condition, let us find the inversion charge density at the drain end of the channel $Q_i(y = L)$ under the biasing condition of Equation 4.78.

We know that at the drain end of MOSFET channel, $y = L$, $V(y) = V_{ds}$. Then substituting for $V(y) = (V_{gs} - V_{th})$, in Equation 4.70, the channel charge $Q_i(y = L)$ at the drain end of the channel is given by

$$Q_i(y = L) = -C_{ox}\left[V_{gs} - V_{th} - (V_{gs} - V_{th})\right] = 0 \tag{4.79}$$

This value of $Q_i(L) = 0$ implies that at $V_{ds} = V_{gs} - V_{th}$, the channel does not exist at the drain end of the device. And, the maximum value of $V(y)$ will be at the drain end of the channel where $V(y) = V_{ds}$; therefore, when $V_{ds} \geq (V_{gs} - V_{th})$, we find $Q_i = 0$ at the drain end of the channel. In other words, once the peak current is reached, the GCA (assumption 1) fails and Equation 4.72 is no longer valid for $V_{ds} \geq (V_{gs} - V_{th})$. And, therefore, we need to derive a separate expression for drain current in the saturation region for $V_{ds} \geq (V_{gs} - V_{th})$.

Device Saturation: The physical understanding of the mathematical interpretations of Equations 4.78 and 4.79 can be achieved by analysis of device operation under varying V_{ds} for a certain value of $(V_{gs} - V_{th}) > 0$, that is, at strong inversion as shown in Figure 4.11. In deriving I_{ds} Equation 4.72, it is assumed that an inversion layer exists along the channel from the source end to drain end as shown in Figure 4.11a. This is only true for $V_{gs} \geq V_{th}$ with very low value of $V_{ds} < 100$ mV. For a given value of V_{gs}, when $V_{ds} = (V_{gs} - V_{th})$, Equation 4.79 shows that the value of Q_i at the drain end drops to zero. This implies that the channel is *pinched off* at the drain end with $Q_i(L)$ approaching to zero as shown in Figure 4.11b. And, consequently, the magnitude of the vertical electric field E_x approaches to that of the lateral electric field E_y at the pinch-off point.

The drain voltage at which the channel pinch-off occurs at the drain end is called the *pinch-off* or *saturation voltage*, V_{dsat}. The corresponding drain current at V_{dsat} is called the saturation drain current I_{dsat} or device on current I_{on}. From Equations 4.78 and 4.79, the condition for pinch-off ($Q_i = 0$) is $(dI_{ds}/dV_{ds}) = 0$, that is, at pinch-off point the slope of $I_{ds} - V_{ds}$ characteristics

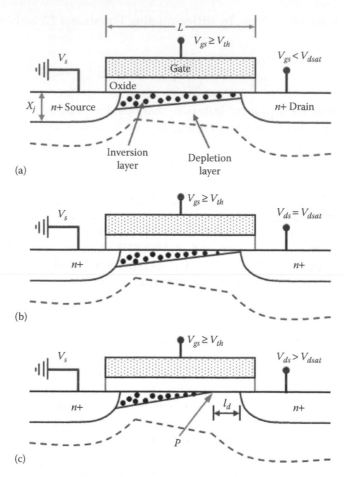

FIGURE 4.11
Schematic diagram of an nMOSFET device at strong inversion showing channel pinch-off as V_{ds} is increased; (a) an inversion layer connects the source and drain, $V_{ds} < V_{dsat}$ and (b) at the onset of saturation, the channel pinches off at the drain end, $V_{ds} = V_{dsat}$, and (c) the pinch-off point P moves toward the source.

becomes zero. At the pinch-off point, $E_x = E_y$, and when $E_y > E_x$, the mobile carriers are pushed off the surface near the drain region creating drain depletion or *pinch-off region* as shown in Figure 4.11c. Then from Equation 4.78, the pinch-off voltage, commonly referred to as the saturation drain voltage, V_{dsat} is given by

$$V_{dsat} = V_{gs} - V_{th} \qquad (4.80)$$

Equation 4.80 shows that the pinch-off voltage V_{dsat} equals the effective gate voltage V_{gst} that increases with increasing V_{gs} as shown in Figure 4.10 by the dash curve. Then substituting for $V_{ds} = V_{dsat}$ from Equation 4.80 in Equation 4.72, we get the drain current I_{dsat} at the pinch-off point as

$$I_{dsat} = \frac{\beta}{2}\left(V_{gs} - V_{th}\right)^2; \quad \text{at } V_{ds} = V_{dsat} = V_{gs} - V_{th} \qquad (4.81)$$

Thus, the simplified compact model requires two separate expressions for I_{ds} given by Equations 4.72 and 4.81 to model the MOSFET device characteristics in strong inversion region in contrast to Pao-Sah and Brews models described earlier.

Saturation Region Operation: Thus, we find that for a given V_{gs}, as V_{ds} increases the channel charge Q_i decreases near the drain end, and when $V_{ds} = V_{dsat} = (V_{gs} - V_{th})$, the channel is pinched off. For $V_{ds} > V_{dsat}$, the pinched-off region moves away from the drain end of the channel, widening the drain depletion region as shown in Figure 4.11c. Thus, as V_{ds} increases beyond pinch-off, the pinched-off region l_d between the channel pinch-off point P and the $n+$ drain region causes the effective channel length to decrease from L to $(L - l_d)$. Since the channel can support only V_{dsat}, any voltage greater than V_{dsat} is absorbed by the l_d region of the channel. Clearly, l_d is bias-dependent parameter, modulating the effective channel length $(L_{eff} = L - l_d)$. This phenomenon is called the *channel length modulation* (CLM). For long channel devices with $L >> l_d$, the drain current I_{ds} remains approximately constant at I_{dsat} for any $V_{ds} > V_{dsat}$. Thus, to a first order, for V_{ds} beyond the pinch-off value, the current $I_{ds} = I_{dsat}$ and is given by Equation 4.81 and is repeated below:

$$I_{dsat} = \frac{\beta}{2}\left(V_{gs} - V_{th}\right)^2; \quad V_{ds} > V_{dsat} \qquad (4.82)$$

The region of operation of the MOSFETs beyond pinch-off ($V_{ds} > V_{dsat}$) is referred to as the saturation region because I_{ds} ideally does not increase in this region. And, the region below V_{dsat} is called the linear region or triode region. Note that Equation 4.82 predicts that I_{ds} in saturation varies as the square of the effective gate voltage and hence often referred to as the square law model of the MOSFETs. Equations 4.72, 4.80, and 4.82 when plotted together result the output characteristics of a MOSFET device as shown by the continuous lines in Figure 4.12.

Figure 4.12 shows that the calculated value of I_{ds} by Equation 4.82 saturates beyond V_{dsat}. This is because Equation 4.82 is based on the assumption that the current is independent of V_{ds}. In reality, I_{ds} depends on $V_{ds} > V_{dsat}$ due to CLM due to the change in the effective channel length $L_{eff} = L - l_d$. Then using L_{eff} for L in Equation 4.82, we get

$$I_{ds} = \mu_s C_{ox} \frac{W}{2\left(L - l_d\right)}\left(V_{gs} - V_{th}\right)^2; \quad V_{ds} > V_{dsat} \qquad (4.83)$$

Equation 4.83 shows that as l_d increases with the increasing $V_{ds} > V_{dsat}$, the drain current increases. We can express Equation 4.83 as

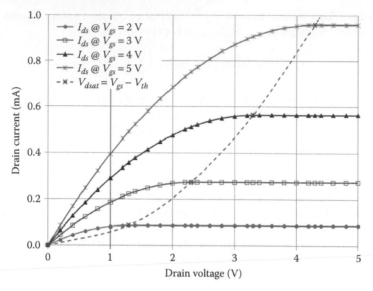

FIGURE 4.12
The current voltage characteristics of an nMOSFET device using Equations 4.72 and 4.82 with $T_{ox} = 20$ nm, $W/L = 1$, $V_{th} = 0.7$ V, and electron mobility $= 600$ cm^2 V^{-1} sec^{-1}; the dashed line separates the linear and the saturation regions of MOSFET operation.

$$I_{ds} = \mu_s C_{ox} \frac{W}{2L\left[1-\left(l_d/L\right)\right]}\left(V_{gs}-V_{th}\right)^2 ; \quad V_{ds} > V_{dsat}$$

$$= \mu_s C_{ox} \frac{W}{2L}\left(V_{gs}-V_{th}\right)^2 \cdot \frac{1}{\left[1-\left(l_d/L\right)\right]} \tag{4.84}$$

$$= \frac{\beta}{2}\left(V_{gs}-V_{th}\right)^2 \cdot \left(1-\frac{l_d}{L}\right)^{-1}$$

Using Equation 4.82 in Equation 4.84, we can show for $V_{ds} > V_{dsat}$

$$I_{ds} = I_{dsat}\left(1-\frac{l_d}{L}\right)^{-1} \tag{4.85}$$

In general, $l_d \ll L$; therefore, by series expansion we get, $\left[1-\left(l_d/L\right)\right]^{-1} \cong 1+\left(l_d/L\right)$. Since l_d increases with the increase of V_{ds}, that is, l_d/L is directly proportional to V_{ds}, we can write $1+\left(l_d/L\right)=1+\lambda V_{ds}$. Then Equation 4.85 becomes

$$I_{ds} = I_{dsat}\left(1+\lambda V_{ds}\right) \tag{4.86}$$

where:
   λ is called the CLM parameter describing the effect of V_{ds} on l_d to model CLM

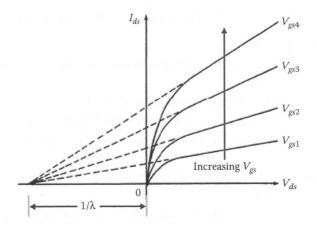

FIGURE 4.13
A typical $I_{ds} - V_{ds}$ characteristics of an nMOSFET device showing the effect of channel length modulation and CLM factor, λ.

We notice from Equation 4.86 that when $V_{ds} = -1/\lambda$, $I_{ds} = 0$. This means that when I_{ds} is extrapolated backward from the saturation region, it will intersect the V_{ds} axis at a value of $-1/\lambda$ as shown in Figure 4.13. However, this is an ideal case and generally the value of λ is obtained by curve fitting the measurement data to Equation 4.86 to minimize the error between the measured data and model.

Equation 4.86 is a first-order approximation for modeling CLM effect in MOSFETs. It provides the basic feature of nonzero slope for the saturated drain current as shown in Figure 4.13. However, the use of Equation 4.86 for calculating I_{ds} in the saturation region results in a discontinuity of the current at $V_{ds} = V_{dsat}$. The SPICE model Level 1 corrects for the discontinuity by multiplying the linear region current by the factor $(1 + \lambda V_{ds})$. Other methods include use of mathematical smoothing functions to make the liner and saturation curve at the transition point at $V_{ds} = V_{dsat}$.

To summarize, we have developed a first-order MOSFET model, which can be described by the following expressions

$$I_{ds} = \begin{cases} 0 \, ; & \left(V_{gs} - V_{th}\right) < 0 \\[2mm] \beta\left(V_{gs} - V_{th} - \dfrac{1}{2}V_{ds}\right)V_{ds} \, ; & 0 < \left(V_{gs} - V_{th}\right) \geq V_{ds} \\[2mm] \dfrac{\beta}{2}\left(V_{gs} - V_{th}\right)^2 \left(1 + \lambda V_{ds}\right); & 0 < \left(V_{gs} - V_{th}\right) \leq V_{ds} \end{cases} \qquad (4.87)$$

In Equation 4.87, β depends on κ [Equation 4.73], W, L, and C_{ox} [Equation 4.74] whereas, V_{th} depends on V_{th0}, $2\phi_B$, and γ [Equation 4.14]. Since C_{ox} depends on T_{ox}, the parameter set of SPICE Level 1 model is $\{V_{th0}, \kappa, \gamma, \lambda, 2\phi_B\}$.

SPICE Level 1 model is derived based on the following assumptions:

1. The GCA is valid
2. Majority carrier current can be neglected (e.g., neglected hole current for nMOSFETs)
3. Recombination and generation are neglected
4. Current flows in the y direction (along the length of channel) only
5. Inversion carrier mobility μ_s is a constant in the y direction along the channel
6. Current flow is due to the drift of minority carriers only (diffusion current is neglected)
7. Bulk charge Q_b is constant at any point in the y direction

The parameters of Level 1 MOSFET model for circuit CAD are shown in Table 4.1.

Although Equation 4.87 is derived for an nMOSFET device, the same expressions apply for a p-channel MOSFET (pMOSFET) device once all polarities of voltages and currents are reversed. The accuracy of Level 1 model is very poor even for long channel (10 µm) devices. However, it is very useful for performing basic circuit analysis and developing design equations for circuit performance.

4.4.4.2 Bulk-Charge Model

The level 1 drain current model is useful for hand calculations; however, it is not accurate for circuit CAD because of the inherent simplifying assumptions in deriving the current model. In order to improve the modeling accuracy, first of all, we examine the effect of bulk charge Q_b on I_{ds}. In level 1, we assumed that Q_b is a constant along the length of the channel. This means that the depletion width X_{dm} under the gate is a constant from the source to drain for all biasing conditions of $V_{ds} > 0$. In reality, when $V_{ds} > 0$, X_{dm} will increase as we move from the source toward the drain as shown in Figure 4.14. Consequently, it is more appropriate to consider the variation of bulk charge along the channel due to the applied drain bias from Equation 4.67. Then from Equations 4.66 and 4.67, the inversion charge density $Q_i(y)$ is given by

TABLE 4.1

Model Parameters for MOS Level 1 Compact Model

Device Parameter	Level 1 Model Parameter	Definition
V_{th0}	VTO	Threshold voltage at zero body bias
κ	KP	Transconductance parameter
γ	GAMMA	Body factor
λ	LAMBDA	CLM factor
$2\lvert \phi_B \rvert$	PHI	Bulk potential

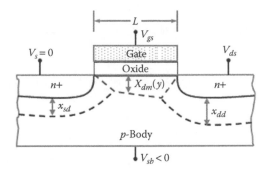

FIGURE 4.14
Depletion region widening at the drain end of the channel of an nMOSFET device due to CLM by applied drain voltage.

$$Q_i(y) = -C_{ox}\left[V_{gs} - V_{fb} - 2\phi_B - V(y) - \gamma\sqrt{2\phi_B + V_{sb} + V(y)}\right] \quad (4.88)$$

Substituting Equation 4.88 in Equation 4.64 and integrating from $V(y) = V_{sb}$ at $y = 0$ to $V(y) = V_{sb} + V_{ds}$ at $y = L$, we get

$$I_{ds} = \mu_s C_{ox}\left(\frac{W}{L}\right)\left\{\left(V_{gs} - V_{th} - \frac{V_{ds}}{2}\right)V_{ds} - \frac{2}{3}\gamma\left[(V_{ds} + 2\phi_B + V_{sb})^{3/2} - (2\phi_B + V_{sb})^{3/2}\right]\right\} \quad (4.89)$$

Equation 4.89 accounts for the bulk-charge variation in the depletion region of MOSFETs. The linear region drain current (Equation 4.89) is sometimes referred to as the Ihantola–Moll model [29] and used as SPICE Level 2 MOS model. Comparing Equations 4.72 and 4.89, we find that Equation 4.89 predicts lower current compared to Equation 4.72. This is because the increasing bulk charge Q_b will reduce the inversion charge Q_i for the same bias condition, resulting in a lower drain current. However, Equation 4.89 is more complex compared to Equation 4.72 and time-consuming for circuit CAD.

In order to derive model equation for I_{dsat}, we calculate V_{dsat} by differentiating Equation 4.89 with respect to V_{ds} and equate the resulting expression to zero. This results in the following expression for V_{dsat}

$$V_{dsat} = V_{gs} - V_{fb} - 2\phi_B + \frac{\gamma^2}{2} - \gamma\sqrt{V_{gs} - V_{fb} + V_{sb} + \frac{\gamma^2}{4}} \quad (4.90)$$

Then substituting $V_{ds} = V_{dsat}$ from Equation 4.90 into Equation 4.89, we can compute the saturation region drain current I_{dsat}.

4.4.4.3 Square Root Approximation of Bulk-Charge Model

In order to develop a computationally efficient drain current model considering $Q_b(y)$, we simplify Equation 4.67 by Taylor series expansion and neglect the higher order terms to get

$$Q_b(y) = -C_{ox}\gamma\left[\sqrt{2\phi_B + V_{sb}} + \frac{1}{2}\frac{V(y)}{\sqrt{2\phi_B + V_{sb}}} - \cdots\right]$$

$$\cong -C_{ox}\gamma\left[\sqrt{2\phi_B + V_{sb}} + \frac{1}{2\sqrt{2\phi_B + V_{sb}}}V(y)\right] \qquad (4.91)$$

$$\cong -C_{ox}\gamma\left[\sqrt{2\phi_B + V_{sb}} + \delta\cdot V(y)\right]$$

Equation 4.91 is called the *square root approximation* of $Q_b(y)$, where δ accounts for the *bulk-charge effect in MOSFETs* and is given by

$$\delta \equiv \frac{1}{2\sqrt{2\phi_B + V_{sb}}} \qquad (4.92)$$

It is found that the value of δ obtained by Equation 4.92 is too large for accurate calculation of I_{ds} at low V_{sb} and high V_{ds}. In order to obtain accurate value of δ, several semi-empirical expressions for δ have been proposed as discussed by Arora [13]. It is found that the more appropriate expressions of δ for circuit CAD are the following semi-empirical relations

$$\delta \equiv \frac{1}{2\sqrt{1 + 2\phi_B + V_{sb}}} \qquad (4.93)$$

and

$$\delta \equiv \frac{1}{2\sqrt{2\phi_B + V_{sb}}}\left[1 - \frac{1}{a_1 + a_2\left(2\phi_B + V_{sb}\right)}\right] \qquad (4.94)$$

where a_1 and a_2 are obtained to minimize the error between the exact function $\sqrt{2\phi_B + V_{sb} + V(y)}$ and its approximation $\left(\sqrt{2\phi_B + V_{sb}} + \delta\cdot V\right)$ within the operating range of V_{sb} and V_{ds}.

With the square root approximation of $Q_b(y)$ from Equation 4.91 in Equation 4.66, we get

$$Q_i(y) = -C_{ox}\left\{V_{gs} - V_{fb} - 2\phi_B - V(y) - \gamma\left[\delta\cdot V(y) + \sqrt{2\phi_B + V_{sb}}\right]\right\}$$

$$= -C_{ox}\left[V_{gs} - \left(V_{fb} + 2\phi_B + \gamma\sqrt{2\phi_B + V_{sb}}\right) - (1 + \delta\cdot\gamma)V(y)\right] \qquad (4.95)$$

$$= -C_{ox}\left[V_{gs} - V_{th} - \alpha V(y)\right]$$

where we have used Equation 4.12 for V_{th} and α is defined as

$$\alpha = 1 + \delta\cdot\gamma \qquad (4.96)$$

α is called the bulk-charge coefficient.

The final expression for $Q_i(y)$ in Equation 4.95 is similar to Equation 4.70 with the difference of the term α, which accounts for the variation of bulk charge along the channel. Using $Q_i(y)$ from Equation 4.95 into Equation 4.64 and after integration and simplification we get the expression for linear current as

$$I_{ds} = \beta \left[V_{gs} - V_{th} - \frac{1}{2} \alpha V_{ds} \right] V_{ds}; \quad V_{gs} > V_{th} \tag{4.97}$$

Comparing Equation 4.97 with Equation 4.89, we see that by approximating the square root term in $Q_b(y)$ we get a much simpler expression for I_{ds}. This current equation is used in most advanced regional drain current models (e.g., BSIM) for circuit CAD [30].

Now, differentiating Equation 4.97 with respect to V_{ds} and equating the resulting expression to zero gives the following simple expression for V_{dsat}

$$V_{dsat} = \frac{V_{gs} - V_{th}}{\alpha} \tag{4.98}$$

Substituting for V_{dsat} from Equation 4.98 into Equation 4.97, we get the drain current model in the saturation region as

$$I_{dsat} = \frac{\beta}{2\alpha} \left(V_{gs} - V_{th} \right)^2; \quad V_{ds} \geq V_{dsat} \tag{4.99}$$

To summarize, we now have a more accurate and compact drain current model that takes into account the bulk-charge variation along the channel region and is represented by the following set of equations

$$I_{ds} = \begin{cases} 0; & \left(V_{gs} - V_{th} \right) < 0 \\[2ex] \beta \left(V_{gs} - V_{th} - \frac{1}{2} \alpha V_{ds} \right) V_{ds}; & 0 < \left(V_{gs} - V_{th} \right) \geq V_{ds} \\[2ex] \frac{\beta}{2\alpha} \left(V_{gs} - V_{th} \right)^2; & 0 < \left(V_{gs} - V_{th} \right) \leq V_{ds} \end{cases} \tag{4.100}$$

Equation 4.100 is simple and has been widely used in circuit CAD prior to the introduction of industry standard compact models.

4.4.4.4 Subthreshold Region Drain Current Model

The regional expressions for I_{ds} in Equations 4.87 and 4.100 are derived assuming that the current flow is due to drift only. This resulted in $I_{ds} = 0$ for $V_{gs} < V_{th}$. In reality, this is not true and I_{ds} has a small but finite value for $V_{gs} < V_{th}$ as shown in Figure 4.9, which shows that I_{ds} is of the order of 10 nA for $V_{gs} \approx V_{th}$ and decreases exponentially below V_{th}. This current below V_{th} is called the *subthreshold* or *weak inversion* current and occurs when $V_{gs} < V_{th}$ or $\phi_B < \phi_s < 2\phi_B$. Unlike the inversion region where drift current dominates, the subthreshold region conduction is dominated by diffusion current as shown in Figure 4.9. The subthreshold region current is important since this is a major contributor to device leakage current that affects the dynamic circuit performance and determines CMOS standby power. In this region of operation, the assumption $I_{ds} = 0$ in Equations 4.87 and 4.100 (by assumption 6) is not valid.

In the subthreshold region of operation, $Q_i \ll Q_b$, and therefore, the surface potential ϕ_s (or band bending) is nearly constant from the source to drain end of the device. This means that we can replace $\phi_s(y)$ in the subthreshold region by some constant value, ϕ_{ss}. Then, the bulk charge Q_b in Equation 4.38 can be expressed as

$$Q_b = -C_{ox}\gamma\sqrt{\phi_s(y)} \cong -C_{ox}\gamma\sqrt{\phi_{ss}} \tag{4.101}$$

Again, since $Q_i \ll Q_b$, we have $Q_s \approx Q_b$, so that Equation 4.6 becomes

$$V_{gb} = V_{fb} + \phi_{ss} - \frac{Q_b}{C_{ox}} \tag{4.102}$$

Substituting for Q_b from Equation 4.101 to Equation 4.93, we get

$$V_{gb} = V_{fb} + \phi_{ss} + \gamma\sqrt{\phi_{ss}}$$

or $\tag{4.103}$

$$\left(\sqrt{\phi_{ss}}\right)^2 + \gamma\sqrt{\phi_{ss}} - \left(V_{gb} - V_{fb}\right) = 0$$

Solving the quadratic Equation 4.103 we can show

$$\phi_{ss} = \left(-\frac{\gamma}{2} + \sqrt{\frac{\gamma^2}{4} + V_{gb} - V_{fb}}\right)^2 \tag{4.104}$$

Equation 4.104 shows that ϕ_{ss} is almost linearly dependent on $V_{gb} = V_{gs}$ for $V_{sb} = 0$. It should be emphasized that ϕ_{ss} is a constant in the subthreshold region for long channel devices only. As the channel length becomes shorter, ϕ_{ss} no longer remains constant over the entire channel length.

Since ϕ_{ss} is a constant, $E_y = -d\phi_{ss}/dy = 0$. Therefore, the only current that can flow is the diffusion current as can be seen from Equation 4.16 and is given by

$$J_n(x,y) = qD_n\frac{dn}{dy} = q\left(\mu_s v_{kT}\right)\frac{dn}{dy} \tag{4.105}$$

where from Einstein relation $D_n = \mu_s v_{kT}$. Integrating $J_n(x,y)$ from $x = 0$ at the Si/SiO$_2$ interface to $x = X_{inv}$ at the end of the inversion layer width, we get for an nMOSFET device of channel length, L, and width, W

$$I_{ds}(y) = W \int_0^{X_{inv}} q\left(\mu_s v_{kT}\right) \frac{dn}{dy} = W\mu_s v_{kT} \frac{d\left(qnX_{inv}\right)}{dy}$$

(4.106)

$$= W\mu_s v_{kT} \frac{dQ_i}{dy}$$

where:

$Q_i = qnX_{inv}$ is the inversion charge per unit area at any point y along the channel in the subthreshold region

Now, consider Q_{is} and Q_{id} are the inversion charge densities at $y = 0$ and $y = L$, respectively. Then integrating Equation 4.106 from $(y = 0, Q_i(y) = Q_{is})$ to $(y = L, Q_i(y) = Q_{id})$, we get

$$I_{ds} = \mu_s \left(\frac{W}{L}\right) v_{kT} \left(Q_{id} - Q_{is}\right)$$

(4.107)

Now, in order to calculate the subthreshold current from Equation 4.107, we need to find the inversion charge in the weak inversion regime, $\phi_B < \phi_s < 2\phi_B$. Again, we solve Poisson's equation to calculate Q_s and find an expression for Q_i in the weak inversion following the procedure in Chapter 3 (Equation 3.68). Then for MOSFETs in the weak inversion region, we can show

$$Q_s \cong -\sqrt{2qK_{si}\varepsilon_0 N_b \phi_{ss}} \left[1 + \frac{v_{kT}}{\phi_{ss}} e^{\left[\phi_{ss} - 2\phi_B - V_{ch}(y)\right]/v_{kT}}\right]^{1/2}$$

(4.108)

Let us assume that the exponential term in Equation 4.108 is much smaller than ϕ_{ss}. Then using series expansion $\sqrt{1+x} \cong 1 + (x/2)$, we get for the total charge Q_s in the substrate at weak inversion as

$$Q_s \cong -\sqrt{2qK_{si}\varepsilon_0 N_b \phi_{ss}} \left[1 + \frac{v_{kT}}{2\phi_{ss}} e^{\left[\phi_{ss} - 2\phi_B - V_{ch}(y)\right]/v_{kT}}\right]$$

$$= Q_b + \left(-\sqrt{\frac{qK_{si}\varepsilon_0 N_b}{2\phi_{ss}}} v_{kT} e^{\left[\phi_{ss} - 2\phi_B - V_{ch}(y)\right]/v_{kT}}\right)$$

(4.109)

where $Q_b = -\sqrt{2qK_{si}\varepsilon_0 N_b \phi_{ss}}$ as shown in Equation 3.64; since, $Q_s = Q_b + Q_i$, from Equation 4.109, the minority carrier charge density at the weak inversion region, $\phi_B < \phi_s < 2\phi_B$, of nMOSFETs is given by

$$Q_i = -\sqrt{\frac{qK_{si}\varepsilon_0 N_b}{2\phi_{ss}}} v_{kT} e^{\left[\phi_{ss} - 2\phi_B - V_{ch}(y)\right]/v_{kT}}$$

(4.110)

Again, from Equation 3.62, the width of the depletion region $X_d = \sqrt{2K_{si}\varepsilon_0 \phi_{ss}/qN_b}$; then the depletion capacitance C_d $(=K_{si}\varepsilon_0/X_d)$ is given by

$$C_d = \sqrt{\frac{qK_{si}\varepsilon_0 N_b}{2\phi_{ss}}} \qquad (4.111)$$

Therefore, the charge density in the weak inversion region of nMOSFETs is given by

$$Q_i = -C_d v_{kT} e^{[\phi_{ss} - 2\phi_B - V_{ch}(y)]/v_{kT}} \qquad (4.112)$$

Now, using the appropriate boundary conditions defined earlier

$$V_{ch}(y) = \begin{cases} V_{sb} & \text{at } y = 0 \quad \text{(source end)} \\ V_{sb} + V_{ds} & \text{at } y = L \quad \text{(drain end)} \end{cases}$$

We can write the expressions for the inversion charges from Equation 4.112 as

$$Q_{is} = -C_d v_{kT} e^{[\phi_{ss} - 2\phi_B - V_{sb}]/v_{kT}}$$

$$Q_{id} = -C_d v_{kT} e^{[\phi_{ss} - 2\phi_B - V_{sb} - V_{ds}]/v_{kT}} \qquad (4.113)$$

Now, substituting for Q_{is} and Q_{id} from Equation 4.113 in Equation 4.107, we get the expression for the subthreshold region current as

$$I_{ds} = \mu_s \left(\frac{W}{L}\right) C_d v_{kT}^2 e^{[\phi_{ss} - 2\phi_B - V_{sb}]/v_{kT}} \left(1 - e^{-(V_{ds}/v_{kT})}\right) \qquad (4.114)$$

Since $\exp(-2\phi_B/v_{kT}) = n_i^2/N_b^2$, Equation 4.114 can also be expressed as

$$I_{ds} = \mu_s \left(\frac{W}{L}\right) C_d \left(v_{kT} \frac{n_i}{N_b}\right)^2 e^{(\phi_{ss} - V_{sb})/v_{kT}} \left(1 - e^{-(V_{ds}/v_{kT})}\right) \qquad (4.115)$$

In order to eliminate ϕ_{ss} from Equation 4.115, we expand V_{gs} in a series around the point $\phi_{ss} = 2\phi_B$ (weak inversion corresponding to $\phi_B < \phi_s < 2\phi_B$). We define $V_{th} = V_{gs}$ @ $\phi_{ss} = 2\phi_B$ and $V_{sb} = 0$; therefore, $V_{gb} = V_{gs}$. Then by series expansion of V_{gs} around the point $\phi_{ss} = V_{sb} + 2\phi_B$ at the onset of inversion

$$V_{gs} = V_{gs}\big|_{\phi_{ss} = V_{sb} + 2\phi_B} + \frac{dV_{gs}}{d\phi_{ss}}(\phi_{ss} - 2\phi_B - V_{sb}) \qquad (4.116)$$

Since $V_{sb} = 0$ at $V_{gs} = V_{th}$, and $\phi_s = 2\phi_B$, by defining $n \equiv dV_{gs}/d\phi_{ss}$, we get from Equation 4.116

$$V_{gs} = V_{th} + n(\phi_{ss} - 2\phi_B - V_{sb})$$

$$\therefore \phi_{ss} - 2\phi_B - V_{sb} = \frac{V_{gs} - V_{th}}{n} \qquad (4.117)$$

Then from Equation 4.114 we get for subthreshold region drain current model as

$$I_{ds} = \mu_s \left(\frac{W}{L}\right) C_d v_{kT}^2 e^{(V_{gs}-V_{th})/nv_{kT}} \left(1 - e^{-(V_{ds}/v_{kT})}\right) \qquad (4.118)$$

where n is the ideality factor that can be determined from Equation 4.103. Using source referencing, we get

$$V_{gs} = V_{fb} + \phi_{ss} + \frac{\sqrt{2qK_{si}\varepsilon_0 N_b \phi_{ss}}}{C_{ox}} \qquad (4.119)$$

Then, from Equation 4.119, we can show

$$\frac{dV_{gs}}{d\phi_{ss}} = 1 + \frac{1}{C_{ox}}\sqrt{\frac{qK_{si}\varepsilon_0 N_b}{2\phi_{ss}}} = 1 + \frac{C_d}{C_{ox}} \qquad (4.120)$$

where we have used Equation 4.111 for C_d. Thus, we have

$$n = 1 + \frac{C_d}{C_{ox}} \qquad (4.121)$$

From Equation 4.121, we get $C_d = (n-1)C_{ox}$. Therefore, we can express the subthreshold region current (Equation 4.118) in terms of C_d as well as C_{ox} as

$$I_{ds} = \begin{cases} \mu_s \left(\frac{W}{L}\right) C_d v_{kT}^2 e^{(V_{gs}-V_{th})/nv_{kT}} \left(1 - e^{-(V_{ds}/v_{kT})}\right) \\ \mu_s \left(\frac{W}{L}\right) (n-1) C_{ox} v_{kT}^2 e^{(V_{gs}-V_{th})/nv_{kT}} \left(1 - e^{-(V_{ds}/v_{kT})}\right) \end{cases} \qquad (4.122)$$

From Equation 4.118 we note that in the subthreshold conduction

1. I_{ds} depends on V_{ds} only for small V_{ds}, that is, $V_{ds} \leq 3v_{kT}$, since $\exp(-V_{ds}/v_{kT}) \to 0$ for larger V_{ds}; therefore, for simplicity of device modeling, Equation 4.118 can be approximated to [31]

$$I_{ds} \cong \mu_s \left(\frac{W}{L}\right) C_d v_{kT}^2 e^{(V_{gs}-V_{th})/nv_{kT}} \qquad (4.123)$$

2. I_{ds} depends exponentially on V_{gs} but with an *ideality factor* $n > 1$ (Equation 4.121); thus, the slope is poorer than a bipolar junction transistor (BJT) but approaches to that of a BJT in the limit $n \to 1$.

3. N_b and V_{bs} enter in the current model through depletion capacitance, C_d.

4. The subthreshold current (Equation 4.122) is strongly dependent on temperature T because of its dependence on the square of the intrinsic concentration n_i through Equation 4.115 and thermal voltage $v_{kT} = kT/q$.

Subthreshold slope: An important characteristic of the subthreshold region is the gate voltage swing required to reduce the current from its ON value to an acceptable OFF value. This gate voltage is also called the *subthreshold slope S* or *SS* or *S-factor*. It is the inverse of the slope of $I_{ds} - V_{gs}$ characteristics and is defined as the change in the gate voltage V_{gs} required to change the subthreshold current I_{ds} by one decade. Clearly, S is a measure of the turn-off characteristics of a MOSFET device. If we take two points (I_{ds1}, V_{gs1}) and (I_{ds2}, V_{gs2}) in the subthreshold region shown in Figure 4.15, then by definition ($V_{gs2} - V_{gs1}$) required to change (I_{ds2}/I_{ds1}) by one decade or 10 can be expressed as

$$S \equiv \frac{V_{gs2} - V_{gs1}}{\log I_{ds2} - \log I_{ds1}} = \frac{dV_{gs}}{d(\log I_{ds})} = 2.3 \frac{dV_{gs}}{d(\ln I_{ds})} \qquad (4.124)$$

where we have used $(\ln I_{ds}) = 2.3(\log I_{ds})$ for the conversion of logarithm base "10" to natural logarithm base "e." In reality, S varies with I_{ds} in the subthreshold region; however, this variation is negligible over one decade of current so that S can be considered as a gate swing per decade of current change. Therefore, from Equation 4.122, we get

$$\ln I_{ds} = \ln\left(\frac{\mu_s W C_d v_{kT}^2}{L}\right) + \frac{V_{gs} - V_{th}}{n v_{kT}} + \ln\left(1 - e^{-(V_{ds}/v_{kT})}\right) \qquad (4.125)$$

Then taking the derivative of Equation 4.125, we get

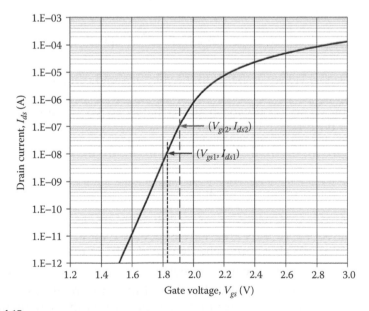

FIGURE 4.15
Log(I_{ds}) versus V_{gs} characteristics of a typical MOSFET device to calculate *S-factor*; the ratio of two data points in the subthreshold current is one decade.

$$d(\ln I_{ds}) = \frac{dV_{gs}}{nv_{kT}}$$

or (4.126)

$$\frac{dV_{gs}}{d(\ln I_{ds})} = nv_{kT}$$

Therefore, combining Equations 4.124 and 4.126, we can show

$$S = 2.3nv_{kT} \qquad (4.127)$$

Using Equation 4.121 for the ideality factor n, we get

$$S = 2.3v_{kT}\left(1 + \frac{C_d}{C_{ox}}\right) \qquad (4.128)$$

Since at room temperature ($T \sim 300$ K), $v_{kT} \cong 26$ mV, Equation 4.128 shows that the theoretical minimum swing S_{min} is given by

$$S_{min} = 2.3v_{kT} \cong 60\,\text{mV per decade} \qquad (4.129)$$

Thus, the minimum attainable S for any device is approximately 60 mV per decade at room temperature. Since, $1 \le n \le 3$, the typical value of $60 \le S \le 180$ mV per decade at room temperature. If there is a substantial interface trap density, then C_d in Equation 4.121 should be replaced by ($C_d + C_{IT}$). Therefore,

$$S = 2.3v_{kT}\left(1 + \frac{C_d + C_{IT}}{C_{ox}}\right) \qquad (4.130)$$

Final notes on subthreshold region conduction:

1. In weak inversion or subthreshold region, MOS devices have exponential characteristics but are less "efficient" than BJTs because $n > 1$.
2. Subthreshold slope S does not scale and is $\approx$ constant. Therefore, V_{th} cannot be scaled as required by the ideal scaling laws.
3. V_{ds} affects V_{th} as well as subthreshold currents.
4. In order to optimize S, the desirable parameters are:
   a. Thin oxide
   b. Low N_b
   c. High V_{bs}

4.4.4.5 Limitations of Regional Drain Current Model

In the regional drain current models developed in Section 4.4.4 we have assumed that in the subthreshold or weak inversion region I_{ds} is due to diffusion component only and in the linear and saturation regions (strong inversion) I_{ds} is due to drift component only. This causes a discontinuity in the device characteristics during transition between these weak and strong inversion regions. This discontinuity is a severe drawback of the simplified model for implementation and usage in circuit CAD. To ensure continuity from weak to strong inversion, we consider that at the transition point, the inversion charge at weak and strong inversions are equal, that is, Q_i(weak inversion) $= Q_i$(strong inversion). Under this condition the gate voltage at the transition point at $V_{gs} = V_{on}$ can be shown to be [32]

$$V_{on} = V_{th} + n v_{kT} \qquad (4.131)$$

From Equation 4.131, we find that the upper limit of subthreshold current is V_{on} instead of V_{th}; then replacing V_{th} by V_{on} in Equation 4.118, we can show

$$I_{ds} \cong I_{on} e^{(V_{gs} - V_{on})/n v_{kT}} \qquad (4.132)$$

where I_{on} is the on current calculated from (4.118) at $V_{gs} = V_{on}$ and is given by

$$I_{on} = \mu_s \left(\frac{W}{L} \right) C_d v_{kT}^2 \left(1 - e^{-(V_{ds}/v_{kT})} \right) \qquad (4.133)$$

Since in the subthreshold region I_{ds} is nearly independent of V_{ds}, we can safely neglect V_{ds} dependence in Equation 4.132 so that

$$I_{on} = \mu_s \left(\frac{W}{L} \right) C_d v_{kT}^2 \qquad (4.134)$$

Thus, for $V_{gs} < V_{on}$, I_{ds} is given by Equation 4.132, whereas for $V_{gs} > V_{on}$, I_{ds} is given by Equation 4.100 with I_{on} from Equation 4.133. Thus, V_{on} acts as a point at which behaviors of strong and weak inversion are pieced together. This is the approach used in SPICE Levels 2 and 3. Combining Equations 4.100 and 4.132 we now have a complete long channel DC MOSFET model for circuit CAD, which is continuous in all regions,

$$I_{ds} = \begin{cases} I_{on} \exp\left(\dfrac{V_{gs} - V_{on}}{n v_{kT}} \right) ; & \left(V_{gs} - V_{on} \right) < 0 \\[2mm] \beta \left(V_{gs} - V_{th} - \dfrac{1}{2} \alpha V_{ds} \right) V_{ds} ; & 0 < \left(V_{gs} - V_{th} \right) \geq V_{ds} \\[2mm] \dfrac{\beta}{2\alpha} \left(V_{gs} - V_{th} \right)^2 ; & 0 < \left(V_{gs} - V_{th} \right) \leq V_{ds} \end{cases} \qquad (4.135)$$

Although Equation 4.135 results in a continuous transition of device characteristics from weak to strong inversion, there are large errors in the I_{ds} calculations around the transitions region, often called the moderate inversion region [25]. However, for most of the digital applications this error is not significant due to the low magnitude of the current in this region. There are other approaches reported, which could be used to achieve better simulation results [30]. However, the improvement is not significant.

4.5 Summary

In this chapter we have introduced the four terminal MOSFET devices. The basic features of MOSFETs are described. A number of simplified assumptions are used to derive threshold voltage model for long channel devices. The fundamental Pao-Sah double-integral model and Brews charge-sheet model are derived to characterize MOSFET devices. We have discussed that the Pao-Sah and Brews models are computationally intensive to use for VLSI circuit analysis with billions of transistors in an IC chip. We have used simplified assumptions to derive the first-generation SPICE models for long channel devices. In these basic models, we have discussed how different equations for different regions of device operation are pieced together using smoothing functions.

The advantages of the simplified regional models include easy implementation of physical effects using empirical relations and fast computation time. On the other hand, the disadvantages include assumption of a constant ϕ_s ($=2\phi_B$) in strong inversion, resulting in an inaccurate modeling of moderate inversion, particularly, capacitances; and the model ignores inversion layer thickness and small geometry effects. However, the basic model is widely used for intuitive analysis of device performance.

Exercises

4.1 Pao-Sah model:

a. Complete the mathematical steps to derive Pao-Sah model given by Equation 4.36. Clearly define all parameters and explain.

b. To calculate MOSFET drain current using Pao-Sah model from Equation 4.36, the surface potential is numerically calculated from Equation 4.37. However, Equation 4.37 is derived assuming strong inversion only. Derive an accurate expression similar to Equation 4.37, which is valid in all regions of MOSFET operation.

4.2 For a device with p-type substrate concentration, $N_a = 2.5 \times 10^{16}$ cm^{-3}; gate oxide thickness, $T_{ox} = 100$ A; and $V_{fb} = -0.97$ V, calculate and plot $\ln(Q_i)$ versus V_{gb} in weak inversion.

4.3 Brews charge-sheet model:

a. Carry out the integration to derive the simplified surface potential based MOSFET drain current (Brews) model Equations 4.48 and 4.51.

b. Derive an expression for ϕ_{s0} in terms of the source-to-body bias V_{sb} to calculate I–V characteristics of the drift and diffusion components of I_{ds} for the above model. Clearly define all parameters and explain.

c. Derive an expression for ϕ_{sL} in terms of drain-to-body bias V_{db} to calculate I–V characteristics of the drift and diffusion components of I_{ds} for the above model.

Clearly define all parameters and explain.

4.4 Consider an nMOSFET device with $N_a = 5 \times 10^{17}$ cm^{-3}, $T_{ox} = 6$ nm, $V_{fb} = -1$ V, $\mu = 600$ cm^2 V^{-1}sec^{-1}, $W = L = 2$ µm, biased with $V_{sb} = 1$ V and $V_{db} = 3$ V, while V_{gb} is varied from 0 to 3 V. Use (Brews model) to calculate the following I_{ds} as a function of V_{gb}:

a. Drift component of I_{ds}, $I_{ds,drift}$

b. Diffusion component of I_{ds}, $I_{ds,diff}$

c. Total current I_{ds}

d. Plot I_{ds}–V_{gb} from part (a)–(c) using the same log drain current I_{ds} axis

e. Plot surface potentials (ϕ_{s0} and ϕ_{sL}) as a function of ($V_{gb} - V_{fb}$)

4.5 Consider Basic MOS models. Explain physically why I–V characteristics of MOSFETs are more sensitive to temperature in the subthreshold region than they are in the strong inversion.

4.6 Show that in the subthreshold region of MOSFETs, the surface potential is given by:

$$\phi_{ss} = \left[-\frac{\gamma}{2} + \sqrt{\frac{\gamma^2}{4} + V_{gb} - V_{fb}} \right]^2$$

4.7 In weak inversion, the drain current I_{ds} is exponentially proportional to an inverse of (60 mV dec I^{-1})(1 + C_d/C_{ox}) at room temperature. Once strong inversion is reached, most of the gate charge resulting from higher V_{gs} value is balanced by channel charge Q_i not depletion charge. Write a simple expression, analogous to the slope expression above, which approximately models the MOSFET devices in strong inversion. State any assumptions you make and explain your results.

4.8 Consider an MOS transistor on a uniformly doped *p*-type silicon substrate with doping concentration $N_a = 1 \times 10^{16} \text{cm}^{-3}$ at room temperature and $W = L = 10 \, \mu\text{m}$. Assume $V_{fb} = 0$, $T_{ox} = 20$ nm; threshold voltage, $V_{th} = 0.7$ V, electron mobility $= 600 \text{ cm}^2 \text{ V}^{-1} \text{ sec}^{-1}$; $\lambda = 0.01 \text{ V}^{-1}$

a. Calculate and plot I_{ds} versus V_{ds} for $0.0 \leq V_{ds} \leq 5.0$ V with $V_{gs} = 2$, 3, 4 and 5 V on the same plot using following equations at different limits of V_{gs} shown:

   i. $I_{ds} = \beta \left[V_{gs} - V_{th} - \left(V_{ds}/2 \right) \right] V_{ds}$; for $V_{gs} > V_{th}$ and $V_{ds} \leq \left(V_{gs} - V_{th} \right)$

   ii. $I_{ds} = \left(\beta/2 \right) \left(V_{gs} - V_{th} \right)^2$; for $V_{gs} > V_{th}$ and $V_{ds} \geq \left(V_{gs} - V_{th} \right)$

   iii. $I_{ds} = \left(\beta/2 \right) \left(V_{gs} - V_{th} \right)^2 \left(1 + \lambda V_{ds} \right)$; for $V_{gs} > V_{th}$ and $V_{ds} \geq \left(V_{gs} - V_{th} \right)$

   iv. Superimpose I_{dsat} versus V_{dsat} on the same plot;

   where $\beta = \mu_s C_{ox} \left(W/L \right)$; clearly, label different operating regimes all different operating regions of MOSFETs and explain

4.9 Measured device data for a silicon nMOSFET are shown in Table E4.1 Considering the bulk-charge effect (α) in drain current I_{ds} calculate:

a. V_{th0}

b. λ

c. γ

d. β

using regional drain current model

4.10 In Section 4.4.4.4, the subthreshold region drain current is modeled using inversion charge at the source and drain ends. In this exercise, formulate the subthreshold region drain current (I_{ds}) model using the inversion carrier density at the source end $n(0)$ and drain end $n(L)$. Clearly state any assumptions you make.

a. Write an expression for the subthreshold region I_{ds} from Fick's first law of diffusion; assume that the concentration gradient of inversion carriers $\left(dn/dy \right)$ is constant along the channel to maintain a constant current flow through the device.

TABLE E4.1

Measurement Data to Extract the Basic nMOSFET Device Model Parameters

V_{gs} (V)	V_{ds} (V)	V_{bs} (V)	I_{ds} (mA)
2	5	0	40
5	5	0	536
5	5	−5	360
5	8	0	644
5	5	−3	420

b. Write down the expressions for the inversion carrier density $n(0)$ and $n(L)$ in terms of the surface potential ϕ_{ss}, bulk potential ϕ_B, and the appropriate channel potential $V_{ch}(y)$ at the respective terminal.

c. Assuming that the depth of the inversion layer is given by $X_{inv} = v_{kT}\sqrt{K_{si}\varepsilon_0/2qN_b\phi_{ss}}$, show that the subthreshold region drain current is given by Equation 4.114; where the parameters have their usual meanings as described in Section 4.4.4.4.

d. Following the procedure in Section 4.4.4.4, show that I_{ds} is given by Equation 4.118.

e. In the subthreshold region, a MOSFET device includes an oxide capacitor C_{ox} in series with a depletion capacitor C_d and any change in gate voltage V_{gs} causes corresponding change in ϕ_{ss}; consider a voltage divider between C_{ox} and C_d, and show that the ideality factor n is given by Equation 4.121.

f. Show that the final I_{ds} in the subthreshold region is given by Equation 4.122.

5

Compact Models for Small Geometry MOSFETs

5.1 Introduction

In this chapter the compact models for small geometry MOSFET (metal-oxide-semiconductor field-effect transistors) devices are presented. The continuous scaling of MOSFET devices toward decananometer regime has resulted in higher device density and faster circuit speed along with higher power dissipation [1–4]. Many new physical phenomena became significant with the device dimension rapidly approaching its physical limit. These include small geometry effects [5–8], channel length modulation (CLM) [9], drain-induced barrier lowering (DIBL) [10], velocity saturation [11], mobility degradation due to high vertical electric field [12], impact ionization [13], band-to-band tunneling [14], velocity overshoot [15], self-heating [16], inversion-layer quantization [17–19], polysilicon depletion [20], and process variability [21,22]. Thus, accurate MOSFET models that include the observed new physical phenomena are crucial to design and optimization of advanced very-large-scale-integrated (VLSI) circuits using nanoscale complementary metal-oxide-semiconductor (CMOS) technologies. In this chapter, we will use regional modeling approach to develop compact MOSFET models to accurately simulate different physical and small geometry effects in advanced VLSI circuits. First of all, we will derive different analytical expressions to model the deviation of long channel V_{th} model derived in Chapter 4 due to geometry and different physical effects and present an accurate V_{th} model for circuit CAD. Then we derive drain current model for short channel MOSFET devices considering high-field effects causing mobility degradation and velocity saturation.

5.2 Threshold Voltage Model

MOSFET threshold voltage model developed in Chapter 4 assumes uniformly doped substrate and neglects geometry effects on device performance. The expression for V_{th} for long channel MOSFETs with uniformly doped substrate is given by Equation 4.12 and can be generalized as

$$V_{th} = V_{fb} + \phi_s + \gamma\sqrt{\phi_s - V_{bs}} \tag{5.1}$$

where:

V_{fb}, ϕ_s, γ, and V_{bs} are the flat band voltage, surface potential, body effect coefficient, and back gate or body bias, respectively

Note that in Equation 5.1, $\phi_s = 2\phi_B$ in strong inversion as shown in Equation 4.12. In Equation 5.1, the body effect coefficient is defined as

$$\gamma = \frac{\sqrt{2qK_{si}\varepsilon_0 N_b}}{C_{ox}} \tag{5.2}$$

where:

q, K_{si}, ε_0, N_b are the electronic charge, permittivity of silicon, permittivity of free space, and substrate concentration, respectively

If we define $V_{TH0} = V_{th}$ @ $V_{bs} = 0$, then we can show

$$V_{th} = V_{TH0} + \gamma\left(\sqrt{\phi_s - V_{bs}} - \sqrt{\phi_s}\right) \tag{5.3}$$

Equation 5.3 models V_{th} for large geometry MOSFET devices of uniformly doped substrate with doping concentration, N_b. In Sections 5.2.1 and 5.2.2, we will derive analytical expressions to consider nonuniform substrate doping and different physical and geometrical effects in modeling V_{th} for advanced MOSFET devices.

5.2.1 Effect of Nonuniform Channel Doping on Threshold Voltage

In nanoscale MOSFET devices, the channel doping concentration N_b varies both vertically and laterally [23–26]. In advanced CMOS technology, the channel doping concentration is vertically nonuniform due to threshold voltage adjust implant dopants and laterally nonuniform due to halo doping implant around the source-drain (S/D) extension (SDE) regions as shown in Figure 5.1a and b.

In a conventional CMOS technology, the type of impurity for V_{th} adjust doping is the same as the channel doping. Thus, the V_{th} adjust implant in the channel increases the channel doping concentration near the surface, that is, provides high–low doping profile [19]. In some advanced technology, the threshold voltage adjust implant creates low–high implant or super-steep-retrograde channel doping profile [19]. The nonuniform vertical channel doping causes a strong dependence of the depletion charge, Q_b, on the applied body bias, V_{bs}, as shown in Figure 5.2a [22]. On the other hand, the nonuniform lateral channel doping causes strong dependence of V_{th} on the channel length (L) as shown in Figure 5.2b [25,26].

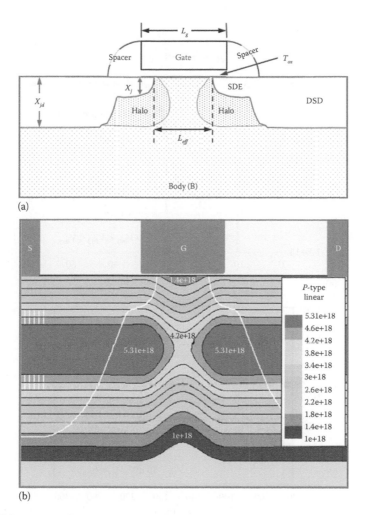

(a)

(b)

FIGURE 5.1

2D cross-section of a MOSFET device: (a) threshold adjust and halo implant causing nonuniform channel doping profile and (b) simulated 2D-doping contours of a typical double-halo *n*MOSFET device with laterally and vertically nonuniform *p*-type channel doping generated using device CAD MEDICI; 2D cross-section shows S, G, and D are the source, gate, and drain terminals, respectively, and the outline of SDE and deep source-drain (DSD) junctions. (Data from S. Saha, *Proc. SPIE Conf.*, 5042, 172–179, 2003.)

5.2.1.1 *Threshold Voltage Modeling for Nonuniform Vertical Channel Doping Profile*

Due to nonuniform channel doping, the body effect coefficient γ depends on the body bias, V_{bs}. For the simplicity of mathematical formulation, let us approximate the nonuniform vertical channel doping profile by a *high–low* step function as shown in Figure 5.3 with uniform concentration N_{CH} from

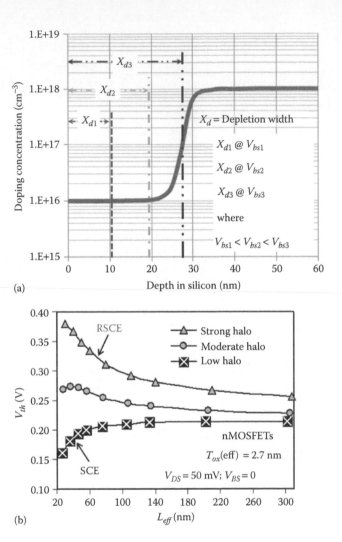

(a)

(b)

FIGURE 5.2
Effect of nonuniform channel doping profile on MOSFET devices: (a) body bias, V_{bs}, dependence of channel depletion widths (X_{d1}, X_{d2}, and X_{d3}) and bulk charge, Q_b, due to nonuniform vertical channel doping profile and (b) channel length dependence of V_{th} due to nonuniform lateral channel doping profile. (Data from S. Saha, *Proc. SPIE Conf.*, 3881, 195–204, 1999.)

the Si/SiO$_2$ interface to a depth X_T and N_{SUB} from X_T to the bottom of the silicon substrate.

With reference to Figure 5.3, let us assume that V_{bx} is the body bias required to fully deplete the region X_T. Then, for the applied body bias V_{bs}, we can show from Equation 5.3

$$V_{th} = V_{TH0} + \gamma_1 \left(\sqrt{\phi_s - V_{bs}} - \sqrt{\phi_s} \right); \quad |V_{bs}| < |V_{bx}| \qquad (5.4)$$

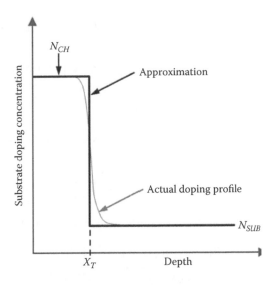

FIGURE 5.3
A typical nonuniform vertical channel doping profile of a MOSFET due to threshold voltage adjust implant approximated to a high–low step profile; N_{CH} and N_{SUB} are the channel doping concentrations at the surface and deep into the substrate, respectively; X_T is the transition depth of doping concentration from the high level to low level.

$$V_{th} = V_{TH0} + \gamma_1 \left(\sqrt{\phi_s - V_{bx}} - \sqrt{\phi_s} \right) + \gamma_2 \left(\sqrt{\phi_s - V_{bs}} - \sqrt{\phi_s - V_{bx}} \right); \quad |V_{bs}| > |V_{bx}| \quad (5.5)$$

It is to be noted that V_{bs} and $V_{bx} < 0$ for n-channel MOSFETs (nMOSFETs) and >0 for p-channel MOSFETs (pMOSFETs). In Equations 5.4 and 5.5, the body effect coefficients γ_1 and γ_2 are given by

$$\gamma_1 = \frac{\sqrt{2qK_{si}\varepsilon_0 N_{CH}}}{C_{ox}} \quad \text{and} \quad \gamma_2 = \frac{\sqrt{2qK_{si}\varepsilon_0 N_{SUB}}}{C_{ox}} \quad (5.6)$$

Equations 5.4 and 5.5 are complex because these require knowledge of the shape of channel doping profile and the exact voltages to deplete different regions of the profile. Therefore, a unified expression for V_{th} is used to model the nonuniform vertical channel doping profile given by [27–29]

$$V_{th} = V_{TH0} + K_1 \left(\sqrt{\phi_s - V_{bs}} - \sqrt{\phi_s} \right) - K_2 V_{bs} \quad (5.7)$$

where K_1 and K_2 are the parameters to model the vertically nonuniform channel doping profile and determined by fitting Equation 5.7 to the measured $I_{ds} - V_{gs}$ data for large geometry devices (e.g., $W/L = 10\ \mu\text{m}/10\ \mu\text{m}$) at low $V_{ds} \approx 50$ mV. The relation between K_1 and K_2 and γ_1 and γ_2 can be determined by solving Equations 5.5 and 5.7 at an intermediate bias $V_{bm} > V_{bx}$. Since Equations 5.5 and 5.7 represent the same V_{th} versus V_{bs} characteristics of a device, at a particular body bias, $V_{bs} = V_{bm}$, we must have the conditions [29]

$$V_{th}\left(\text{Equation 5.5}\right)\Big|_{V_{bs}=V_{bm}} = V_{th}\left(\text{Equation 5.7}\right)\Big|_{V_{bs}=V_{bm}}$$

$$\frac{d}{dV_{bs}}V_{th}\left(\text{Equation 5.5}\right)\Big|_{V_{bs}=V_{bm}} = \frac{d}{dV_{bs}}V_{th}\left(\text{Equation 5.7}\right)\Big|_{V_{bs}=V_{bm}}$$

Using the above conditions, we can show

$$\gamma_1\left(\sqrt{\phi_s - V_{bx}} - \sqrt{\phi_s}\right) + \gamma_2\left(\sqrt{\phi_s - V_{bm}} - \sqrt{\phi_s - V_{bx}}\right) = K_1\left(\sqrt{\phi_s - V_{bm}} - \sqrt{\phi_s}\right) - K_2 V_{bm} \quad (5.8)$$

$$-\frac{\gamma_2}{2\sqrt{\phi_s - V_{bm}}} = -\frac{K_1}{2\sqrt{\phi_s - V_{bm}}} - K_2 \quad (5.9)$$

Solving Equations 5.8 and 5.9 simultaneously, we can show that

$$K_1 = \gamma_2 - \frac{2\left(\gamma_1 - \gamma_2\right)\left(\sqrt{\phi_s - V_{bx}} - \sqrt{\phi_s}\right)\left(\sqrt{\phi_s - V_{bm}}\right)}{2\sqrt{\phi_s}\left(\sqrt{\phi_s - V_{bm}} - \sqrt{\phi_s}\right) + V_{bm}};$$

$$(5.10)$$

$$K_2 = \frac{\left(\gamma_1 - \gamma_2\right)\left(\sqrt{\phi_s - V_{bx}} - \sqrt{\phi_s}\right)}{2\sqrt{\phi_s}\left(\sqrt{\phi_s - V_{bm}} - \sqrt{\phi_s}\right) + V_{bm}}$$

If K_1 and K_2 are not given, they can be computed from Equation 5.10 using the channel doping concentration [27–29]. From Equation 5.6, we note that for a conventional high–low channel doping profile, $\gamma_1 > \gamma_2$; then, from Equation 5.10, the value of $K_2 > 0$. And, therefore, the devices with conventional high–low channel doping profile are sensitive to strong body bias. On the other hand, in the devices with low–high channel doping profile, $\gamma_1 < \gamma_2$; therefore, from Equation 5.10, $K_2 < 0$ and the devices are less sensitive to strong body bias.

5.2.1.2 Threshold Voltage Modeling for Nonuniform Lateral Channel Doping Profile

In advanced CMOS technologies, localized high doping concentration regions of the same doping type as the channel are used near the source and drain ends of the channel [3,4,23–26] to suppress SCEs [5,6]. This localized concentration of additional channel doping near the source and drain ends of the channel is referred to as the *halo* or *pocket* doping as shown in Figure 5.1a and is a critical technology parameter for optimizing the performance of MOSFET devices. Due to halo doping, the channel doping profile becomes laterally nonuniform with higher concentration at the source and drain ends and low concentration near the channel as shown in Figure 5.1b. This nonuniform lateral channel doping profile can also be modeled by two step functions as shown in Figure 5.4.

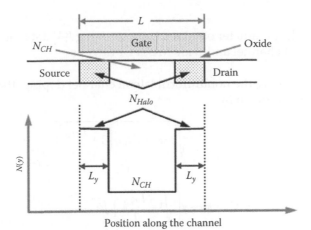

FIGURE 5.4
Nonuniform lateral channel doping profile due to halo/pocket implant approximated by two step profiles at the source and drain ends overlapping the gate; N_{Halo} and N_{CH} are the halo and channel doping concentrations of the same type of dopants, respectively; L_y is the width of the halo doping concentration inside the channel region.

In Figure 5.4, L is the channel length, L_y is the length of each halo region, N_{Halo} is the uniform concentration in each halo region L_y, and N_{CH} is the uniform concentration in the region $L - 2L_y$. If N_{eff} is the average channel doping concentration, then the channel charge per unit area is given by

$$Q = qN_{eff}L = \left[qN_{CH}L + q\left(N_{Halo} - N_{CH}\right)\left(2L_y\right) \right] \tag{5.11}$$

After simplifying Equation 5.11, we can show that the effective channel concentration due to lateral channel doping profile is given by

$$N_{eff} = N_{CH}\left[1 + 2L_y\left(\frac{N_{Halo} - N_{CH}}{N_{CH}} \right)\frac{1}{L} \right]$$

$$= N_{CH}\left[1 + \frac{L_{PE0}}{L} \right] \tag{5.12}$$

where L_{PE0} is defined as

$$L_{PE0} = 2L_y \frac{N_{Halo} - N_{CH}}{N_{CH}} \tag{5.13}$$

L_{PE0} is a model parameter and is obtained by optimizing *I–V* data at $V_{bs} = 0$ for short channel and wide MOSFET devices. Similarly, we can show that the effective channel doping concentration with applied V_{bs} is

$$N_{eff}(V_{bs}) = N_{CH}\left[1 + \frac{L_{PEB}}{L} \right] \tag{5.14}$$

where:

L_{PEB} is a model parameter obtained by optimizing *I–V* data for different V_{bs} for short channel and wide MOSFET devices

Now, by introducing the nonuniform lateral channel doping, the expression for V_{th} at $V_{bs} = 0$ is given by

$$
\begin{aligned}
V'_{TH0} &= V_{fb} + \phi_s + \frac{\sqrt{2qK_{si}\varepsilon_0 N_{CH}\left(1+\dfrac{L_{PE0}}{L}\right)}}{C_{ox}}\sqrt{\phi_s} \\[2ex]
&= V_{fb} + \phi_s + \gamma\left(\sqrt{1+\frac{L_{PE0}}{L}}\right)\sqrt{\phi_s} \\[2ex]
&= V_{fb} + \phi_s + \gamma\sqrt{\phi_s} + \gamma\left(\sqrt{1+\frac{L_{PE0}}{L}}-1\right)\sqrt{\phi_s} \\[2ex]
&= V_{TH0} + \gamma\left(\sqrt{1+\frac{L_{PE0}}{L}}-1\right)\sqrt{\phi_s}
\end{aligned}
\tag{5.15}
$$

where we have added and subtracted $\gamma\sqrt{\phi_s}$ and used Equation 5.1 at $V_{bs} = 0$ along with $\gamma = \sqrt{2qK_{si}\varepsilon_0 N_{CH}}/C_{ox}$ to obtain Equation 5.15 for threshold voltage at $V_{bs} = 0$ due to the halo doping. Similarly, the expression for V_{th} at V_{bs} with halo doping can be shown as

$$
V_{th} = V'_{TH0} + \gamma\left(\sqrt{\phi_s - V_{bs}} - \sqrt{\phi_s}\right)\cdot\left(\sqrt{1+\frac{L_{PEB}}{L}}\right)
\tag{5.16}
$$

Then combining Equations 5.15 and 5.16, we get the general expression for threshold voltage for modeling nonuniform lateral channel doping profile of MOSFETs as

$$
V_{th} = V_{TH0} + \gamma\left(\sqrt{\phi_s - V_{bs}} - \sqrt{\phi_s}\right)\cdot\left(\sqrt{1+\frac{L_{PEB}}{L}}\right) + \gamma\left(\sqrt{1+\frac{L_{PE0}}{L}}-1\right)\sqrt{\phi_s}
\tag{5.17}
$$

Now, considering the vertically nonuniform channel doping profile and K_1 and K_2 parameters, we can write the combined expression for substrate doping effect on V_{th} as

$$
V_{th} = V_{TH0} + K_1\left(\sqrt{\phi_s - V_{bs}} - \sqrt{\phi_s}\right)\cdot\left(\sqrt{1+\frac{L_{PEB}}{L}}\right) + K_1\left(\sqrt{1+\frac{L_{PE0}}{L}}-1\right)\sqrt{\phi_s} - K_2 V_{bs}
\tag{5.18}
$$

Thus, the effect of nonuniform substrate doping introduces the set of compact model parameters $\{K_1, K_2, L_{PE0}, L_{PEB}\}$. In Equation 5.18,

- K_1 and K_2 model the effect of nonuniform vertical channel doping profile on V_{th};
- L_{PE0} and L_{PEB} model the nonuniform lateral channel doping profile on V_{th} at $V_{bs} = 0$ and $|V_{bs}| > 0$, respectively;
- At $V_{bs} = 0$, as L decreases, Equation 5.18 shows that V_{th} increases, showing reverse short channel effect due to halo-doping profile as shown in Figure 5.2b.

In long channel devices, halo/pocket implant causes a significant *drain-induced threshold voltage shift* (DITS) [30,31]. The applied V_{ds} reduces the drain barrier in the long channel MOSFET devices with halo implant. For large V_{ds}, the shift ΔV_{th}(DITS) due to DITS is given by [30]

$$\Delta V_{th}(\text{DITS}) \cong nv_{kT}.\ln\left(\frac{L}{L + \text{DVTP0}.\left(1 + e^{-\text{DVTP1}.V_{ds}}\right)}\right) \qquad (5.19)$$

where:

  DVTP0 and *DVTP1* are fitting parameters

  nv_{kT} depends on subthreshold slope as discussed in Chapter 4

5.2.2 Small Geometry Effect on Threshold Voltage Model

The MOSFET threshold voltage is sensitive to both the channel length (Figure 5.2b) and the channel width [32]. Experimental data show that V_{th} decreases with the decrease of channel length called SCE whereas it increases with the decrease of channel width referred to as the narrow width effect (NWE). Therefore, it is critical to determine the shift in the long channel threshold voltage due to SCE and NWE and develop an expression for threshold voltage that accurately models the nanoscale device technology for circuit CAD. In Sections 5.2.2.1 and 5.2.2.2, we will develop mathematical expressions of the shift in the threshold voltage due to small geometry effects and present a threshold voltage expression to model all geometries in an advanced technology.

5.2.2.1 Threshold Voltage Model for Short Channel MOSFET Devices

For short channel devices, SCE or the decrease in V_{th} with the decreases in L is caused by the *bulk-charge sharing* between the gate and S/D *pn*-junctions as shown in Figure 5.5. Figure 5.5 shows that a significant amount of the bulk charge Q_b near the source and drain ends is controlled by reversed bias S/D *pn*-junctions. As a result, gate-induced Q_b decreases as channel length decreases (i.e., less V_{gs} is used to induce the same amount of Q_b). Since $Q_s = Q_b + Q_i$, for the same V_{gs}, Q_i increases as the devices are scaled down. Thus, less gate voltage is required to turn on the device, causing V_{th} decrease

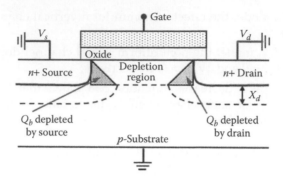

FIGURE 5.5
Short channel effect in MOSFETs caused by bulk-charge sharing by the gate and S/D *pn*-junctions; a significant part of channel depletion is caused by S/D regions; the source and drain each contributes an amount of channel charge Q_b to the total channel charge in silicon; X_d is the width of the S/D depletion region at zero bias condition.

as channel length decreases. The physics of SCE can be understood by a simple mathematical model based on charge sharing [32,33]. However, this model is not suitable for circuit CAD. Therefore, compact models are developed to calculate the shift in V_{th} due to SCE for circuit CAD.

Again, for a particular $V_{gs} > V_{th}$, as the drain voltage increases, the depletion region near the drain end of the channel gradually increases and extends toward the source end of the channel. As a result, the potential barrier to the inversion charge near the source end is reduced so that more carriers are injected from the source to the channel as V_{ds} increases. Thus, for a particular value of V_{gs} more inversion charges are injected as L decreases. This is referred to as the DIBL, causing V_{th} fall with the increase in V_{ds} as shown in Figure 5.6.

In order to model SCE, we solve Poisson's equation in the y direction along the channel. It can be shown that the shift in V_{th} due to SCE and DIBL is given by [34]

$$\Delta V_{th}\left(\text{SCE, DIBL}\right) = -\theta_{th}\left(L_{eff}\right)\left[2\left(V_{bi} - \phi_s\right) + V_{ds}\right] \tag{5.20}$$

where

$$\theta_{th}(L_{eff}) = \frac{1}{2\left[\cosh\left(L_{eff}/2l_t\right) - 1\right]} \tag{5.21}$$

V_{bi} is the built-in potential of the S/D *pn*-junctions and is given by (Equation 2.84)

$$V_{bi} = v_{kT} \ln\left(\frac{N_{CH}N_{SD}}{n_i^2}\right) \tag{5.22}$$

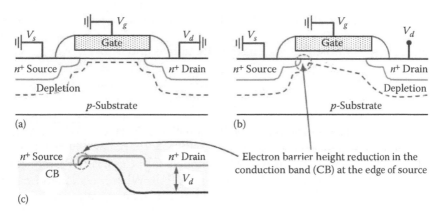

FIGURE 5.6
Short channel effect in MOSFETs due to drain voltage V_{ds}–DIBL in an n-channel device: (a) $V_{gs} = 0$ and $V_{ds} = 0$, (b) $V_{gs} = 0$ and $V_{ds} =$ supply voltage, V_{dd}, and (c) plot of conduction bands along the length of the device under zero bias (top curve) and at drain bias conditions (bottom curve).

where N_{CH} and N_{SD} are the effective channel and S/D doping concentrations, respectively, and l_t represents the characteristic length given by

$$l_t = \sqrt{\frac{K_{si}T_{ox}W_d}{K_{ox}\eta}} \qquad (5.23)$$

With depletion width $W_d = \sqrt{2K_{si}\varepsilon_0\left(\phi_s - V_{bs}\right)/qN_{CH}}$ and η (ETA) is a fitting parameter so that W_d/η = average width of the depletion region along the length of the channel.

Equation 5.20 shows that ΔV_{th} depends linearly on V_{ds} showing that V_{th} decreases as V_{ds} increases due to DIBL. In order to improve modeling flexibility for different technologies, different model parameters are introduced to get

$$\theta_{th}(\text{SCE}) = \frac{0.5\text{DVT0}}{\cosh\left(\text{DVT1}.L_{eff}/l_t\right) - 1} \qquad (5.24)$$

$$\Delta V_{th}(\text{SCE}) = -\theta_{th}(\text{SCE})\left(V_{bi} - \phi_s\right) \qquad (5.25)$$

$$l_t = \sqrt{\frac{K_{si}T_{ox}W_d}{K_{ox}}}\left(1 + \text{DVT2}.V_{bs}\right) \qquad (5.26)$$

Similarly, the shift in threshold voltage due to DIBL is described by

$$\theta_{th}(\text{DIBL}) = \frac{0.5}{\cosh\left(\text{DSUB}.L_{eff}/l_{t0}\right) - 1} \qquad (5.27)$$

$$\Delta V_{th}(\text{DIBL}) = -\theta_{th}(\text{DIBL}) \cdot (\text{ETA0} + \text{ETAB} \cdot V_{bs}) \cdot V_{ds} \tag{5.28}$$

where l_{t0} is the characteristics length without bias and is given by

$$l_{t0} = \sqrt{\frac{K_{si} T_{ox} W_{d0}}{K_{ox}\eta}} \tag{5.29}$$

with zero bias depletion width is given by (Equation 2.97), $W_{d0} = \sqrt{2K_{si}\varepsilon_0\phi_s/qN_{CH}}$.

In summary, the parameters used in BSIM4 (Berkeley Short Channel IGFET Model, version 4) compact MOS models for *SCE* and *DIBL* modeling are *DVT0*, *DVT1*, *DVT2*, *DSUB*, *ETA0*, and *ETAB*. Where, *DVT2* and *ETAB* account for substrate bias effect on *SCE* and *DIBL*, respectively.

5.2.2.2 Threshold Voltage Modeling for Narrow Channel MOSFET Devices

In addition to channel length effect on V_{th}, narrow channel widths also affect V_{th}. These effects can be understood physically with reference to local oxidation of silicon (LOCOS) isolation process in CMOS technology as shown in Figure 5.7. LOCOS isolation process has been used prior

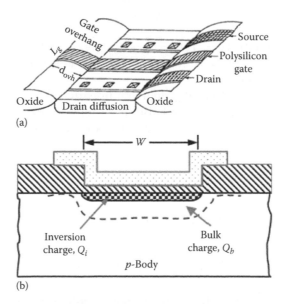

(a)

(b)

FIGURE 5.7

Narrow channel effect in MOSFETs: (a) gate overlap over isolation oxide and (b) additional bulk charge, Q_b, controlled by gate bias due to gate overlap.

to shallow trench isolation (STI) techniques used in advanced CMOS technology.

Figure 5.7b shows two-dimensional (2D) cross-section of a MOSFET device along the channel width direction from the layout shown in Figure 5.7a. As shown in Figure 5.7b, the depletion layer does not abruptly change from deep to shallow at the edge of gate oxide. Therefore, there is a transition region and some spreading of field lines outside W. Thus, the gate charge Q_g supports some charge outside W. Since $Q_g = Q_b + Q_i$, for the same gate bias, Q_b is higher for narrow devices (i.e., gate is required to induce more Q_b out of the same Q_g), resulting in lower Q_i and consequently higher V_{th}.

For STI devices, the fringing field from the gate regions beyond the channel edges support channel depletion charges. This fringing field causes deeper depletion resulting in higher band bending and, therefore, an increase in the surface potential ϕ_s near the STI channel edge. The higher ϕ_s induces channel inversion near the STI at a lower V_{gs} than the rest of the channel. Thus, it takes lower effective V_{gs} to reach maximum channel depletion and the formation of inversion layer. Since the percentage contribution of the fringing field increases as the channel width W decreases, V_{th} tends to decrease with decreasing W in MOSFETs using STI technology (in contrast to LOCOS isolation technology), resulting in *inverse NWE*.

The physics of NWE can be understood by charge-sharing model similar to SCE [32]. However, these models are not suitable for compact modeling of billions of transistors in a VLSI circuit. Besides, NWE depends on the isolation technology. Therefore, universally accurate physical model is not available. For compact modeling, an empirical model can be developed, based on the observation of NWE from the experimental data. We know that V_{th} is directly proportional to gate oxide thickness T_{ox} and surface potential ϕ_s and experimentally it is found that V_{th} is inversely proportional to the channel width, W; therefore, for a long channel device, the shift in V_{th} due to NWE can be expressed as

$$\Delta V_{thW} \propto \frac{T_{ox}}{W_{eff}} \phi_s$$

or (5.30)

$$\Delta V_{thW} = K_3 \frac{T_{ox}}{W_{eff}} \phi_s$$

where:
K_3 is the constant of proportionality and is a W-dependent model parameter extracted from the measurement data
W_{eff} is the effective channel width

In order to model V_{bs} dependence of NWE, a model parameter K_{3B} can be used. Thus, Equation 5.30 can be expressed to include the effect of body bias as

$$\Delta V_{thW} = \left(K_3 + K_{3B}V_{bs}\right)\frac{T_{ox}}{W'_{eff} + W_0}\phi_s \qquad (5.31)$$

Thus, three fitting parameters K_3, K_{3B}, and W_0 are required to model NWE. Here W'_{eff} is the effective channel width with the additional fitting parameter W_0 for accurate fitting of the measured data. Equation 5.31 models NWE in MOSFETs; however it does not model SCE of the narrow devices. In order to model SCE in narrow channel devices, we use Equations 5.24 and 5.25 to obtain the shift in V_{th} for narrow- and short channel devices as

$$\Delta V_{thWL} = \frac{0.5DVT0W}{\cosh\left(DVT1W.L_{eff}W_{eff}/l_{tw}\right)-1}\left(V_{bs}-\phi_s\right) \qquad (5.32)$$

where:

$l_{tw} = \sqrt{K_{si}T_{ox}W_d/K_{ox}}\left(1+DVT2W.V_{bs}\right)$ is the characteristic length for short and narrow devices

Equation 5.32 models V_{th} shift in the short- and narrow channel devices whereas Equation 5.31 models that in narrow- and long channel devices. The final expression for V_{th} including nonuniform substrate concentration and small geometry effects is given by

$$V_{th} = V_{th}(non-uniform\ substarate) + \Delta V_{th}(NWE) + \Delta V_{th}(NWE, SCE)$$
$$+ \Delta V_{th}(SCE) + \Delta V_{th}(DIBL) + \Delta V_{th}(DITS) \qquad (5.33)$$

Thus, combining Equations 5.18, 5.19, 5.25, 5.28, 5.31, and 5.32, we can show the expression for V_{th} as used in BSIM4 model for circuit CAD.

$$V_{th} = V_{TH0} + K_{1ox}\left(\sqrt{\phi_s - V_{bseff}} - \sqrt{\phi_s}\right).\left(\sqrt{1+\frac{L_{PEB}}{L_{eff}}}\right)$$

$$+ K_{1ox}\left(\sqrt{1+\frac{L_{PE0}}{L_{eff}}} - 1\right)\sqrt{\phi_s} - K_{2ox}V_{bseff}$$

$$+ \left(K3 + K3B.V_{bseff}\right)\frac{T_{ox}}{W'_{eff} + W_0}\phi_s \qquad (5.34)$$

$$- \frac{1}{2}\left[\frac{DVT0W}{\cosh\left[DVT1W\left(L_{eff}.W_{eff}/l_{tw}\right)\right]-1} + \frac{DVT0}{\cosh\left[DVT1\left(L_{eff}/l_t\right)\right]-1}\right]\left(V_{bseff}-\phi_s\right)$$

$$- \frac{1}{2}\frac{\left(EAT0+EATB.V_{bseff}\right)}{\cosh\left(DSUB\left(L_{eff}/l_{t0}\right)\right)-1}.V_{ds} - nv_{kT}.\ln\left[\frac{L_{eff}}{L_{eff}+DVTP0\left(1+e^{-DVTP1.V_{ds}}\right)}\right]$$

where we have used the effective channel length (L_{eff}) and effective channel width (W_{eff}) in Equation 5.34. Equation 5.34 shows the overall V_{th} expression for MOSFET devices to accurately model geometry and substrate doping dependence on device performance. In real CAD implementation the following modifications are made [28].

1. Electrical oxide thickness, TOXE dependence is introduced in model parameters $K1$ and $K2$ to improve scalability of V_{th} model over TOXE as

$$K_{1ox} = K1 \cdot \frac{TOXE}{TOXM}$$

and (5.35)

$$K_{2ox} = K2 \cdot \frac{TOXE}{TOXM}$$

where:
TOXM is the model parameter required to fit device characteristics

2. In order to set a lower bound for the body bias during circuit simulations to prevent occurrence of unreasonable values during iterations in CAD environment, V_{bs} is implemented as [28]

$$V_{bseff} = V_{bc} + \frac{1}{2} \cdot \left[(V_{bs} - V_{bc} - \delta_1) + \sqrt{(V_{bs} - V_{bc} - \delta_1)^2 - 4\delta_1 \cdot V_{bc}} \right] \quad (5.36)$$

where $\delta_1 = 1$ mV and V_{bc} is the maximum allowable V_{bs} and found from $dV_{th}/dV_{bs} = 0$ to be

$$V_{bc} = 0.9 \left(\phi_s - \frac{K1^2}{4K2^2} \right) \quad (5.37)$$

For positive V_{bs}, there is need to set an upper bound for the body bias as [28]

$$V_{bseff} = 0.95\phi_s - \frac{1}{2} \cdot \left[(0.95\phi_s - V'_{bseff} - \delta_1) + \sqrt{(0.95\phi_s - V'_{bseff} - \delta_1)^2 - 4\delta_1 \cdot 0.95\phi_s} \right] \quad (5.38)$$

Effective Channel Length and Width: The effective channel length (L_{eff}) and width (W_{eff}) used in Equation 5.34 are given by

$$L_{eff} = L_{drawn} - 2\Delta L \quad (5.39)$$

$$W_{eff} = W_{drawn} - 2\Delta W \quad (5.40)$$

where ΔL and ΔW are model parameters that include S/D overlap under the gate and poly overlap along the width direction, respectively, and are given by

$$\Delta L = L_{INT} + \frac{L_L}{L^{L_{LN}}} + \frac{L_W}{L^{L_{WN}}} + \frac{L_{WL}}{L^{L_{LN}} L^{L_{WN}}} \tag{5.41}$$

$$\Delta W = W_{INT} + DWG.V_{gsteff} + DWB\left(\sqrt{\phi_s - V_{bseff}} - \sqrt{\phi_s}\right)$$

$$+ \frac{W_L}{L^{W_{LN}}} + \frac{W_W}{W^{W_{WN}}} + \frac{W_{WL}}{L^{W_{LN}} W^{W_{WN}}} \tag{5.42}$$

where L_{INT}, W_{INT}, DWG, and DWB are extracted from the measured data. Other parameters in Equations 5.41 and 5.42 are fitting parameters to improve the modeling accuracy (and rarely used). In Equation 5.42, V_{gsteff} is the effective value of $(V_{gs} - V_{th})$ used to ensure the channel charge continuity at the weak and strong inversion regions in the regional model. V_{gsteff} is obtained by equating channel charge at the weak inversion and at strong inversion at the transition point.

5.3 Drain Current Model

The total current density (J) in a MOSFET is the sum total of the *electron* and *hole* current densities J_n and J_p, respectively. And, the total J_n and J_p are the sum of the drift component of the respective carriers due to electric field E and diffusion component of the respective carriers due to the concentration gradient along the channel as discussed in Chapter 4 (Section 4.4) and is given by

$$J_n = qn\mu_n E + qD_n \nabla n$$

$$J_p = qp\mu_p E - qD_p \nabla p \tag{5.43}$$

where:
  q is the electronic charge
  n and p are the electron and hole concentrations, respectively
  ∇n and ∇p are the electron and hole concentration gradient, respectively
  μ_n and μ_p are the electron and hole surface mobility, respectively

The accuracy of MOSFET drain current model depends on the accuracy of inversion layer mobility model. Therefore, in the following section, we will derive the surface mobility model used in circuit CAD for small geometry MOSFETs.

5.3.1 Surface Mobility Model

In Chapter 4, we have assumed a constant surface mobility, μ_s for modeling MOSFET drain current, I_{ds}. This assumption is not valid under high electric field operation of the devices. As the vertical electric field E_x and lateral electric field E_y increase with increasing gate voltage V_{gs} and drain voltage

V_{ds}, respectively, the inversion carriers suffer increased scattering. Therefore, μ_s strongly depends on E_x and E_y. Let us consider the effect of E_x only on the surface mobility, that is, $V_{ds} \sim 0$. For the simplicity of I_{ds} modeling, let us define an *effective mobility* as the average mobility of carriers given by

$$\mu_{eff} = \frac{\int_0^{X_{inv}} \mu_s(x,y) \cdot n(x,y)dx}{\int_0^{X_{inv}} n(x,y)dx} \tag{5.44}$$

Using the definition of mobility from Equation 5.44 in Equation 4.64, we can write

$$I_{ds} = \frac{W}{L}\mu_{eff} \int_0^{V_{ds}} Q_i dV \tag{5.45}$$

In reality, μ_{eff} is highly reduced by large vertical electric field due to the high applied V_{gs}. The vertical electric field E_x pulls the inversion layer electrons in nMOSFETs toward the surface causing higher surface scattering as well as Coulomb scattering due to the interaction of electrons with oxide charges (Q_f, N_{it}) discussed in Chapter 2. Since the electric field varies vertically through the inversion layer, the average field in the inversion layer is given by

$$E_{eff} = \frac{E_{x1} + E_{x2}}{2} \tag{5.46}$$

where:
  E_{x1} is the vertical electric field at the Si/SiO$_2$ interface
  E_{x2} is the vertical electric field at the channel/depletion layer interface as
      shown in Figure 5.8

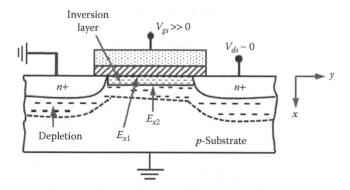

FIGURE 5.8
Effective vertical electric field on MOSFET inversion carriers due to the large applied gate bias V_{gs}; E_{x1} is the vertical electric field at the Si/SiO$_2$ interface and E_{x2} is the vertical electric field at the channel/depletion layer interface.

Now, from Gauss's law we can show that

$$E_{x1} - E_{x2} = \frac{Q_i}{K_{si}\varepsilon_0}$$

and   (5.47)

$$E_{x2} = \frac{Q_b}{K_{si}\varepsilon_0}$$

where:
   Q_i and Q_b are the inversion and bulk-charge densities, respectively, due to
   the applied V_{gs}

Substituting for E_{x1} and E_{x2} from Equation 5.47 into Equation 5.46, we can show

$$E_{eff} = \frac{1}{K_{si}\varepsilon_0}\left(\frac{1}{2}Q_i + Q_b\right)$$ (5.48)

In order to represent both electrons and holes, the general expression for the
effective electric field is expressed as

$$E_{eff} = \frac{1}{K_{si}\varepsilon_0}\left(\eta Q_i + Q_b\right)$$ (5.49)

where:
   the constant $\eta = 1/2$ for electrons and $\eta = 1/3$ for holes [35–37]

The measured μ_{eff} versus E_{eff} plots show a *universal behavior* independent of
doping concentration at high effective vertical electrical fields and dependence on the substrate doping concentration and interface charge at low
effective vertical electric fields as shown in Figure 5.9a.

The experimentally observed universal mobility behavior is due to the
relative contributions of different scattering mechanisms [38,39] set by the
strength of vertical electrical fields as shown in Figure 5.9b. As shown in
Figure 5.9b, μ_{eff} is determined by Coulomb scattering of the ionized impurities and oxide charges, *phonon* scattering due to thermal vibration, and
surface roughness scattering at the Si/SiO$_2$ interface. At high vertical electric
fields, *surface roughness scattering* dominates as the carrier confinement is
close to the interface, resulting in a decrease of μ_{eff} with the increase of E_{eff} as
observed in Figure 5.9a.

The deviation from the *universal behavior* observed in Figure 5.9a, particularly in the heavily doped substrates at low effective electric fields, is
due to the ionized impurity scattering, Coulomb scattering, and phonon
scattering. At low effective vertical electric fields, Q_i is low and $<<Q_b$. As a
result, the ionized impurity scattering and Coulomb scattering by ionized
impurities and oxide charges become dominant scattering mechanisms in

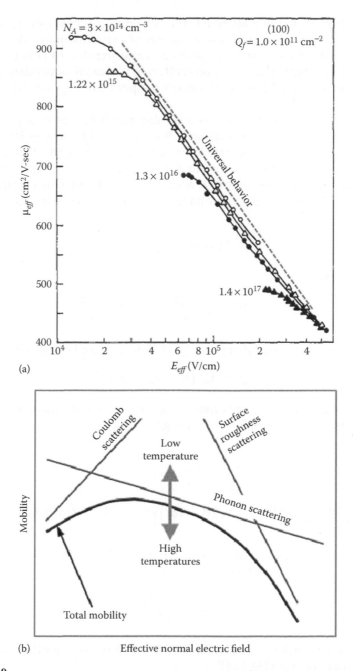

(a)

(b)

FIGURE 5.9
Low field mobility of inversion carriers in MOSFETs: (a) universal mobility behavior of inversion layer electrons in nMOSFET devices (Data from S.C. Sun and J.D. Plummer, *IEEE Trans. Electron Dev.*, 27, 1497–1508, 1980.) and (b) physical mechanisms showing the dependence of the inversion layer mobility on the effective vertical electric field.

the depletion region of a MOSFET device and μ_{eff} becomes strong function of channel doping concentration as observed experimentally. As the effective electric field increases, the phonon scattering due to lattice vibration becomes important. Thus, phonon scattering is weakly dependent on vertical electric fields and has the strongest temperature dependence on μ_{eff} as shown in Figure 5.9b.

The previous physical analysis describes the behavior of μ_{eff} versus E_{eff}. However, we need to develop an effective mobility model that can be used in drain current calculation to account for the vertical field effects on device performance. In order to develop μ_{eff} model for circuit CAD, we substitute the expressions for Q_b and Q_i for a MOSFET in Equation 5.48. For MOSFETs with threshold voltage V_{th} at strong inversion, the inversion charge is given by

$$Q_i = -C_{ox}\left(V_{gs} - V_{th}\right) \tag{5.50}$$

Again, we know,

$$V_{th} = V_{fb} + 2\phi_B + \frac{Q_s}{C_{ox}} \cong V_{fb} + 2\phi_B - \frac{Q_b}{C_{ox}} \tag{5.51}$$

where we have assumed that $Q_s \cong Q_b$. Therefore, from Equation 5.51 we get

$$Q_b = -C_{ox}\left(V_{th} - V_{fb} - 2\phi_B\right) \tag{5.52}$$

Now, substituting the expressions for Q_i and Q_b from Equations 5.50 and 5.51, respectively, in Equation 5.48, we get

$$E_{eff} = \frac{C_{ox}}{K_{si}\varepsilon_0}\left[\frac{V_{gs} - V_{th}}{2} + V_{th} - V_{fb} - 2\phi_B\right]$$

$$= \frac{K_{ox}\varepsilon_0}{T_{ox}} \cdot \frac{1}{2K_{si}\varepsilon_0}\left[V_{gs} + V_{th} - \left(2V_{fb} + 4\phi_B\right)\right] \cong \frac{1}{6T_{ox}}\left[V_{gs} + V_{th} - \left(2V_{fb} + 4\phi_B\right)\right] \tag{5.53}$$

In the above expression, we have used $K_{ox}/K_{si} \cong 1/3$. Typically $\left(V_{gs} + V_{th}\right) >> \left(2V_{fb} + 4\phi_B\right)$; therefore, after simplification of Equation 5.53 we get

$$E_{eff} \cong \frac{V_{gs} + V_{th}}{6T_{ox}} \tag{5.54}$$

Now, we know that the unified formulation of effective mobility is given by the empirical relation [34,40,41]

$$\mu_{eff} = \frac{\mu_0}{\left[1 + \left(E_{eff}/E_0\right)\right]^\nu} \tag{5.55}$$

where:

μ_0 is concentration-dependent surface mobility

E_0 is the critical electric field

v is a constant

Since the parameter $v \ll 1$, we can use Taylor's series expansion of the denominator and neglect the higher order terms to obtain

$$\left(1+\frac{E_{eff}}{E_0}\right)^v = 1 + v\frac{E_{eff}}{E_0} + \frac{v(v-1)}{2!}\left(\frac{E_{eff}}{E_0}\right)^2 + \cdots \qquad (5.56)$$

Now substituting for E_{eff} from Equation 5.54 to the right-hand side of Equation 5.56, we get

$$\left(1+\frac{E_{eff}}{E_0}\right)^v \cong 1 + \frac{v}{E_0}\left(\frac{V_{gs}+V_{th}}{6T_{ox}}\right) + \frac{v(v-1)}{2E_0^2}\left(\frac{V_{gs}+V_{th}}{6T_{ox}}\right)^2 + \cdots$$

$$= 1 + \frac{v}{6E_0}\left(\frac{V_{gs}+V_{th}}{T_{ox}}\right) + \frac{v(v-1)}{72E_0^2}\left(\frac{V_{gs}+V_{th}}{T_{ox}}\right)^2 + \cdots \qquad (5.57)$$

$$= 1 + U_a\left(\frac{V_{gs}+V_{th}}{T_{ox}}\right) + U_b\left(\frac{V_{gs}+V_{th}}{T_{ox}}\right)^2$$

where we have defined $U_a \equiv v/6E_0$ and $U_b \equiv v(v-1)/72E_0^2$ as the model parameters to be extracted from the measured I_{ds} *versus* V_{gs} characteristics of MOSFET devices at low drain bias, V_{ds}. Therefore, combining Equations 5.55 and 5.57, the simplified low lateral field mobility model for MOSFET inversion carriers can be shown as [27,28]

$$\mu_{eff} = \frac{\mu_0}{1 + U_a\left[(V_{gs}+V_{th})/T_{ox}\right] + U_b\left[(V_{gs}+V_{th})/T_{ox}\right]^2} \qquad (5.58)$$

In order to improve the modeling accuracy at high body bias, a term U_cV_{bs} is introduced in the denominator of Equation 5.58 so that

$$\mu_{eff} = \frac{U_0}{1 + (U_a + U_cV_{bs})\cdot\left[(V_{gs}+V_{th})/T_{ox}\right] + U_b\left[(V_{gs}+V_{th})/T_{ox}\right]^2} \qquad (5.59)$$

where:

$U_0 \equiv \mu_0$

The alternative expression to include the body bias dependence on μ_{eff} is

$$\mu_{eff} = \frac{U_0}{1 + \left\{U_a\left[(V_{gs}+V_{th})/T_{ox}\right] + U_b\left[(V_{gs}+V_{th})/T_{ox}\right]^2\right\}(1+U_cV_{bs})} \qquad (5.60)$$

The mobility Equations 5.58 through 5.60 have been derived assuming strong inversion condition. In the strong inversion regime, the inversion carrier mobility is a function of gate bias. In the subthreshold region the accuracy of the mobility is not critical since Q_{inv} varies with V_{gs} and cannot be modeled accurately. Therefore, in subthreshold regime, the mobility is usually modeled as a constant concentration dependent mobility.

To ensure the continuity of the mobility model, BSIM mobility model is modified based on the V_{gsteff} expression to obtain the basic empirical models as [28]

$$\mu_{eff} = \frac{U_0}{1 + \left(U_A + U_C V_{bseff}\right) \cdot \left[\left(V_{gsteff} + 2V_{th}\right)/T_{OX}\right] + U_B\left[\left(V_{gsteff} + 2V_{th}\right)/T_{OX}\right]^2} \tag{5.61}$$

or

$$\mu_{eff} = \frac{U_0}{1 + \left[U_A \cdot \left[\left(V_{gsteff} + 2V_{th}\right)/T_{OX}\right] + U_B\left[\left(V_{gsteff} + 2V_{th}\right)/T_{OX}\right]^2\right]\left(1 + U_C V_{bseff}\right)} \tag{5.62}$$

where:

V_{bseff} is the effective value of body bias to set the upper limit of computation as defined earlier

The BSIM4 model parameter set for the basic mobility model is $\{U_0, U_A, U_B, U_C\}$ and is extracted from the $I_{ds} - V_{gs}$ characteristics at low V_{ds} with body bias. Different options of Equation 5.59 have been implemented in BSIM4 model and readers are encouraged to look at the users' manual to use the appropriate model and extract the appropriate model parameters for circuit CAD [28]. It can be observed from the earlier defined mobility models that μ_{eff} approaches a constant value of U_0 for $V_{gs} < V_{th}$ as used in the subthreshold regime.

The expression for V_{gsteff} is obtained by equating the channel charge of weak and strong inversions at the transition point for model continuity in the entire range of device operation and can be shown as [28]

$$V_{gsteff} = \frac{n v_{kT} \ln\left\{1 + \exp\left[m^*\left(V_{gs} - V_{th}\right)/n v_{kT}\right]\right\}}{m^* + n C_{ox}\sqrt{2\phi_s/q K_{si}\varepsilon_0 N_{CH}}\exp\left\{-\left[\left(1 - m^*\right)\left(V_{gs} - V_{th} - V_{off}\right)/2 n v_{kT}\right]\right\}} \tag{5.63}$$

It should be pointed out that all of the mobility models given earlier account for only the influence of the vertical electrical field due to V_{gs} at low lateral electric field and often referred to as the *low-field mobility* model. The influence of the lateral electric field due to the applied V_{ds} on device performance is modeled in drain current by considering the velocity saturation in MOSFET devices under high lateral electric field.

5.3.2 Subthreshold Region Drain Current Model

The subthreshold current model is the same as derived for the long channel devices in Chapter 4 with minor change for improving the accuracy of data fitting and is given by [27,28]

$$I_{ds} = I_{s0}e^{(V_{gs}-V_{th}-V_{OFF})/nv_{kT}}\left[1-e^{-(V_{ds}/v_{kT})}\right]; \quad V_{gs} < V_{th} \tag{5.64}$$

where V_{OFF} is the model parameter to account for the difference between V_{th} in the strong inversion and the subthreshold region and I_{s0} is given by (see Equation 4.118)

$$I_{s0} = \mu_s\left(W_{eff}/L_{eff}\right)C_d v_{kT}^2 \tag{5.65}$$

In Chapter 4 (Equation 4.127), we have shown that the subthreshold slope is given by

$$S = 2.3nv_{kT} \tag{5.66}$$

where the ideality factor is given by

$$n = 1 + \frac{C_d}{C_{ox}} + \frac{C_{IT}}{C_{ox}} \tag{5.67}$$

In BSIM [27,28] compact models, a parameter called *NFACTOR* is introduced to ensure accurate calculation of C_d and is extracted from the measured data. Again, in short channel devices the surface potential in the channel is determined by both V_{gs} and V_{ds} through the coupling of C_{ox} and C_{dsc} as shown in Figure 5.10. The coupling capacitance $C_{dsc}(L)$ is an exponential function of L. Therefore, in BSIM4 the parameter n is modeled as

$$n = 1 + NFACTOR\frac{C_d}{C_{ox}} + \frac{C_{IT}}{C_{ox}}$$
$$+ \frac{\left(C_{DSC} + C_{DSCD}.V_{ds} + C_{DSCB}.V_{bseff}\right)\left(0.5/\cosh(DVT1.L_{eff}/l_t)-1\right)}{C_{ox}} \tag{5.68}$$

where:

C_{DSC}, C_{DSCD}, and C_{DSCB} are the model parameters that describe the coupling between the channel and the drain

C_{DSCD} and C_{DSCB} represent the drain bias and body bias dependence of channel/drain coupling, respectively

5.3.3 Linear Region Drain Current Model

The high lateral electric field along the channel due to the applied V_{ds} significantly effects device performance. As we observe from Figure 5.11 that for electrons in silicon, the drift velocity v_d saturates near $E \sim 10^4$ V cm^{-1}.

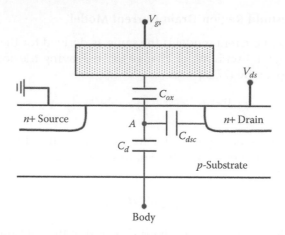

FIGURE 5.10
MOSFET device showing gate capacitance C_{ox}, bulk capacitance C_d, and source drain to channel coupling capacitances C_{dsc}; all the capacitances have an effect on the channel potential and subthreshold conduction.

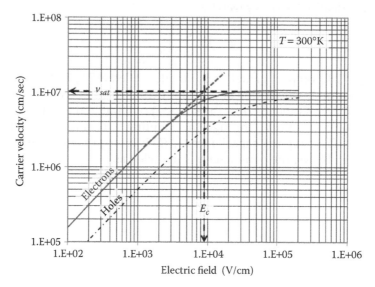

FIGURE 5.11
Drift velocity versus electric field showing carrier velocity saturation in silicon at an electric field near 1×10^4 V cm^{-1}.

As a result, the relation $v_d = \mu E$ does not hold at high electric field. Since average electric field for short channel devices $> 10^4$ V cm^{-1}, small geometry MOSFET devices will operate at $v_d = v_{sat} \cong 1 \times 10^7$ cm sec^{-1}, that is, the saturation velocity of electrons.

We discussed earlier that the mobility is not a constant at high electric field; therefore, we must account for the high lateral electric field effects in

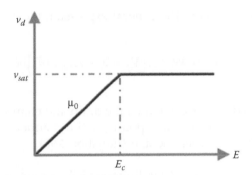

FIGURE 5.12
Drift velocity, v_d versus lateral electrical field, E; piecewise linear mobility behavior of inversion layer electrons due to high E along the channel of MOSFETs; v_{sat}, μ_0, and E_c are the saturation velocity of inversion carriers, concentration-dependent mobility of inversion carriers, and critical electric field at which carrier velocity saturates, respectively.

the expression for I_{ds} derived from simple theory (Chapter 4). Thus, at high electric field along the channel, MOSFET devices operate at a drift velocity, $v_d = v_{sat}$ [11]. Then with reference to Figure 5.11, we assume a v_d versus E piecewise linear model for I–V modeling as shown in Figure 5.12. Thus, at a particular lateral electric field, E_y, we can write [11]

$$v_d = \begin{cases} \dfrac{\mu_{eff}E_y}{1+\left(E_y/E_c\right)}, & \left(E_y < E_c\right) \\[3mm] v_{sat}, & \left(E_y > E_c\right) \end{cases}$$

(5.69)

As shown in Figure 5.12, we assume that v_d saturates abruptly at a critical lateral electric field E_c along the channel.

In Figure 5.12, E_c is the field at which carriers are velocity saturated, that is, at $E_y = E_c$, $v_d = v_{sat}$. Then from Equation 5.69 we can show [11]

$$v_{sat} = \frac{\mu_{eff}E_c}{2}$$

or

(5.70)

$$E_c = \frac{2v_{sat}}{\mu_{eff}}$$

We will use Equation 5.69 to derive linear region drain current expression to account for the high lateral field along the channel due to V_{ds}.

Now, we know that the current density at any point y along the channel in the y direction of an nMOSFET is given by $J_n(y) = nqv(y) = Q_iv(y)$, where n, q, and $v(y)$ are the inversion carrier density, electronic charge, and drift velocity of inversion layer electrons, respectively; $Q_i = nq$ is the inversion carrier charge per unit area. Using the expression for Q_i from Chapter 4,

Equation 4.95, we can write the general expression for drain current in the linear regime as

$$I_{ds} = I(y) = W_{eff}C_{ox}\left[V_{gs} - V_{th} - A_{bulk}V(y)\right]v(y) \tag{5.71}$$

where:
$V(y)$ = potential difference between the drain and channel at y
$v(y)$ is the carrier velocity at any point y in the channel
A_{bulk} is the body effect coefficient, α (Equation 4.96)

Then substituting Equation 5.69 in Equation 5.71, we get for $E_y < E_c$

$$I_{ds} = W_{eff}C_{ox}\left[V_{gs} - V_{th} - A_{bulk}V(y)\right]\frac{\mu_{eff}E(y)}{1+\left[E(y)/E_c\right]} \tag{5.72}$$

After simplification, we can show from (5.72),

$$E(y) = \frac{I_{ds}}{W_{eff}\mu_{eff}C_{ox}\left[V_{gs} - V_{th} - A_{bulk}V(y)\right]-\left(I_{ds}/E_c\right)} = \frac{dV(y)}{dy}$$

$$\text{or} \tag{5.73}$$

$$I_{ds}dy = \left(W_{eff}\mu_{eff}C_{ox}\left[V_{gs} - V_{th} - A_{bulk}V(y)\right]-\frac{I_{ds}}{E_c}\right)dV(y)$$

Integrating Equation 5.73 from ($y = 0$, $V(y) = 0$) to ($y = L_{eff}$, $V(y) = V_{ds}$) and after simplification, we get the linear region ($V_{ds} < V_{dsat}$) current as

$$I_{ds} = \frac{W_{eff}}{L_{eff}\left[1+\left(V_{ds}/L_{eff}E_c\right)\right]}\mu_{eff}C_{ox}\left(V_{gs} - V_{th} - \frac{1}{2}A_{bulk}V_{ds}\right)V_{ds} \tag{5.74}$$

From Equation 5.74 note that the effect of high lateral electric field is the apparent increase in L_{eff} for higher V_{ds}, thus decreasing the linear current. Also, note that Equation 5.74 is valid when parasitic S/D series resistance, $R_{ds} = 0$. For $R_{ds} > 0$, the drain current is modified as [28]

$$I_{ds} = \frac{I_{ds0}}{1+\left(R_{ds}I_{ds0}/V_{ds}\right)} \tag{5.75}$$

where:
I_{ds0} is the drain current at $R_{ds} = 0$ and is given by Equation 5.74

5.3.4 Saturation Region Drain Current Model

Let us assume that V_{dsat} is the drain saturation voltage at which the inversion carriers attain saturation velocity v_{sat}, that is, at $E_y = E_c$. Using the condition,

$v(y) = v_{sat}$ at $E_y = E_c$, in Equation 5.71, we get the saturation region ($V_{ds} \geq V_{dsat}$) drain current as

$$I_{ds} = W_{eff}C_{ox}\left(V_{gs} - V_{th} - A_{bulk}V_{dsat}\right)v_{sat} \qquad (5.76)$$

Using the expression for v_{sat} from Equation 5.70, we get from Equation 5.76 an alternate expression for drain current in the saturation region of MOSFETs as

$$I_{ds} = \frac{1}{2}W_{eff}\mu_{eff}C_{ox}\left(V_{gs} - V_{th} - A_{bulk}V_{dsat}\right)E_c \qquad (5.77)$$

Again, Equations 5.76 and 5.77 are valid when $R_{ds} = 0$ and must be modified for $R_{ds} > 0$.

In order to derive the expression for saturation drain voltage V_{dsat}, we recognize that I_{ds} given by Equations 5.74 and 5.77 must be continuous at $V_{ds} = V_{dsat}$; therefore, equating Equation 5.74 to Equation 5.77, we get

$$\frac{W_{eff}}{L_{eff}\left[1 + \left(V_{dsat}/L_{eff}E_c\right)\right]}\mu_{eff}C_{ox}\left(V_{gs} - V_{th} - \frac{1}{2}A_{bulk}V_{dsat}\right)$$

$$= \frac{1}{2}W_{eff}\mu_{eff}C_{ox}\left(V_{gs} - V_{th} - A_{bulk}V_{dsat}\right)E_c \qquad (5.78)$$

or

$$\frac{2}{E_cL_{eff} + V_{dsat}}\left(V_{gs} - V_{th} - \frac{1}{2}A_{bulk}V_{dsat}\right)V_{dsat} = \left(V_{gs} - V_{th} - A_{bulk}V_{dsat}\right)$$

After simplification of Equation 5.78, we can show

$$V_{dsat} = \frac{E_cL_{eff}\left(V_{gs} - V_{th}\right)}{A_{bulk}E_cL_{eff} + \left(V_{gs} - V_{th}\right)} \qquad (5.79)$$

For $R_{ds} > 0$, V_{dsat} is higher than that given by Equation 5.79 and can be calculated from Equations 5.75 and 5.77 with two model parameters, $A1$ and $A2$, to account for the nonsaturating effect of I–V characteristics [28,41].

The I_{dsat} model described in Equations 5.76 and 5.77 must be corrected for output resistance, R_{out}, due to (1) CLM, (2) DIBL, and (3) substrate current–induced body effect (SCBE).

5.3.5 Bulk-Charge Effect

When V_{ds} is large and/or when the channel length is long, the depletion region thickness of the channel is not uniform along the channel length. This will cause V_{th} to vary along the channel. This effect is called the bulk-charge effect as discussed in Chapter 4, defining the parameter called, α (Equation 4.96). In BSIM4, the parameter A_{bulk} is used to model the bulk-charge effect including both short channel effects and narrow channel effects and is given by

$$A_{bulk} = \left\{ 1 + F\_doping \cdot \left[\begin{array}{c} \dfrac{A0 \cdot L_{eff}}{L_{eff} + 2\sqrt{XJ \cdot X_{dep}}} \left[1 - AGS \cdot V_{gsteff} \left(\dfrac{L_{eff}}{L_{eff} + 2\sqrt{XJ \cdot X_{dep}}} \right)^2 \right] \\ + \dfrac{B0}{W'_{eff} + B1} \end{array} \right] \right\} \cdot \dfrac{1}{1 + KETA \cdot V_{bseff}} \quad (5.80)$$

where, F_doping models nonuniform doping profiles and is given by

$$F\_doping = \dfrac{\sqrt{1 + \left(LPEB/L_{eff} \right) K_{1ox}}}{2\sqrt{\phi_s - V_{bseff}}} + K_{1ox} - K3B \dfrac{TOXE}{W'_{eff} + W_0} \quad (5.81)$$

where:
K_{1ox} and K_{2ox} are defined in Equation 5.35

In Equation 5.80, the model parameters introduced to characterize A_{bulk} are A0, AGS, B0, B1, and KETA. These parameters are extracted from the measured I–V data. It is found that the value of A_{bulk} increases with the increase in L and approaches 1 for shorter devices. This is due to the fact that for short channel devices, the depletion width is almost uniform from source to drain, whereas for long channel devices the depletion near the drain end is much wider than that near the source end of the channel.

5.3.6 Output Resistance

I_{ds}–V_{ds} plot along with the output resistance (R_{out}), which is reciprocal of its first-order derivative, is shown in Figure 5.13 [27,28]. As shown in Figure 5.13, the behavior of R_{out} is characterized by four separate regions based on different physical mechanisms. These regions are (1) triode or linear, (2) CLM, (3) DIBL, and (4) SCBE. Three mechanisms CLM, DIBL, and SCBE affect R_{out} in the saturation region; however, each of them dominates in one of the three distinct regions as shown in Figure 5.13.

We know that I_{ds} depends on both V_{gs} and V_{ds}, and from Figure 5.13, we find that I_{ds} is weakly dependent on V_{ds} in the saturation region (CLM and DIBL).

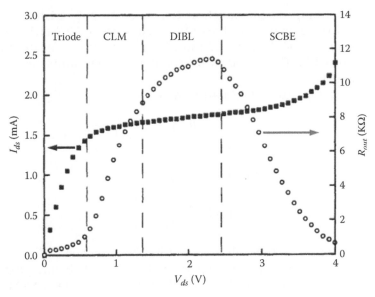

FIGURE 5.13
MOSFET output characteristics: drain current, I_{ds}, and output resistance, R_{out}, of an nMOS-FET device divided into different operating regions based on different physical mechanisms. (Reproduced with permission from J.H. Huang et al., *International Electron Devices Meeting 1992, Technical Digest*, pp. 569–572, IEEE, 1992. Copyright 1992 IEEE.)

Since the saturation region I_{ds} depends weakly on V_{ds}, we can use Taylor series expansion of I_{ds} @ $V_{ds} = V_{dsat}$ and neglect the higher order terms to get

$$I_{ds}\left(V_{gs},V_{ds}\right) = I_{ds}\left(V_{gs},V_{dsat}\right) + \frac{dI_{ds}\left(V_{gs},V_{ds}\right)}{dV_{ds}}\left(V_{ds} - V_{dsat}\right)$$

$$= I_{dsat}\left(1 + \frac{1}{I_{dsat}}\frac{dI_{ds}\left(V_{gs},V_{ds}\right)}{dV_{ds}}\left(V_{ds} - V_{dsat}\right)\right) \qquad (5.82)$$

$$= I_{dsat}\left(1 + \frac{V_{ds} - V_{dsat}}{V_A}\right)$$

where I_{dsat} and V_A are given by

$$I_{dsat} = I_{ds}\left(V_{gs},V_{dsat}\right)$$

$$V_A = I_{dsat}\left(\frac{dI_{ds}}{dV_{ds}}\right)^{-1} \qquad (5.83)$$

In Equation 5.83, the expression for I_{dsat} is given by Equations 5.76 and 5.77. In Equation 5.82, V_A is called the *early voltage* (following the original term used in describing bipolar junction transistor output resistance) and is introduced for the analysis of the output resistance of MOSFET devices in the saturation region. In order to model V_A, we have to consider the contributions of CLM, DIBL, and SCBE components on output resistance as described next:

The early voltage due to CLM is given by

$$V_{ACLM} = I_{dsat} \left(\frac{dI_{ds}}{dL} \cdot \frac{dL}{dV_{ds}} \right)^{-1} = C_{clm} \cdot \left(V_{ds} - V_{dsat} \right) \tag{5.84}$$

The early voltage due to DIBL is given by

$$V_{ADIBL} = I_{dsat} \left(\frac{dI_{ds}}{dV_{th}} \cdot \frac{dV_{th}}{dV_{ds}} \right)^{-1} \tag{5.85}$$

The early voltage due to SCBE is caused by the reduction of V_{th} due to the substrate current induced forward biasing of the source/channel pn-junction (as discussed in Section 5.4). Therefore, the early voltage due to SCBE can be defined as

$$V_{ASCBE} = I_{dsat} \left(\frac{dI_{ds}}{dV_{ds}} \right)^{-1} = \frac{L_{eff}}{PSCBE2} \exp\left(\frac{PSCBE1.l_c}{V_{ds} - V_{dsat}} \right) \tag{5.86}$$

where:

PSCBE1 and PSCBE2 are model parameters extracted from $I_{ds} - V_{ds}$ plots in the saturation region

l_c is the characteristic length of the impact ionization region at the drain-end of MOSFETs

In addition, for long channel devices with halo implant we have to consider the component of early voltage, V_{ADITS} due to DITS.

5.3.7 Unified Drain Current Equation

In the regional modeling approach, separate model expressions for each region of MOSFET device operation such as subthreshold and strong inversion as well as the linear and saturation regions are developed. Although these expressions can accurately describe device behavior within their own respective region of operation, problems are likely to occur in the transition region between two well-described regions. In order to address this persistent problem, a unified model should be synthesized to preserve the region-specific accuracy and to ensure continuity of current and conductance and their derivatives in all transition regions.

In order to ensure this continuity, a unified current expression based on continuous channel charge and mobility is used in BSIM4 model. Thus, a single I–V equation is obtained and is given by [28]

$$I_{ds} = \frac{I_{ds0}}{1+\left(R_{ds}I_{ds0}/V_{dseff}\right)}\left[1+\frac{1}{C_{lm}}\ln\left(\frac{V_A}{V_{Asat}}\right)\right]\cdot\left(1+\frac{V_{ds}-V_{dseff}}{V_{ADIBL}}\right)\cdot$$

$$\left(1+\frac{V_{ds}-V_{dseff}}{V_{ASCBE}}\right)\cdot\left(1+\frac{V_{ds}-V_{dseff}}{V_{ADITS}}\right) \tag{5.87}$$

where V_{Asat} = early voltage @ $V_{ds}=V_{dsat}$; $V_A = V_{Asat} + V_{ACLM}$; and I_{ds0} is given by

$$I_{ds0} = \frac{W_{eff}}{L_{eff}\left[1+\left(V_{dseff}/E_cL_{eff}\right)\right]}C_{ox}\mu_{eff}V_{gsteff}V_{dseff}\left(1-\frac{V_{dseff}}{2V_{bs}}\right) \tag{5.88}$$

Also, an effective drain voltage, V_{dseff}, is a function that guarantees continuity of I_{ds} and its derivatives at V_{dsat} with a user defined parameter δ (DELTA) and is given by

$$V_{dseff} = V_{dsat} - \frac{1}{2}\left[V_{dsat} - V_{ds} - \delta + \sqrt{\left(V_{dsat} - V_{ds} - \delta\right)^2 - 4\delta.V_{dsat}}\right] \tag{5.89}$$

V_{dseff} along with the optimized value of δ ensures continuity of *I–V* plot and its derivatives from linear to saturation regimes. It is shown that the unified Equation 5.87 addresses the continuity from the subthreshold to linear region also by the introduction of the parameter V_{gsteff} given in Equation 5.63.

5.3.8 S/D Parasitic Series Resistance

The S/D parasitic series resistance, R_{ds}, of advanced MOSFET devices is modeled as

$$R_{ds} = \frac{R_{DSW} + \left[1 + P_{RWG}V_{gsteff} + P_{RWB}\left(\sqrt{\phi_s - V_{bseff}} - \sqrt{\phi_s}\right)\right]}{\left(10^6 W'_{eff}\right)^{WR}} \tag{5.90}$$

where:
R_{DSW}, P_{RWG}, P_{RWB}, and WR are model parameters
P_{RWG} and P_{RWB} are gate- and body bias–dependent parameters
WR is empirical fitting parameters to improve the accuracy of the model

5.3.9 Polysilicon Gate Depletion

When a gate voltage is applied to a heavily doped polysilicon gate, for example, nMOSFETs with *n*+ polysilicon (poly-Si) gate, a thin depletion layer in the poly-Si can be formed at the interface between the poly-Si and the gate oxide. This depletion layer is very thin because of the high doping concentration in the poly-Si gate. However, its effect cannot be ignored for devices with gate oxides thinner than 10 nm [28].

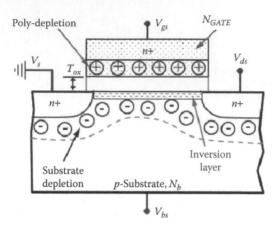

FIGURE 5.14
Charge distribution in an nMOSFET device due to polysilicon gate depletion effect as the device operates in the strong inversion region.

Figure 5.14 shows an nMOSFET device with the depletion region in the $n+$ poly-Si gate. The doping concentration in the poly-Si gate is N_{GATE} and the doping concentration in the substrate is N_{SUB}. The gate oxide thickness is T_{ox}. If we assume that the doping concentration in the gate is infinite, then no depletion region will exist in the gate, and there would be no one sheet of positive charge at the interface between the poly-Si gate and gate oxide. In reality, the doping concentration is finite. The positive charge near the interface of the poly-Si gate and the gate oxide is distributed over a finite depletion region with thickness X_p. The depletion width in the substrate is X_d. In the presence of the depletion region, the voltage drop across the gate oxide and the substrate will be reduced, because part of the gate voltage will be dropped across the depletion region in the gate. That means the effective gate voltage will be reduced.

Let us assume that the potential drop in the depletion layer X_p in the polysilicon gate is ϕ_p; following the procedure discussed in Section 3.4.2.1 [Equation 3.62], we can show

$$\phi_p = \frac{qN_{GATE}}{2K_{si}\varepsilon_0} X_p^2 \tag{5.91}$$

where:
N_{GATE} is the effective concentration in the poly-depletion region

If E_p is the electric field at the poly-Si/SiO$_2$ interface, then the depletion charge in the poly is given by (Equation 3.64)

$$Q_{GATE} = \sqrt{2qK_{si}\varepsilon_0 N_{GATE}\phi_p} \tag{5.92}$$

Again, from Gauss's law we get $K_{ox}\varepsilon_0 E_{ox} = Q_{GATE}$; therefore, from Equation 5.92 we get

$$E_{ox} = \frac{1}{K_{ox}\varepsilon_0}\sqrt{2qK_{si}\varepsilon_0 N_{GATE}\phi_p} \tag{5.93}$$

Now, the applied gate voltage with additional voltage drop in the poly-depletion region is given by

$$V_{gs} = V_{fb} + \phi_s + \phi_p + V_{ox} \tag{5.94}$$

Since $V_{ox} = E_{ox}T_{ox}$, we can simplify Equation 5.94 using Equation 5.93 as

$$V_{gs} = V_{fb} + \phi_s + \phi_p + \frac{T_{ox}}{K_{ox}\varepsilon_0}\sqrt{2qK_{si}\varepsilon_0 N_{GATE}\phi_p} \tag{5.95}$$

After simplification we can show from Equation 5.95

$$a\left(V_{gs} - V_{fb} - \phi_s - \phi_p\right)^2 - \phi_p = 0 \tag{5.96}$$

where we defined

$$a = \frac{K_{ox}^2\varepsilon_0^2}{2qK_{si}\varepsilon_0 N_{GATE}T_{ox}^2} \tag{5.97}$$

Now let us define that the effective gate voltage due to additional voltage drop in the poly is given by $V_{gseff} = (V_{gs} - \phi_p)$; then rearranging Equation 5.96 we get

$$a\left[\left(V_{gs} - \phi_p\right) - \left(V_{fb} + \phi_s\right)\right]^2 + \left(V_{gs} - \phi_p\right) - V_{gs} = 0$$

$$\text{or} \tag{5.98}$$

$$a\left[V_{gseff} - \left(V_{fb} + \phi_s\right)\right]^2 + V_{gseff} - V_{gs} = 0$$

After simplification of Equation 5.98, we can show

$$aV_{gseff}^2 - \left[2a\left(V_{fb} + \phi_s\right) - 1\right]V_{gseff} + \left[a\left(V_{fb} + \phi_s\right)^2 - V_{gs}\right] = 0 \tag{5.99}$$

Now, we solve the quadratic Equation 5.99 on V_{gseff} due to poly gate depletion. Solving V_{gseff} we get

$$V_{gseff} = \left(V_{fb} + \phi_s\right) - \frac{1}{2a} \pm \frac{1}{2a}\sqrt{\left(2a\left(V_{fb} + \phi_s\right) - 1\right)^2 - 4a^2\left(V_{fb} + \phi_s\right)^2 + 4aV_{gs}} \tag{5.100}$$

Since $(V_{gs} - \phi_p) > 0$, we consider the positive sign of Equation 5.100, to get

$$V_{gseff} = \left(V_{fb} + \phi_s\right) - \frac{1}{2a} + \frac{1}{2a}\sqrt{\left(2a\left(V_{fb} + \phi_s\right) - 1\right)^2 - 4a^2\left(V_{fb} + \phi_s\right)^2 + 4aV_{gs}}$$

$$= V_{fb} + \phi_s - \frac{1}{2a}$$

$$+ \frac{1}{2a}\sqrt{4a^2\left(V_{fb} + \phi_s\right)^2 - 4a\left(V_{fb} + \phi_s\right) + 1 - 4a^2\left(V_{fb} + \phi_s\right)^2 + 4aV_{gs}} \qquad (5.101)$$

$$= V_{fb} + \phi_s - \frac{1}{2a} + \frac{1}{2a}\sqrt{1 - 4a\left(V_{fb} + \phi_s\right) + 4aV_{gs}}$$

$$= V_{fb} + \phi_s + \frac{1}{2a}\left(\sqrt{1 + 4a\left(V_{gs} - V_{fb} - \phi_s\right)} - 1\right)$$

Now, substituting the expression for a from Equation 5.97 in Equation 5.101, we can show

$$V_{gseff} = V_{fb} + \phi_s + \frac{qK_{si}\varepsilon_0 N_{GATE}T_{ox}^2}{K_{ox}^2\varepsilon_0^2}\left(\sqrt{1 + \frac{2K_{ox}^2\varepsilon_0^2\left(V_{gs} - V_{fb} - \phi_s\right)}{qK_{si}\varepsilon_0 N_{GATE}T_{ox}^2}} - 1\right) \quad (5.102)$$

For metal gate $K_{si} = 0$; therefore, Equation 5.91 shows that there are no gate depletion and $V_{gs} = V_{gseff}$.

Due to polysilicon gate depletion, the effective gate voltage can be reduced by about 10%. We can estimate the drain current reduction in the linear region as a function of V_{gs}. Assume that V_{ds} is very small (e.g., 50 mV). The linear drain current is proportional to $C_{ox}(V_{gs} - V_{th})$. The ratio of the linear drain current with and without polysilicon gate depletion is equal to

$$\frac{I_{ds}\left(V_{gseff}\right)}{I_{ds}\left(V_{gs}\right)} \cong \frac{V_{gseff} - V_{th}}{V_{gs} - V_{th}} \qquad (5.103)$$

Since $V_{gs} > V_{gseff}$, Equation 5.103 shows that $I_{ds}(V_{gseff})$ is reduced due to polysilicon depletion effect. A significant capacitance reduction has been observed in MOSFETs with oxide thickness less than 5 nm. Thus, the polysilicon depletion effect has to be accounted for in modeling the capacitance characteristics of devices with very thin oxide thickness.

5.3.10 Temperature Dependence

The temperature dependence of the major BSIM model parameters are briefly described next with reference to the reference temperature T_{NOM}.

At any temperature T, the temperature dependence of threshold voltage is modeled by

$$V_{th}(T) = V_{th}(T_{NOM}) + \left(KT_1 + \frac{KT_{1L}}{L_{eff}} + KT_2 V_{bseff} \right)\left(\frac{T}{T_{NOM}} - 1 \right) \qquad (5.104)$$

where:

KT_1, KT_{1L}, and KT_2 are the model parameters to characterize the temperature dependence of threshold voltage for different channel lengths and body biases

The temperature dependence of carrier mobility is given by

$$U_0(T) = U_0 \left(\frac{T}{T_{NOM}} \right)^{UTE}$$

$$U_A(T) = U_A + U_{A1}\left[\frac{T}{T_{NOM}} - 1 \right]$$

$$U_B(T) = U_B + U_{B1}\left[\frac{T}{T_{NOM}} - 1 \right] \qquad (5.105)$$

$$U_C(T) = U_C + U_{C1}\left[\frac{T}{T_{NOM}} - 1 \right]$$

where:

UTE is the parameter to model the temperature dependence of concentration dependent mobility

U_{A1}, U_{B1}, and U_{C1} are the parameters to model the temperature dependence of mobility parameters U_A, U_B, and U_C, respectively, as discussed in Section 5.3.1

The temperature dependence of the saturation velocity is defined by model parameter AT as

$$v_{sat}(T) = v_{sat} - AT\left(\frac{T}{T_{NOM}} - 1 \right) \qquad (5.106)$$

The temperature dependence of S/D series resistance is modeled by a parameter PRT such that

$$R_{DSW}(T) = R_{DSW} - PRT\left(\frac{T}{T_{NOM}} - 1 \right) \qquad (5.107)$$

The temperature coefficients are optimized to fit the measurement data obtained at the target range of operating temperatures.

5.4 Substrate Current Model

The channel electrons traveling through high electric field near the drain end of the channel can become highly energetic. These high energetic electrons are called *hot electrons* and can cause impact ionization generating electrons and holes [42–44]. The holes go into the substrate creating substrate current I_{sub} as shown in Figure 5.15. Some of the electrons have enough energy to overcome the Si/SiO$_2$ energy barrier generating gate current I_g as shown in Figure 5.15. And, some are collected to the drain, contributing to the drain current. The maximum electric field E_m near the drain has the greatest control of hot carrier effects.

Figure 5.16 shows the detailed mechanism of hot carrier effects on nMOS-FET device performance.

Figure 5.16 shows the effect of high drain bias $V_{ds} > V_{dsat}$ on nMOSFET devices at strong inversion, $V_{gs} > V_{th}$. As shown in Figure 5.16, the inversion layer electrons traveling under high electric field cause the following:

1. High energetic electrons traveling along the channel acquire energy from the electric field and become hot;

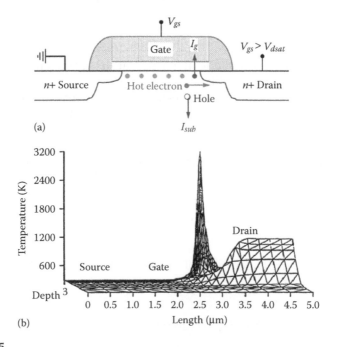

(a)

(b)

FIGURE 5.15
Hot carrier effect in MOSFETs: (a) channel hot electrons in an nMOSFET device contributing to the drain current and generating gate current and (b) electron temperature near the drain end of the channel of the nMOSFET.

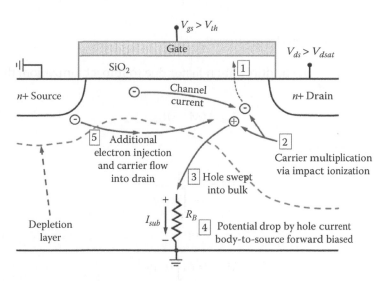

FIGURE 5.16

Cross section of an nMOSFET device in saturation showing hot carrier effects: different physical mechanisms include (1) electron injection into the oxide generating gate current, (2) carrier multiplication by impact ionization, (3) hole flow in the bulk, (4) substrate current flow due to holes, and (5) secondary impact ionization generating additional drain current; the substrate current flow causes a potential drop on the substrate due to the finite substrate resistance R_B, thus forward biasing the source-body *pn*-junction.

2. These hot electrons cause carrier multiplication due to impact ionization by collision with the silicon atoms and breaking covalent bonds, thus creating electrons and holes;

3. Holes are swept into the substrate due to the favorable electric field producing substrate current, I_{sub};

4. I_{sub} flowing through the bulk causes a potential drop in the body, which forward biases the source channel *pn*-junction, thus reducing the source channel potential barrier, $\phi_{bi}(s)$, and enabling more carrier injection from the source to channel;

5. Additional carrier injection due to reduced $\phi_{bi}(s)$ causes more carrier flow in the drain, thus increasing I_{ds} referred to the SCBE discussed earlier.

From the above discussions, we find that the substrate current in an nMOSFET device is due to the holes that are generated by impact ionization of channel hot electrons as they travel from the source to drain. The total drain current, I_{ds}, including the substrate current due to impact ionization is given by

$$I_{ds} = I_{dsat} + I_{sub} \tag{5.108}$$

where:

I_{dsat} is the saturation drain current

If M is the avalanche multiplication factor due to impact ionization, then I_{sub} can be expressed as

$$I_{sub} = (M-1)I_{dsat} \tag{5.109}$$

where M is given by

$$M = \frac{1}{1 - \int \alpha_n dy}$$

or   (5.110)

$$M - 1 = M \int \alpha_n dy$$

where:

α_n is the electron impact ionization coefficient per unit length and is a strong function of the channel electric field E

In order to derive a generalized expression for I_{sub}, we replace I_{dsat} by I_{ds}. Then from Equations 5.109 and 5.110, we can show

$$I_{sub} = I_{ds} M \int \alpha_n dy \tag{5.111}$$

Since I_{sub} resulting from the channel hot electrons impact ionization process is 3–5 orders of magnitude smaller than the drain current I_{ds}, it can be considered as a low-level avalanche current. For low-level multiplication $M \approx 1$, and therefore, Equation 5.111 becomes

$$I_{sub} = I_{ds} \int_0^{l_i} \alpha_n dy \tag{5.112}$$

where y is the distance along the channel with $y = 0$ representing the start of the impact ionization region, and l_i is the length of the drain section where impact ionization takes place as shown in Figure 5.17. Several forms for α_n have been proposed but most commonly used form is

$$\alpha_n = A_i \exp\left[-\frac{B_i}{E}\right] \tag{5.113}$$

where:

A_i and B_i are called the impact ionization coefficients

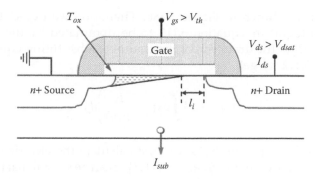

FIGURE 5.17
Hot carrier current effect in nMOSFETs showing the impact ionization region, l_i, at the drain end of the device.

Most of the reported data on α_n have been measured in bulk silicon and the constants A_i and B_i show a wide range of values [42–44]. Slotboom et al. [44] have measured α_n at the surface and in the bulk silicon and reported the values for the constants, which are provided in Table 5.1.

Due to the exponential dependence of α_n on electric field as shown in Equation 5.113, it is easy to see that the impact ionization will dominate at the position of the maximum electric field. In a MOSFET, the maximum electric field E_m is present at the drain end as shown in Figure 5.18a. The sharp maximum E_m shown in Figure 5.18a can be reduced by device

TABLE 5.1

Surface and Bulk Impact Ionization Coefficients in Silicon

α_n	A_i (cm^{-1})	B_i (V cm^{-1})
Surface	2.45×10^6	1.92×10^6
Bulk	7.03×10^5	1.23×10^6

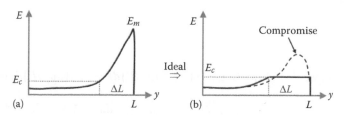

FIGURE 5.18
Hot carrier effect in nMOSFETs: (a) maximum electric field, E_m, at the drain end of the channel and (b) smoother E_m to reduce the effect of substrate current on device performance.

optimization as shown in Figure 5.18b. Therefore, we expect the impact ionization integral in Equation 5.112 to be dominated by the maximum electric field E_m at the drain end of the channel. Substituting Equation 5.113 in Equation 5.112 we get

$$I_{sub} = I_{ds}A_i \int_0^{l_i} \exp\left(-\frac{B_i}{E(y)}\right) dy \tag{5.114}$$

In order to solve Equation 5.114, we first calculate the electric field in the channel. Based on a pseudo-2D analysis [45], it can be shown that the channel electric field E can be expressed as

$$E(y) = -\frac{dV}{dy} = \sqrt{\frac{\left(V(y) - V_{dsat}\right)^2}{l_i^2} + E_c^2} \tag{5.115}$$

where E_c represents the channel electric field at which the carriers reach velocity saturation (at $y = 0$ and $E = E_c$) and the corresponding voltage at that point is the saturation voltage V_{dsat}. E_c is about 2×10^4 V cm^{-1} for electrons. The parameter l_i can be treated as an effective impact ionization length and is given by

$$l_i^2 = \frac{\varepsilon_{si}}{\varepsilon_{ox}} T_{ox} X_j \tag{5.116}$$

where:
T_{ox} is the gate oxide thickness
X_j is the S/D junction depth

Although Equations 5.115 and 5.116 were derived for conventional S/D junctions, they are still valid for lightly-doped drain (LDD) as well as SDE MOSFET structures. For LDD and SDE devices, X_j is the junction depth of the LDD or SDE region. The maximum field E_m, which occurs at the drain end, can easily be obtained replacing $V(y)$ by V_{ds} in Equation 5.115. Again, since $E_c^2 \ll \left(V_{ds} - V_{dsat}\right)^2 / l_i^2$ in Equation 5.115, neglecting E_c results in the following approximate expression for E_m, we get

$$E_m \cong \frac{\left(V_{ds} - V_{dsat}\right)}{l_i} \tag{5.117}$$

Now, we replace dy in Equation 5.114 by $\left(dy/dE\right)dE = -E^2\left(dy/dE\right)d\left(1/E\right)$ to get

$$I_{sub} = -I_{ds}A_i \int_{E_c}^{E_m} \exp\left(-\frac{B_i}{E(y)}\right) E^2 \frac{dy}{dE} d\left(\frac{1}{E}\right) \tag{5.118}$$

From Pseudo-2D analysis of the velocity saturation region, we can show

$$E(y) = E_c \cosh\left(\frac{y}{l_i}\right) = E_c \frac{\exp(y/l_i) - \exp(-y/l_i)}{2} \cong E_c \frac{1}{2} \exp\left(\frac{y}{l_i}\right) \quad (5.119)$$

Since l_i is very small and y/l_i is a very large number, $\exp(-y/l_i)$ is negligibly small; then differentiating Equation 5.119 we get

$$\frac{dE}{dy} = E_c \frac{1}{2} \exp\left(\frac{y}{l_i}\right) \cdot \left(\frac{1}{l_i}\right) = \frac{E}{l_i} \quad (5.120)$$

Therefore,

$$-E^2\left(\frac{dy}{dE}\right) = -E^2\left(\frac{l_i}{E}\right) = l_i E \quad (5.121)$$

Substituting Equation 5.121 in Equation 5.118, we get

$$I_{sub} = -I_{ds} A_i \int_{E_c}^{E_m} l_i E \exp\left(-\frac{B_i}{E(y)}\right) d\left(\frac{1}{E}\right) \quad (5.122)$$

Since the exponential term in Equation 5.122 has a pronounced peak at $E = E_m$, we evaluate it at $E = E_m$ and let it be constant over the region so that we can remove it from the integral. After this simplification, Equation 5.122 can be solved for I_{sub} as

$$I_{sub} = -I_{ds} A_i l_i E_m \int_{E_c}^{E_m} \exp\left(-\frac{B_i}{E(y)}\right) d\left(\frac{1}{E}\right) \quad (5.123)$$

After integration and simplification, we can show assuming $E_c \ll E_m$,

$$I_{sub} \cong I_{ds} \frac{A_i}{B_i} l_i E_m \exp\left(-\frac{B_i}{E_m}\right) \quad (5.124)$$

Substituting for E_m from Equation 5.117 and Equation 5.124 can be expressed in terms of drain voltage as

$$I_{sub} \cong I_{ds} \frac{A_i}{B_i} (V_{ds} - V_{dsat}) \exp\left(-\frac{l_i B_i}{V_{ds} - V_{dsat}}\right) \quad (5.125)$$

Equation 5.125 is used for substrate current modeling. Note that Equation 5.125 is independent of device geometry. In order to model channel length dependence of I_{sub}, the ratio A_i/B_i can be replaced by $(\alpha_0 + \alpha_1/L_{eff})$ to express

$$I_{sub} \cong \left(\alpha_0 + \frac{\alpha_1}{L_{eff}}\right)(V_{ds} - V_{dsat}) \exp\left(-\frac{\beta}{V_{ds} - V_{dsat}}\right) \cdot I_{dsa} \quad (5.126)$$

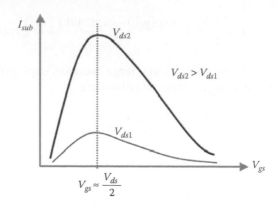

FIGURE 5.19
Impact ionization induced substrate current I_{sub} versus gate voltage V_{gs} characteristics of nMOSFET devices for two different values of V_{ds}; typically, for any value of V_{ds}, the value of I_{sub} attains a maximum value at $V_{gs} \approx V_{ds}/2$.

where:
$$\beta = l_i B_i$$
I_{dsa} is the drain current without the impact ionization

Thus, the basic parameter set for modeling I_{sub} is $\{\alpha_0, \alpha_1, \beta\}$ which is obtained by optimizing the measurement data for MOSFET devices.

Figure 5.19 shows a typical I_{sub} versus V_{gs} plot for two values of V_{ds}. It is found that for a given value of V_{ds}, initially I_{sub} increases with increasing V_{gs} due to an increase in the drain current (i.e., increase in the inversion charge density from weak to strong inversion regime as V_{gs} increases from 0 to strong inversion). Further increase in V_{gs} eventually results in a decrease in I_{sub} due the reduction in the effective width of the pinch-off region, resulting in an increase in V_{dsat}, which in turn reduces the electric field along the channel. Thus, as V_{gs} increases, I_{sub} increases first, reaches its peak value at a certain V_{gs}, and then decreases resulting in a bell-shaped curve with its maximum occurring at a gate voltage, $V_{gs} \approx 0.5V_{ds}$. However, in nanoscale devices, the lateral electric field along the direction of current flow is extremely high and due to local carrier heating, the entire channel length is velocity saturated. Therefore, for any nanoscale MOSFETs, the impact ionization occurs at a lower value of $V_{gs} > V_{th}$ and the value of V_{gs} at I_{sub}(peak) is almost independent of V_{ds} [43].

In order to extract the impact ionization parameters A_i, B_i, and l_i, the general Equation 5.125 can be expressed as [46,47]

$$\ln(Y) = mX + c \qquad (5.127)$$

where

$$Y = \frac{I_{sub}}{I_{ds}(V_{ds} - V_{dsat})}$$

$$X = \frac{1}{(V_{ds} - V_{dsat})} \qquad (5.128)$$

$$m = -l_i B_i$$

and

$$c = \ln\left(\frac{A_i}{B_i}\right)$$

Equation 5.127 represents a straight line with the slope, m, and intercept, c, given by Equation 5.128. Thus, $\ln\left[I_{sub}/I_{ds}(V_{ds} - V_{dsat})\right]$ versus $1/(V_{ds} - V_{dsat})$ plot is a straight line with a slope $m = -l_i B_i$ and the intercept $c = \ln(A_i/B_i)$. From such plots for MOSFETs with different processing parameters, the value of l_i can be determined [47] as shown in Figure 5.20.

As discussed in Section 5.3.6, substrate current I_{sub} flowing into the substrate increases drain current significantly, resulting in lower output resistance as shown in Figure 5.13. This is due to the fact that I_{sub} flowing to the substrate causes an *IR* drop in the substrate, resulting in a body bias; this body bias forward biases the source/body *pn*-junction thus lowering the source to chain potential barrier for carriers. As a result, more carriers are injected from the source to the inversion channel, causing a significant increase in I_{ds}, which is referred to as the SCBE. The SCBE causes V_{th} drop and manifold increase in I_{sub} and consequently, I_{dS} as shown in Figure 5.13

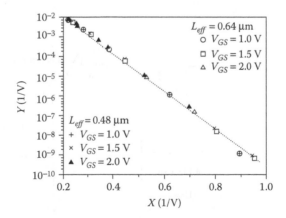

FIGURE 5.20
Plot of $Y = I_{sub}/\left[I_{ds}(V_{ds} - V_{dsat})\right]$ versus $X = (V_{ds} - V_{dsat})^{-1}$ with different V_{gs} for LDD type nMOS-FETs of different channel length; all data are obtained under $V_{bs} = 0$ for $W = 40$ μm devices and $T_{ox} = 150$ A. (Data from S. Saha, *Solid-State Electron.*, 37, 1786–1788, 1994.)

5.4.1 Gate-Induced Drain Leakage Body Current Model

When $V_{gs} < 0$ (or $V_{gs} = 0$) and high V_{ds} is applied to the device as shown in Figure 5.21, the electric field is very high in the drain region. This high electric field causes a large band bending, which results in *band-to-band tunneling* (BTBT). As a result a significant amount of drain leakage current is observed.

The drain leakage current due to BTBT is related to the generation of carriers in the drain overlap region under the gate as shown in Figure 5.21. From the basic device physics, we know that a positive gate bias tends to invert the *p*-type channel. Similarly, a negative gate bias tends to invert the *n*-type drain junction in the overlap region. The inversion of the drain does not easily take place, since the drain is doped more heavily than the channel. Nevertheless, when V_{gd} is fairly negative, the applied drain bias at least causes the overlap region to be depleted of carriers. As the minority carriers, generated either by BTBT or trap-assisted tunneling, arrive at the surface to attempt to form the inversion layer, they immediately get swept laterally to the substrate. The current that flows as a result of the carriers being swept from the overlap region constitutes the *gate-induced drain leakage* (GIDL) current, I_{gidl}. In the framework of this explanation, we see that GIDL is not an SCE. The leakage current tends to be significant in LDD devices where the overlapped region is lightly doped. GIDL is, generally, less a severe in nanometer-scale devices whose drain extension forms a heavily doped junction.

Similar current is also observed at the source end of the device. The components of body current observed are gate-induced drain leakage and gate-induced source leakage (GISL). The general expressions to model GIDL and GISL are given by

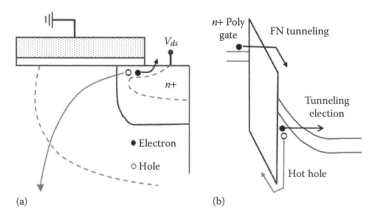

(a)   (b)

FIGURE 5.21
GIDL current in an nMOSFET device: (a) gated diode, at the drain MOSFET only, showing electron–hole pair generation and transport and (b) Fowler-Nordheim (FN) tunneling due to high lateral electric field by applied drain voltage.

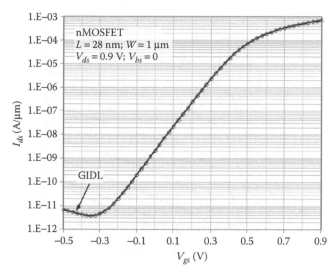

FIGURE 5.22

GIDL in MOSFETs: I_{ds} versus V_{gs} characteristics of an nMOSFET device showing the effect of GIDL on a 28 nm nMOSFET performance for $V_{gs} < 0$.

$$I_{gidl} = NF.AGIDL.W_{eff}\left(\frac{V_{ds} - V_{gseff} - EGIDL}{3TOXE}\right)\exp\left(\frac{-3TOXE.BGIDL}{V_{ds} - V_{gseff} - EGIDL}\right)\frac{V_{DB}^3}{CGIDL + V_{DB}^3}$$

and $\qquad$ (5.129)

$$I_{gisl} = NF.AGISL.W_{eff}\left(\frac{-V_{ds} - V_{gseff} - EGISL}{3TOXE}\right)\exp\left(\frac{-3TOXE.BGISL}{-V_{ds} - V_{gseff} - EGISL}\right)\frac{V_{SB}^3}{CGISL + V_{SB}^3}$$

The model parameters (AGIDL, AGISL), (BGIDL, BGISL), (CGIDL, CGISL), and (EGIDL, EGISL) are obtained from the measured $I_{ds} - V_{gs}$ data obtained for $-V_{gs}$ to $+V_{gs}$ at $V_{ds} = V_{dd}$ (supply voltage); NF is the number of fingers used in the layout for MOSFETs. GIDL must be accounted if the standby current of a circuit is an important specification. Figure 5.22 shows GIDL effect in a 28 nm channel length nMOSFET device.

5.4.2 Gate Current Model

As the oxide becomes progressively thinner in each generation of IC technology, the magnitude of the direct tunneling currents through the oxide becomes more significant. In direct tunneling, the carriers from the inversion layer of silicon surface can tunnel directly through the energy gap of the SiO_2 layer instead of tunneling into the conduction band of the SiO_2 layer.

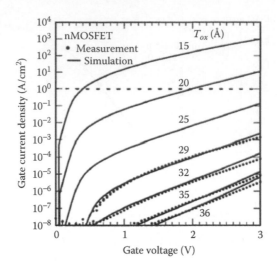

FIGURE 5.23
Measured and simulated tunneling currents in thin oxide polysilicon gate MOSFET devices. The horizontal broken line indicates a tunneling current level of 1 A cm⁻². (Data from S.-H. Lo et al., *IEEE Electron Device Lett.*, 18, 209–211, 1997.)

The direct tunneling current can be very large for advanced CMOS technologies with oxide thickness of about 1 nm. Figure 5.23 shows the plots of measured and simulated tunneling current versus gate voltage in polysilicon-gate MOSFETs with different gate oxide thicknesses [48]. Figure 5.23 shows that I_{gate} is extremely high for thinner $T_{ox} < 2$ nm due to direct tunneling gate leakage current. Therefore, it is critical to model gate current of advanced MOSFETs for circuit design.

There are five tunneling components of gate current, I_g, as shown in Figure 5.24. They are

1. I_{gd} = gate-to-drain current between the gate and the heavily doped drain junction
2. I_{gcd} = gate-to-channel current and to the drain
3. I_{gs} = gate-to-source current between the gate and the heavily doped source diffusion
4. I_{gcs} = gate-to-channel current and to the source
5. I_{gb} = gate-to-substrate tunneling current (accumulation and inversion)

The detailed analysis of these tunneling currents unavoidably involves quantum mechanical analysis [48–56]; however, the analytical expressions for compact gate current modeling are described in BSIM4 [28].

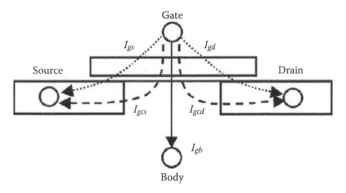

FIGURE 5.24
Gate current model: different components of gate tunneling current in nanometer-scale MOSFETs.

5.5 Summary

This chapter presents compact MOSFET models for small geometry devices. In order to develop accurate small geometry compact model, the different structural and physical effects are modeled in device threshold voltage. First of all, the nonuniform substrate doping is modeled in device threshold voltage. Then the model for small geometry effects such as short channel effect, reverse short channel effect, narrow width and reverse-NWEs are included in the threshold voltage model. In this chapter, the accurate mobility model is derived to account for the effect of high gate bias on device performance. With accurate mobility model, the regional drain current models for the linear and saturation regions are developed to model high lateral electric field and velocity saturation due to high drain bias. Finally, the compact models for hot carrier–induced substrate current for MOSFETs devices are presented.

Exercises

5.1 Consider an nMOSFET device with channel doping concentration $N_a = 1 \times 10^{18}$ cm^{-3} and $T_{ox} = 3$ nm. Assume $Q_f = 0$, $V_{sb} = 0$, and $n+$ degenerately doped poly gate.

a. Calculate the value of long channel threshold voltage V_{th0}.

b. Derive the expressions for $K1$ and $K2$ in terms of the channel and substrate body effect coefficients and an intermediate substrate bias.

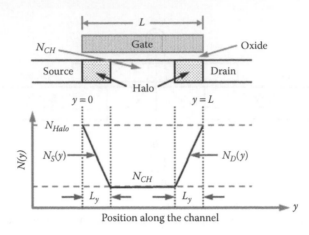

FIGURE E5.2.1
Triangular halo/pocket doping profiles for MOSFET device structure.

Discuss the impact of the model parameters $K1$ and $K2$ on V_{th} of an MOS transistor.

5.2 In this problem you will use the triangular halo doping profiles shown in Figure E5.2.1 to model the halo doping distribution near the source and drain ends of a MOSFET channel. Given: L = channel length, L_y = halo spread inside L at the source and drain ends, N_{Halo} = maximum halo concentration, and N_{CH} = channel doping concentration:

a. Show that the halo doping profile $N_S(y)$ at any point y near the source end of the channel is given by

$$N_S(y) = N_{CH}\left(\frac{y}{L_y}\right) + N_{Halo}\left[1 - \left(\frac{y}{L_y}\right)\right]$$

b. Show that the halo doping profile $N_D(y)$ at any point y near the drain end of the channel is given by

$$N_D(y) = N_{CH}\left[\left(\frac{L}{L_y}\right) - \left(\frac{y}{L_y}\right)\right] + N_{Halo}\left[1 - \left\{\left(\frac{L}{L_y}\right) - \left(\frac{y}{L_y}\right)\right\}\right]$$

5.3 In order to develop V_{th}-model for nonuniform lateral channel doping profile, we used piecewise box-shaped step functions for N_{Halo} to represent a constant channel doping concentration near the source and drain ends of the channel while N_{CH} to represent a constant channel concentration, where $N_{Halo} > N_{CH}$. In reality, the halo doping profile near the source and drain ends can be more accurately modeled by a triangular-shaped function. Use triangular profiles [Figure E5.2.1] to represent the halo doping, model V_{th} for nonuniform lateral channel doping. Given L = channel length, and L_y = halo spread inside L at the source/drain ends:

a. Derive an expression for the average channel doping concentration to account for the halo doping in the channel. Clearly define all parameters and explain any assumptions you make.

b. Show the expressions for model parameters from your work in part (a).

c. How would you extract the model parameters obtained in part (b)?

d. Compare the model parameters in part (b) with that derived using box-shaped profiles given by Equation 5.12. Explain.

5.4 In order to derive an effective inversion carrier mobility model, it is shown that the effective channel electrical field, $E_{eff} = [0.5Q_{inv} + Q_b]/\varepsilon_{si}$, where Q_{inv} and Q_b are the inversion charge and bulk (depletion) charge under the gate, respectively, and ε_{si} is the dielectric constant of silicon. The dependence of surface mobility μ_s on process parameters such as T_{ox} and N_{sub} and the terminal voltages are lumped in E_{eff}. Assume $V_{gs} > V_{th}$ and small V_{ds}:

a. Show that $E_{eff} \cong (V_{gs} + V_{th})/6T_{ox}$.

b. If the effective mobility is modeled by: $\mu_{eff} = \mu_0/[1 + E_{eff}/E_0)]^\nu$, where $\mu_s = \mu_0$ @ $V_{gs} - 0$ and E_0 and ν are parameters determined from the measured data. Use the expression for E_{eff} in part (a) to show that:

$$\mu_{eff} = \frac{\mu_0}{1 + U_a\left[\left(V_{gs} + V_{th}\right)/T_{ox}\right] + U_b\left[\left(V_{gs} + V_{th}\right)/T_{ox}\right]^2}$$

where:

U_a and U_b are the model parameters that are determined experimentally from *I–V* data of MOSFET devices

Clearly state any assumptions you make.

5.5 An nMOSFET device is designed with a gate oxide thickness of 5 nm and a uniformly doped substrate with $N_a = 5 \times 10^{17}$ cm^{-3}. Assuming that the "ON" state of this device is characterized by $\phi_s = 2\phi_B$ and the "OFF" state by $\phi_s = \phi_B$, estimate the ratio of ON to OFF currents flowing in the device.

5.6 Complete the mathematical steps to show that the MOSFET drain current expression in the linear region is given by Equation 5.74.

5.7 Complete the mathematical steps to show that the general expression for MOSFET saturation drain voltage is given by Equation 5.79.

6

MOSFET Capacitance Models

6.1 Introduction

This chapter presents the dynamic compact MOSFET (metal-oxide-semiconductor field-effect transistor) models for analyzing the device performance under time-varying terminal voltages in circuit operation. The MOSFET device models developed in Chapters 4 and 5 are applicable to devices under DC or steady-state biasing condition, that is, when the terminal voltages do not vary with time. However, the real circuit operates under time-varying terminal voltages. Under such biasing condition, the device behavior is described by dynamic models. If the rate of change of terminal voltages is sufficiently small, the device operation can be described by a *small signal dynamic model*. On the other hand, if the rate of change of terminal voltages is large, the device is represented by a *large signal dynamic model*. In dynamic models, the device is represented by capacitors, resistors, current sources, and so on. The dynamic MOSFET models are essential part of circuit CAD (computer-aided design).

The dynamic operations of MOSFET devices are due to the capacitive effects of the device, resulting from the stored charges in the device. Thus, a capacitance model describing the intrinsic and extrinsic components of the device capacitance is an essential part of a compact model for circuit simulation besides DC model. In most circuit simulators, the same capacitance model is used for both the small signal AC analysis and the large signal transient analysis. A capacitance model is always based on *quasistatic assumptions*, that is, charge in a device can follow the varying terminal voltage instantaneously without any delay. In this chapter, first of all, a large signal dynamic model is described by developing models for intrinsic charges and capacitances of a large geometry device (large L and wide W). Then the models for short channel devices are discussed. Finally, the small signal linear model parameters required for small signal analysis are discussed.

6.2 Basic MOSFET Capacitance Model

The various capacitances present within an *n*-channel MOSFET are shown in Figure 6.1. The MOS transistor capacitances are categorically divided into two components: intrinsic and extrinsic, as shown in Figure 6.1. The region between the metallurgical source and drain junctions where the gate to source-drain (S/D) region is at flat band voltage is referred to as the intrinsic region.

- Intrinsic capacitances are between S/D metallurgical junctions.
- Extrinsic capacitances are outside the intrinsic part.

The extrinsic capacitances are divided into five components as shown in Figure 6.1. These are:

1. Outer fringing capacitances between the gate and the S/D region, C_{FO}
2. Inner fringing capacitances between the gate and the S/D region, C_{FI}
3. The overlap capacitances between the gate and heavily doped S/D regions C_{GSO} and C_{GDO} and between the gate and bulk region, C_{GBO} (not shown in Figure 6.1)
4. Overlap capacitances between the gate and the lightly doped S/D regions C_{GSL}, C_{GDL} (not shown in Figure 6.1)
5. S/D junction capacitances C_{JS} and C_{JD}

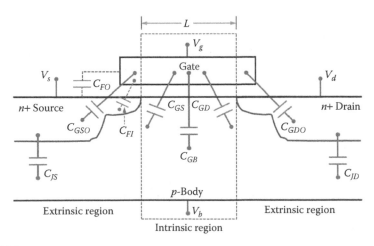

FIGURE 6.1
MOSFET capacitances: intrinsic capacitances between the S/D metallurgical junctions; extrinsic capacitances include overlap, fringing, and junction capacitances outside the active area of the device.

6.2.1 Intrinsic Charges and Capacitances

In a typical steady-state operation, the current flow through a MOSFET device is due to the transport of the mobile carriers (e.g., electrons in n-channel MOSFETs or nMOSFETs and holes in p-channel MOSFETs or pMOSFETs) from the source to drain under the applied drain voltage. This current is referred to as the *transport current* in transient analysis. In a dynamic operation, additional currents flow through the device due to the stored charges at the device terminals and are called the charging currents as shown in Figure 6.2.

Figure 6.2a shows the transient or dynamic currents i_g, i_s, i_d, and i_b flowing through the gate (g), source (s), drain (d), and bulk (b) terminals, respectively, of a MOSFET device. In Figure 6.2a, Q_G, Q_S, Q_D, and Q_B are the total gate, source, drain, and bulk charges, respectively, corresponding to four terminals of the MOSFET. These terminal charges are functions of the gate, source, drain, and bulk terminal voltages V_g, V_s, V_d, and V_b, respectively. Thus, in general

$$Q_j = f(V_g, V_s, V_d, V_b), \quad \text{where } j = G, S, D, B \qquad (6.1)$$

From Kirchhoff's current law (KCL) for the total current, we have

$$i_g + i_s + i_d + i_b = 0 \qquad (6.2)$$

and from the law of conservation of charge, we have

$$Q_G + Q_S + Q_D + Q_B = 0 \qquad (6.3)$$

In order to calculate various charges of a MOSFET device, we assume *quasi-static operation* of the device [1]. In quasistatic operation, the terminal voltages are assumed to vary sufficiently slowly so that the distribution in the stored

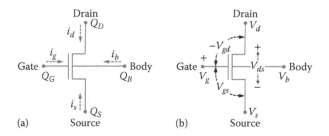

(a)　Source　　　　(b)　Source

FIGURE 6.2
Schematic of a MOSFET as a circuit element: (a) transient currents i_g, i_s, i_d, and i_b flowing through the gate, source, drain, and body terminals, respectively and (b) terminal DC voltages, V_g, V_d, V_s and V_b at the gate, drain, source, and body terminals, respectively, where $V_{ds} = V_{gs} - V_{gd}$.

charges Q_G, Q_S, Q_D, and Q_B can follow the voltage variations. This implies that the terminal currents vary instantaneously with the terminal voltages. Thus, at any time t the charge per unit area is due to dynamic and DC operation is the same. The dynamic model developed by quasistatic assumption is called the *quasistatic model*. In practice, the quasistatic model works quite well for much of the circuit CAD. However, *this approach may fail, especially with long channel devices operating at high switching speeds, or when the load capacitance is very small.*

Assuming quasistatic operation, the total transient current at each terminal can be expressed as the sum of the time-dependent transport current and a charging current as

$$i_s(t) = -I_s\left[V(t)\right] + \frac{dQ_S}{dt}$$

$$i_d(t) = -I_d\left[V(t)\right] + \frac{dQ_D}{dt}$$

$$i_g(t) = \frac{dQ_G}{dt} \tag{6.4}$$

$$i_b(t) = \frac{dQ_B}{dt}$$

where we assumed that no transport current is flowing to the gate ($I_g = 0$) and substrate ($I_b = 0$). In Equation 6.4, we have assumed that Q_S and Q_D are known. However, we only know the total inversion or channel charge Q_I so that

$$i_s(t) = -I_s\left[V(t)\right] + \frac{dQ_S}{dt}$$

$$i_s + i_d = I_{ds}\left[V(t)\right] + \frac{dQ_I}{dt} \tag{6.5}$$

However, Equation 6.5 is unsuitable for circuit simulation, since circuit CAD requires separate expressions for i_s and i_d. Thus, in order to develop a dynamic MOSFET model for circuit CAD, it is necessary to derive expressions for Q_G, Q_B, and Q_I as functions of terminal voltages.

In order to derive the expressions for Q_G, Q_B, and Q_I as functions of terminal voltages, we use the corresponding known steady-state charges $Q_g(y)$, $Q_b(y)$, and $Q_i(y)$ per unit area at any point y along the length of the channel. By integrating these charges over the area of the active gate region we can obtain the corresponding total charge Q_G, Q_B, and Q_I. Now, the gate charge contained in a small area of device width W and length dy is $Q_g \cdot W \cdot dy$. Then integrating this charge over the channel length L gives the total gate charge Q_G as

$$Q_G = W \int_0^L Q_g(y)dy \qquad (6.6)$$

Similarly, we can show

$$Q_I = W \int_0^L Q_i(y)dy$$

$$\qquad (6.7)$$

$$Q_B = W \int_0^L Q_b(y)dy$$

Again, from the charge conservation principle,

$$Q_G + Q_I + Q_B = 0 \qquad (6.8)$$

In Equation 6.8, we have neglected the total oxide charge (Q_o) since $Q_G \gg Q_o$. In Equations 6.6 and 6.7, Q_G, Q_I, and Q_B are distributed charges. Therefore, the corresponding intrinsic capacitances must be modeled as distributed capacitances. However, such a model is not suitable for circuit CAD. Thus, for the simplicity of circuit CAD, these distributed capacitances are usually modeled as lumped two-terminal capacitances appearing between the gate, source, drain, and bulk or substrate terminals of a MOSFET. The Meyer model is one of such lumped capacitance model that is widely implemented in many circuit simulation tools [2].

The Meyer model was derived for long channel MOSFET devices. The most serious error in the model is that it violates the law of charge conservation [3]. However, due to the inherent simplicity of the Meyer model, it has been extensively used in simulating circuits that do not have charge conservation problems. The Meyer model is the default capacitance model for SPICE (Simulation Program with Integrated Circuit Emphasis) Levels 1–4. In order to overcome the deficiencies in the Meyer model, charge is used as a state variable in capacitance modeling. This is known as charge-based capacitance models [4–9]. We will first discuss the Meyer model and then develop a more accurate charge-based capacitance model.

6.2.2 Meyer Model

In Meyer model the distributed gate-channel capacitances are split into three lumped capacitances: gate to source (C_{GS}), gate to drain (C_{GD}), and gate to bulk (C_{GB}). These are defined as the derivative of the total gate charge Q_G with respect to the source, drain, and bulk [Figure 6.2b], respectively, as given next:

$$C_{GS} = \left.\frac{\partial Q_G}{\partial V_{gs}}\right|_{V_{gd}, gb}$$

$$C_{GD} = \left.\frac{\partial Q_G}{\partial V_{gd}}\right|_{V_{gs}, gb} \tag{6.9}$$

$$C_{GB} = \left.\frac{\partial Q_G}{\partial V_{gb}}\right|_{V_{gs}, gd}$$

where:

$V_{gd} = (V_{gs} - V_{ds})$
$V_{gb} = (V_{gs} - V_{bs})$

It is seen that the capacitances defined in Equation 6.9 imply that these capacitances are reciprocal; that is, both terminals of a capacitor are equivalent and the capacitance is symmetric, for example: $C_{GD} = C_{DG}$. In this case, the change in the charge Q_G due to V_{gd} may be due to the change either in the gate voltage V_g or in the drain voltage V_d. In Meyer model, the following assumptions are used to derive the capacitances:

1. MOSFET capacitances are reciprocal, that is, $C_{GB} = C_{BG}$, $C_{GD} = C_{DG}$, and $C_{GS} = C_{SG}$.
2. The bulk charge Q_b is constant along the length of the channel depending only on the applied bias V_{gb} and independent of V_{ds}. Thus, bulk-source (C_{BS}) and bulk-drain (C_{BD}) capacitances are zero.

From the law of conservation of charge given in Equation 6.8, we can express the total gate charge as

$$Q_G = -(Q_I + Q_B) = -W \int_0^L Q_i(y)dy - W \int_0^L Q_b(y)dy \tag{6.10}$$

where we have used the expressions for Q_I and Q_B from Equation 6.7. By assumption 2, the bulk charge density Q_b is a constant along the length of the channel and can be taken out of the integral. Thus, Equation 6.10 becomes

$$Q_G = -W \int_0^L Q_i(y)dy - Q_B \tag{6.11}$$

where:

$$Q_B = WLQ_b$$

Equation 6.11 is the generalized expression for Q_G in a MOSFET device. In order to calculate Q_G from Equation 6.11, any expression for Q_i used to calculate I_{ds} can be used. However, in deriving the Meyer intrinsic capacitance model, long channel expressions for Q_i and Q_b from Chapter 4 are used to derive Q_G and the capacitances in the different mode of operations of MOSFET devices as described next.

6.2.2.1 Strong Inversion

From Equations 4.68 and 4.70, the expressions for Q_b and Q_i, respectively, for long channel MOSFETs are given by

$$Q_b(y) = -\gamma C_{ox}\sqrt{2\phi_B + V_{sb}} \tag{6.12}$$

$$Q_i(y) = -C_{ox}\left[V_{gs} - V_{th} - V(y)\right] \tag{6.13}$$

where:
  C_{ox} is the gate oxide capacitance per unit area
  V_{th} is the threshold voltage
  $V(y)$ is the voltage at any point y along the length of the channel from the
    source to drain

Since Q_i is a function of V, to integrate Equation 6.11 we first change the variable of integration from dy to dV using Equation 4.63 so that

$$dy = -\frac{W\mu_s}{I_{ds}}Q_i(y)dV \tag{6.14}$$

Now, combining Equations 6.11 through 6.14, we can show

$$Q_G = \frac{W^2\mu_s C_{ox}^2}{I_{ds}}\int_0^{V_{ds}}\left(V_{gs} - V_{th} - V\right)^2 dV - Q_B \tag{6.15}$$

where the limits of integration change from $y = 0$ to $V(y) = 0$, and $y = L$ to $V(y) = V_{ds}$. Again, using Q_i from Equations 6.13 through 6.14 and integrating the resulting expression from source to drain, we get the expression for I_{ds} (Equation 4.72)

$$I_{ds} = \mu_s C_{ox} \left(\frac{W}{L}\right)\left[V_{gs} - V_{th} - \frac{V_{ds}}{2}\right]V_{ds}; \quad V_{gs} > V_{th} \tag{6.16}$$

Now, from Figure 6.2b we get, $V_{ds} = (V_{gs} - V_{gd})$; then substituting for $V_{ds} = (V_{gs} - V_{gd})$ in Equation 6.16 we can write

$$
\begin{aligned}
I_{ds} &= \mu_s C_{ox}\left(\frac{W}{2L}\right)\left[2V_{gs} - 2V_{th} - \left(V_{gs} - V_{gd}\right)\right]\left(V_{gs} - V_{gd}\right) \\
&= \mu_s C_{ox}\left(\frac{W}{2L}\right)\left[\left(V_{gs} - V_{th}\right) + \left(V_{gd} - V_{th}\right)\right]\cdot\left[\left(V_{gs} - V_{th}\right) - \left(V_{gd} - V_{th}\right)\right] \\
&= \frac{W\mu_s C_{ox}}{2L}\left[\left(V_{gs} - V_{th}\right)^2 - \left(V_{gd} - V_{th}\right)^2\right]
\end{aligned}
\tag{6.17}
$$

Now, substituting for I_{ds} from Equation 6.17 to Equation 6.15, we get

$$
\begin{aligned}
Q_G &= \frac{2WLC_{ox}}{\left(V_{gs} - V_{th}\right)^2 - \left(V_{gd} - V_{th}\right)^2}\int_0^{V_{ds}}\left(V_{gs} - V_{th} - V(y)\right)^2 dV - Q_B \\
&= \frac{2}{3}WLC_{ox}\left[\frac{\left(V_{gs} - V_{th} - V_{ds}\right)^3 - \left(V_{gs} - V_{th}\right)^3}{\left(V_{gd} - V_{th}\right)^2 - \left(V_{gs} - V_{th}\right)^2}\right] - Q_B \\
&= \frac{2}{3}WLC_{ox}\left[\frac{\left(V_{gd} - V_{th}\right)^3 - \left(V_{gs} - V_{th}\right)^3}{\left(V_{gd} - V_{th}\right)^2 - \left(V_{gs} - V_{th}\right)^2}\right] - Q_B
\end{aligned}
\tag{6.18}
$$

where we have used $(V_{gs} - V_{ds}) = V_{gd}$ from Figure 6.2b. Then differentiating Equation 6.18 with respect to V_{gs}, V_{gd}, and V_{gb}, we obtain the intrinsic capacitance C_{GS}, C_{GD}, and C_{GB}, respectively, in the different operation regions of MOSFETs.

In the *linear region*, we get the expressions for the intrinsic capacitances from Equation 6.18 as

$$
C_{GS} = \frac{\partial Q_G}{\partial V_{gs}} = \frac{2}{3}WLC_{ox}\left[1 - \frac{\left(V_{gd} - V_{th}\right)^2}{\left(V_{gd} + V_{gs} - 2V_{th}\right)^2}\right]
$$

$$
C_{GD} = \frac{\partial Q_G}{\partial V_{gd}} = \frac{2}{3}WLC_{ox}\left[1 - \frac{\left(V_{gs} - V_{th}\right)^2}{\left(V_{gd} + V_{gs} - 2V_{th}\right)^2}\right] \tag{6.19}
$$

$$
C_{GB} = \frac{\partial Q_G}{\partial V_{gb}} = 0
$$

Note that $C_{GB} = 0$ in the strong inversion is expected since the inversion charge in the channel from S to D shields the gate from the bulk and, therefore, prevents any response of Q_G due to substrate bias V_{bs}. Let us define $V_{gt} = V_{gs} - V_{th}$; then using $V_{gs} - V_{ds} = V_{gd}$ (Figure 6.2b), Equation 6.19 can be expressed as

$$C_{GS} = \frac{\partial Q_G}{\partial V_{gs}} = \frac{2}{3} WLC_{ox} \left[1 - \left(\frac{V_{gt} - V_{ds}}{2V_{gt} - V_{ds}} \right)^2 \right]$$

$$C_{GD} = \frac{\partial Q_G}{\partial V_{gd}} = \frac{2}{3} WLC_{ox} \left[1 - \left(\frac{V_{gt}}{2V_{gt} - V_{ds}} \right)^2 \right] \qquad (6.20)$$

$$C_{GB} = \frac{\partial Q_G}{\partial V_{gb}} = 0$$

In the saturation regime, we can obtain the expression for Q_G by replacing V_{ds} in Equation 6.18 by V_{dsat}. We know that for a long channel device in saturation, $V_{dsat} = V_{gs} - V_{th}$, and from Figure 6.2b, we get: $V_{ds} = V_{gs} - V_{gd} = V_{dsat}$ ($= V_{gs} - V_{th}$). Therefore, in the saturation region, $V_{gd} = V_{th}$. Then, substituting for $V_{gd} = V_{th}$ in Equation 6.18, we get

$$Q_G = \frac{2}{3} WLC_{ox} \left[\frac{(V_{th} - V_{th})^3 - (V_{gs} - V_{th})^3}{(V_{th} - V_{th})^2 - (V_{gs} - V_{th})^2} \right] - Q_B$$

$$= \frac{2}{3} WLC_{ox} (V_{gs} - V_{th}) - Q_B \qquad (6.21)$$

From Equation 6.21, we get the *saturation region* intrinsic capacitances at $V_{ds} > V_{dsat}$

$$C_{GS} = \frac{\partial Q_G}{\partial V_{gs}} = \frac{2}{3} WLC_{ox}$$

$$C_{GD} = \frac{\partial Q_G}{\partial V_{gd}} = 0 \qquad (6.22)$$

$$C_{GB} = \frac{\partial Q_G}{\partial V_{gb}} = 0$$

Note that the saturation region capacitances are independent of V_{ds}. Since, in saturation, the channel is pinched off at the drain end, it is electrically isolated from the drain. Thus, Q_G is not influenced by a change in V_{ds} and the capacitances are independent of V_{ds}.

6.2.2.2 Weak Inversion

In the weak inversion region ($V_{gs} < V_{th}$), $Q_i \ll Q_b$ so that Equation 6.11 becomes

$$Q_G = -Q_B = W \int_0^L Q_b(y)dy = WLQ_b \tag{6.23}$$

Under the depletion approximation, the depletion charge density in the bulk for long channel devices is given by (Equation 4.101)

$$Q_b = -C_{ox}\gamma\sqrt{\phi_{ss}} \tag{6.24}$$

where the surface potential ϕ_{ss} in weak inversion is given by Equation 4.104

$$\phi_{ss} = \left[-\left(\frac{\gamma}{2}\right) + \sqrt{\frac{\gamma^2}{4} + V_{gb} - V_{fb}} \right]^2 \tag{6.25}$$

Thus, ϕ_{ss}, is practically independent of the position y along the channel. This means that Q_b is independent of position along the channel. Therefore, using for Q_b from Equation 6.24 and ϕ_{ss} from Equation 6.25, we get the expression for the gate charge in weak inversion from Equation 6.23 as

$$Q_G = -Q_B = -\frac{1}{2}WLC_{ox}\gamma^2 \left[1 - \sqrt{1 + \frac{4}{\gamma^2}\left(V_{gb} - V_{fb}\right)} \right] \tag{6.26}$$

Now, differentiating Equation 6.26 with respect V_{gb} gives the gate-to-bulk capacitance C_{GB} in the subthreshold or weak inversion region as

$$C_{GB} = \frac{\partial Q_G}{\partial V_{gb}} = \frac{WLC_{ox}}{\sqrt{1 + \left(4/\gamma^2\right)\left(V_{gb} - V_{fb}\right)}} \tag{6.27}$$

In deriving Equation 6.27, we assumed that γ is constant independent of V_{bs}. This is true only for a uniformly doped substrate. In reality, MOSFETs are nonuniformly doped and γ is bias dependent as discussed in Chapter 4. Therefore, appropriate value of γ and its derivative must be used for accurate modeling of C_{GB} in the weak inversion regime of MOSFETs. Since in weak inversion, Q_G does not depend on V_{ds}, we can safely write

$$C_{GS} = \frac{\partial Q_G}{\partial V_{gs}} = 0$$

$$C_{GD} = \frac{\partial Q_G}{\partial V_{gd}} = 0 \tag{6.28}$$

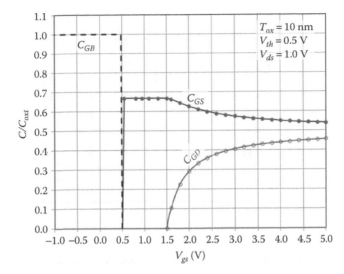

FIGURE 6.3

Plots of the intrinsic capacitances C_{GS}, C_{GD}, and C_{GB} associated with the gate terminal of MOSFET devices as a function of gate voltage V_{gs} for $V_{ds} = 1$ V; the plots are obtained by Equations 6.20, 6.22, and 6.27.

At $V_{gs} = V_{fb}$, the calculated value of C_{GB} from Equation 6.27 is not accurate due to the failure of the depletion approximation used in deriving Equation 6.27. However, because of the simplicity of calculation, Equation 6.27 is used for computing C_{GB} at weak inversion.

Figure 6.3 shows normalized plots of three capacitances as a function of V_{gs} for $V_{ds} = 1$. The capacitances are normalized with respect to the total gate capacitance given by

$$C_{oxt} = WLC_{ox} \tag{6.29}$$

Finally, the *accumulation* region capacitances are given by: $C_{GB} = C_{oxt}$, and $C_{GS} = 0 = C_{GD}$.

The gate capacitance C_{oxt} is the maximum capacitance of a MOSFET device that occurs in the accumulation condition. In the inversion region, that is, in the active mode of operation of the device, the maximum capacitance occurs in saturation and is equal to $(2/3)C_{oxt}$ as shown in Equation 6.22.

The Meyer model can be represented by a simple equivalent circuit as shown in Figure 6.4.

In Figure 6.4, C_{JS} and C_{JD} are the source-body and drain-body *pn*-junction capacitances, respectively.

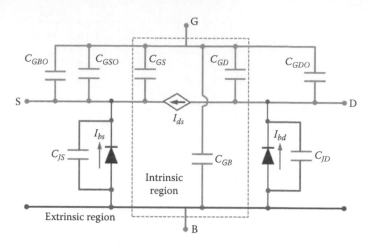

FIGURE 6.4
Complete equivalent circuit of a MOSFET device showing the extrinsic and Meyer's intrinsic capacitances.

6.2.3 Limitations of Meyer Model

The Meyer model is simple and predicts acceptable simulation results for most circuit analysis since its implementation in SPICE [10]. However, it is found to generate nonphysical simulation results when used to model circuits with charge storage nodes. The model incorrectly predicts the charge built up on these nodes in circuit simulation. It is found that the Meyer model is inadequate in predicting accurate capacitances in circuits such as MOS (metal-oxide-semiconductor) charge pumps [11], silicon on sapphire [4], dynamic random access memory, and switched-capacitor circuits [8]. This inaccuracy in simulation results when using the Meyer model is due the (1) charge nonconservation and (2) nonphysical reciprocity assumption.

The charge nonconservation problem has been extensively analyzed [8,11,12]. The detailed investigation of the Meyer model reveals that the incorrect implementation of the model in circuit CAD causes charge nonconservation [11]. However, in order to ensure charge conservation in modeling MOSFET capacitances, it is required to assign charges at each terminal of the device. With quasistatic assumption, the charges at any time t only depend on the values of the terminal voltages at the same time so that we can write

$$Q_j = Q_j\left(V_{gs}, V_{gd}, V_{gb}\right), \quad \text{where } j = G, S, D, B \tag{6.30}$$

Thus, the capacitance C_{ji} with $i = G$ (e.g., C_{GG}, C_{DG}, C_{SG}, and C_{BG}) in a MOSFET must satisfy the relation

$$C_{ji}\left(V_{gs}, V_{gd}, V_{gb}\right) \equiv \frac{dQ_j}{dV_g}, \quad \text{where } j = G, S, D, B; \text{ and } i = G \tag{6.31}$$

and the sum of the charges in the device must satisfy the law of charge conservation given by Equation 6.3, that is,

$$\sum_j Q_j = 0, \quad \text{where } j = G, S, D, B \tag{6.32}$$

In addition to the charge nonconservation problem, the assumption of capacitance reciprocity, $C_{ij} = C_{ji}$, in the Meyer model is more critical. It is shown that the assumption of reciprocity is inconsistent with the charge conservation law [13,14]. The detailed analysis shows that in order to ensure charge conservation principle, the reciprocity of the Meyer model requires Q_S to depend only on V_{gs} and Q_D to depend only on V_{gd}. This implies that $C_{GS} = C_{SG} \equiv dQ_S/dV_{gd}$ cannot be a function of V_{ds} or V_{bs} [14]. In reality, the channel charge can be modulated by both V_{ds} and V_{bs}. Therefore, the assumption of capacitance reciprocity is nonphysical. The nonreciprocal effect in MOSFETs is due to the fact that the channel charge is controlled by three or more terminal voltages. And, the reciprocal capacitors simply cannot be used to model the capacitive effects in a MOSFET device.

6.3 Charge-Based Capacitance Model

The charge-based capacitance modeling is one of the approaches to solve charge nonconservation problem in MOSFET capacitance modeling [4,15,16]. In this approach, the charges in the drain, gate, source, and bulk of a MOSFET are determined to use them as state variables in circuit simulation. Transient currents and the capacitances are obtained by differentiating the charges with respect to time and voltage, respectively. The charge-based capacitance model automatically ensures the charge conservation, as long as Equation 6.3 is satisfied, that is,

$$Q_G + Q_S + Q_D + Q_B = 0 \tag{6.33}$$

Since the terminal charge Q_j ($j = G, D, S, B$) is a function of terminal voltages V_g, V_s, V_d, and V_b, we can write the terminal current, i_j, as

$$i_j = \frac{dQ_j}{dt} = \frac{\partial Q_j}{\partial V_g}\frac{\partial V_g}{\partial t} + \frac{\partial Q_j}{\partial V_d}\frac{\partial V_d}{\partial t} + \frac{\partial Q_j}{\partial V_s}\frac{\partial V_s}{\partial t} + \frac{\partial Q_j}{\partial V_b}\frac{\partial V_b}{\partial t} \tag{6.34}$$

Equation 6.34 shows that each terminal of a MOSFET device has a capacitance with respect to the remaining three terminals. Thus, a four-terminal device has 16 capacitances that include 4 self-capacitances corresponding to its four terminals and 12 nonreciprocal intrinsic capacitances. The 16 capacitances form the so called indefinite admittance matrix. Each element C_{ij} of this capacitance matrix describes the dependence of the charge at the terminal i with respect to the voltage applied at the terminal j with all other voltages held constant. For example, C_{GS} specifies the rate of change of Q_G with respect to the source voltage V_s keeping the voltages at the other terminals (V_g, V_d, and V_b) constant. Thus, in general

$$
C_{ij} = \begin{cases} -\dfrac{\partial Q_i}{\partial V_j}, & i \neq j; \quad i,j = G, S, D, B \\[3mm] \dfrac{\partial Q_i}{\partial V_j}, & i = j \end{cases} \tag{6.35}
$$

In Equation 6.35, the sign of any C_{ij} is chosen to keep all of the capacitance terms positive for well-behaved devices, that is, devices for which the charge at a node increases with an increase in the voltage at that node whereas decreases with an increase in the voltage at any other node. All 16 capacitances of the matrix C_{ij}, shown here, are not independent.

$$
C_{ij} = \begin{bmatrix} C_{GG} & -C_{GD} & -C_{GS} & -C_{GB} \\ -C_{DG} & C_{DD} & -C_{DS} & -C_{DB} \\ -C_{SG} & -C_{SD} & C_{SS} & -C_{SB} \\ -C_{BG} & -C_{BD} & -C_{BS} & C_{BB} \end{bmatrix} \tag{6.36}
$$

In Equation 6.36, each row must sum to zero for the matrix to be reference-independent and each column must sum to zero for the device description to be charge-conservative, which is equivalent to obeying KCL. One of these four capacitances, corresponding to each terminal of the device, is the self-capacitance, which is the sum of the remaining three capacitances. Thus, for example, the gate capacitance C_{GG} is given by

$$
C_{GG} = C_{GS} + C_{GD} + C_{GB} \tag{6.37}
$$

The 12 *inter-nodal* or *intrinsic capacitances* of a MOSFET device are also called the *trans-capacitances*. And, these capacitances are nonreciprocal. Thus, for example, C_{DG} and C_{GD} differ both in value and physical interpretation. Out of the 12 trans-capacitances, only 9 are independent: C_{GB}, C_{GS}, C_{GD}, C_{BG}, C_{BS}, C_{BD}, C_{DG}, C_{DS}, and C_{DB}. Therefore, if we evaluate the independent nine capacitances, then the other three capacitances C_{SG}, C_{SD}, and C_{SB} can be determined from the following relations

$$C_{SG} = C_{GB} + C_{GD} + C_{GS} - C_{BG} - C_{DG}$$

$$C_{SD} = C_{BG} + C_{BD} + C_{BS} - C_{GB} - C_{DB} \qquad (6.38)$$

$$C_{SB} = C_{DG} + C_{DB} + C_{DS} - C_{GD} - C_{BD}$$

Thus, it is evident from Equation 6.36 that to calculate MOSFET intrinsic capacitances we need to calculate the charges Q_G, Q_D, Q_S, and Q_B as a function of terminal voltages, and if we take these charges as independent state variables, then charge conservation will be guaranteed. Thus, charge-based capacitance model is obtained by integrating the terminal charges Q_G and Q_B given in Equations 6.6 and 6.7 over the length of the channel under the charge conservation principle given by Equation 6.8. Thus, Q_G and Q_B can be easily obtained by integrating the corresponding charge per unit area over the active gate region. However, Q_S and Q_D can only be determined from the channel charge Q_I, because both source and drain terminals are in intimate contact with the channel region. Therefore, it is necessary to partition the channel charge into charge Q_D associated with the drain terminal and a charge Q_S associated with the source terminal, such that

$$Q_I = Q_S + Q_D \qquad (6.39)$$

Although this partition of Q_I into $(Q_S + Q_D)$ is not physically accurate, it does lead to MOSFET capacitance model, which agrees with the experimental results.

Channel Charge Partition: There are various approaches to partition Q_I into Q_S and Q_D [4 9,15–19]. These approaches vary from an equal division of Q_I across both terminals ($Q_S = Q_D = 0.5Q_I$) [7] to a Q_I multiplied by a "linear partitioning" or "weighted function" [4]. However, the channel-charge partition scheme proposed by Ward and Dutton [4] agrees very well with the experimental results.

The Ward–Dutton partition is derived from 1D (one-dimensional) continuity equation. Neglecting the generation-recombination in the channel region, 1D continuity equation (Equation 2.80 or 2.81) at a point y along the channel at any instant t can be expressed as

$$\frac{\partial I(y,t)}{\partial y} = -W \frac{\partial Q_i(y,t)}{\partial t} \qquad (6.40)$$

Integrating Equation 6.40 along the channel from the source ($y = 0$) to an arbitrary point y along the channel, we get

$$\int_0^y \frac{\partial I(y',t)}{\partial y'} dy' = -W \int_0^y \frac{\partial Q_i(y',t)}{\partial t} dy'$$

or   (6.41)

$$I(y,t) - I(0,t) = -W \int_0^y \frac{\partial Q_i(y',t)}{\partial t} dy'$$

Again, integrating Equation 6.41 along the entire length of the channel, we get

$$\int_0^L I(y,t) dy - \int_0^L I(0,t) dy = -W \int_0^L \int_0^y \frac{\partial Q_i(y',t)}{\partial t} dy' dy \qquad (6.42)$$

Since the integration is at any instant t, the right-hand side of the above equation can be rewritten by taking the time derivative outside the integral. Then integrating by parts and simplifying the resulted expression, we can show

$$I(0,t) = \frac{1}{L} \int_0^L I(y,t) dy + \frac{W}{L} \frac{\partial}{\partial t} \int_0^L \left(1 - \frac{y}{L}\right) Q_i dy \qquad (6.43)$$

Equation 6.43 is the expression for the channel current at the position $y = 0$ at any time t, that is, the total current flowing through the source contact. The first term on the right-hand side is the average transport current in the channel at time t, that is, the DC current under quasistatic operation. Comparing Equation 6.43 with the expression for $i_s(t)$ in Equation 6.4, we find that the charge Q_S associated with the source is

$$Q_S = -W \int_0^L \left(1 - \frac{y}{L}\right) Q_i dy \qquad (6.44)$$

An expression similar to Equation 6.43 can be derived for the drain current and the charge Q_D associated with the drain can be shown as

$$Q_D = -W \int_0^L \frac{y}{L} Q_i dy \qquad (6.45)$$

Thus, we can now calculate the terminal charges Q_G, Q_B, Q_S, and Q_D from Equations 6.6, 6.7, 6.44, and 6.45, respectively, using the expression for Q_I from Equation 6.8 to ensure charge conservation. First of all, we will derive the charge expressions for the long channel devices and then modify those charge expressions for short channel devices. In general, the expressions for Q_i and Q_b required for deriving the charge expressions (Equations 6.6 and 6.7) can be used from any DC current model for a MOSFET. However, in the following section, the widely used regional DC current model in circuit CAD tools is used to derive the expressions for charge-based capacitance model.

6.3.1 Long Channel Charge Model

In this section, the terminal charges are derived using the regional DC current model discussed in Section 4.4.4. Thus, similar to drain current model, the charge-based model also consists of different expressions for terminal charges for different regions of device operations.

6.3.1.1 Strong Inversion

In Equation 4.95, the channel charge density Q_i for a long channel MOSFET device is shown as

$$Q_i(y) = -C_{ox}\left[V_{gs} - V_{th} - \alpha V(y)\right] \qquad (6.46)$$

and in Equation 4.91, the bulk-charge density for a long channel device is shown as

$$Q_b(y) = -C_{ox}\gamma\left[\delta V(y) + \sqrt{2\phi_B + V_{sb}}\right] \qquad (6.47)$$

Since the total charge in the system must be zero, that is $Q_g + Q_i + Q_b = 0$, using Q_i and Q_b from Equations 6.46 and 6.47, respectively, we get

$$\begin{aligned}Q_g(y) &= C_{ox}\left[V_{gs} - V_{th} - \alpha V(y) + \delta\gamma V(y) + \gamma\sqrt{2\phi_B + V_{sb}}\right] \\ &= C_{ox}\left[V_{gs} - \left(V_{th} - \gamma\sqrt{2\phi_B + V_{sb}}\right) - (\alpha - \delta\gamma)V(y)\right]\end{aligned} \qquad (6.48)$$

Now, from the threshold voltage (V_{th}) Equation 4.10 of a MOSFET device, we can show that $V_{th} - \gamma\sqrt{2\phi_B + V_{sb}} = V_{fb} + 2\phi_B$, and from Equation 4.96, we get, $(\alpha - \delta\gamma) = 1$. Then, Equation 6.48 can be expressed as

$$Q_g(y) = C_{ox}\left[V_{gs} - V_{fb} - 2\phi_B - V(y)\right] \qquad (6.49)$$

Similarly, using Equations 4.10 and 4.96, Equation 6.47 can be expressed as

$$Q_b(y) = -C_{ox}\left[V_{th} - 2\phi_B - V_{fb} - (1 - \alpha)V(y)\right] \qquad (6.50)$$

Equations 6.46, 6.49, and 6.50 are used to calculate the terminal charges using Equations 6.6 and 6.7 along with Equations 6.44 and 6.45 for charge partitioning. Let us first calculate Q_S and Q_D using Equations 6.44 and 6.45, respectively. Since $Q_i(y)$ is known as a function of V, we first change the variable of integration dy in Equations 6.44 and 6.45 to dV using Equation 6.14 to get

$$Q_S = -\frac{\mu_s W^2}{I_{ds}} \int_{V_s}^{V_d} \left(1 - \frac{y}{L}\right) Q_i \cdot Q_i dV$$

$$Q_D = -\frac{\mu_s W^2}{I_{ds}} \int_{V_s}^{V_d} \frac{y}{L} Q_i \cdot Q_i dV$$

(6.51)

To express y in terms of V_{ds} in Equation 6.51, we integrate Equation 6.14 from $(y = 0, V = V_s = 0)$ to an arbitrary point (y, V) along the length of the channel using Equation 6.46 for Q_i. This yields

$$y = -\frac{W\mu_s}{I_{ds}} \int_0^V Q_i dV = \frac{\mu_s W C_{ox}}{I_{ds}} \left(V_{gs} - V_{th} - \frac{1}{2}\alpha V\right) V$$

(6.52)

At the drain end $y = L$ and $V = V_{ds}$ so that we have

$$I_{ds} = \mu_s C_{ox} \left(\frac{W}{L}\right) \left[V_{gs} - V_{th} - \frac{1}{2}\alpha V_{ds}\right] V_{ds}; \quad V_{gs} > V_{th}$$

(6.53)

Now combining Equation 6.51 with Equations 6.46 and 6.52 and carrying out the integration, we get after simplification the terminal charges in the *linear region* of device operation as

$$Q_D = -WLC_{ox} \left[\frac{1}{2}(V_{gs} - V_{th}) - \frac{1}{3}\alpha V_{ds} + AB\right]$$

$$Q_S = -WLC_{ox} \left[\frac{1}{2}(V_{gs} - V_{th}) - \frac{1}{6}\alpha V_{ds} + A(1 - B)\right]$$

(6.54)

where the parameters A and B are defined as

$$A = \frac{\alpha^2 V_{ds}^2}{12\left((V_{gs} - V_{th}) - (1/2)\alpha V_{ds}\right)}$$

$$B = \frac{5(V_{gs} - V_{th}) - 2\alpha V_{ds}}{10\left((V_{gs} - V_{th}) - (1/2)\alpha V_{ds}\right)}$$

(6.55)

When $V_{ds} = 0$, it is found from Equations 6.54 and 6.55, $Q_S = Q_D = (1/2)WLC_{ox}(V_{gs} - V_{th})$, which is obvious because of the symmetry.

The total gate charge Q_G can be obtained by integrating the gate charge density Q_g over the area of the active gate region as

$$Q_G = W \int_0^L Q_g(y)dy = \frac{\mu_s W^2}{I_{ds}} \int_0^{V_d} Q_g \cdot Q_i dV \qquad (6.56)$$

where we have replaced the differential channel length dy with the corresponding differential potential drop dV using Equation 6.14. Substituting for Q_i and Q_g from Equations 6.46 and 6.49, respectively, and carrying out the integration results in the following expression for the charge Q_G, we get

$$Q_G = WLC_{ox}\left[V_{gs} - V_{fb} - 2\phi_B - \frac{1}{2}V_{ds} + \frac{A}{\alpha}\right] \qquad (6.57)$$

Similarly, the total bulk charge Q_B can be written as

$$Q_B = W \int_0^L Q_b(y)dy = -\frac{\mu_s W^2}{I_{ds}} \int_0^{V_{ds}} Q_b \cdot Q_i dV \qquad (6.58)$$

Again, substituting Q_i and Q_b from Equations 6.46 and 6.47 (or 6.50), respectively, and carrying out the integration yields

$$Q_B = -WLC_{ox}\left[\gamma\sqrt{2\phi_B + V_{sb}} - (1-\alpha)V_{ds}D\right] \qquad (6.59)$$

where the parameter D is defined as

$$D = \frac{3(V_{gs} - V_{th}) - 2\alpha V_{ds}}{6\left[(V_{gs} - V_{th}) - (1/2)\alpha V_{ds}\right]} \qquad (6.60)$$

It is seen from the first expression in Equation 6.59 that the bulk charge consists of two terms. The first term gives the total bulk charge due to the back bias V_{sb} and is related to the threshold voltage. The second term describes the additional charge induced by the drain bias. The second term reduces to zero when $V_{ds} = 0$.

It is very easy to verify that the sum of Q_G, Q_S, Q_D, and Q_B is zero.

Equations 6.54, 6.57, and 6.59 are the terminal charges for the *linear region* of the device operation. The corresponding charges in the *saturation region* are obtained by replacing V_{ds} by $V_{dsat} = (V_{gs} - V_{th})/\alpha$ (Equation 4.98), in the expressions for terminal charges in the linear region. Thus, the expressions for terminal charges Q_S, Q_D, Q_G, and Q_B in the *saturation region* are given by

$$Q_D = -\frac{4}{15} WLC_{ox} \left(V_{gs} - V_{th} \right)$$

$$Q_S = -\frac{2}{5} WLC_{ox} \left(V_{gs} - V_{th} \right)$$

$$Q_G = WLC_{ox} \left[V_{gs} - V_{fb} - 2\phi_B - \frac{1}{3\alpha} \left(V_{gs} - V_{th} \right) \right] \tag{6.61}$$

$$Q_B = -WLC_{ox} \left[V_{th} - V_{fb} - 2\phi_B + \frac{1}{3\alpha} (1-\alpha) \left(V_{gs} - V_{th} \right) \right]$$

$$\therefore Q_I = Q_s + Q_D = -\frac{2}{3} WLC_{ox} \left(V_{gs} - V_{th} \right)$$

It is observed from Equation 6.61 that the terminal charges in the saturation region are independent of V_{ds}. This is due to the fact that because of the channel pinch-off near the drain end of the device in saturation, the drain has no influence on the behavior of the device. Also, it is observed that the mobility degradation factor due to the gate field does not appear in the charge expressions. This is because of the global way of modeling the mobility, which cancels out while deriving the charges. Numerical device simulation results show that the mobility degradation has little effect on the terminal charges, thus validating the results obtained by analytical charge-based model [13].

6.3.1.2 Weak Inversion

In the weak inversion region of a MOSFET device, though the number of mobile charges at the interface is small, these charges are important for modeling the switching behavior of the device. Also, in this region, $Q_b \gg Q_i$, and therefore, the bulk charges are not shielded by the inversion charge and behave differently compared to the strong inversion condition.

In order to derive expressions for the terminal charges in weak inversion, we assume that the current transport occurs by diffusion only as discussed in deriving the subthreshold drain current expression in Chapter 4. Indeed, this is a valid approximation for low gate voltages as discussed in Section 4.4.4.4. Then from Equation 4.106, the drain current at any point y along the channel is given by

$$I_{ds} = \mu_s W v_{kT} \frac{dQ_i}{dy} \tag{6.62}$$

Integrating Equation 6.62 from ($y = 0$, $Q_i = Q_{is}$) to any point (y, Q_i) along the channel and after simplification we can show

$$y = \frac{\mu_s W}{I_{ds}} v_{kT} \left(Q_i - Q_{is} \right) \tag{6.63}$$

where:

v_{kT} is the thermal voltage

Q_{is} is the mobile charge density at the source end

At the drain end of the channel $Q_i = Q_{id}$.

Let us first calculate the source and drain charge Q_S and Q_D, respectively. Substituting for dy and y from Equations 6.62 and 6.63, respectively, to the expression for Q_D in Equation 6.45, we get

$$Q_D = \frac{W}{L} \left(\frac{\mu_s W}{I_{ds}} \right)^2 v_{kT}^2 \int_{Q_{is}}^{Q_{id}} Q_i \left(Q_i - Q_{is} \right) dQ_i \tag{6.64}$$

which on integration and after simplification using Equation 6.62 for I_{ds} can be shown as

$$Q_D = \frac{1}{6} WL \left(2Q_{id} - Q_{is} \right) \tag{6.65}$$

Now, substituting for the charge densities Q_{is} and Q_{id} from Equation 4.113, we get the expression for the drain charge Q_D as

$$Q_D = -\frac{1}{6} WLC_d v_{kT} \exp\left(\frac{V_{gs} - V_{th}}{n v_{kT}} \right) \cdot \left[2\exp\left(-\frac{V_{ds}}{v_{kT}} \right) + 1 \right] \tag{6.66}$$

where we have used Equation 4.117 to eliminate ϕ_B from the expressions for Q_{is} and Q_{id} in Equation 4.113. Equation 6.66 can also be expressed by using Equation 4.121 relating the depletion capacitance C_d and the ideality factor $n = \left[1 + \left(C_d / C_{ox} \right) \right]$ as

$$Q_D = -\frac{1}{6} WLC_{ox} (n-1) v_{kT} \exp\left(\frac{V_{gs} - V_{th}}{n v_{kT}} \right) \cdot \left[2\exp\left(-\frac{V_{ds}}{v_{kT}} \right) + 1 \right] \tag{6.67}$$

Using similar procedures we can show that the expression for the source charge Q_S in the weak inversion region is given by

$$Q_S = -\frac{1}{6} WLC_{ox} (n-1) v_{kT} \exp\left(\frac{V_{gs} - V_{th}}{n v_{kT}} \right) \cdot \left[\exp\left(-\frac{V_{ds}}{v_{kT}} \right) + 2 \right] \tag{6.68}$$

From Equations 6.67 and 6.68, we observe that at $V_{ds} = 0$ and $V_{gs} = V_{th}$, $Q_D = Q_S = 0.5\ WLC_{ox}(n - 1)v_{kT}$. It is also observed from Equations 6.67 and 6.68 that Q_S and Q_D depend weakly on V_{ds}. This is due to fact that for V_{ds} greater than a few v_{kT}, the terms involving V_{ds} become negligible and therefore, $Q_S = 2Q_D$.

Since in weak inversion, the bulk charge Q_B is virtually independent of the S/D voltage V_{ds}, we can use Equation 6.24 for Q_B, which at the boundary of the strong inversion can be rewritten as

$$Q_B = -WLC_{ox}\gamma\sqrt{2\phi_B + V_{sb}} \tag{6.69}$$

Equation 6.69 is the same as the first term of the first expression in Equation 6.59. If the channel charge is assumed zero ($Q_I = 0$) in the subthreshold region, the gate charge becomes equal to the bulk charge. Thus, $Q_G = -Q_B$.

6.3.1.3 Accumulation

In the accumulation region of a MOSFET device operation, $V_{gb} < V_{fb}$; thus a thin layer of majority carriers are formed at the interface, forming a parallel plate capacitor with the gate. In this case, the bulk charge Q_B is simply written as

$$Q_B = -WLC_{ox}\left(V_{gs} + V_{sb} - V_{fb}\right) \tag{6.70}$$

Since there is no current flow, the gate charge is given by

$$Q_G = -Q_B = WLC_{ox}\left(V_{gs} + V_{sb} - V_{fb}\right) \tag{6.71}$$

6.3.2 Long Channel Capacitance Model

We can now derive the expressions for capacitances associated with a MOSFET using the equations derived for various charges in different regions of device operation and the definition of C_{ij} in Equation 6.35. The mathematics to derive 12 capacitances is basic, however involved and long. It is left as an exercise for the readers.

The expressions for C_{GD} and C_{DG} in the *linear region* are obtained by differentiating Q_G (Equation 6.57) with respect to V_d (or V_{ds}) and Q_D (Equation 6.54) with respect to V_g or V_{gs}, respectively, and using A and B defined in Equation 6.55, that is,

$$C_{GD} = -\frac{\partial Q_G}{\partial V_d} = \frac{1}{2}WLC_{ox}\left[-1 + \frac{1}{\left(V_{gs} - V_{th}\right) - (1/2)\alpha V_{ds}}\left(A + \frac{1}{3}\alpha V_{ds}\right)\right]$$

$$C_{DG} = -\frac{\partial Q_D}{\partial V_g} = \frac{1}{2}WLC_{ox}\left[1 + \frac{A}{\left(V_{gs} - V_{th}\right) - (1/2)\alpha V_{ds}}(1 - 4B)\right] \tag{6.72}$$

The corresponding capacitances in the *saturation* region are obtained, either by differentiating the corresponding charges derived for saturation region (Equation 6.61) or by replacing V_{ds} with $V_{dsat} = (V_{gs} - V_{th})/\alpha$, in Equation 6.72. Thus, in the saturation we can show

$$C_{GD} = 0$$

$$C_{DG} = -\frac{4}{15} WLC_{ox}$$

(6.73)

Equations 6.72 and 6.73 clearly show the nonreciprocal nature of MOSFET terminal capacitances. It should be pointed out that C_{GD} is most important among the gate capacitances because its effect is multiplied by the voltage gain between the drain and gate nodes due to the Miller effect.

The expressions for C_{GS} and C_{SG} are obtained by differentiating Q_G (Equation 6.57) with respect to V_s and Q_S (Equation 6.54) with respect to V_g (or V_{gs}), respectively, and using A and B defined in Equation 6.55, that is,

$$C_{GS} = -\frac{\partial Q_G}{\partial V_s}$$

$$= WLC_{ox}\left[-\frac{1}{2} - A\left(\frac{\partial \alpha}{\partial V_{bs}} \frac{1}{\alpha^2} + \frac{2}{\alpha V_{ds}} \right) - \frac{A}{\alpha\left[V_{gs} - V_{th} - (1/2)\alpha V_{ds} \right]} \right.$$

(6.74)

$$\left. \left(-1 + \frac{\partial V_{th}}{\partial V_{bs}} + \frac{\alpha}{2} + \frac{1}{2} \frac{\partial \alpha}{\partial V_{bs}} V_{ds} \right) \right]$$

$$C_{SG} = -\frac{\partial Q_S}{\partial V_g} = \frac{1}{2} WLC_{ox}\left[1 - \frac{A}{V_{gs} - V_{th} - (1/2)\alpha V_{ds}}(3 - 4B) \right]$$

(6.75)

The corresponding capacitances in the saturation region can be shown as

$$C_{GS} = WLC_{ox}\left[-1 - \frac{1}{3\alpha}\left(-1 + \frac{\partial V_{th}}{\partial V_{bs}} \right) - \frac{V_{gs} - V_{th}}{3\alpha^2} \frac{\partial \alpha}{\partial V_{bs}} \right]$$

(6.76)

$$C_{SG} = \frac{1}{5} WLC_{ox}$$

(6.77)

Again, the nonreciprocal nature of the capacitance is self-evident. The detailed model equations with discussions can be found in the literature [3]. Interested readers are encouraged to read the relevant references.

6.3.3 Short Channel Charge Model

In the derivation of long channel terminal charges and capacitances in the previous section, we have neglected the effects of velocity saturation, channel length modulation, and series resistance, since these effects are important only for short channel devices (Chapter 5). As in the case of drain current modeling, we need to consider these in modeling terminal charges for short channel devices. However, the final charge equations including these short channel effects become more complex.

For the simplicity of modeling capacitances for short channel MOSFETs, the long channel charge model has been used by modifying the body effect coefficient, α [8]. However, for accurate modeling of terminal charges and capacitances in short channel devices, short channel effects including carrier velocity saturation, channel length modulation, and S/D series resistance must be considered. In order to include the short channel effects in modeling charges and hence capacitances for short channel devices, I_{ds} expression (Equation 5.72) for short channel devices in the linear region is used. Repeating Equation 5.72, I_{ds} for short channel devices that includes SCE is given by

$$I_{ds} = WC_{ox}\left(V_{gs} - V_{th} - \alpha V\right) \cdot \frac{\mu_{eff}E_y}{1 + \left(E_y/E_c\right)} \tag{6.78}$$

Replacing E_y by $-dV/dy$ and rearranging, we get

$$dy = \left[\frac{\mu_{eff}WC_{ox}}{I_{ds}}\left(V_{gs} - V_{th} - \alpha V\right) - \frac{1}{E_c}\right]dV \tag{6.79}$$

where:
$E_c = 2v_{sat}/\mu_{eff}$ (Equation 5.70)

After integrating Equation 6.79, we get

$$y = \left[\frac{\mu_{eff}WC_{ox}}{I_{ds}}\left(V_{gs} - V_{th} - \frac{1}{2}\alpha V\right) - \frac{1}{E_c}\right]V \tag{6.80}$$

Substituting $y = L$ and $V = V_{ds}$ in Equation 6.80, we get the expression for linear region I_{ds} Equation 5.74. Equations 6.78 through 6.80 account for velocity saturation whereas μ_{eff} accounts for S/D series resistance.

Now, following the procedure used to derive terminal charges and capacitances for long channel MOSFET devices in Sections 6.3.1 and 6.3.2, we get the expressions for the (Q_D) drain and source (Q_S) charges in the linear region of device operation as

$$Q_D = -WLC_{ox}\left[\frac{1}{2}(V_{gs}-V_{th})-\frac{1}{3}\alpha V_{ds}+A'B'\right]$$

$$Q_S = -WLC_{ox}\left[\frac{1}{2}(V_{gs}-V_{th})-\frac{1}{6}\alpha V_{ds}+A'(1-B')\right]$$

(6.81)

where

$$A' = A\cdot\left(1+\frac{V_{ds}}{LE_c}\right)$$

$$B' = B\cdot\left(1+\frac{V_{ds}}{LE_c}\right)$$

(6.82)

and, A and B are defined in Equation 6.55.

Comparing the expressions for Q_D and Q_S in Equation 6.81 for short channel devices with the corresponding expressions for long channel devices in Equation 6.54, we notice that the two equations have the same form differing only in parameters A' and B'. As can be seen from Equation 6.82, A' and B' include the velocity saturation factor. Thus, in the case for a long channel device, the product LE_c is very large, then $A' = A$, $B = B'$, and Equation 6.81 converges to Equation 6.54 as is expected.

Again, using the procedure for deriving Q_G for long channel devices, we can show for short channel devices

$$Q_G = \frac{\mu_s W^2}{I_{ds}}\int_0^{V_d} Q_g\cdot Q_i dV - \frac{W}{E_c}\int_0^{V_d} Q_g dV$$

(6.83)

Substituting for Q_i and Q_g from Equations 6.46 and 6.49, respectively, and carrying out the integration, we get after simplification

$$Q_G = WLC_{ox}\left[V_{gs}-V_{fb}-2\phi_B-\frac{1}{2}V_{ds}+\frac{A'}{\alpha}\right]$$

(6.84)

Here again, for long channel devices Equation 6.84 converges to Equation 6.57. Similarly, we can show the bulk charge expression for short channel devices as

$$Q_B = -WLC_{ox}\left[\gamma\sqrt{2\phi_B+V_{sb}}+(\alpha-1)V_{ds}D'\right]$$

(6.85)

where

$$D' = D - \frac{1}{12\left[V_{gs}-V_{th}-(1/2)\alpha V_{ds}\right]}\cdot\frac{\alpha V_{ds}}{LE_c}$$

(6.86)

and, D is given by Equation 6.60.

In the case of short channel MOSFETs, the terminal charges and capacitances cannot be calculated just by substituting $V_{ds} = V_{dsat}$. However, for short channel devices, where velocity saturation and channel length modulation (CLM) become important, charge near the saturation consists of two components. One is the charge near the source region where the gradual channel approximation can be applied and the other is the charge near the pinch-off region at the drain-end where carrier velocity saturates. This two-section model creates a discontinuity in the capacitances from the linear to saturation regions, similar to the case of drain current modeling. Therefore, often the charge in the pinch-off is ignored for short channel modeling.

The effect of including velocity saturation in the charge expressions is a reduction in the amount of charge from its long channel value, which intuitively makes sense because carriers are velocity saturated. Although the effect of S/D resistance is not taken into account it is possible to include its effect externally.

In weak inversion, Q_I, and hence Q_S and Q_D, are assumed zero, similar to the long channel case. This means that $Q_G = -Q_B$ in weak inversion. For short channel devices, the bulk charge Q_B is still given by Equation 6.24, however, the long channel body factor γ is replaced by an effective value of γ, to account for the reduction in the bulk charge density due to short channel and narrow width effects as discussed in Chapter 5.

6.3.4   Short Channel Capacitance Model

The expressions for the terminal charges for short channel devices given in Section 6.3.3 are used to calculate the corresponding capacitances using the procedure discussed for the long channel devices. The mathematics is basic, however involved. Thus, we will not derive the final expressions for the capacitances.

It is difficult to accurately measure the capacitances for short channel MOSFETs unlike the long channel devices. This is attributed to very small value of capacitances ($\sim 1 \times 10^{-18}$ F) and the difficulty in separating the small transient currents due to the capacitances associated with the source and drain terminals by the large steady-state current (I_{ds}) in small devices. Thus, most reported data on short channel capacitances are on the gate capacitances C_{GS}, C_{GD}, and C_{GB}.

The measured capacitances include the overlap capacitances, and as such, they do not entirely describe intrinsic capacitances. This is particularly true for short channel devices with lightly doped drain (LDD) regions. However, no such bias-dependent overlap is generally observed in short channel conventional S/D pn-junctions. The bias dependence of the overlap capacitance is due to the modulation of the lightly doped regions (n-region for nMOSFETs and p-region for pMOSFETs).

6.4 Gate Overlap Capacitance Model

The S/D overlap capacitances are parasitic elements that originate due to the encroachment of S/D implant profile under the gate region during IC (integrated circuit) fabrication processes. The postimplant thermal processing steps cause lateral diffusion of dopants under the gate and overlap of the S/D regions in the final device structure. Since in MOSFETs, S/D regions are normally symmetrical, the source overlap distance l_{ov} is same as that of the drain (Figure 6.5). Assuming the parallel plate formulation, the overlap capacitance C_{GSO} and C_{GDO} for the source and drain regions, respectively, can be approximated as

$$C_{GSO} = C_{GDO} = \frac{\varepsilon_0 K_{ox}}{T_{ox}} \cdot Wl_{ov} = C_{ox} Wl_{ov} \tag{6.87}$$

From Equation 6.87, the source and drain overlap capacitances per unit width C_{gso} and C_{gdo}, respectively, are given by

$$C_{gso} = C_{gdo} = C_{ox} l_{ov} \tag{6.88}$$

A third overlap capacitance that can be significant is due to the overlap between the gate and bulk as shown in Figure 5.7. This is the capacitance C_{GBO} that occurs due to the overhang of the transistor gate required at one end and is a function of the effective polysilicon width that is equivalent to the drawn channel lengths. Thus, if C_{gbo} is the gate to bulk overlap capacitance per unit length, then the total gate-to-bulk overlap capacitance becomes

$$C_{GBO} = C_{gbo} L_g \tag{6.89}$$

where:
  L_g is the final physical gate length of MOSFET devices

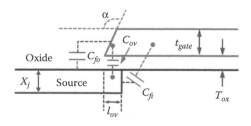

FIGURE 6.5
Different components of MOSFET gate overlap capacitance: C_{ov} due to S/D encroachment length l_{ov} under the gate, C_{fo} outer fringing, and C_{fi} inner fringing.

Typically, C_{GBO} is much smaller than C_{GSO}/C_{GDO} and, therefore, is often neglected.

In a MOSFET, in addition to the outer fringing capacitance, there is another parasitic capacitance that must be taken into account while calculating the overlap capacitance. Thus, MOSFET overlap capacitance can be approximated by the parallel combination of the (1) direct overlap capacitance C_{ov} between the gate and the S/D, (2) outer fringing capacitance C_{fo} on the outer side between the gate and S/D, and (3) inner fringing capacitance C_{fi} on the channel side (inner side) between the gate and side wall of the S/D junction such that [20]

$$C_{ov} = \frac{\varepsilon_0 K_{ox}}{T_{ox}}\left(l_{ov} + \Delta\right)$$

$$C_{fo} = \frac{\varepsilon_0 K_{ox}}{\alpha}\ln\left(1 + \frac{t_{gate}}{T_{ox}}\right) \tag{6.90}$$

$$C_{fi} = \frac{2\varepsilon_0 K_{si}}{\pi}\ln\left(1 + \frac{X_j}{T_{ox}}\sin\beta\right)$$

where:

X_j is the S/D junction depth

Therefore, the total overlap capacitance is given by

$$C_{gso} = C_{gdo} = \frac{\varepsilon_0 K_{ox}}{T_{ox}}\left(l_{ov} + \Delta\right) + \frac{\varepsilon_0 K_{ox}}{\alpha}\ln\left(1 + \frac{t_{gate}}{T_{ox}}\right) + \frac{2\varepsilon_0 K_{si}}{\pi}\ln\left(1 + \frac{X_j}{T_{ox}}\sin\beta\right) \tag{6.91}$$

In Equation 6.91, C_{ov} is the parallel plate component of the effective overlap distance $(l_{ov} + \Delta)$, where Δ accounts for the fact that polysilicon thickness has a slope of angle α. It is a correction factor of higher order and is given by

$$\Delta = \frac{T_{ox}}{2}\left[\frac{1 - \cos\alpha}{\sin\alpha} + \frac{1 - \cos\beta}{\sin\beta}\right] \tag{6.92}$$

where

$$\beta = \left(\frac{\pi}{2}\cdot\frac{K_{ox}}{K_{si}}\right)$$

It is observed from Equation 6.90 that the inner fringing component C_{fi} (channel side) is much larger than the outer fringing component C_{fo} because $K_{si} \cong 3K_{ox}$, and in general, $\alpha > \pi/2$. Thus, it is clear from Equation 6.91

that if the overlap distance l_{ov} approaches zero, there will be an *overlap* capacitance present in MOSFET devices due to the fringing components C_{fo} and C_{fi}.

Although, the inner fringing capacitance C_{fi} is found to be gate and drain bias dependent, C_{fi} calculated from Equation 6.90 is its maximum value. The value of C_{fi} decreases with the increase in the gate voltage from the sub-threshold to strong inversion and approaches to zero in strong saturation. The overlap capacitance is bias dependent, particularly for advance CMOS (complementary metal-oxide-semiconductor) technology and thin gate oxide devices.

In advanced capacitance models [21,22], the bias dependence of overlap capacitance is considered by analytical expressions. Thus, the source overlap capacitance is given by the source charge overlap as

$$\frac{Q_{overlap,s}}{W_{active}} = \begin{cases} C_{GSO}V_{gs} - \dfrac{1}{2}C_{KAPPA}C_{GSL}\left(-1+\sqrt{1-\dfrac{4V_{gs}}{C_{KAPPA}}}\right), & V_{gs} < 0; \\ \\ (C_{GSO}+C_{KAPPA}C_{GSL})\cdot V_{gs}, & V_{gs} \geq 0 \end{cases} \tag{6.93}$$

where:

C_{GSL} and C_{KAPPA} are the model parameters that account for the gate-bias dependence of the gate charge due to the source-body overlap region

Similarly, the drain overlap capacitance is given by

$$\frac{Q_{overlap,d}}{W_{active}} = \begin{cases} C_{GDO}V_{gd} - \dfrac{1}{2}C_{KAPPA}C_{GDL}\left(-1+\sqrt{1-\dfrac{4V_{gd}}{C_{KAPPA}}}\right), & V_{gd} < 0; \\ \\ (C_{GDO}+C_{KAPPA}C_{GDL})\cdot V_{gd}, & V_{gd} \geq 0 \end{cases} \tag{6.94}$$

where:

C_{GDL} is a model parameter that accounts for the gate-bias dependence of the gate charge due to the drain-body overlap region

Then the total charge for the gate overlap over the source and drain regions is given by

$$Q_{overlap,g} = -\left(Q_{overlap,s}+Q_{overlap,d}\right) \tag{6.95}$$

A single equation for the overlap capacitance in both the accumulation and depletion regions is found through the smoothing functions $V_{gs,overlap}$ and $V_{ds,overlap}$ for the source and drain side, respectively.

$$V_{gs,overlap} = \frac{1}{2}\left[\left(V_{gs}+\delta_1\right)-\sqrt{\left(V_{gs}+\delta_1\right)+4\delta_1}\right]$$

$$V_{gd,overlap} = \frac{1}{2}\left[\left(V_{gd}+\delta_2\right)-\sqrt{\left(V_{gd}+\delta_2\right)+4\delta_2}\right]$$

(6.96)

where:
$$\delta_1 = \delta_2 = 0.02 \text{ V}$$

And, the source overlap capacitance is given by the charge in the gate/source overlap region as

$$\frac{Q_{overlap,s}}{W_{active}} = C_{GSO}V_{gs} - C_{GSL}\left[V_{gs}-V_{gs,overlap}-\frac{1}{2}C_{KAPPA}\left(-1+\sqrt{1-\frac{4V_{gs,overlap}}{C_{KAPPA}}}\right)\right]$$

(6.97)

Similarly, the drain overlap capacitance is given from the charge in the gate/drain overlap region

$$\frac{Q_{overlap,d}}{W_{active}} = C_{GDO}V_{gd} - C_{GDL}\left[V_{gd}-V_{gd,overlap}-\frac{1}{2}C_{KAPPA}\left(-1+\sqrt{1-\frac{4V_{gd,overlap}}{C_{KAPPA}}}\right)\right]$$

(6.98)

Finally, the total charge for the gate overlap over the source and drain regions is given by

$$Q_{overlap,g} = -\left(Q_{overlap,s}+Q_{overlap,d}\right)$$

(6.99)

6.5 Limitations of the Quasistatic Model

The analytical expressions derived in Sections 6.2 and 6.3 for modeling the terminal charges and capacitances of a MOSFET device are based on the quasistatic assumption; that is, the terminal voltages vary sufficiently slowly so that the stored charges (Q_G, Q_S, Q_D, and Q_B) can follow the variation in terminal voltages. It has been found that for most of the digital circuits the quasistatic model predicts acceptable results if the rise time τ_r of the waveforms of the applied signal and the transit time τ_t associated with the DC operation of the device satisfy the relation [1]

$$\tau_r > N_{qs}\tau_t \tag{6.100}$$

where:

N_{qs} is a factor with a value between 15 and 25, depending on the application
τ_t is defined as the average time taken by an inversion carrier to travel the length of the channel

that is,

$$\tau_t = \frac{|Q_I|}{I_{ds}} \tag{6.101}$$

Now using Q_I from Equation 6.61 and saturation region I_{ds} from Equation 4.100, we get

$$\tau_t = \frac{4}{3}\alpha\frac{L^2}{\mu_s\left(V_{gs}-V_{th}\right)} = \frac{4}{3}\alpha\frac{L^2}{\mu_s V_{dsat}} \tag{6.102}$$

Equation 6.102 shows that the transit time is proportional to L^2. Thus, the transit time decreases with the decrease in L, resulting in higher speed of device operation. If the carriers are velocity saturated, then Equation 6.102 becomes invalid and the expressions for Q_I and I_{ds} discussed in Section 5.3 must be used to derive τ_t. However, a simple estimate for τ_t can be made by assuming that carriers are moving from source to drain with their scattering limited saturation velocity v_{sat} for the entire length of the channel rather than only a part of the channel. Since carriers cannot move faster than v_{sat}, the time required for the drain current to respond to the changes in the gate voltage is simply v_{sat}/L. Thus, in general

$$\tau_t > \frac{L}{v_{sat}} \tag{6.103}$$

For long channel devices it can be shown from Equation 6.103 that the switching is limited by the parasitic capacitances rather than the time required for the charge redistribution within the transistor itself. Thus, quasistatic operation is valid for modeling intrinsic capacitances of the most long channel MOSFET devices.

It should be pointed out that Equation 6.100 is only a rough rule of thumb and often, due to the significant extrinsic parasitic capacitances, this rule is not restrictive. For modeling nanoscale devices the time dependence in the basic charge equations must be considered. The resulting analysis is called non-quasistatic analysis.

6.6 S/D *pn*-Junction Capacitance Model

Source-drain *pn*-junction capacitances consist of three components: the bottom junction capacitance, sidewall junction capacitance along the isolation edge, and sidewall junction capacitance along the gate edge. An analogous set of equations are used for both sides but each side has a separate set of model parameters.

In Chapter 2, we have shown that the expression for junction diode capacitance is given by

$$C_j = \frac{C_{j0}}{\left[1 - \left(V_d / \phi_{bi}\right)\right]^{mj}} \tag{6.104}$$

For IC *pn*-junctions, the value mj is in the range of $0.2 < mj < 0.6$. Plots of Equation 6.104 show that the capacitance C_j decreases as the reverse-biased $|V_d|$ increases ($V_d < 0$) as shown in Figure 2.28. However, Equation 6.104 shows that when the *pn*-junction is forward biased ($V_d > 0$), the capacitance C_j increases and becomes infinite at $V_d = \phi_{bi}$ as observed in Figure 2.28 (curve 1). This is because Equation 6.104 no longer applies due to depletion approximation becoming invalid. For simplicity of modeling forward-biased *pn*-junction capacitances, we can make a series expansion of Equation 6.104. Thus, for modeling the forward-biased *pn*-junction, Equation 6.104 is simplified by a series expansion of the denominator and by neglecting the higher order terms we can show

$$\left(1 - \frac{V_d}{\phi_{bi}}\right)^{-mj} = 1 + mj\frac{V_d}{\phi_{bi}} + \cdots \tag{6.105}$$

Then Equation 6.104 can be written as

$$C_j = \begin{cases} \dfrac{C_{j0}}{\left(1 - \dfrac{V_d}{\phi_{bi}}\right)^{mj}}; & V_d < 0 \\[4ex] C_{j0}\left(1 + mj\dfrac{V_d}{\phi_{bi}}\right); & V_d > 0 \end{cases} \tag{6.106}$$

6.6.1 Source-Body *pn*-Junction Diode

The source-side *pn*-junction capacitance can be calculated by

$$C_{bs} = A_{seff}C_{jbs} + P_{seff}C_{jbssw} + W_{effcj}NF \cdot C_{jbsswg} \tag{6.107}$$

where:

C_{jbs} is the unit-area bottom source-body junction capacitance

C_{jbssw} is the unit-length source-body junction sidewall capacitance along the isolation edge

C_{jbsswg} is the unit-length source-body junction sidewall capacitance along the gate edge

A_{seff} and P_{seff} are the effective junction area and perimeter, on the source-side of the S/D diffusion, respectively

W_{effcj} and NF are the effective width of the S/D diffusions and the number of fingers, respectively

The components of the S/D pn-junction capacitances are obtained using the following expressions:

- C_{jbs} is calculated by

$$C_{jbs} = \begin{cases} CJS(T) \cdot \left(1 - \dfrac{V_{bs}}{PBS(T)}\right)^{-MJS}, & V_{bs} < 0 \\[4mm] CJS(T) \cdot \left(1 + MJS\dfrac{V_{bs}}{PBS(T)}\right), & V_{bs} \geq 0 \end{cases} \tag{6.108}$$

- C_{jbssw} is calculated by

$$C_{jbssw} = \begin{cases} CJSWS(T) \cdot \left(1 - \dfrac{V_{bs}}{PBSWS(T)}\right)^{-MJSWS}, & V_{bs} < 0 \\[4mm] CJSWS(T) \cdot \left(1 + MJSWS\dfrac{V_{bs}}{PBSWS(T)}\right), & V_{bs} \geq 0 \end{cases} \tag{6.109}$$

- C_{jbsswg} is calculated by

$$C_{jbsswg} = \begin{cases} CJSWGS(T) \cdot \left(1 - \dfrac{V_{bs}}{PBSWGS(T)}\right)^{-MJSWGS}, & V_{bs} < 0 \\[4mm] CJSWGS(T) \cdot \left(1 + MJSWGS\dfrac{V_{bs}}{PBSWGS(T)}\right), & V_{bs} \geq 0 \end{cases} \tag{6.110}$$

6.6.2 Drain-Body Junction Diode

The drain-side pn-junction capacitance can be calculated by

$$C_{bd} = A_{deff}C_{jbd} + P_{deff}C_{jbdsw} + W_{effcj}NF \cdot C_{jbdswg} \tag{6.111}$$

where:

C_{jbd} is the unit-area bottom drain-body junction capacitance

C_{jbdsw} is the unit-length drain-body junction sidewall capacitance along the isolation edge

C_{jbdswg} is the unit-length drain-body junction sidewall capacitance along the gate edge

A_{deff} and P_{deff} are the effective junction area and perimeter on the drain-side of the source-drain diffusion, respectively

- C_{jbd} is calculated by

$$C_{jbd} = \begin{cases} CJD(T) \cdot \left(1 - \dfrac{V_{bs}}{PBD(T)}\right)^{-MJD}, & V_{bd} < 0; \\[4mm] CJD(T) \cdot \left(1 + MJD\dfrac{V_{bs}}{PBD(T)}\right), & V_{bd} \geq 0 \end{cases} \tag{6.112}$$

- C_{jbdsw} is calculated by

$$C_{jbdsw} = \begin{cases} CJSWD(T) \cdot \left(1 - \dfrac{V_{bs}}{PBSWD(T)}\right)^{-MJSWD}, & V_{bd} < 0; \\[4mm] CJSWD(T) \cdot \left(1 + MJSWD\dfrac{V_{bs}}{PBSWD(T)}\right), & V_{bd} \geq 0 \end{cases} \tag{6.113}$$

- C_{jbdswg} is calculated by

$$C_{jbdswg} = \begin{cases} CJSWGD(T) \cdot \left(1 - \dfrac{V_{bs}}{PBSWGD(T)}\right)^{-MJSWGD}, & V_{bd} < 0; \\[4mm] CJSWGD(T) \cdot \left(1 + MJSWGD\dfrac{V_{bs}}{PBSWGD(T)}\right), & V_{bd} \geq 0 \end{cases} \tag{6.114}$$

In the above model equations, the compact model parameters are represented by *upper-case* letters.

6.7 Summary

This chapter presented the dynamic MOSFET models for small signal analysis of MOSFETs in very-large-scale-integrated circuits using quasistatic assumptions. MOSFET device capacitances are separated into intrinsic and extrinsic or parasitic capacitances. First of all, the widely used simple Meyer intrinsic capacitance model is presented. We have derived the expressions for the terminal charges and capacitances and discussed the merits and demerits of the Meyer model. In order to overcome the limitations of Meyer model, more accurate charge-based capacitance model for both the long and short channel devices are presented. The validity and limitations of quasi-static assumptions in capacitance modeling are also discussed. Finally, the extrinsic capacitances such as the MOSFET S/D overlap capacitance and S/D junction capacitances are presented.

Exercises

6.1 Complete the mathematical steps to show that the linear region capacitances of MOSFETs are given by Equation 6.20.

6.2 Plot the normalized capacitance C/C_{oxt} versus $(V_{gs} - V_{th})$ characteristics for each of the components of gate capacitance C_{GB}, C_{GS}, and C_{GD} for an nMOSFET device with $W = 1$ μm, $L = 250$ nm, and $T_{ox} = 10$ nm; consider the biasing condition $-2.0\ V < V_{gs} < 3.0\ V$ and $V_{ds} = 1, 2,$ and $3\ V$. Clearly state any assumptions you make and explain your plots.

6.3 Show that the channel charge partition for a MOSFET device with channel length L and channel width W at the drain end is given by Equation 6.45.

7

Compact MOSFET Models
for RF Applications

7.1 Introduction

For analog and radio frequency (RF) applications of metal-oxide-semiconductor field-effect transistors (MOSFETs), it is critical to understand the behavior of these devices under applied signal. In Chapter 6, the analytical expressions are derived for modeling the terminal charges and capacitances of a MOSFET device based on the quasistatic (QS) assumption. For analog/RF applications of MOSFETs, the effect of noise, non-quasistatic (NQS) effect, and resistive network of these devices must be properly characterized and modeled for circuit CAD (computer-aided design). This chapter presents the basic understanding of modeling device noise, high-frequency characteristics such as NQS effect, and the resistive network of the devices.

7.2 MOSFET Noise Models

A careful observation of the drain current of a MOSFET reveals minute random fluctuations, referred to as *noise*. Noise is inherently present in electronic devices with or without the presence of an applied external signal. Noise could be in output current as well as in output voltage. In an MOS (metal-oxide-semiconductor) analog circuits, noise in MOSFET devices can interfere with weak signals and significantly impact the accurate characterization of circuit performance. Therefore, it is very important to model and characterize noise in devices [1–3].

The amount of noise depends on the bandwidth of the characterizing system. A common noise characterization technique involves a very narrow bandwidth, centered on a frequency *f*. The current noise spectral components within this bandwidth have a certain mean square value. The ratio of this value to the bandwidth, as the latter is allowed to approach zero, tends to what is called the power spectral density (PSD) of the current noise, denoted by $S_i(f)$. This quantity has units of square amperes per hertz (A^2/Hz). Often the square root of the PSD is used instead, given by $A/\sqrt{Hz}$. For a noise voltage v_n, one can similarly define a PSD $S_v(f)$ as V^2/Hz or its square root in $V/\sqrt{Hz}$ [4].

The total mean square noise current within an arbitrary bandwidth extending from $f = f_1$ to $f = f_2$ can be found by summing the mean square values of the individual components within each sub-bandwidth Δf. More precisely, using the PSD concept, we have

$$\overline{i_n^2} = \int_{f1}^{f2} S_i(f)df \tag{7.1}$$

A similar result can be obtained for voltage noise. Detailed characterization and modeling techniques of low- [5] and high-frequency [6] noise in advanced MOSFET devices are available in the literature. In this chapter, we have presented the basic understanding of modeling noise in MOSFET devices.

7.2.1 Fundamental Sources of Noise

Random fluctuations in the current (or voltage) in a device are generated by some fundamental processes in the device. The various types of noise present in an electronic device include (1) thermal (Johnson/Nyquist) noise, (2) shot noise, (3) generation–recombination noise, (4) random telegraph signal (burst/popcorn) noise, and (5) $1/f$ or flicker noise. The detailed description of these sources can be found in the review articles [6,7]. This chapter presents only the basic models of the thermal and flicker noise in MOSFET devices.

7.2.2 Thermal Noise

7.2.2.1 Physical Mechanism of Thermal Noise

Thermal noise arises from the random thermal motion of electrons in a material. When an electron gets scattered, its velocity is randomized. Thus, at a particular instant, the number of electrons moving in a certain direction may be more than that in another direction and small net current flows. This current fluctuates in magnitude and direction, but the average over a long time is always zero. The PSD of thermal noise current in a material of resistance R and temperature T is not white or flat up to infinitely high frequencies. It exists in every resistive medium and is unavoidable. However, it may be minimized by proper circuit design technique. For instance, input matching techniques using reactive elements can be used to lower the noise in amplifiers since reactive elements do not generate thermal noise. Also, system bandwidth should be kept as small as possible to pass the desired signal since unused portions of the bandwidth cause unnecessary noise.

In order to understand thermal noise in a MOSFET, we will discuss first the thermal noise model of a resistor. It is known that the thermal noise of a resistor is directly proportional to temperature T. The spectral noise power density $S_i(f)$ (mean square value of current per frequency bandwidth) of a resistor, R, can be given by [8]

$$S_i(f) = \frac{\overline{i^2}}{\Delta f} 4kT \frac{1}{R} \tag{7.2}$$

where:
k is the Boltzmann's constant

The equivalent circuit of the thermal noise can be represented by a shunt current source $\overline{i^2}$, as shown in Figure 7.1.

The thermal noise characteristics in a MOSFET operating in strong inversion region have been studied for over two decades. The origin of thermal noise in a MOSFET has been found to be related to the random thermal motion of carriers in the channel of the device [9]. Depending on this understanding, noise models have been developed and implemented in circuit simulators [4]. Even though using the thermal noise model of a resistor can qualitatively explain the thermal noise in a MOSFET, it is not quantitatively accurate even at low drain bias [10,11]. Furthermore, as the moderate inversion region becomes important for low power applications, there is an increasing need for accurate noise modeling in this region. Therefore, the noise behavior of a transistor should be well modeled from strong inversion through moderate inversion, into weak inversion.

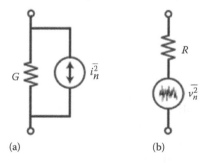

(a) (b)

FIGURE 7.1
Equivalent circuit for the thermal noise of a resistor: (a) noise power as a current source in shunt with mean square value $\overline{i_n^2}$ and (b) noise power as a voltage source with mean square value $\overline{v_n^2}$.

7.2.2.2 Thermal Noise Model

Basic Model: The thermal noise model originally implemented in SPICE2 (Simulation Program with Integrated Circuit Emphasis) [12] is given by

$$Si_d(f) = \frac{8kT}{3} g_m \tag{7.3}$$

where:
g_m is the gate transconductance of the device

Equation 7.3 is found to be inadequate in the linear region, especially when $V_{ds} = 0$, where the transconductance is zero so that the calculated noise density is zero. However, in reality, the noise power density is not zero. To resolve this problem, the SPICE2 noise model is modified to the following form:

$$Si_d(f) = \frac{8kT}{3}\left(g_m + g_{ds} + g_{mb}\right) \tag{7.4}$$

where:
g_{ds} and g_{mb} are the output conductance and the bulk transconductance, respectively

Advanced Thermal Noise Model: An advanced thermal noise model is used in the industry standard compact modeling tools. We will derive the thermal noise model following the steps described by Tsividis and McAndrew [4]. It is well known that PSD of the noise voltage, generated across a resistor of value R, is $4kTR$ [2]. If a small element in the MOSFET channel has a resistance, ΔR, the noise voltage power of this element is

$$\overline{\left(\Delta v_t\right)^2} = 4kT\Delta R\Delta f \tag{7.5}$$

Assuming the length of the small element of the channel is Δy, then ΔR is

$$\Delta R = \frac{\Delta y}{\mu W_{eff} Q_i} \tag{7.6}$$

where:
W_{eff} is the effective channel width
μ is the electron mobility
Q_i is the channel charge per unit area

Substituting Equation 7.6 into Equation 7.5 gives

$$\overline{\left(\Delta v_t\right)^2} = \frac{4kT\Delta f\Delta y}{\mu W_{eff} Q_i} \tag{7.7}$$

The current change caused by the voltage change Δv_t in a device of effective channel length, L_{eff} is given by

$$\Delta i_t = \frac{W_{eff}}{L_{eff}} \mu Q_i \Delta v_t \qquad (7.8)$$

Then the mean square value of Δi_t is

$$\overline{\left(\Delta i_t\right)^2} = \left(\frac{W_{eff}}{L_{eff}} \mu Q_i\right)^2 \overline{\left(\Delta v_t\right)^2} \qquad (7.9)$$

Substituting Equation 7.7 into Equation 7.9, we have

$$\overline{\left(\Delta i_t\right)^2} = 4kT \frac{W_{eff}}{L_{eff}^2} \mu Q_i \Delta y \Delta f \qquad (7.10)$$

The total noise current power in a bandwidth Δf can be obtained by integrating Equation 7.10 along the channel

$$\overline{\left(\Delta i_t\right)^2} = 4kT \Delta f \frac{\mu}{L_{eff}^2} \int_0^{L_{eff}} Q_i W_{eff} dy$$

$$= 4kT \frac{\mu}{L_{eff}^2} Q_I \Delta f \qquad (7.11)$$

where:
   Q_I is the total inversion layer charge in the channel

Then from Equation 7.11, the PSD of thermal noise in a MOSFET can be expressed as

$$Si_d(f) = \frac{\overline{\Delta i_t^2}}{\Delta f} = 4kT \frac{\mu}{L_{eff}^2} Q_I \qquad (7.12)$$

Equation 7.12 can be used for modeling thermal noise in MOSFETs with model-specific computation of the total inversion charge Q_I. Equation 7.12 shows that the channel thermal noise PSD is independent of frequency where the assumption of QS behavior is valid. It is observed from Equation 7.12 that the thermal noise increases with the increasing gate voltage V_{gs} since Q_I increases with V_{gs} and it increases with the decreasing channel lengths [6]. However, the experimental data show that thermal noise depends weakly on the drain voltage V_{ds}. This indicates that the noise contribution from the velocity saturation region of the channel is negligible. This is theoretically justified from the thermal noise model since Q_I saturates in the velocity saturation region.

7.2.3 Flicker Noise

7.2.3.1 Physical Mechanism of Flicker Noise

The low-frequency noise, commonly referred to as the *flicker noise* or $1/f$ *noise*, is characterized by $1/f$ dependency of its spectral density. There has been a continuous effort to understand the physical origin of flicker noise [13–18], leading to three different theories: (a) carrier density fluctuation model [19], (b) mobility fluctuation model [20], and (c) correlated carrier and mobility fluctuation model or the unified theory [21]. In the *carrier density fluctuation* model, the noise is explained by the fluctuation of channel free carriers due to the random capture and emission of carriers by interface traps at the Si-SiO$_2$ interface. According to this model, the input noise is independent of the gate bias, and the magnitude of the noise spectrum is proportional to the density of the interface traps. A $1/f$ noise spectrum is predicted if the trap density is uniform in the oxide. The experimental results show a $1/f^\gamma$ spectrum where the value of γ is in the range of $0.7 < \gamma < 1.2$ [17,22]. Experimental data also show that γ decreases with increasing gate bias in p-channel MOSFETs [23]. These experimental results are explained using modified charge density fluctuation model whereas the technology and the gate bias dependence of γ are explained assuming nonuniform spatial distribution of active traps in the oxide [19,23]. In the *mobility fluctuation model*, the flicker noise is considered to be the result of fluctuations in the carrier mobility given by Hooge's empirical relation for the spectral density of the flicker noise in a homogeneous device [24]. It has been proposed that the fluctuation of the bulk mobility in MOSFETs is induced by changes in phonon population [25]. The mobility fluctuation models predict a gate bias–dependent noise. However, they cannot always account for the magnitude of the noise [26]. In the *unified theory*, the origin of $1/f$ noise is assumed to be due to the capture and emission of carriers by the interface traps causing fluctuation in both the carrier number and mobility [21]. The unified theory can explain most of the experimental data and has been implemented in most compact-model extraction and circuit CAD tools [27–29].

7.2.3.2 Flicker Noise Model

The basic flicker noise model implemented in SPICE2 is given by

$$S_{id}(f) = \frac{K_F \cdot I_{ds}^{AF}}{f^{EF} C_{ox} L^2} \tag{7.13}$$

where:
 S_{id} is the drain current noise PSD
 I_{ds} is the drain current
 AF is the flicker noise exponent
 EF is the flicker noise frequency coefficient
 K_F is the flicker noise coefficient

The basic model given by Equation 7.13 is not adequate for the characterization of noise in advanced MOSFET devices. Therefore, in this section, the unified flicker noise model that explains most of the observed behavior of low-frequency noise is presented.

Let us assume that y is the distance along the direction of channel length, z is the distance along the direction of channel width, and x is the coordinate along the direction of oxide thickness perpendicular to both the y and z directions.

For a section of channel width W_{eff} and length Δy in a MOSFET, fluctuations in the amount of trapped interface charge will introduce correlated fluctuations in the channel carrier concentration and mobility. The resulting fractional change in the local drain current can be expressed as [29]

$$\frac{\delta I_{ds}}{I_{ds}} = \left[\frac{1}{\Delta N} \frac{\delta \Delta N}{\delta \Delta N_t} \pm \frac{\delta \mu_{eff}}{\delta \Delta N_t} \right] \cdot \delta \Delta N_t \qquad (7.14)$$

where:

$\Delta N = N W_{eff} \Delta y$

$\Delta N_t = N_t W_{eff} \Delta y$

N is the number of channel carriers per unit area

N_t is the number of occupied traps per unit area

The sign in front of the mobility term in Equation 7.14 is dependent on whether the trap is neutral or charged when filled [29]. The ratio of the fluctuations in the carrier number to the fluctuations in occupied trap number, $R_n = \delta \Delta N / \delta \Delta N_t$, is close to unity at strong inversion but assumes smaller values at other bias conditions [30]. A general expression for R_n is

$$R_n = \frac{\delta \Delta N}{\delta \Delta N_t} = -\frac{C_{inv}}{C_{ox} + C_{inv} + C_{dep} + C_{it}} \qquad (7.15)$$

where:

C_{inv} is the inversion layer capacitance

C_{dep} is the depletion layer capacitance

C_{it} is the interface trap capacitance

The relationship between C_{inv} and N can be approximated as

$$C_{inv} \cong \frac{q}{v_{kT}} N \qquad (7.16)$$

where:

v_{kT} is the thermal voltage

Thus, Equation 7.15 can be rewritten as,

$$R_n = -\frac{N}{N + N*} \qquad (7.17)$$

where

$$N^* = \frac{v_{kT}}{q}\left(C_{ox} + C_{dep} + C_{it}\right) \tag{7.18}$$

In order to evaluate $\delta\Delta\mu_{eff}/\delta\Delta N_t$, Matthiessen's rule is used so that

$$\frac{1}{\mu_{eff}} = \frac{1}{\mu_n} + \frac{1}{\mu_{ox}}$$

$$= \frac{1}{\mu_n} + \alpha N_t \tag{7.19}$$

where:
    μ_{ox} is the mobility limited by oxide charge scattering
    α is the scattering coefficient and is a function of the local carrier density
        due to channel charge screening effect [31]

It can be shown from Equation 7.19 that

$$\frac{\delta\mu_{eff}}{\delta\Delta N_t} = -\frac{\alpha\mu_{eff}^2}{W_{eff}\Delta y} \tag{7.20}$$

Substituting Equations 7.15 and 7.20 in Equation 7.14 yields

$$\frac{\delta I_{ds}}{I_{ds}} = \left[\frac{R_n}{N} \pm \alpha\mu_{eff}\right] \cdot \frac{\delta\Delta N_t}{W_{eff}\Delta y} \tag{7.21}$$

Thus, PSD of the local current fluctuations is

$$S_{\Delta Id}(y,f) = \left(\frac{I_{ds}}{W_{eff}\Delta y}\right)^2 \left(\frac{R_n}{N} \pm \alpha\mu_{eff}\right)^2 S_{\Delta Nt}(y,f) \tag{7.22}$$

where:
    $S_{\Delta Nt}(y,f)$ is the PSD of the fluctuations in the number of the occupied traps
        over the area $W_{eff}\Delta y$

and is given by

$$S_{\Delta Nt}(y,f) = \int_{E_v}^{E_c}\int_0^{W_{eff}}\int_0^{T_{ox}} 4N_t(E,x,y,z)\Delta yf_t\left(1-f_t\right)\cdot\frac{\tau(E,x,y,z)}{1+\omega^2\tau(E,x,y,z)}dx\cdot dz\cdot dE \tag{7.23}$$

where:
    $N_t(E,x,y,z)$ is the distribution of the traps in the oxide and over the energy
        band
    $\tau(E,x,y,z)$ is the trapping time constant

$f_t = \left\{1 - \exp\left[\left(E - E_{fn}\right)/kT\right]\right\}^{-1}$ is the trap occupancy function where E_{fn} is the electron quasi-Fermi level

$\omega = 2\pi f$ is the angular frequency

T_{ox} is the oxide thickness

$E_c - E_v$ is the silicon energy gap

In order to evaluate the integral in Equation 7.23, the following assumptions are used:

1. The oxide traps have a uniform spatial distribution near the interface, that is, $N_t(E,x,y,z) = N_t(E)$
2. The probability of an electron penetrating into the oxide decreases exponentially with the distance from the interface

   As a result the trapping time constant is given by

$$\tau = \tau_0(E)\exp\left(\gamma \cdot x\right) \tag{7.24}$$

where:

$\tau_0(E)$ is the time constant at the interface

γ is the attenuation coefficient of the electron wave function in the oxide

Since $f_t\left(1 - f_t\right)$ in Equation 7.23 behaves like a delta function around the quasi-Fermi level, the major contribution to the integral is from the trap level around E_{fn}. Thus, $N_t(E)$ can be approximated by $N_t(E_{fn})$ and taken out of the integral. Replacing $f_t\left(1 - f_t\right)$ in Equation 7.23 by $kT\left(df_t/dE\right)$ and carrying out the integration yields

$$S_{\Delta Id}(y, f) = N_t\left(E_{fn}\right)\frac{kTW_{eff}\Delta y}{\gamma \cdot f} \tag{7.25}$$

The total drain current noise power spectrum density can be derived as

$$S_i(f) = \frac{1}{L_{eff}^2}\int_0^{L_{eff}} S_{\Delta Ids}(y, f)\Delta y \, dy$$

$$= \frac{kTI_{ds}^2}{\gamma \cdot fW_{eff}L_{eff}^2}\int N_t\left(E_{fn}\right)\left[\frac{R_n}{N(y)} + \alpha\mu_{eff}\right]^2 dy \tag{7.26}$$

$$= \frac{qkTI_{ds}\mu_{eff}}{\gamma \cdot fL_{eff}^2}\int_0^{V_{ds}} N_t\left(E_{fn}\right)\left[1 \pm \frac{\alpha\mu_{eff}N}{R_n}\right]^2 \frac{R_n^2}{N} dV$$

Since α and μ_{eff} are functions of the local carrier density N, Equation 7.26 can be expressed as

$$S_i(f) = \frac{qkTI_{ds}\mu_{eff}}{\gamma \cdot f \cdot L_{eff}^2} \int_0^{V_{ds}} N_t^*\left(E_{fn}\right) \frac{R_n^2}{N} dV \qquad (7.27)$$

where:

$N_t^*\left(E_{fn}\right)$ is the equivalent oxide-trap density that produces the same noise power if there were no contributions from the mobility fluctuations

In the present model $N_t^*\left(E_{fn}\right)$ is approximated as a three-parameter function of the channel carrier density such as

$$N_t^*\left(E_{fn}\right) = A + BN + CN^2 \qquad (7.28)$$

where:
A, B, and C are technology-dependent model parameters

Using Equation 7.27, the flicker noise power spectrum density in the different operation regions of MOSFETs can be found.

1. Linear region in strong inversion ($V_{gs} > V_{th}$ and $V_{ds} < V_{dsat}$)

   In the strong inversion region, the charge density of carrier is given by

$$qN(y) = C_{ox}\left[V_{gs} - V_{th} - \alpha V(y)\right] \qquad (7.29)$$

Thus, we have

$$qN_0 = qN(0) = C_{ox}\left[V_{gs} - V_{th}\right] \qquad (7.30)$$

$$qN_L = qN\left(L_{eff}\right) = C_{ox}\left[V_{gs} - V_{th} - \alpha V_{ds}\right] \qquad (7.31)$$

where:
N_0 and N_L are carrier densities at the source and drain ends of the channel, respectively

Now, using Equations 7.30 and 7.31, Equation 7.27 can be rearranged as

$$S_i(f) = \frac{q^2kTI_{ds}\mu_{eff}}{\gamma \cdot f \cdot L_{eff}^2 C_{ox}} \int_{N_L}^{N_0} N_t^*\left(E_{fn}\right) \frac{R_n^2}{N} dN \qquad (7.32)$$

Substituting Equations 7.15 and 7.28 into Equation 7.32 and performing the integration, we can show

$$S_i(f) = \frac{q^2kTI_{ds}\mu_{eff}}{\alpha\gamma \cdot f \cdot L_{eff}^2 C_{ox}}\left[A\ln\left(\frac{N_0 + N^*}{N_L + N^*}\right) + B\left(N_0 - N_L\right) + \frac{1}{2}C\left(N_0^2 - N_L^2\right)\right] \qquad (7.33)$$

2. Saturation region in strong inversion ($V_{gs} > V_{th}$ and $V_{ds} \geq V_{dsat}$)

In the saturation region, the channel can be divided into two parts: one part is from the source to the velocity-saturation or *pinched-off* point discussed in Chapter 4 and the other part is from the velocity-saturation point to the drain end given by the length l_i. Then the total flicker noise PSD in the saturation region can be shown as

$$S_i(f) = \frac{q^2 kTI_{ds}\mu_{eff}}{\alpha \cdot \gamma \cdot f \cdot L_{eff}^2 C_{ox}} \cdot \left[A\ln\left(\frac{N_0 + N^*}{N_L + N^*}\right) + B(N_0 - N_L) + \frac{1}{2}C\left(N_0^2 - N_L^2\right) \right]$$

$$+ \frac{kTI_{ds}^2\mu_{eff}}{\gamma \cdot f \cdot L_{eff}^2 W_{eff}} \frac{A + BN_L + CN_L^2}{\left(N_L + N^*\right)^2} l_i \tag{7.34}$$

where:

$$N_L = \frac{1}{q}C_{ox}\left(V_{gs} - V_{th} - \alpha V_{dsat}\right)$$

3. Subthreshold region ($V_{gs} < V_{th}$)

From Equations 4.112 and 4.117, we can show that the channel charge density in the subthreshold region is given by

$$qN(V) = C_d v_{kT} \exp\left[\frac{\left(V_{gs} - V_{th}\right)}{n v_{kT}} - \frac{V}{v_{kT}}\right] \tag{7.35}$$

where:

n is the subthreshold swing factor

V_{th} is the voltage when the surface potential is equal to $2\phi_B$

Substituting Equation 7.35 into Equation 7.27 and rearranging, we get the following expression for the spectral flicker noise power density in the subthreshold region:

$$Si(f) = \frac{q^2 v_{kT}^2 I_{ds}^2 \mu_0}{\gamma \cdot f \cdot L_{eff}^2} \int_{N_L}^{N_0} \frac{N_t^*(E_{fn})}{\left(N + N^*\right)^2} dN \tag{7.36}$$

where

$$qN_0 = C_d v_{kT} \exp\left(\frac{V_{gs} - V_{th}}{n v_{kT}}\right)$$

$$qN_L = qN_0 \left[1 - \exp\left(-\frac{V_{ds}}{v_{kT}}\right)\right] \tag{7.37}$$

In the subthreshold region, it is reasonable to assume that $N \ll N^*$ and $N^*_t = A + BN + CN^2 \cong A$. Thus, the flicker noise power in the subthreshold region can be simplified to

$$Si(f) = \frac{Aqv_{kT}I_{ds}^2}{\gamma \cdot f \cdot W_{eff}L_{eff}N^{*2}} \quad (7.38)$$

However, the flicker noise model of p-channel MOSFETs is not clear, especially in the weak inversion region [32].

Figure 7.2 shows the measured and simulated low-frequency noise behavior for both nMOSFET and pMOSFET devices of a typical 130 nm CMOS (complementary metal-oxide-semiconductor) technology.

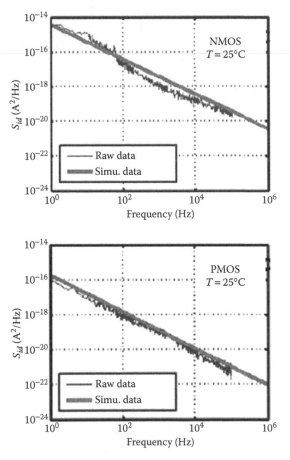

FIGURE 7.2
Simulated and measured low-frequency noise characteristics of nMOSFET and pMOSFET devices as a function of frequency with $L = 0.28$ μm and $W = 10$ μm; the data are obtained under the biasing condition $V_{gs} = 1.25$ V and $V_{ds} = 2.5$ V at room temperature.

Figure 7.3 shows noise PSD as a function of frequency. The plot shows thermal noise and flicker noise for both nMOSFETs and pMOSFETs.

Figure 7.4 shows the small-signal equivalent circuit of a MOSFET device with different noise sources and parasitic resistance-capacitance components.

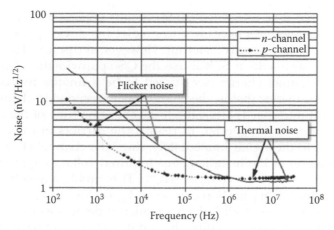

FIGURE 7.3
Noise spectra of an nMOSFET and a pMOSFET device measured at $I_{ds} = 500\ \mu A$ at room temperature; dimensions of both devices are $W = 2000\ \mu m$ and $L = 0.5\ \mu m$.

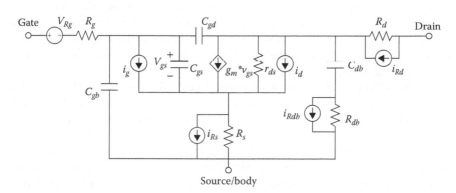

FIGURE 7.4
Small signal equivalent circuit model of a MOSFET suitable for RF and microwave frequencies including the noise sources ignoring body effect; v_{Rg}, i_{Rs}, i_{Rd}, and i_{Rdb} are the noise sources that model thermal noise from parasitic resistaces R_g, R_s, R_d, and R_{db}, respectively; i_g and i_d are the noise sources that model the gate current noise and channel noise (thermal and flicker noise), respectively; C_{gs}, C_{gd}, C_{gb}, and C_{db} are the gate-source, gate-drain, gate-body, and drain-body capacitances, respectively; and i_g, i_d, g_m, and v_{gs} are small-signal parameters. (Data from M.J. Deen et al., *IEEE Trans. Electron Dev.*, 53, 2062–2081, 2006.)

7.3 NQS Effect

In Chapter 6, the charge and C–V models are derived based on the QS approximation; that is, the inversion charge responds to the changes in the applied signal instantaneously. However, the QS approximation breaks down when signal changes occur on a timescale comparable to the device transit time.

As very-large-scale-integrated (VLSI) circuits become more performance-driven, it is sometimes necessary to predict the device performance for operation near the device transit time. However, as discussed in Chapter 5, most models available in SPICE use the QS approximation [33]. In a QS model, the channel charge is assumed to be a unique function of the instantaneous biases; that is, the charge has to respond to change in voltages with infinite speed. Thus, the finite charging time of the carriers in the inversion layer is ignored. In reality, the carriers in the channel do not respond to the signal immediately; thus, the channel charge is not a unique function of the instantaneous terminal voltages (QS) but a function of the history of the voltages (NQS). This problem may become pronounced in the RF applications, or when V_{gs} is close to V_{th}, or when long channel devices coexist with deep submicron devices as in many mixed signal circuits. In these circuits, the input signals may have rise or fall times comparable to or even smaller than the channel transit time. For long channel devices, the channel transit time is approximately inversely proportional to $(V_{gs} - V_{th})$ and directly proportional to L^2. Since the carriers in these devices cannot follow the changes of the applied signal, the QS models may give inaccurate or anomalous simulation results that cannot be used to guide circuit design. Two-dimensional (2D) numerical simulation results show that the most common QS model that uses 40/60 drain/source charge partitioning [34] results in an unrealistic large drain current spike during a fast turn-on [35].

Besides affecting the accuracy of simulation, the nonphysical results can also cause oscillation and convergence problems in the numerical iterations in circuit CAD. It is common among circuit designers to circumvent the convergence problem by using a 0/100 drain/source charge partitioning ratio [36], which attributes all transient charges to the source side. However, the numerical device simulation results show that this nonphysical solution merely shifts the current-spike problem to the source current; thus it only works when the source is grounded.

Moreover, none of these QS models can be used to accurately predict the high-frequency transadmittance of a MOSFET as pointed out by Tsividis and Masetti [37]. It is a common practice in high-frequency circuit designs to break a long channel MOSFET into N equal parts in series (N-lumped model) due to the lack of NQS models. The accuracy increases with N, at the expense of simulation time [38]. However, this method becomes impractical when the device channel length is small because the short channel effects in the subtransistors may be activated. As an example, it is found that the results of modeling a 200 μm long MOSFET device in strong inversion (saturation) is completely different from modeling two 100 μm long MOSFETs in series.

It has been found that for RF applications the NQS model is necessary to fit the measured high-frequency characteristics of devices with even short channel length where the operation frequency is above 1 GHz [39,40]. Thus, NQS model is desirable in some mixed signal IC (integrated circuit) and RF applications. Therefore, a compact model that accounts for the NQS effect is highly desirable. Some NQS models based on solving the current continuity equation have been proposed [41–43]. They are complex and require long simulation time, making them unattractive for use in circuit simulation. In this chapter, an NQS model based on the Elmore-equivalent resistance-capacitance (RC) circuit is described [35]. It uses a physical relaxation time approach to account for the finite channel charging time. This NQS model applicable for both the large signal transient and small signal AC analysis is discussed in the next section.

7.3.1 Modeling NQS Effect in MOSFETs

Typically, the channel of a MOSFET is analogous to a bias-dependent RC-distributed transmission line [44]. In QS approach, the gate capacitors are lumped with the intrinsic source and drain nodes [35]. This ignores the fact that the charge build-up in the center portion of the channel does not follow a change in V_g as readily as it does at the source or drain edge of the channel. Breaking the transistor into N devices in series offers a good approximation for the RC network but has the disadvantages discussed in Section 7.3. A physical and efficient approach to model the NQS effect would be to formulate an estimate for the delay time through the channel RC network, and incorporate this time constant into the model equations.

One of the most widely used methods to approximate the RC delay was proposed by Elmore [45], considering the mean, or the first moment, of the impulse response. Utilizing Elmore's approach, the RC distributed channel can be approximated by a simple RC equivalent circuit that retains the lowest frequency pole of the original RC network. The new equivalent circuit is shown in Figure 7.5. The Elmore resistance (R_{Elmore}) in strong inversion is calculated from the channel resistance and is given by [35].

$$R_{Elmore} = \frac{L_{eff}}{E_{LM}\mu W_{eff}Q_i}$$

$$= \frac{L_{eff}}{E_{LM}\mu W_{eff}C_{ox}\left(V_{gs} - V_{th}\right)}$$

(7.39)

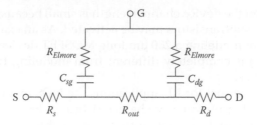

FIGURE 7.5
Transient and small signal equivalent circuit model for a MOSFET device. (Data from M. Chan et al., *IEEE Trans. on Electron Dev.*, 45, 834–841, 1998.)

where:
   Q_i is the amount of channel inversion charge per unit area
   E_{LM} is a parameter, referred to as the Elmore constant (=5) that is used to match the lowest frequency pole

It is reported that the value of E_{LM} required to match the output of different possible equivalent circuits is about 3 and it is invariant with respect to W and L [35]. The comparison of the time and frequency domain responses of the Elmore equivalent network shown in Figure 7.5 with the conventional distributed channel representation of a device shows a reasonable match between the Elmore's equivalent circuit and the distributed RC network [35]. However, direct implementation of the model shown in Figure 7.5 requires two additional nodes that increase the computational time. In addition, the change in the device topology may require modifications of the existing compact model formulations. Therefore, to improve the computational efficiency within the framework of an existing compact model, simplifying assumptions are made to develop a simplified NQS model. For example, we assume that the bulk charging current is negligibly small. Then the gate, drain, and source terminal currents can be described by the expression

$$I_j(t) = I_j(t)\big|_{DC} + j_{xpart} \frac{dQ_i(t)}{dt}; \quad \text{where } j = G, D, S \tag{7.40}$$

where:
   $I_j(t)$ represents the total gate, drain, and source currents
   $I_j(t)\big|_{DC}$ represents the DC gate, drain, and source currents
   $Q_i(t)$ is the actual channel charge at any given time t
   j_{xpart} represents the channel charge partitioning ratios [46,47] for the gate (G_{xpart}), drain (D_{xpart}), and source (S_{expart})

so that

$$D_{xpart} + S_{xpart} = -G_{xpart} = 1 \tag{7.41}$$

In the 40/60 partitioning scheme, D_{xpart} varies from 0.5 at $V_d = 0$ to 0.4 in the saturation region, and S_{xpart} varies from 0.5 to 0.6 [33,48]. However, since the 0.4/0.6 scheme covers a wide range of voltage and the error introduced by using a constant $D_{xpart} = 0.4$ and $S_{xpart} = 0.6$ is less than 5%, these values can be adopted to simplify the model.

In the QS approach, it is assumed that

$$\frac{dQ_i(t)}{dt} = \frac{dQ_{iq}(t)}{dt}$$
$$= \frac{dQ_{iq}}{dV}\frac{dV}{dt} \tag{7.42}$$

where:

$Q_{iq}(t)$ is the equilibrium, or QS, channel charge under the instantaneous bias at any time t

The assumption of equilibrium at all times gives an error in calculating the NQS currents. To account for the NQS current, a new state variable Q_{def} is introduced to keep track of the amount of deficit (or surplus) channel charge relative to the QS charge at a given time.

$$Q_{def}(t) = Q_{iq}(t) - Q_i(t) \tag{7.43}$$

and

$$\frac{dQ_{def}(t)}{dt} = \frac{dQ_{iq}(t)}{dt} - \frac{dQ_i(t)}{dt} \tag{7.44}$$

Q_{def} is allowed to decay exponentially to zero after a step change in bias with a bias-dependent NQS relaxation time, τ. Thus, the charging current can be approximated by

$$\frac{dQ_i(t)}{dt} \cong \frac{dQ_{def}(t)}{\tau} \tag{7.45}$$

$Q_{def}(t)$ can be calculated from Equation 7.44, and a subcircuit, shown in Figure 7.6, has been introduced to obtain the solution. The subcircuit is a direct translation from Equation 7.44. The node voltage gives the value of $Q_{def}(t)$. The total charging current is given by the current going through the resistor of value τ. With this approach, only one additional node is needed and the topology of the original transistor model is not affected.

The value of the channel relaxation time constant τ is composed of the terms related to the diffusion and drift currents (calculated from the RC Elmore equivalent circuit discussed earlier). The components of τ are given by

$$\tau_{diffusion} = \frac{(L_{eff}/4)^2}{\mu v_{kT}} \tag{7.46}$$

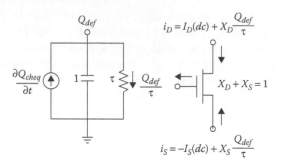

FIGURE 7.6
A simplified representation of Elmore's equivalent circuit for modeling NQS effect in MOSFETs. (Data from M. Chan et al., *IEEE Trans. on Electron Dev.*, 45, 834–841, 1998.) In text, $Q_{cheq} \equiv Q_{iq}$.

$$\tau_{drift} = \frac{1}{2} R_{Elmore} C_{ox} W_{eff} L_{eff} \tag{7.47}$$

$$\frac{1}{\tau} = \frac{1}{\tau_{diffusion}} + \frac{1}{\tau_{drift}} \tag{7.48}$$

In order to derive a simplified NQS model, C_{ox} is used in Equation 7.47 instead of the bias-dependent parameters ($C_{sg} + C_{dg}$). It is found that the simulated relaxation time τ obtained by Equation 7.48 agrees very well with that obtained by a 2D numerical device simulation under various biasing conditions [35].

In reality, NQS effects are important for long channel devices driven by fast switching inputs. However, NQS effects have also been observed in short channel devices [39,40]. When a MOSFET is operated in the velocity saturation regime, the channel conductivity is reduced, thus increasing the value of τ [35]. The circuit simulation data using NQS model show that the error resulting from the velocity saturation effect in a MOSFET is usually less than 20%. This error can be further reduced by optimizing Elmore constant to achieve an acceptable simulation data for both linear and saturation regions. However, accurate results can be obtained using an empirical model for the relaxation time [35] such as

$$\tau'_{drift} = \begin{cases} \tau_{drift}\left(1 + \frac{3}{8}\left(\frac{V_{ds}}{V_{dsat}}\right)^2\right); & \text{for } 0 < \left(V_{gs} - V_{th}\right) \geq V_{ds} \\ 1.375\tau_{drift}; & \text{for } 0 < \left(V_{gs} - V_{th}\right) \leq V_{ds} \end{cases} \tag{7.49}$$

Again, the simulated relaxation time obtained by empirical expressions in Equation 7.49 and 2D numerical device simulation agree very well [35].

For the simplicity of NQS modeling, it is assumed that the bulk charging current is zero [35]. This assumption is justified since for most applications the NQS effect from the bulk charge is negligible and does not significantly impact small signal simulation. However, the body current can be included by partitioning Q_{def} between the gate and the body [35,49].

7.4 Modeling Parasitic Elements for RF Applications

With the continuous scaling down of CMOS technologies to the nanoscale regime, RF circuits are realized in a standard CMOS process [50]. Therefore, a compact model for circuit CAD that is valid for a broad range of bias conditions, device sizes, and operating frequencies is of utmost importance. The widely used RF modeling approach is to build subcircuits based on MOSFET models that are suitable for analog/digital applications [40,51–55]. In the subcircuit, parasitic elements around gate, source, drain, and substrate as shown in Figure 7.7 are added to improve the accuracy of the model at high frequencies [40]. An important part of RF modeling is to establish physical and scalable model equations for the parasitic elements at the source, drain, gate, and substrate. The scalability of the intrinsic device is ensured by the core model library developed using the target compact model discussed in Chapters 4 and 5. We will now discuss the techniques to model the gate and substrate resistances for RF and analog applications.

7.4.1 Modeling Gate Resistance

At any low frequency, the gate resistance of a MOSFET can be calculated from the sheet resistance of the gate material and is given by

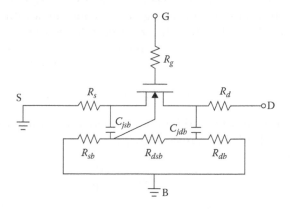

FIGURE 7.7
A subcircuit with parasitic elements added to an intrinsic MOSFET model for RF analysis. (Data from Y. Cheng et al., *Proceedings of the ICSICT*, 416–419, 1998.)

$$R_g = \frac{W_{eff}}{L_{eff}} \rho_{sh,gate} \qquad (7.50)$$

where:

W_{eff} and L_{eff} are the effective channel width and channel length of the device, respectively

$\rho_{sh,gate}$ is the gate sheet resistance per square [Chapter 2, Section 2.2.5.3]. The typical sheet resistance for a polysilicon gate ranges between 20 and 40 Ω per square and is significantly lower for silicide as well as for metal stack processes

At high frequencies, the accurate modeling of the gate resistance is very complex due to the distributed transmission-line effect. Therefore, a lumped equivalent gate resistance α times the end-to-end gate resistance given in Equation 7.50 is used such that [56]

$$R_g = \frac{\alpha_g W_{eff}}{L_{eff}} \rho_{sh,gate} \qquad (7.51)$$

where:

$\alpha_g = 1/3$ to account for the distributed RC effects when the gate electrode is contacted at one end and $\alpha_g = 1/12$ when the electrode is contacted on both ends [57]

It is found that the distributed RC effect of the gate as well as the NQS effect, that is, the distributed RC effect of the channel, affects the high-frequency characteristics of MOSFET devices. Therefore, an additional component of gate resistance must be considered to account for the distributed RC effect in the channel or NQS effect. Thus, at high-frequency operation of a MOSFET device, the distributed channel resistance *seen* by the signal applied to the gate also contributes to the effective gate resistance in addition to the resistance of the gate electrode. Then the effective gate resistance consists of two parts: the distributed gate electrode resistance ($R_{g,eltd}$) and the distributed channel resistance seen from the gate (R_{gch}), as shown in Figure 7.8 [58].

$$R_{g,eff} = R_{g,eltd} + R_{gch} \qquad (7.52)$$

Typically, $R_{g,eltd}$ is insensitive to bias and frequency and, therefore, is obtained from the gate electrode sheet resistance ($\rho_{sh,geltd}$)

$$R_{g,eltd} = \rho_{sh,geltd} \frac{\alpha_g W}{L + \beta} \qquad (7.53)$$

where:

α_g is 1/3 when the gate terminal is brought out from one side and 1/12 when connected on both sides

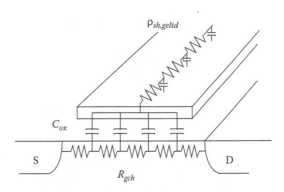

FIGURE 7.8
Distributed nature of gate electrode resistance, channel resistance network, and gate capacitance for modeling the efective gate resistance for RF analysis of MOSFET devices. (Data from X. Jin et al., *IEDM Technical Digest*, 961–964, 1998.)

L is the channel length
the parameter β models the external gate resistance

Now, R_{gch} includes the static channel resistance (R_{st}) that accounts for the DC channel resistance and the excess-diffusion channel resistance (R_{ed}) due to the change of channel charge distribution by the AC excitation of the gate voltage. Both R_{st} and R_{ed} together determine the time constant of the NQS effect. R_{st} is modeled by integrating the resistance along the channel under the QS assumption [58]

$$R_{st} = \int dR = \int \frac{dV}{I_d} = \begin{cases} \dfrac{V_{ds}}{I_{ds}}; & 0 < \left(V_{gs} - V_{th}\right) \geq V_{ds} \\[3mm] \dfrac{V_{dsat}}{I_{ds}}; & 0 < \left(V_{gs} - V_{th}\right) \leq V_{ds} \end{cases} \tag{7.54}$$

where:
  $V_{dsat} \cong (V_{gs} - V_{th})$ is the saturation drain voltage

Both I_{dsat} and V_{dsat} can be calculated from the target compact model. And, R_{ed} can be derived from the diffusion current (Equation 4.122) as

$$R_{ed} = \frac{L_{eff}}{\eta W_{eff} \mu C_{ox} v_{kT}} \tag{7.55}$$

where:
  η is a technology-dependent constant
  C_{ox} is the gate capacitance as shown in Figure 7.8

Thus, the overall channel resistance seen from the gate is

$$\frac{1}{R_{gch}} = \gamma\left(\frac{1}{R_{st}} + \frac{1}{R_{ed}}\right) \tag{7.56}$$

where:

γ is a parameter accounting for the distributed nature of the channel resistance

It is shown that if the resistance is uniformly distributed along the channel, the value of γ is 12 [58]. However, this assumption is true only in the saturation region; therefore, γ is used as a fitting parameter. Thus, from Equations 7.53 and 7.56 (along with Equation 7.52), we can model the effective gate resistance network for RF application. The reported simulation results using the above-described model show very good agreement with the data obtained by numerical device simulation.

7.4.2 Modeling Substrate Network

Modeling of the substrate parasitic elements and substrate network is very important for RF applications of MOSFET devices [40,59]. In order to obtain the desired scalable RF model, it is critical to develop scalable model for each component of the substrate network. A three-resistance substrate network equivalent circuit shown in Figure 7.9 is used for modeling the substrate parasitic elements at high frequencies [40].

In Figure 7.9, C_{JSB} and C_{JDB} are the capacitances between the source-body and drain-body pn-junctions, respectively, R_{SB} and R_{DB} are the resistances between the source-body and drain-body to account for the resistive losses at the source-drain signal coupling, and R_{BDS} is the substrate resistance. Using a two-port substrate network, the above model has been verified using 2D-numerical device simulation [40,59].

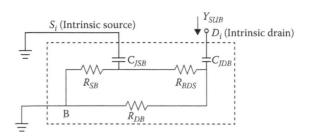

FIGURE 7.9

A three-resistance equivalent circuit for the substrate network: C_{JSB} and C_{JDB} are the capacitances between the source-body and drain-body pn-junctions, respectively.

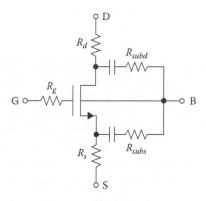

FIGURE 7.10
A MOSFET RF model developed used in conjunction with standard compact model. (Data from J.-J. Ou et al., *IEEE Symposium on VLSI Technology*, 94–95, 1998.)

7.4.3 MOSFET RF Model for GHz Applications

Using the lumped gate resistance and the substrate resistance network models described earlier, a unified compact device model can be realized for simulations of both RF and baseband analog circuits. The equivalent circuit model is shown in Figure 7.10 [54]. The RF MOSFET model can be realized with the addition of three resistors R_g, R_{subd}, and R_{subs} to the existing compact model. In Figure 7.10, R_g models both the physical gate resistance and the NQS effect, whereas R_{subd} and R_{subs} are the lumped substrate resistances between the S/D junctions and the substrate contacts. The values of R_{subd} and R_{subs} may not be equal as they are functions of the transistor layout [54].

7.5 Summary

In this chapter, the basic modeling techniques for analog and RF applications are presented. These approaches include modeling of noise, NQS effect, and the parasitic gate and substrate networks for high-frequency applications. The primary objective of this chapter is to expose readers to these RF models for the sake of completeness of the compact modeling activities and understanding of the advanced VLSI circuits. The readers would benefit from the basic understanding presented in this chapter to use the state-of-the-art RF compact models that are continuously developed and updated.

Exercises

7.1 Use the expression for the inversion charge from the MOS Level 1 model to calculate and plot the thermal noise PSD as a function of the channel length at room temperature for nMOSFET and pMOSFET devices of channel length, $100 < L < 1000$ nm; $W = 1000$ nm. Given that the surface mobilities for electrons and holes are 600 cm^2 V^{-1} sec^{-1} and 400 cm^2 V^{-1} sec^{-1}, respectively.

7.2 Complete the mathematical steps to derive the flicker noise PSD given in Equation 7.27. Clearly state any assumptions you make.

7.3 Use the subthreshold region drain current expression (Equation 4.122) to derive the diffusion resistance in the channel given by Equation 7.55. Clearly state any of the assumptions you make.

8

Modeling Process Variability
in Scaled MOSFETs

8.1 Introduction

This chapter presents compact MOSFET (metal-oxide-semiconductor field-effect transistor) modeling approaches for process variability-aware VLSI (very-large-scale-integrated) circuit CAD. The circuit design for advanced VLSI technology is severely constrained by random and systematic process variability [1]. With continued miniaturization of MOSFET devices [2–8], performance variability induced by process variability has become a critical issue in the design of VLSI circuits using advanced CMOS (complementary metal-oxide-semiconductor) technologies. Process variability in scaled CMOS technologies severely impacts the delay and power variability in VLSI devices, circuits, and chips, and this impact keeps increasing as MOSFET devices and CMOS technologies continue to scale down [1,9–13]. The increasing amount of within-die process variability on the yield of VLSI circuits, such as static random access memory (SRAM), has imposed an enormous challenge in the conventional VLSI design methodologies. Similarly, the chip mean variation due to across-the-chip systematic process variability also imposes serious challenge in the conventional VLSI circuit design methodologies. Because of process variability constraints, an advanced VLSI circuit, optimized using the conventional design methodology, is more susceptible to random performance fluctuations. Thus, new circuit design techniques to account for the impact of process variability in VLSI circuits have become essential [1,9]. And, compact model addressing the impact of random and systematic process variability in scaled MOSFET devices is crucial for the simulation and analysis of advanced VLSI circuits. Process variability in manufacturing technology includes *front-end* or intrinsic process variability due to various dopant implant and thermal processing steps and *back-end* variability of metal lines for interconnecting the devices in the VLSI circuits. Both the front-end and interconnection process variabilities are important for circuit analysis. Over the years, different approaches have been used to develop statistical models for circuit analysis to account for intrinsic process variability [9,14–21]. In this chapter, we will present different approaches to develop statistical compact MOSFET device models to account for the front-end process variability in VLSI circuit design.

8.2 Sources of Front-End Process Variability

The intrinsic sources of variability in VLSI device performance arise from the random variability of fabrication-processing steps [1,9–11]. Typically, the intrinsic process variability is grouped as *systematic* and *stochastic* or random as shown in Figure 8.1.

8.2.1 Systematic or Global Process Variability

Systematic variation in IC (integrated circuit) device and chip performance is caused by inherent random character of IC-processing steps. The systematic component is defined as the global or inter-die process variability [1,9–13]. The global process variability causes die-to-die, wafer-to-wafer, or lot-to-lot systematic parametric fluctuations between identical devices [1,9–13]. Global variability causes a shift in the mean value of the sensitive design parameters, including the channel length (L), channel width (W), gate oxide thickness (T_{ox}), resistivity, doping concentration, and body effect as shown in Figure 8.2. Systematic differences may lead to longer channel length transistors than the nominal devices, causing them to switch more slowly due to reduced drive current, resulting in slower ICs with lower leakage current. On the other hand, the shorter (than the nominal) channel length devices would lead to faster die easily meeting clock-frequency specifications; however, these devices may exhibit excessive leakage current and fail leakage

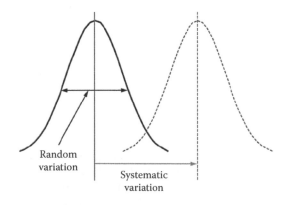

FIGURE 8.1
Types of process variability: random variation of a parameter around its mean value and systematic variation of the mean value of a parameter.

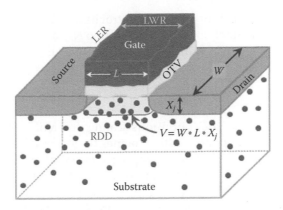

FIGURE 8.2
The critical sources of variability in CMOS devices; here L, W, and X_j are the channel length, channel width, and S/D junction depth of MOSFET devices, respectively; and V is the volume of charge in the channel region. (Data from S.K. Saha, *IEEE Access*, 2, 104–115, 2014.)

current specifications. In the semiconductor industry, the systematic process variation has been the major interest to IC chip manufacturers in maintaining competitive yield over multiple technology nodes [4]. The systematic process variability in manufacturing technology has been accounted in compact modeling by defining process corners, that is, boundaries in parameter variation that account for process tolerances [1,9].

8.2.2 Random or Local Process Variability

The random variation in IC device and chip performance arises from stochastic variations inherent to the discrete nature of dopant impurities and point defects as well as random variations in the complex processing steps of nanometer scale CMOS technology. Random variability is defined as the *local* or *intra-die* process variability [1,9]. Local process variability causes parametric fluctuations or mismatch between identically designed devices within a die as shown in Figure 8.1. The major sources of intrinsic process variability in advanced CMOS technologies include random discrete doping (RDD), line-edge roughness (LER), line-width roughness (LWR), and oxide thickness variation (OTV) as shown in Figure 8.2 [1,9–13].

The front-end process variability in CMOS technology has been extensively studied and the major sources of process variability along with their impact on device and VLSI circuit performance have been reported [1,9–11]. In the following section, a brief overview of the sources of front-end process variability is presented.

8.2.2.1 Random Discrete Doping

In the channel region of a MOSFET device, RDD results from the discreteness of dopant atoms as shown in Figure 8.2. In a MOSFET device, the channel region is doped with dopant atoms to control its threshold voltage (V_{th}) [5–8]. For a device with channel doping concentration N_{CH} and source/drain (S/D) junction depth X_j, the total number of dopant atoms in the channel region is given by [1,9]:

$$N_{CHtotal} \cong N_{CH} \cdot W \cdot L \cdot X_j \qquad (8.1)$$

Equation 8.1 shows that the continuous scaling down of L, W, and X_j causes the total number of dopants in the channel to decrease, despite the corresponding increase in the channel-doping concentration according to the CMOS scaling rule [2,3]. Using Equation 8.1 and the target specifications for advanced CMOS technology scaling by *International Technology Roadmap for Semiconductors* [22], the estimated decrease in $N_{CHtotal}$ over the scaled technology nodes is shown in Figure 8.3. Figure 8.3 implies that the number of dopants in a transistor channel is a discrete statistical quantity with probability to occupy any random location. Therefore, in an advanced CMOS technology, two identical transistors next to each other have different electrical characteristics because of the randomness in a few dopant atoms, resulting in intra-die device and circuit performance variability.

The major effects of RDD include significant variability in V_{th}, variability in the overlap capacitance (C_{ov}) due to the uncertainty in the position of S/D dopants under the gate, and variability in the effective S/D series resistance (R_{DS}). The impact of RDD-induced process variability on V_{th} mismatch between two identically designed within-die devices is given by [9]

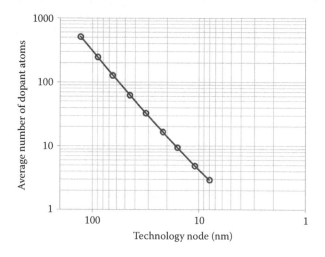

FIGURE 8.3
Estimated average channel doping concentration with scaling bulk CMOS devices in the nanoscale regime; the calculation is performed following ITRS. (Data from S.K. Saha, *IEEE Access*, 2, 104–115, 2014.)

$$\sigma V_{th,RDD} \cong C \cdot \left(\sqrt[4]{q^3 \varepsilon_{si} \phi_B}\right) \frac{T_{ox}}{\varepsilon_{ox}} \left(\frac{\sqrt[4]{N_{CH}}}{\sqrt{W_{eff} L_{eff}}}\right) \tag{8.2}$$

where:

C is a number and is given by 0.8165 [23] or 0.7071 [24] with or without the dopant variation along the depth of the channel region, respectively

q is the electronic charge

ε_{si} and ε_{ox} are the permittivity of silicon and silicon-dioxide (SiO_2), respectively

$\phi_B = v_{kT} \ln(N_{CH}/n_i)$ is the bulk potential of the channel region of MOSFETs and v_{kT} and n_i are the thermal voltage and intrinsic carrier concentration, respectively

W_{eff} and L_{eff} represent the effective dimension of W and L, respectively

Since the device area ($W_{eff} \cdot L_{eff}$) decreases with each new technology generation, it is obvious from Equation 8.2 that the net result of RDD is a significant increase in process variability for scaled CMOS technology as shown in Figure 8.4. In fact, RDD is a major contributor to mismatch (σV_{th}) in advanced MOSFETs [25]. As the device size scales down, the total number of channel dopants decreases [1,9], resulting in a larger variation of dopant numbers, and significantly impacting V_{th} as shown in Figure 8.4.

Equation 8.2 is the generalized analytical expressions for σV_{th} in planar devices due to RDD that represents σV_{th} equations derived by Stolk et al. [23] and Mizuno et al. [24] with appropriate value of the parameter C [9]. For devices of a particular process technology, Equation 8.2 can be expressed as

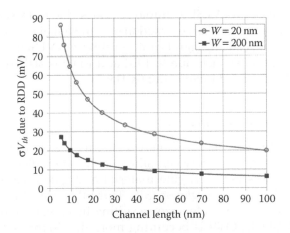

FIGURE 8.4
Estimated threshold voltage variation for a typical 20-nm bulk CMOS technology as a function of device channel length for different channel width following ITRS (Data from S.K. Saha, *IEEE Access*, 2, 104–115, 2014.); parameters used in Equation 8.2 are $N_{CH} = 6 \times 10^{18}$ cm^{-3}; SiO_2 equivalent oxide thickness (EOT) = 1.1 nm; and $C = 0.8165$.

$$\sigma V_{th,RDD} = \frac{C_{vt}}{\sqrt{W_{eff}L_{eff}}} \qquad (8.3)$$

where C_{vt} is a process technology dependent constant given by

$$C_{vt} \cong C \cdot \left(\sqrt[4]{q^3 \varepsilon_{si} \phi_B N_{CH}} \right) \frac{T_{ox}}{\varepsilon_{ox}} \qquad (8.4)$$

Equation 8.3 shows that V_{th} variability due to RDD is inversely proportional to the square root of device area. Thus, Equation 8.3 can be used to model the impact of RDD-induced V_{th} variability and the parameter C_{vt} is the slope of $\sigma V_{th,RDD}$ versus $1/\sqrt{W_{eff}L_{eff}}$ plots obtained from a large set of measurement data on a set of paired devices with varying W and L.

8.2.2.2 Line-Edge Roughness

In CMOS technology, LER results from sub-wavelength lithography and etching processes that cause variation in the critical dimension of the transistor feature size, as shown in Figure 8.2 [26]. The impact of LER includes variation in V_{th} and higher subthreshold current. LER-induced V_{th} mismatch depends on the variability in W_{eff} of MOSFETs and is given by [1,9,26,27]

$$\sigma V_{th,LER} \propto \frac{1}{\sqrt{W_{eff}}} < \sigma V_{th,RDD} \qquad (8.5)$$

Thus, LER increases as VLSI technology scales down. In scaled MOSFET devices, LER has become a larger fraction of L and a major source of intrinsic statistical variation, causing significant variability in VLSI device and circuit performance. The mismatch due to LER and RDD is statistically independent and can be modeled independently [1,9,26].

8.2.2.3 Oxide Thickness Variation

In CMOS technologies, OTV, shown in Figure 8.2, is caused by atomic level interface roughness between silicon and gate dielectric and remote interface roughness between gate material and gate dielectric, hereafter referred to as the *surface roughness (SR)*. This SR causes fluctuations of the voltage drop across the oxide layer, resulting in V_{th} variation [1,9,28,29]. In nanoscale MOSFETs, OTV is becoming more dominant as T_{ox} approaches the length of a few silicon atoms and is comparable to the thickness of interface roughness.

In nano-MOSFET devices, OTV causes significant device parameter variability. In polysilicon gate MOSFETs, OTV introduces a gate current (I_g)

variation. This I_g variation induces a voltage drop in the polysilicon gate and significantly changes V_{th}. In addition, the device transconductance g_m changes significantly because of the reduction in the gate voltage V_{gs} due to the voltage drop in the polysilicon gate. In high-k gate dielectric and metal gate devices, OTV introduces significant mobility degradation [1,9].

8.2.2.4 Other Sources Process Variability

Other sources of process variability include variation associated with polysilicon as well as metal gates granularity [30,31]; variation in fixed charge [32] and defects and traps in gate dielectric [33]; variation associated with patterning proximity effects such as optical proximity correction [34]; variation associated with polish such as shallow trench isolation [35] and gate [36]; variation associated with the strain such as in wafer-level biaxial strain [37], high-stress capping layers [38], and embedded silicon germanium (SiGe) [39]; and variation associated with implants and anneals due to implant tools, the implant profile, and millisecond annealing [40,41].

Thus, from the above discussions, it is clear that the advanced CMOS process technologies introduce within-die random performance variability, which causes severe variability in the performance of advanced VLSI circuits and systems. Therefore, it is critical to accurately model process variability when predicting the performance of advanced VLSI circuits and systems.

8.3 Characterization of Parametric Variability in MOSFETs

The random parametric variation such as threshold voltage variation (σV_{th}) is a key factor in determining the yield of memory elements such as SRAM and register file cells. Equation 8.3 can be used to characterize random V_{th} variation in devices.

8.3.1 Random Variability

In Figure 8.1, the random variability of a parameter is defined as the variation around its mean value. Therefore, random variability can be characterized by monitoring the differences in the value of a parameter of two closely spaced identical transistors, that is, paired transistor. Thus, the random V_{th} variation of identical transistor pairs can be determined by measuring the difference in V_{th} (i.e., ΔV_{th}) between a number of sets of closely spaced paired transistors (e.g., all the transistor pairs on a wafer) and computing the standard deviation of the difference ΔV_{th} (i.e., $\sigma \Delta V_{th}$). Thus

$$\sigma_{random-pair} = \sigma(V_{th1} - V_{th2}) = \sigma\left(\Delta V_{th}\right) \equiv \sigma\left(\sigma V_{th}\right) \tag{8.6}$$

In order to determine the local V_{th} variability (also referred to as the *mismatch*) in devices, V_{th} shifts (ΔV_{th}) between the closely spaced identical paired transistors are measured for a large number of pairs. Any standard procedure can be used to extract V_{th} for individual devices. Typically, the V_{th} extraction procedure is used on a set of device geometries (L, W) for both n-channel and p-channel MOSFETs. Then, $\sigma\Delta V_{th}$ is plotted as a function of $1/\sqrt{W_{eff}L_{eff}}$ as shown in Figure 8.5, which is known as Pelgrom plot [42]. Pelgrom plot presents the V_{th} variability ($\sigma\Delta V_{th}$) extracted for various (L, W) gate dimensions.

The slope A_{vt} of Pelgrom plot as shown in Figure 8.5 is called the *mismatch coefficient* and describes the mismatch between closely spaced identical transistor pairs [42,43]. Thus, we can write

$$A_{vt} = \sigma\left(\Delta V_{th}\right) \cdot \sqrt{W_{eff} \cdot L_{eff}} \tag{8.7}$$

In order to determine the local V_{th} variation of an individual transistor in the pair, we consider

$$\sigma_{\Delta V_{th}}^2 = \sigma_{V_{th1}}^2 + \sigma_{V_{th2}}^2 - 2 \cdot \rho \cdot \sigma_{V_{th1}} \cdot \sigma_{V_{th2}} \tag{8.8}$$

where:
V_{th1} and V_{th2} are the threshold voltages of the transistors 1 and 2 in the pair
ρ is the correlation coefficient between V_{th1} and V_{th2} fluctuations

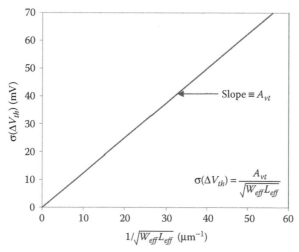

FIGURE 8.5
Plot of $\sigma(\Delta V_{th})$ versus $\left(\sqrt{W_{eff}L_{eff}}\right)^{-1}$ of identically designed paired transistors; A_{vt} is the slope of the plot.

Note that the fluctuations on one transistor of the pair cannot induce fluctuations on the second one (i.e., σV_{th1} and σV_{th2} are independent); thus, $\rho = 0$. In addition, $\sigma V_{th1} = \sigma V_{th2} \equiv \sigma V_{th}$ (i.e., $V_{th1} - V_{th2} = V_{th2} - V_{th1}$). Therefore, defining $\sigma V_{th1} = \sigma V_{th2} \equiv \sigma V_{th}$ from Equation 8.8, we can show that

$$\sigma V_{th} = \frac{\sigma \Delta V_{th}}{\sqrt{2}} \tag{8.9}$$

Equation 8.9 describes σV_{th} of individual transistors of the closely spaced pair. Equation 8.9 can be experimentally verified by comparing the local V_{th} variability obtained either in paired transistors (which give a value of $\sigma(\Delta V_{th})$) or in dense transistor arrays (which give a value of (σV_{th})) [44]. Note that the A_{vt} factor is defined historically from $\sigma(\Delta V_{th})$. Therefore, in order to develop compact variability model to simulate mismatch between identical devices, we get σV_{th} from Equations 8.7 and 8.9, as

$$\sigma V_{th} = \frac{A_{vt}}{\sqrt{2}} \cdot \frac{1}{\sqrt{W_{eff} L_{eff}}} \tag{8.10}$$

Comparing Equations 8.3 and 8.10 we get, $C_{vt} = A_{vt}/\sqrt{2}$; Thus, we can estimate the mismatch coefficient A_{vt} for any technology from Equation 8.4 using the technology parameters T_{ox} and N_{CH}. However, A_{vt} is extracted from the measured data from a set of closely spaced identical paired transistors.

The same procedure is used to determine the mismatch σP of any parameter P between closely spaced identical devices with mismatch coefficient A_p such that

$$\sigma P = \frac{A_p}{\sqrt{2}} \frac{1}{\sqrt{W_{eff} L_{eff}}} \tag{8.11}$$

8.3.2 Systematic Variability

As shown in Figure 8.1, the systematic or global variability is the shift of the mean value of a parameter. Therefore, global variability is obtained simply by calculating the standard deviation (σ) of any parameter P causing systematic variability. Thus, the systematic variability of V_{th} is characterized by calculating σV_{th} of the total V_{th} population, that is, V_{th} data from the target MOSFET test structures distributed across the wafer. The total V_{th} population could include devices from several wafers of a lot or from several lots collected over a period of time. The same procedure is used to determine the systematic variation σP of any parameter causing global device performance variability.

8.4 Conventional Process Variability Modeling for Circuit CAD

In order to account for process variability in circuit performance, typically corner models are used to set the lower and upper limits of process variation. These models are implemented in the process design kit to support process variability-aware VLSI circuit design.

8.4.1 Worst-Case Fixed Corner Models

In conventional circuit design technique, process variability is modeled by *four* worst-case corners: two for analog applications and two for digital [1,9]. The corners for analog applications are generated from slow NMOS (*p*-type body with *n*+ source-drain) and slow PMOS (*n*-type body with *p*+ source-drain; SS) to model the worst-case speed and from fast NMOS and fast PMOS (FF) to model the worst-case power. The corners for digital applications are generated from fast NMOS and slow PMOS (FS) to model the worst-case "1" and from slow NMOS and fast PMOS (SF) to model the worst-case "0". A standard set of model parameters (e.g., V_{th}) is used to account for process variability and model worst-case corner performance of the devices and circuits for the target CMOS technology [1,9].

In this modeling approach, the standard deviation (σ) limits are preset pessimistically to include any potential process variability over a wide range. The worst-case corner models are generated by offsetting the selected compact-model parameter, P, of the typical (TT) compact model by $\pm dP = n\sigma$ to account for the window of process variability, where n is the number of σ for P. Typically, $3 \leq n \leq 6$ is selected to set the fixed *lower limit* (LL) and *upper limit* (UL) of the worst-case models; and TT is the typical compact model extracted from the *golden die* of *golden wafer*, representing the centerline process technology [9]. For example, the TT model parameter V_{TH0} of BSIM4 [45] corner models is defined as $V_{TH} = V_{TH0} \pm dvth$, where $dvth$ is used to set the target LL and UL of the worst-case models.

To obtain the worst-case corner of drain current I_{ds}, let us consider the basic I_{ds} expression in the ON state (saturation regime) of a large MOSFET device [46] (Equation 4.87)

$$I_{ds} \cong \left(\frac{W}{2L}\right)\mu_{eff}C_{ox}\left(V_{gs} - V_{th}\right)^2; \quad 0 < \left(V_{gs} - V_{th}\right) < V_{ds} \tag{8.12}$$

where:
  μ_{eff}, C_{ox}, and V_{ds} are the inversion carrier mobility, gate oxide capacitance, and drain-to-source voltage, respectively
  $(V_{gs} - V_{th}) \equiv V_{dsat}$
  the remaining parameters have their usual meanings as defined in Chapters 4 and 5

Then, the UL is set by taking the appropriate maximum or minimum off-set of model parameters to maximize the value of I_{ds}. Thus, the UL of ION, defined at $V_{ds} = V_{dsat}$ for nMOSFETs is given by:

$$IONN(UL) \cong \mu_{eff}\left(\frac{W+dW}{2(L-dL)}\right)\left(\frac{\varepsilon_{ox}}{T_{ox}-dT_{ox}}\right)\left[V_{gs}-\left(V_{th}-dV_{th}\right)\right]^2 \qquad (8.13)$$

In Equation 8.13, W is increased by dW, L is reduced by dL, T_{ox} is reduced by dT_{ox}, and V_{th} is reduced by dV_{th} to achieve the UL of ION specification. Similarly, the LL for ION is set by

$$IONN(LL) \cong \mu_{eff}\left(\frac{W-dW}{2(L+dL)}\right)\left(\frac{\varepsilon_{ox}}{T_{ox}+dT_{ox}}\right)\left[V_{gs}-\left(V_{th}+dV_{th}\right)\right]^2 \qquad (8.14)$$

The FF corner is obtained using the UL values of the selected model parameters for both NMOS and PMOS devices whereas SS corner is obtained considering the LL values of the selected model parameters for both NMOS and PMOS devices. The SF corner is derived using LL values of NMOS and UL values of PMOS model parameters. Similarly, The FS corner is derived using UL values of NMOS and LL values of PMOS model parameters.

Figure 8.6 shows ION plots for both nMOSFET and pMOSFET devices obtained by fixed corner models along with the distribution of electrical test (ET) data. It is observed from Figure 8.6 that the simulation results obtained by fixed corner models are too wide, so they could end up rejecting a valid

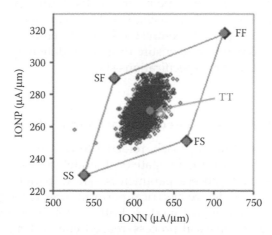

FIGURE 8.6
Distribution of measurement and simulation data generated using fixed corner models: NMOS ON current (IONN) versus PMOS ON current (IONP). (FF: fast NMOS and fast PMOS; FS: fast NMOS and slow PMOS; SF: slow NMOS and fast PMOS; SS: slow NMOS and slow PMOS). (Data from S.K. Saha, *IEEE Access*, 2, 104–115, 2014.)

design, causing yield loss. The major problems with the worst-case corner models are that in most cases the existing correlations between the device parameters are ignored and the models include pessimistic corner values. As a result, the models generate a large spread of data during analog circuit simulation.

The worst-case corner models offer designers capability to simulate the pass/fail results of a typical design and are usually pessimistic.

8.4.2 Statistical Corner Models

During IC chip manufacturing, a large set of ET data on critical device and process parameters are collected for process monitoring. Therefore, unlike fixed corner models, statistical corner models can be generated using ET data from different die, wafers, and wafer lots collected over a certain period of time to represent realistic process variability of a target technology [14–21].

In one approach, ET data are collected from a large number of sites of the target technology. And, for each site of ET data a compact model file is generated. Thus, a large number of compact model files, referred to as the *performance-aware model* (PAM) cards, are generated for the target technology [17,18]. In this approach about 1000 PAM cards or model files are generated for realistic statistical analysis of circuit performance.

In another approach, ET data are used to determine the depth of the location of device parameters in the distribution to generate corner models, referred to as the *location depth corner modeling* (LDCM) [19]. In LDCM, the wafers corresponding to the extreme data points in the distribution are used to extract separate compact models. Thus, using LDCM, the number of model cards is reduced significantly (<20) in contrast to PAM. An enhanced LDCM is used with proper guard banding to ensure design validation against future process shift from the baseline specifications [19].

8.4.3 Process Parameters–Based Compact Variability Modeling

The statistical modeling approach, referred to as the *backward propagation of variance* (BPV) [20], formulates statistical models as a set of independent, normally distributed process parameters. These parameters control the variations seen in the device electrical performances through the behavior described in the TT compact models. With recent extensions [21], BPV is used to characterize physical process–related compact model parameters. For an accurate analysis of process variability–induced circuit performance variability using BPV, the TT model file must be physical, the sensitivity matrix must be well-conditioned, and the variances of parameters must be physically consistent.

8.5 Statistical Compact Modeling

In the conventional variability modeling approaches, a standard set of model parameters are used for fixed corner modeling or a large number of model files are generated from ET data. The fixed corner models are inadequate, whereas, ET data–based modeling is resource-intensive. Therefore, an analytical technique to obtain the process-sensitive compact model parameters of any industry standard compact model to generate compact variability model library for circuit analysis is crucial for variability-aware circuit design as described in the following section.

A generalized approach for process variability modeling is shown in Figure 8.7. The method includes selection of target compact model, consideration of basic I_{ds} expression, derivation of a generalized expression of I_{ds} variance, selection of device parameters causing process-induced I_{ds} variation,

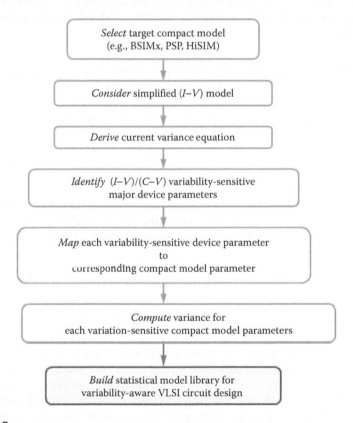

FIGURE 8.7
Generalized modeling approach for process variability-aware VLSI circuit design; here, BSIM*x*, PSP, and HiSIM represent industry standard MOSFET compact models; where x = 3, 4, and 6. (Data from S.K. Saha, *IEEE Access*, 2, 104–115, 2014.)

mapping process-sensitive device parameters to corresponding compact model parameters, determination of variances for mismatch modeling and global variability modeling, and finally, building compact variability model.

The modeling methodology outlined in Figure 8.7 is described in the following section.

8.5.1 Determination of Process Variability-Sensitive MOSFET Device Parameters

It is clear from our discussions in Section 8.2 that process variability causes variability in MOSFET device performance, which in turn causes variability in VLSI circuit performance. Since, the MOSFET device performance is determined by I_{ds}, in order to determine the impact of process variability on circuit performance, we determine the process variability-sensitive device parameters causing I_{ds} variability. For the selection of major process variability-sensitive device parameters, we consider the basic I_{ds} model in the subthreshold, linear, and saturation regions of MOSFETs (Equations 4.122 and 4.135)

$$I_{ds} \cong \begin{cases} \left(\dfrac{W}{L}\right)\mu_{eff}C_{ox}(n-1)v_{kT}^2 e^{\left(V_{gs}-V_{th}\right)/nv_{kT}}\left(1-e^{-\left(V_{ds}/v_{kT}\right)}\right); & \left(V_{gs}-V_{th}\right)<0 \\[2ex] \left(\dfrac{W}{L}\right)\mu_{eff}C_{ox}\left(V_{gs}-V_{th}-\dfrac{V_{ds}}{2}\right)V_{ds}; & 0<\left(V_{gs}-V_{th}\right)>V_{ds} \quad (8.15) \\[2ex] \left(\dfrac{W}{2L}\right)\mu_{eff}C_{ox}\left(V_{gs}-V_{th}\right)^2; & 0<\left(V_{gs}-V_{th}\right)\le V_{ds} \end{cases}$$

where the parameters have their usual meanings as defined in Chapter 4. From Equation 8.15, we can determine the major device parameters most sensitive to process variability in each region of MOSFET device operation.

8.5.1.1 Selection of Local Process Variability-Sensitive Device Parameters

The local process variability or mismatch between identically designed transistors is caused by microscopic process that makes every transistor different from its neighbors [1,9–13]. As a result, a device parameter P can be considered as consisting of a fixed component P_0 and a randomly varying component p resulting in different values of P for closely spaced identical paired transistors. Then the difference ΔP between two identical transistors within a die is a randomly varying parameter and is defined as the "mismatch" in P between two identical paired transistors. For a large number of samples, ΔP converges to a Gaussian distribution with zero mean. Then the mismatch in relative drain current, $\Delta I_{ds}/I_{ds}$, between paired transistors due to P is given by [47]:

$$\sigma^2_{\Delta I_{ds}/I_{ds}} = \sum_{i=1}^{l} \left(\frac{1}{I_{ds}} \frac{\partial I_{ds}}{\partial P_i} \right)^2 \sigma^2_{\Delta P_i} + \frac{2}{I^2_{ds}} \sum_{i=1}^{l} \frac{\partial I_{ds}}{\partial P_i} \frac{\partial I_{ds}}{\partial P_{i+1}} \rho(\Delta P_i, \Delta P_{i+1}) \qquad (8.16)$$

where l is the total count of ΔP contributing to I_{ds} mismatch; ΔP_i is the ith count of ΔP with standard deviation $\sigma_{\Delta P_i}$; and $\rho(\Delta P_i, \Delta P_{i+1})$ is the correlation between ΔP_i and ΔP_{i+1}. Since ΔP_i is random and independent, the correlation $\rho(\Delta P_i, \Delta P_{i+1}) = 0$ as discussed in Section 8.3.1. In order to model I_{ds} mismatch between paired transistors, we determine the major local process variability-sensitive device parameters P.

From Equation 8.15, we find that for all regions of MOSFET device operation, the value of I_{ds} depends on a common set of parameters $\{V_{th}, W, L, C_{ox}, \mu_{eff}, V_{gs}, V_{ds}\}$. We know $C_{ox} = f(T_{ox})$; then considering only parametric variation in Equation 8.16, ΔP represents any of the mismatch parameters of the set $\{\Delta V_{th}, \Delta W, \Delta L, \Delta T_{ox}, \Delta \mu_{eff}\}$. It is to be noted that the parameter set $\{\Delta W, \Delta L, \Delta T_{ox}, \Delta \mu_{eff}\}$ describes the mismatch in current gain, $\beta = \left[(W/L)C_{ox}\mu_{eff}\right]$, defined in Equation 4.74.

Again, V_{th} can be expressed as $V_{th} = f(V_{th0}, \gamma, \phi_s, V_{bs})$, where V_{bs} is the applied body bias and $V_{th0} = V_{th}$ at $V_{bs} = 0$ whereas γ and ϕ_s are the body effect coefficient and channel surface potential, respectively. Here, ΔV_{th0} describes the mismatch $\Delta I_{ds}(V_{bs} = 0)$ due to RDD of the channel doping concentration N_{CH} of MOSFETs whereas, $\Delta \gamma$ describes the mismatch in $\Delta I_{ds}(V_{bs})$ due to the variation in N_{CH} in the depletion region under the gate. We know that $\gamma = f(N_{CH})$ (Equation 4.11) and with the change in the value of V_{bs}, the depth of the depletion layer under the gate changes due to nonuniform channel doping profile [1,9,48–51]. As a result, the amount of bulk charge qN_{CH} changes with the change in V_{bs} as shown in Figure 8.8 for the graded retrograde channel doping profile [49]. Thus, RDD of the vertical channel doping profile under the gate contributes to the mismatch in $I_{ds}(V_{bs})$. Hence, $I_{ds}(V_{bs})$ mismatch between the identical paired transistors due to variation in the vertical channel doping concentration must be modeled by γ.

Thus, the set of major local process variability-sensitive device parameters contributing to the mismatch between identically designed paired transistors within a die is $\{V_{th0}, W, L, T_{ox}, \mu_{eff}, \gamma\}$ as shown in Table 8.1. Here, ΔV_{th0} describes the variation in ΔI_{ds} due to RDD; ΔW and ΔL describe ΔI_{ds} due to LER and LWR; ΔT_{ox} defines ΔI_{ds} due to OTV; $\Delta \mu_{eff}$ defines ΔI_{ds} due to mobility variation caused by SR scattering; and γ models $\Delta I_{ds}(V_{bs})$ due to RDD in the vertical channel doping profile. Therefore, we have used the basic I–V relation to determine the major process variability-sensitive device parameters for modeling mismatch in VLSI circuit performance.

8.5.1.2 Selection of Global Process Variability-Sensitive Device Parameters

The global process variability is caused by nonuniform processing temperature as well as by the variation of implant doses across wafers and relative

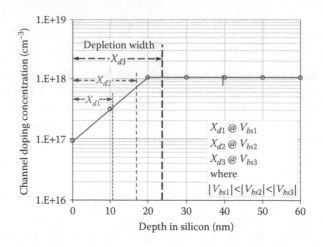

FIGURE 8.8
A piecewise graded-retrograde MOSFET channel doping profile from the silicon/SiO$_2$ interface at depth $x = 0$ into the substrate; here X_{d1}, X_{d2}, and X_{d3} are the depletion width due to the applied body bias V_{bs1}, V_{bs2}, and V_{bs3}, respectively, causing $V_{th}(V_{bs})$ variability due to RDD along the depth of the channel. (Data from S.K. Saha, *IEEE Access*, 2, 104–115, 2014.)

TABLE 8.1

Process Variability-Sensitive Local Device Parameters Mapped to the Corresponding BSIM4 Compact Model Parameters

Device Parameter		Compact Model Parameter	
Symbol	**Definition**	**Symbol**	**Definition**
V_{th0}	Threshold voltage	VTH0	V_{th} at $V_{bs} = 0$
W	Channel width	XW	W offset due to masking and lithography
L	Channel length	XL	L offset due to masking and lithography
T_{ox}	Gate oxide thickness	TOXE/TOXM	Equivalent T_{ox}
μ_{eff}	Inversion carrier mobility	U0	Low field mobility
γ	Body bias coefficient	K1	1st order body bias coefficient

Source: S.K. Saha, *IEEE Access*, 2, 104–115, 2014.

location of devices [7,8]. The global variation shifts the average or mean value of device performance. As a result, a device parameter within a chip varies for two identically designed devices. For a large count of P from a large number of on-chip measurement data, P converges to a Gaussian distribution with mean value P_0 and standard deviation $\sigma = \Delta P$. Then the chip mean variation in I_{ds} due to global process variability-sensitive parameter P is given by [9]

$$\sigma_{I_{ds}}^2 = \sum_{i=1}^{l}\left(\frac{\partial I_{ds}}{\partial P_i}\right)^2 \sigma_{P_i}^2 + 2\sum_{i=1}^{l}\frac{\partial I_{ds}}{\partial P_i}\frac{\partial I_{ds}}{\partial P_{i+1}}\rho(P_i,P_{i+1}) \tag{8.17}$$

where l is the total number of occurrence (total count of data) of the device parameter P contributing to global I_{ds} variation; P_i is the ith count of P with standard deviation σ_{P_i} from its mean value P_0; and $\rho(P_i,P_{i+1})$ is the correlation between the occurrence P_i and P_{i+1}. In order to model the variation of I_{ds} around its mean value, we determine the major global process variability-sensitive parameters P.

Again, from Equation 8.15, the chip mean variation in I_{ds} due to global process variability can be described by the parameter set $\{V_{th0}, W, L, C_{ox}, \mu_{eff}, \gamma\}$. In addition, the I_{ds} variability due to the variation in the S/D dopant implantation dose and processing temperature across wafers are described by the variation in the S/D series resistance R_{DS} of MOSFET devices. Furthermore, the gate delay, $\tau_{pd} \propto C_{load}$, where C_{load} is the load capacitance of the inverter circuit. Therefore, for an accurate simulation of digital circuits, the across-the-chip variation in MOSFET gate capacitance (C_g) along with the S/D junction capacitance (C_j) must be modeled. Now, the variability in the mean value of C_g is described by the gate overlap capacitance (C_{ov}) whereas that in C_j is described by S/D area as well as S/D sidewall and isolation-edge sidewall capacitances. Thus, the variation in the AC and transient performance of VLSI digital circuits are also described by an additional parameter set $\{C_{ov}, C_j\}$. Therefore, the set of major MOSFET device parameters sensitive to global process variability can be represented by $\{V_{th0}, W, L, T_{ox}, \mu_{eff}, \gamma, R_{DS}, C_{ov}, C_j\}$ as shown in Table 8.2.

8.5.2 Mapping Process Variability-Sensitive Device Parameters to Compact Model Parameters

In order to develop compact MOSFET model to analyze the impact of process variability in advanced VLSI circuits, the process variability-sensitive device parameters $\{P\}$ selected in Section 8.5.1 are mapped to the corresponding compact model parameter $\{M\}$ of the selected compact model. In this study, we select BSIM4 [45] compact model to describe the methodology of generating compact MOSFET variability model library for VLSI circuit CAD.

8.5.2.1 Mapping Local Process Variability-Sensitive Device Parameters to Compact Model Parameters

In Section 8.5.1.1, we have described an analytical approach to select the randomly variable set of device parameters, $\{V_{th0}, W, L, T_{ox}, \mu_{eff}, \gamma\}$, causing mismatch between identically designed paired transistors. The corresponding set of BSIM4 MOS model parameters, shown in Table 8.1, is

TABLE 8.2

Process Variability-Sensitive Global Device Parameters Mapped to the
Corresponding Compact Model Parameters

Device Parameter		Compact Model Parameter	
Symbol	Definition	Symbol	Definition
V_{th}	Threshold voltage	VTH0	V_{th} at $V_{bs} = 0$
W	Channel width	XW	W offset due to masking and lithography
L	Channel length	XL	L offset due to masking and lithography
T_{ox}	Gate oxide thickness	TOXE/TOXM	Equivalent T_{ox}
μ_{eff}	Inversion carrier mobility	U0	Low field mobility
γ	Body bias coefficient	K1	1st order body bias coefficient
R_{DS}	SDE resistance	RDSW	Zero bias R_{DS}
C_{ov}	Gate overlap capacitance	CGSL/CGDL	SDE C_{ov}
		CGSO/CGDO	Non-SDE region C_{ov}
C_j	S/D junction capacitance	CJS/CJD	Area component of C_j
		CJSWS/CJSWD	Isolation-edge sidewall C_j
		CJSWGS/CJSWGD	Gate-edge sidewall C_j

Source: S.K. Saha, *IEEE Access*, 2, 104–115, 2014.

$\{V_{TH0}, XW, XL, T_{ox}, U0, K1\}$; where, XW and XL are the channel width and length offset parameters due to masking and photolithography, respectively, and account for the mismatch due to LER and LWR, whereas $U0$ and $K1$ account for the variation in μ_{eff} and N_{CH} under V_{bs}, respectively. In order to build the compact model, the variance $\sigma_{\Delta M_{mismatch}}$ is computed for each M from a large set of data to account for the mismatch in identical paired transistors.

8.5.2.2 Mapping Global Process Variability-Sensitive Device Parameters to Compact Model Parameters

In Section 8.5.1.2, we have shown an analytical approach to determine the critical set of device parameters, $\{V_{th0}, W, L, T_{ox}, \mu_{eff}, \gamma, R_{DS}, C_{ov}, C_j\}$, impacting MOSFET device performance due to global process variability. The corresponding set of BSIM4 compact model parameters is {VTH0, XW, XL, TOX, U0, K1, RDSW, CGSO, CGDO, CGSL, CGDL, CJS, CJD, CJSWS, CJSWD, CJSWGS, CJSWGD}, where the parameter set {CGSO, CGDO, CGSL, CGDL} defines C_{ov}; {CJS, CJD} defines S/D junction area capacitances and {CJSWS, CJSWD, CJSWGS, CJSWGD} defines S/D *pn*-junction sidewall capacitances as shown in Table 8.2. For each M, the variance $\sigma_{M_{global}}$ is obtained from a large set of ET data and added to M_0 to analyze the impact of chip mean variation in VLSI circuits.

8.5.3 Determination of Variance for Process Variability-Sensitive Compact Model Parameters

The variance σM of the compact model parameter M due to process variability is included to the mean (TT) value M_0 to model the impact of process variability on VLSI circuit performance.

8.5.3.1 Variance of Local Process Variability-Sensitive Compact Model Parameters

For a large number of samples $\Delta M_{mismatch}$ between paired transistors is described by standard normal distribution, $N\left(0, \sigma_{\Delta M_{mismatch}}\right)$, where the variance $\sigma_{\Delta M_{mismatch}}$ is given by: $\sigma_{\Delta M_{mismatch}}\big|_{pair} \cong A_M / \sqrt{WL}$ as described in Section 8.3.1 [43,44], where the parameter A_M is a technology-dependent constant of ΔM and is extracted from ΔM_i versus $\left(1/\sqrt{WL}\right)$ plot for a large number ($i = 1, 2, 3, \ldots l$) of sample ET data [1,43,44]. Thus, for the compact model parameter V_{TH0}, the variance of ΔV_{TH0} between two paired transistors is given by:

$$\sigma_{\Delta V_{TH0}}\big|_{pair} \cong \frac{A_{vt}}{\sqrt{WL}} \tag{8.18}$$

where:
A_{vt} is the area dependent constant of ΔV_{TH0}

Typically, each mismatch parameter $\Delta VTH0$, ΔXW, ΔXL, ΔT_{ox}, $\Delta U0$, and $\Delta K1$ can be represented by an expression similar to Equation 8.16. Again, since ΔM_i is random and independent, the correlation $\rho\left(\Delta M_i, \Delta M_{i+1}\right) = 0$ [43]. Then, for a single device we get

$$\sigma M_{mismatch} = \frac{1}{\sqrt{2}} \sigma_{\Delta M_i} = \frac{1}{\sqrt{2}} \frac{A_M}{\sqrt{WL}} \tag{8.19}$$

In Equation 8.19, $\sigma M_{mismatch}$ represents the variance of ΔM due to within-die stochastic process variability. Thus, the variance of ΔV_{TH0} is given by

$$\sigma V_{TH0,mismatch} = \frac{1}{\sqrt{2}} \sigma_{\Delta VTH0} = \frac{1}{\sqrt{2}} \frac{A_{vt}}{\sqrt{WL}} \tag{8.20}$$

For statistical compact modeling, $\sigma M_{mismatch}$ for each variability-sensitive parameter is added to the corresponding M_0 to compute mismatch between paired transistors. Typically, for each M, A_M is extracted from Pelgrom's plot from a large set of measurement data. For next generation technology development, a large set of data can be obtained by numerical process and device CAD to compute $\sigma M_{mismatch}$ for each variability-sensitive compact model parameters [51–55].

8.5.3.2 Variance of the Global Process Variability-Sensitive Compact Model Parameters

For Monte Carlo (MC) statistical modeling, M_{global} is described by normal distribution $N\left(M_0, \sigma M_{global}\right)$, around its mean (TT) value M_0. The global variance σM_{global} is obtained from the statistical distribution of ET data for each M measured from multiple die, wafers, and lots over a period of time [1,9]. However, for the next-generation technology, the ET data are scarcely available for statistical analysis. In this case, the numerical simulation data can be used for the computation of σM_{global} and generate rev0 compact model for circuit analysis of the target technology [51–55]. Typically, $n\sigma M_{global}$ is used to model global process variability with $3 \leq n \leq 6$.

8.5.4 Formulation of Compact Model for Process Variability-Aware Circuit Design

As described in Section 8.4.1, the TT model for circuit CAD consists of a set of parameters $\{M_0\}$ that models the device and circuit performance of centerline process of the target technology node. The set $\{M_0\}$ represents the nominal device specifications of the target technology. The local and global components of the variability-sensitive compact model parameter are included in the nominal set $\{M_0\}$ to generate compact variability model library for circuit CAD. The final model library includes the nominal parameters with the components of process variability. Thus, a process variability-sensitive model parameter M including both local and global process variability components is given by

$$M = M_0 + \sigma M_{mismatch} + n\sigma M_{global} \tag{8.21}$$

Equation 8.21 is used to build the compact model of the target technology for process variability-aware circuit analysis. Thus, for the compact model parameter V_{TH}, Equation 8.21 yields

$$V_{TH} = V_{TH0} + \sigma V_{TH0,mismatch} + n\sigma V_{TH0,global} \tag{8.22}$$

Equation 8.22 is used to build statistical corner model for realistic analysis of process variability in scaled MOSFETs. Table 8.3 shows FF and SS corner limit of a set of process variability-sensitive model parameters obtained by analytical approach discussed in Section 8.5.2.2.

For MC statistical compact modeling, the *probability distribution function* (PDF) of the mismatch component of M for HSPICE (see Section 1.2.2.1) [56] circuit CAD is obtained using 1-σ variation between paired transistors

$$PDF\left(\sigma M_{mismatch}\right) = \left(\sigma M_{mismatch}\right)agauss(0,1,1) \tag{8.23}$$

TABLE 8.3

Typical Parameter Limits for Worst-Case BSIM4
Fixed Corner Model Generation

Compact Model Parameter	FF	SS
TOXE	Minimum	Maximum
TOXM		
XL	Minimum	Maximum
XW	Maximum	Minimum
VTH0	Minimum	Maximum
U0	Maximum	Minimum
K1	Minimum	Maximum
RDSW	Minimum	Maximum
CGSL	Maximum	Minimum
CGDL		
CGSO	Maximum	Minimum
CGDO		
CJS	Minimum	Maximum
CJD		
CJSWS	Minimum	Maximum
CJSWD		
CJSWGS	Minimum	Maximum
CJSWGD		

Source: S.K. Saha, *IEEE Access*, 2, 104–115, 2014.
FF: fast NMOS and fast PMOS; SS: slow NMOS and slow
PMOS.

Similarly, the PDF for the global component of M is expressed as

$$PDF\left(\sigma_{global}\right) = \left(\sigma_{global}\right)agauss(0,1,3) \tag{8.24}$$

Equations 8.22 through 8.24 are used to formulate the variability-sensitive compact model parameters to develop the final model library for HSPICE circuit CAD. Table 8.4 shows the formulation of variability-sensitive BSIM4 model parameters determined in Section 8.5.2 in the model library. Thus, for the variability-sensitive V_{TH}, we have

$$V_{TH} = V_{TH0} + \frac{1}{\sqrt{2}}\frac{Avt}{\sqrt{WL}}agauss(0,1,1) + \sigma V_{TH0}agauss(0,1,3) \tag{8.25}$$

The above procedure is used to build a BSIM4 MOSFET compact model library for the advanced CMOS technology [5–7]. In order to show the basic function-ality of the present modeling approach, all mismatches are lumped into V_{th} mismatch and the correlation between global model parameters is ignored.

Compact Models for Integrated Circuit Design

TABLE 8.4

Process Variability-Sensitive BSIM4 Model Parameters

Compact Model Parameter	Typical Value (M_0)	Local Component of Model Parameter		Global Component of Model Parameter		Compact Model Parameter for MC Statistical Analysis
		$\sigma M_{mismatch\_z}$	PDF for MC Analysis	σM_{global_z}	PDF for MC Analysis	
TOXE	$toxe0$	$\sigma_{toxm\_z} = \left(1/\sqrt{2}\right)\left(A_{tox}/\sqrt{WL}\right)$	$\sigma_{toxm\_z}\,agauss(0,1,1)$	σ_{tox_z}	$\sigma_{tox\_z}\,agauss(0,1,3)$	$toxe0 + \sigma_{toxm\_z}\,agauss(0,1,1)$ $+ \sigma_{tox\_z}\,agauss(0,1,3)$
TOXM	$toxm0$	$\sigma_{toxm\_z} = \left(1/\sqrt{2}\right)\left(A_{tox}/\sqrt{WL}\right)$	$\sigma_{toxm\_z}\,agauss(0,1,1)$	σ_{tox_z}	$\sigma_{tox\_z}\,agauss(0,1,3)$	$toxm0 + \sigma_{toxm\_z}\,agauss(01,1)$ $+ \sigma_{tox\_z}\,agauss(0,1,3)$
XL	$xl0$	$\sigma_{xlm\_z} = \left(1/\sqrt{2}\right)\left(A_{xl}/\sqrt{WL}\right)$	$\sigma_{xlm\_z}\,agauss(0,1,1)$	σ_{xl_z}	$\sigma_{xl\_z}\,agauss(0,1,3)$	$xl0 + \sigma_{xlm\_z}\,agauss(0,1,1)$ $+ \sigma_{xl\_z}\,agauss(0,1,3)$
XW	$xw0$	$\sigma_{xwm\_z} = \left(1/\sqrt{2}\right)\left(A_{xw}/\sqrt{WL}\right)$	$\sigma_{xwm\_z}\,agauss(0,1,1)$	σ_{xw_z}	$\sigma_{xw\_z}\,agauss(0,1,3)$	$xw0 + \sigma_{xwm\_z}\,agauss(0,1,1)$ $+ \sigma_{xw\_z}\,agauss(0,1,3)$
VTH0	$vth0$	$\sigma_{vthm\_z} = \left(1/\sqrt{2}\right)\left(A_{vt}/\sqrt{WL}\right)$	$\sigma_{vthm\_z}\,agauss(0,1,1)$	σ_{vth_z}	$\sigma_{vth\_z}\,agauss(0,1,3)$	$vth0 + \sigma_{vthm\_z}\,agauss(0,1,1)$ $+ \sigma_{vth\_z}\,agauss(0,1,3)$
U0	$u00$	$\sigma_{u0m\_z} = \left(1/\sqrt{2}\right)\left(A_{u0}/\sqrt{WL}\right)$	$\sigma_{u0m\_z}\,agauss(0,1,1)$	σ_{u0_z}	$\sigma_{u0\_z}\,agauss(0,1,3)$	$u00 + \sigma_{u0m\_z}\,agauss(0,1,1)$ $+ \sigma_{u0\_z}\,agauss(0,1,3)$
K1	$k10$	$\sigma_{k1m\_z} = \left(1/\sqrt{2}\right)\left(A_{k1}/\sqrt{WL}\right)$	$\sigma_{k1m\_z}\,agauss(0,1,1)$	σ_{k1_z}	$\sigma_{k1\_z}\,agauss(0,1,3)$	$k10 + \sigma_{k1m\_z}\,agauss(0,1,1)$ $+ \sigma_{k1\_z}\,agauss(0,1,3)$

(Continued)

TABLE 8.4 (Continued)

Process Variability-Sensitive BSIM4 Model Parameters

Compact Model Parameter	Typical Value (M_0)	Local Component of Model Parameter		Global Component of Model Parameter		Compact Model Parameter for MC Statistical Analysis
		$\sigma M_{mismatch\_z}$	PDF for MC Analysis	σM_{global_z}	PDF for MC Analysis	
RDSW	rdsw0			σ_{rdsw_z}	$\sigma_{rdsw\_z}\,agauss(0,1,3)$	$rdsw0 + \sigma_{rdsw\_z}\,agauss(0,1,3)$
CGSL/CGDL	cgl0	*	*	σ_{cgl_z}	$\sigma_{cgl\_z}\,agauss(0,1,3)$	$cgl0 + \sigma_{cgl\_z}\,agauss(0,1,3)$
CGSO/CGDO	cgo0	*	*	σ_{cgo_z}	$\sigma_{cgo\_z}\,agauss(0,1,3)$	$cgo0 + \sigma_{cgo\_z}\,agauss(0,1,3)$
CJS/CJD	cj0	*	*	σ_{cj_z}	$\sigma_{cj\_z}\,agauss(0,1,3)$	$cj0 + \sigma_{cj\_z}\,agauss(0,1,3)$
CJSWS/CJSWD	cjsw0	*	*	σ_{cjsw_z}	$\sigma_{cjsw\_z}\,agauss(0,1,3)$	$cjsw0 + \sigma_{cjsw\_z}\,agauss(0,1,3)$
CJSWGS/CJSWGS	cjswg0	*	*	σ_{cjswg_z}	$\sigma_{cjswg\_z}\,agauss(0,1,3)$	$cjswg0 + \sigma_{cjswg\_z}\,agauss(0,1,3)$

Source: Data from S.K. Saha, *IEEE Access*, 2, 104–115, 2014.

8.5.5 Simulation Results and Discussions

The model library developed in Section 8.5.4 is used for MC statistical analysis of advanced MOSFET devices [5–7]. Since RDD is the dominant contributor in mismatch, Figure 8.9 is obtained using only RDD in mismatch model. Figure 8.9 shows the distribution of IONN and IONP obtained by HSPICE circuit simulation. Here, ION is defined as $|V_{gs}| = |V_{ds}| = 1$ V. The IONN versus IONP distribution in Figure 8.9 clearly shows the impact of local process variability or mismatch, global process variability or chip mean variation, and the local and global process variability combined. In Figure 8.9, the simulation data from statistical corner values of IONN and IONP are also superimposed on the plot for reference. In Figure 8.9, FF and SS corners enclose the MC distribution of ON currents. Thus, in contrast to fixed pessimistic corners, shown in Figure 8.6, the statistical corners offer realistic analysis of process variability similar to MC analysis as shown in Figure 8.9.

Figure 8.9 shows that global variability is dominated mainly by local fluctuations, as observed for advanced bulk technologies [57]. It would indicate that global variability is dominated by local random fluctuations or that most of the systematic process variations are present already within the distance between two mismatch transistors [57].

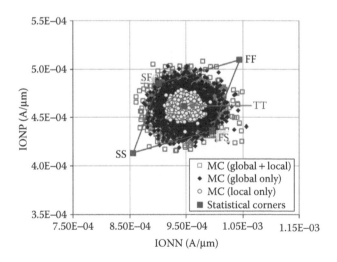

FIGURE 8.9
MC simulation data obtained by HSPICE circuit CAD for an advanced CMOS technology; simulation data show the distribution of ON currents for pMOSFETs (IONP) and nMOSFETs (IONN) for local only, global only, and both local and global process variability. The simulated statistical corners (SS, FF, SF, and FS) along with the nominal (TT) values of drain currents are also superimposed on the plot using solid rectangular symbols. (Data from S.K. Saha, *IEEE Access*, 2, 104–115, 2014.)

8.6 Mitigation of the Risk of Process Variability in VLSI Circuit Performance

Techniques to mitigate the risk of process variability include (1) pure process optimization such as targeting key transistor properties to reduce RDD, improve patterning techniques to reduce LER, and improve polishing techniques to reduce systematic cross-wafer variation; (2) combination of process and design techniques such as optimization of topology, use of OPC to reduce random and systematic variations, and adding dummy features to reduce systematic variations; and (3) pure design techniques such as common-centroid layout to compensate for systematic variation.

As we discussed in Section 8.2.2.1, RDD is a major contributor to random variation and is modeled by Equation 8.2. From Equation 8.2, it is found that we can reduce the impact of RDD by reducing channel doping, N, and gate oxide thickness, T_{ox}. In advanced CMOS technologies, T_{ox} is scaled appropriately using Hi-K dielectric with metal gate to mitigate the risk of process variability due to OTV. However, due to the scaling constraint of N_{CH}, RDD cannot be controlled in nanoscale planar CMOS technology.

Recently, advanced channel engineering has been used to design nanoscale MOSFET devices with undoped or lightly doped channel to mitigate the risk of RDD [48]. The channel is formed on undoped epitaxial layer grown on silicon substrate followed by standard CMOS processing steps [58]. Also, it has been shown that the double-halo MOSFET device architecture [5–8] controls the V_{th} variation in nanoscale devices. Recently, an enhanced double-halo MOSFET [7] device architecture is proposed to design undoped or lightly doped channel MOSFETs and mitigate the risk of process variability in planar CMOS technology [59]. This enhanced double-halo structure is referred to as the buried-halo MOSFET (BH-MOSFET), which is shown in Figure 8.10. The simulation results shown in Figure 8.11 show a significant reduction of threshold voltage variation due to RDD in nanoscale BH-MOSFETs compared to the conventional MOSFET devices.

In order to further mitigate the risk of process variability in nonplanar devices and technologies including Fin field-effect transistors (FinFETs) and ultrathin body (UTB) silicon-on-insulator field-effect transistors referred to as the UTB-SOI MOSFETs [60] have emerged as the most promising alternatives to MOSFET devices and CMOS technology. An overview of the compact models for these devices is presented in Chapter 9.

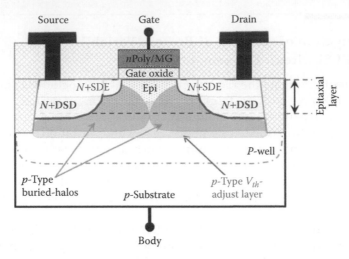

FIGURE 8.10
A variability-tolerant buried-halo MOSFET (BH-MOSFET) device structure; multiple halo implants are buried under the epitaxial layer to obtain undoped or lightly doped channel region.

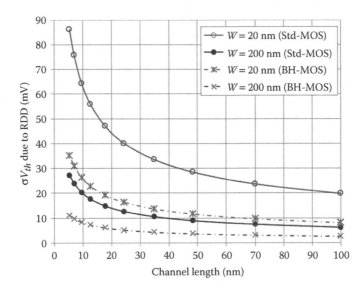

FIGURE 8.11
Comparison of the simulated threshold voltage variation of the conventional (Std-MOS) and BH-MOSFET (BH-MOS) for a typical 20-nm bulk CMOS technology as a function of device channel length for channel width 20 and 200 nm following ITRS (Data from International Technology Roadmap for Semiconductors. http://www.itrs.net/) and using Equation 8.2; parameters used are $N_{CH} = 6 \times 10^{18}$ cm^{-3}; SiO$_2$ equivalent oxide thickness (EOT) = 1.1 nm; and C = 0.8165.

8.7 Summary

This chapter presented the intrinsic process variability in CMOS technology and different approaches to model process variability in VLSI circuit CAD. A brief overview of the systematic and stochastic front-end process variability and sources of process variability is described. A methodology to characterize the random process variability that causes mismatch in the performance of identical MOSFETs in a die is discussed. Conventional approaches to generate compact MOSFET variability models are overviewed and a detailed statistical MOSFET compact modeling approach is discussed. The basic steps to generate statistical compact MOSFET models including selection of expressions defining device performance, selection of device parameters sensitive to process variability, mapping process-variability sensitive device parameters to corresponding compact model parameters, and model formulation are described. The results obtained by MC statistical model and statistical corner model are presented. The basic statistical modeling methodology can be used to generate statistical compact MOSFET models using any compact models considering the basic equations for device performance. Finally, different approaches to mitigate the risk of process variability in VLSI circuits are briefly discussed.

Exercises

8.1 Write an expression for variance $\left(\sigma_{\Delta\beta}^2/\beta^2\right)$ for mismatch in current factor β (a) in terms of the variance of its mutually independent components described in Section 8.5.1.1 and (b) in terms of the mismatch coefficient of each component.

8.2 Consider an nMOSFET device with channel length $L = 100$ nm, channel width $W = 200$ nm, channel doping concentration $N_a = 5 \times 10^{17}$ cm^{-3}, $T_{ox} = 1$ nm, and S/D junction depth $X_j = 50$ nm of the 100 nm CMOS technology node; use $C = 0.8165$ to solve the following problems:

    a. Scale down the above technology by 70% up to five times and calculate the total number of dopants (N_{total}) in the channel for all the technology nodes and plot N_{total} versus L. (Scaling: multiply all geometry parameters by 0.7 and divide doping by 0.7.)

    b. Considering the device with $W = 200$ nm, calculate and plot $\sigma V_{th,RDD}$ as a function of L calculated in part (a); assume $L = L_{eff}$ and $W = W_{eff}$.

c. Repeat part (b) for $W = 30$ nm to calculate and plot $\sigma V_{th,RDD}$ as a function of L on the same graph (b).

d. Compare your results in parts (b) and (c) and explain.

e. Repeat parts (b) and (c) for $C = 0.7071$; explain the difference, if any.

8.3 Use the given technology parameters in exercise 8.2 to solve the following problems (consider only RDD):

a. Estimate mismatch coefficient A_{vt} for the nMOSFET devices of the technology using $C = 0.8165$.

b. Use the estimated A_{vt} number from part (a) to calculate $\sigma(\Delta V_{th})$ for a set of devices with varying W and L and plot $\sigma(\Delta V_{th})$ *versus* $1/\sqrt{W \cdot L}$. Explain your plot.

c. Repeat part (a) to calculate σV_{th} for a set of devices with varying W and L and plot σV_{th} versus $1/\sqrt{W \cdot L}$. Explain your results.

d. Compare results from part (b) and part (c) and explain the significance of each plot in compact MOSFET modeling.

First of all, select a wide W (~2 μm) and keeping W constant vary L from the nominal geometry to a long (~250 nm) device and calculate the area $W.L$; then select a long L (=200 nm) and keeping L constant vary W from the nominal geometry to a wide device (~1 μm) and calculate $W.L$.

8.4 If the distance between the identical paired transistors in the x direction is D_x, write an expression for the variance $\sigma_{\Delta P}^2$ of the stochastic parameter P showing the correction factor due to separation between the transistors of the pair.

8.5 Following references [23,24] derive Equation 8.2. Clearly state any assumptions you make.

9

Compact Models for Ultrathin Body FETs

9.1 Introduction

This chapter presents compact models for the emerging ultrathin-body (UTB) field-effect-transistors (FETs). The UTB FETs include *multiple-gate* or *multigate* FinFETs and silicon-on-insulator (SOI) multigate UTB-FETs (UTB-SOI FETs) [1,2]. FinFETs and UTB-SOI FETs have emerged as the real alternatives to MOSFETs (metal-oxide-semiconductor field-effect transistors) and planar CMOS (complementary metal-oxide-semiconductor) technology to surmount the continuous scaling challenges of MOSFET devices. The continuous miniaturization of the conventional planar MOSFET devices has become more challenging at the same rate of Moore's law [3–6] due to several fundamental device-physics constraints such as short channel effects (SCEs). Shrinking the gate length, L, in the decananometer regime degrades the transfer characteristics of planar MOSFETs, degrades the subthreshold swing (S), and decreases V_{th} (e.g., V_{th} roll-off) [3] as discussed in Chapter 5. This implies that the scaled MOSFETs cannot be turned off easily by lowering the gate voltage V_g due to SCEs [7]. Because of SCEs, the device characteristics become increasingly sensitive to L variations and process-induced variability imposes a serious challenge in continued scaling of bulk MOSFETs as discussed in Chapter 8 [8,9]. The early theoretical and modeling approaches on SCEs [10–12] suggest increasing the gate control by reducing the gate dielectric thickness in proportion to L, which increases manufacturing process complexity. Another constraint for the continuous scaling of conventional bulk MOSFETs is controlling leakage current in scaled devices [12]. It is observed that at gate length below 20 nm, the leakage paths several nanometers below the silicon-dielectric interface (subsurface leakage paths) are primarily responsible for the leakage current. These leakage paths are weakly controlled by the gate irrespective of gate oxide thickness and their potential barriers can be easily lowered by drain bias V_d through the enhanced electric field coupling to the drain, referred to as the drain-induced barrier lowering [12]. This new challenge to scaling L led to engineering efforts on channel-profile engineering, shallow source-drain extensions (SDE), and halo implants around SDEs as discussed in Chapter 5 [13–19].

In order to overcome the increasing challenges in continuous scaling of the conventional planar MOSFETs, the major research and development efforts for the last two decades have been exploring alternative device architectures and materials [20–28]. Among the exploratory devices, FinFETs [29–36] and UTB-SOI MOSFETs [37–40] have emerged as the most promising devices for advanced nanometer scale VLSI (very-large-scale-integrated) technology and beyond. The multiple-gates of multigate FETs offer strong electrostatic control over the channel and reduce the coupling between the source and drain in the subthreshold region, thus enabling continuous scaling of FETs. Multigate FETs have a great potential to mitigate the risk of process variability by using undoped channel. The efforts are under way to enable large-scale manufacturing of multigate FETs [41–44]. A reduction of four orders of magnitude in the leakage current over the 32 nm planar manufacturing process has been reported [29]. UTB-SOI FETs [45], deeply depleted channel MOSFETs [46], and BH-halo MOSFETs [47] are close competitors to the FinFET architecture along with IBM's aggressively scaled planar MOSFET down to the 10 nm node [48]. Thus, ultrathin body enables continuous scaling down of FETs by overcoming the major scaling constraints such as SCE and random discrete doping (RDD) of the conventional bulk MOSFETs discussed in Chapters 5 and 8. For computer analysis of the performance of these emerging multigate FETs in VLSI circuits, compact models are critical. This chapter presents surface potential–based compact models for multigate FET devices.

9.2 Multigate Device Structures

The desirables from any alternative device structure include surmounting the impending L scaling barrier, preserving today's CMOS technology as much as possible, and using innovative device architectures to eliminate major problems in scaled planar MOSFETs including undesirable leakage currents and excessive static power. Among the alternative architectures, FinFETs [29–36] and UTB-MOSFETs [37–40] are found to offer solutions to major issues for the continuous scaling of FETs. Both of these structures show potential to eliminate the leakage paths that are far from the gate(s) by limiting the thickness of semiconductor body in the immediate vicinity of the gate(s) [29].

9.2.1 Bulk-Multigate Device Structure

Figure 9.1 shows a 3D cross section of an ideal double-gate MOSFET (DG-MOSFET) device structure [49]. As shown in Figure 9.1, the structure consists of a thin film of undoped silicon body, referred to as the *fin*, a front

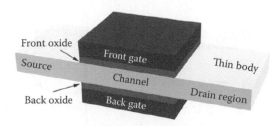

FIGURE 9.1
3D cross section of an ideal DG-MOSFET device structure with an undoped thin film silicon body; all leakage paths are close to the gates due to thin body, thus suppressing the short-channel effects. (Data from N. Paydavosi et al., *IEEE Access*, 1, pp. 201–215, 2013.)

and a back gate oxide layers, a source and a drain regions, and front and back gates. If the body is sufficiently thin, any line drawn between the source and drain including possible leakage paths would not be far from one of the gates. In this structure, channel doping is not required for suppressing SCEs. Thus, RDD, a major contributor to the variation in the performance of IC devices and VLSI circuits, is eliminated [8,9].

Figure 9.2 shows a typical manufacturable version of the multiple-fin FinFET device structure commonly referred to as the *multigate* structure [50]. The fin can be fabricated on SOI or cost-effective bulk silicon substrates using the standard patterning and etching technologies.

Let us consider an ideal symmetric double-gate FinFET (DG-FinFet) structure with channel length L and the channel thickness defined by fin

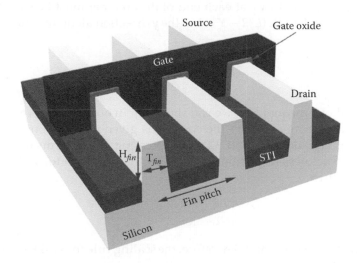

FIGURE 9.2
3D cross section of a typical multifin FinFET structure used in manufacturing; in the structure, W is the channel width, H_{fin} is the fin height, and $T_{fin} \equiv t_{fin}$ is the fin thickness. (Data from N. Paydavosi et al., *IEEE Access*, 1, pp. 201–215, 2013.)

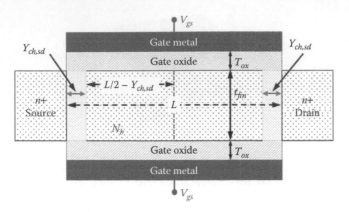

FIGURE 9.3
A typical symmetric DG-nMOSFET device structure: t_{fin}, T_{ox}, and N_b are the fin thickness, gate oxide thickness, and body doping concentration, respectively; $Y_{ch,sd}$ is the depletion width in the y direction along the channel due to the applied drain bias V_{ds}.

thickness t_{fin} as shown in Figure 9.3. In order to ensure a complete gate control of the channel, it is required that t_{fin} is completely depleted by gate bias V_{gs} so that the fin depletion width ($X_{ch,g}$) satisfies the relation

$$X_{ch,g} \geq \frac{t_{fin}}{2} \tag{9.1}$$

and, in order to suppress source-drain punchthrough, the lateral channel depletion ($Y_{ch,d}$) due to V_{ds} at each end of the channel must be such that the neutral channel length ($L/2 - Y_{ch,sd}$) in the y direction along the channel must satisfy

$$X_{ch,g} \ll \frac{L}{2} - Y_{ch,sd} \tag{9.2}$$

From the above inequalities, we can show that in order to suppress SCE, the device structure must satisfy the conditions given in Equations 9.1 and 9.2. Combining Equations 9.1 and 9.2 and expressing $Y_{ch,d}$ in terms of equivalent gate oxide thickness (Equation 3.82), we get the condition for *scaling* FinFET device structure

$$\frac{t_{fin}}{2} + \frac{K_{si}}{K_{ox}}T_{ox} \ll \frac{L}{2} \tag{9.3}$$

Typically, $Y_{ch,d}$ is very small. Therefore, the scaling rule for FinFETs can safely be defined by

$$\frac{t_{fin}}{2} \ll \frac{L}{2} \tag{9.4}$$

Thus, if the fin is sufficiently thin with a thickness, t_{fin}, smaller than L, then SCEs are suppressed and subthreshold slope (S) is expected to be near its ideal value of about 60 mV per decade (at room temperature) [29]. Thus, the new device architecture results in a new *scaling rule* given in Equation 9.3; that is, L can be scaled by maintaining the condition $t_{fin} < L$, relaxing the scaling of gate dielectric and body doping.

In 1988 and 1999, 45 and 18 nm working DG-FinFETs, respectively, were reported [30,31]. Subsequently, 10 nm double-gate [32], 10 nm triple-gate (Q gate) [33], 5 nm nanowire [34], and 3 nm all-around gate [35] FinFETs were reported.

9.2.2 UTB-SOI Device Structure

Figure 9.4 shows 3D cross section of an ideal UTB-SOI transistor structure. If t_{fin} in an SOI-MOSFET is only several nanometers (e.g., thinner than about one-half of L), the leakage paths far from the gate will be eliminated and SCEs can be significantly suppressed. It is found that the transistor leakage current is reduced by about ten times for every nanometer drop in t_{fin} [37]. The UTB-SOI MOSFETs require SOI substrates with extremely uniform silicon films (sub-nanometer uniformity). In 2009, SOI wafer supplier, Soitec, developed SOI wafers with a desired tolerance of ±0.5 nm using a process called *smart cut* [51]. It is reported that UTB-SOI MOSFETs with $t_{fin} \approx 3$ nm have been experimentally realized [52]. The most attractive channel materials for UTB-SOI MOSFETs are the *monolayer* semiconductors such as graphene [22], MoS$_2$ [23], and WSe$_2$ monolayer [53].

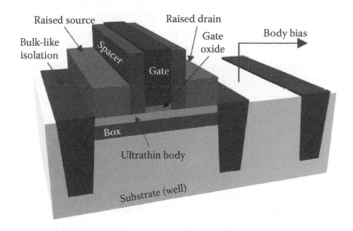

FIGURE 9.4

3D cross section of an ultrathin body SOI MOSFET device structure: the body can be a thin film of silicon, or any monolayer semiconductors; appropriate thickness of the buried oxide, BOx can be used as the back gate oxide to bias the body for the target dynamic V_{th} shift. (Data from N. Paydavosi et al., *IEEE Access*, 1, pp. 201–215, 2013.)

In a UTB transistor, the thickness of the buried oxide (BOx) layer is reduced to use the substrate immediately below the BOx as a back gate to bias the body of the device and to enable a multi-V_{th} technology, especially for system-on-chip design [54–56].

The multiple-gate FET structures can be classified as (1) common multigate (CMG) structure where a common gate terminal is used to bias the device and the gate dielectric thicknesses is the same and (2) independent multigate (IMG) structure where gates are independently biased and the gate dielectric thickness is different for each gate.

9.3 Common Multiple-Gate FinFET Model

The term common gate defines all gates in the multigate (double-gate or triple-gate or quadruple-gate) FinFET, which are electrically interconnected and are biased at the same electrical terminal voltage. It is also assumed that the gate work functions and the dielectric thicknesses on all sides to the silicon fin are the same. However, the carrier mobilities in the inversion are dependent on crystal orientations and/or strain.

9.3.1 Core Model: Poisson-Carrier Transport

The core CMG model is formulated using gradual channel approximation (GCA) [57], described in Chapter 4, and assuming physical effects such as mobility degradation can safely be neglected. Several basic models have been proposed for the FinFET, where charge [58] and surface potential [59,60] modeling approaches have been mainly used for model formulations. The core model described in the following section is based on the solution of Poisson's drift/diffusion equations for a long channel DG-FinFET assuming a finite doping in the channel [29]. The reported simulation data obtained by the core model agree very well with the numerical device simulation data [60,61].

9.3.1.1 Electrostatics

For the simplicity of model formulation, let us consider 2D (two-dimensional) cross section of an ideal n-type FinFET device structure as a common double-gate transistor as shown in Figure 9.5. First of all, we obtain surface potential ϕ_s within the device by solving 1D Poisson's equation given by (Equation 3.30)

$$\frac{d^2\phi(x,y)}{dx^2} = -\frac{q}{K_{si}\varepsilon_0}\left[p(x,y)-n(x,y)+N_d^+(x,y)-N_a^-(x,y)\right] \qquad (9.5)$$

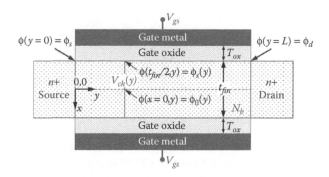

FIGURE 9.5
Schematic of an idealized symmetric common DG-nMOSFET device used to derive device equations: T_{ox}, t_{fin}, and N_b are the gate oxide thickness, fin or body thickness, and body doping concentration, respectively; the origin of the coordinate system (0,0) is at the center at $(L = 0, t_{fin}/2)$; ϕ_s and ϕ_d are the surface potentials at the source and drain ends of the device, respectively.

where:

$\phi(x, y)$ is the electrostatic potential at any point (x, y) in the channel

q is the magnitude of the electronic charge

K_{si} and ε_0 are the dielectric constant of the silicon channel (fin) and permittivity of free space, respectively

$p(x, y)$, $n(x, y)$, $N_d^+(x,y)$, and $N_a^-(x,y)$ are the hole, electron, ionized donor, and ionized acceptor concentrations at any point (x, y) of the semiconductor substrate, respectively

For a p-type substrate, the minority carrier concentration at any point (x, y) of the substrate is given by (Equation 3.40)

$$n(x,y) \cong \frac{n_i^2}{p(x)} = \frac{n_i^2}{N_a} \exp\left[\frac{\phi(x,y)}{v_{kT}} \right] \tag{9.6}$$

where:

N_a is the acceptor doping concentration in a p-type substrate (assuming complete ionization)

n_i is the intrinsic carrier concentration

v_{kT} is the thermal voltage given by kT/q

k and T are the Boltzmann constant and ambient temperature, respectively

Again, from Equation 3.35, we can show that for a p-type substrate

$$\phi_B = v_{kT} \ln\left(\frac{N_a}{n_i} \right) \tag{9.7}$$

and

$$\frac{n_i^2}{N_a} = n_i \exp\left(-\frac{\phi_B}{v_{kT}}\right) \tag{9.8}$$

where:
ϕ_B is the bulk potential

Typically, FinFETs are undoped or lightly doped channel devices; therefore, we consider only the inversion carrier electron concentration $n(x, y)$ at any point (x, y) given by Equation 9.6 and uniformly doped p-type body doping concentration, $N_a(x, y) \equiv N_b$. Let us assume that $V_{ch}(y)$ is the channel potential at any point y and GCA as described in Chapter 4 is valid [57]. Then for a double-gate FET (DG-FET) shown in Figure 9.5, we can express Poisson's Equation 9.5 as

$$\frac{d^2\phi(x,y)}{dx^2} = \frac{q}{K_{si}\varepsilon_0}\left\{n_i \exp\left[\frac{\phi(x,y)-\phi_B-V_{ch}(y)}{v_{kT}}\right]+N_b\right\} \tag{9.9}$$

where:
$V_{ch}(y)$ is given by $V_{ch}(0) = V_s$ at the source and $V_{ch}(L) = V_d$ at the drain

From Equation 9.9, the electrostatic potential $\phi(x, y)$ at any point (x, y) in the channel can be written as

$$\phi(x,y) \cong \phi_1(x,y) + \phi_2(x,y) \tag{9.10}$$

In Equation 9.10, $\phi_1(x, y)$ is the contribution to $\phi(x, y)$ due to the inversion carriers without the effect of the ionized body dopants, and $\phi_2(x, y)$ is the contribution to $\phi(x, y)$ due to body dopants, N_b. Therefore, we have

$$\frac{d^2\phi_1(x,y)}{\partial x^2} = \frac{qn_i}{K_{si}\varepsilon_0}\exp\left[\frac{\phi(x,y)-\phi_B-V_{ch}(y)}{v_{kT}}\right] \tag{9.11}$$

$$\frac{d^2\phi_2(x,y)}{\partial x^2} = \frac{qN_b}{K_{si}\varepsilon_0} \tag{9.12}$$

If t_{fin} is less than the width of the depletion region, then for a certain gate bias V_g, the silicon fin is fully depleted and consequently the inversion carriers are spread throughout the entire body. Thus, $Q_i \gg Q_b$, and therefore, we can safely neglect the term containing N_b in Equation 9.9 and the channel potential is obtained by solving Equation 9.11.

We know that for a symmetric double-gate structure, the vertical component of the electric field E_x is zero at the center, that is, at $x = 0$, $d\phi_1/dx = 0$

and $\phi_1(x = 0, y) = \phi_0(y)$; then using Equation 3.47 and following the procedure described in Section 3.4.2, we get $\phi_1(x, y)$ by integrating Equation 9.11 twice as

$$\phi_1(x,y) = \phi_0(y) - 2v_{kT} \ln\left(\cos\left\{\sqrt{\frac{q}{2K_{si}\varepsilon_0 v_{kT}} \frac{n_i^2}{N_b} \exp\left[\frac{\phi_0(y) - V_{ch}(y)}{v_{kT}}\right]} \cdot \frac{x}{2}\right\}\right) \quad (9.13)$$

where:

$\phi_0(y)$ is the potential at the center of the body as shown in Figure 9.5 and we have used Equation 9.8 to express ϕ_B in terms of n_i and N_b

Similarly, in order to solve for $\phi_2(x, y)$, we apply the boundary conditions: $E_x = 0$ at the center of the channel ($x = 0$) and $\phi_2(x = 0, y) = 0$, and integrate Equation 9.12 twice. Again, using Equation 3.47, we can express Equation 9.12 as

$$\frac{d}{dx}\left(\frac{\partial\phi_2(x,y)}{\partial x}\right)^2 = 2\frac{qN_b}{K_{si}\varepsilon_0}\frac{d\phi_2(x,y)}{dx} \quad (9.14)$$

Now, integrating Equation 9.14 from $d\phi_2(x = 0, y)/dx = 0$, $\phi_2(x = 0, y) = 0$ to any point $d\phi_2(x, y)/dx$, $\phi_2(x, y)$ we get

$$\int_0^{d\phi_2/dx} d\left(\frac{\partial\phi_2(x,y)}{\partial x}\right)^2 = 2\frac{qN_b}{K_{si}\varepsilon_0}\int_0^{\phi_2(x,y)} d\phi_2(x,y) \quad (9.15)$$

After integration and simplification, we get from Equation 9.15

$$\frac{d\phi_2(x,y)}{\sqrt{\phi_2(x,y)}} = \sqrt{\frac{2qN_b}{K_{si}\varepsilon_0}} \cdot dx \quad (9.16)$$

Integrating Equation 9.16, from $x = 0$, $\phi_2(x = 0, y) = 0$ to any point x, $\phi_2(x, y)$, we can show

$$\phi_2(x,y) = \frac{qN_bx^2}{2K_{si}\varepsilon_0} \quad (9.17)$$

The surface potential $\phi_s(y)$ at any point y along the surface is obtained by evaluating the sum of $\phi_1(x,y)$ and $\phi_2(x,y)$ at the surface ($x = -t_{fin}/2$) such that

$$\phi_s(y) \cong \phi_1\left(-\frac{t_{fin}}{2}, y\right) + \phi_2\left(-\frac{t_{fin}}{2}, y\right) \quad (9.18)$$

In Equation 3.23 we have shown that

$$V_{gs} = V_{fb} + \phi_s - \frac{Q_s}{C_{ox}} \quad (9.19)$$

where:
  V_{gs} is the gate voltage
  V_{fb} is the flat-band voltage
  Q_s is the total charge in the body
  C_{ox} is the gate oxide capacitance per unit area, given by $K_{ox}\varepsilon_0/T_{ox}$, with K_{ox}
    and T_{ox} are the permittivity of oxide and oxide thickness, respectively

Then from Gauss's law at the channel/oxide interface, we get

$$Q_s = -K_{si}\varepsilon_0 E_{xs} \tag{9.20}$$

where:
  E_{xs} is the vertical component of the electric field at the surface

Substituting Equation 9.20 in Equation 9.19, we get

$$V_{gs} = V_{fb} + \phi_s(y) + \frac{K_{si}\varepsilon_0 E_{xs}}{C_{ox}} \tag{9.21}$$

Now, following the procedure to obtain E_{xs} for bulk MOS (metal-oxide-semiconductor) capacitor system in Equation 3.51, we can show for a DG-FET device

$$\frac{d}{dx}\left(\frac{d\phi}{dx}\right)^2 = \frac{2q}{K_{si}\varepsilon_0}\left\{n_i \exp\left[\frac{\phi(x,y)-\phi_B-V_{ch}(y)}{v_{kT}}\right]+N_b\right\}\frac{d\phi}{dx} \tag{9.22}$$

We integrate Equation 9.22 from center potential $\phi(x=0,y)\equiv\phi_0(y), d\phi(x=0,y)/dx=0$ to any point $\phi(x,y)$ and $d\phi(x,y)/dx$ to get

$$\int_0^{d\phi/dx} d\left(\frac{d\phi}{dx}\right)^2 = \frac{2qn_i}{K_{si}\varepsilon_0}\int_{\phi_0(y)}^{\phi(x,y)}\left\{\exp\left[\frac{\phi(x,y)-\phi_B-V_{ch}(y)}{v_{kT}}\right]+\frac{N_b}{n_i}\right\}d\phi \tag{9.23}$$

After integration and simplification, we can express Equation 9.23 as

$$\left[\frac{d\phi(x,y)}{dx}\right]^2 = \frac{2qn_i}{K_{si}\varepsilon_0}\left(\begin{array}{l} v_{kT}\left\{\exp\left[\frac{\phi_s(y)}{v_{kT}}\right]-\exp\left[\frac{\phi_0(y)}{v_{kT}}\right]\right\}\cdot\exp\left[\frac{-\phi_B-V_{ch}(y)}{v_{kT}}\right] \\[3mm] +\exp\left(\frac{\phi_B}{v_{kT}}\right)\left[\phi(y)-\phi_0(y)\right] \end{array}\right) \tag{9.24}$$

$$= E_s^2(x) \equiv E_{xs}^2$$

where in Equation 9.24, we have used Equation 9.7 for N_b/n_i. Thus, the vertical electric field at any point y along the surface of the channel is given by

$$E_{xs} = \sqrt{\frac{2qn_i}{K_{si}\varepsilon_0}\left[\begin{array}{c} v_{kT}\left\{\exp\left[\dfrac{\phi_s(y)}{v_{kT}}\right]-\exp\left[\dfrac{\phi_0(y)}{v_{kT}}\right]\right\}\cdot\exp\left[\dfrac{-\phi_B-V_{ch}(y)}{v_{kT}}\right] \\ +\exp\left(\dfrac{\phi_B}{v_{kT}}\right)\cdot\left[\phi_s(y)-\phi_0(y)\right] \end{array}\right]} \tag{9.25}$$

Combining Equations 9.21 and 9.25, we get

$$V_{gs} = V_{fb} + \phi_s(y)$$

$$+\frac{K_{si}\varepsilon_0}{C_{ox}}\sqrt{\frac{2qn_i}{K_{si}\varepsilon_0}\left(\begin{array}{c} v_{kT}\left\{\exp\left[\dfrac{\phi_s(y)}{v_{kT}}\right]-\exp\left[\dfrac{\phi_0(y)}{v_{kT}}\right]\right\}\cdot\exp\left[\dfrac{-\phi_B-V_{ch}(y)}{v_{kT}}\right] \\ +\exp\left(\dfrac{\phi_B}{v_{kT}}\right)\cdot\left[\phi_s(y)-\phi_0(y)\right] \end{array}\right)} \tag{9.26}$$

Equations 9.18 and 9.26 represent a self-consistent system of equations that can be solved to obtain $\phi_0(y)$ and $\phi_s(y)$ for a fully depleted DG-FET structure under a set of external biases.

In the partially depleted DG-FETs, the depletion width X_d is bias dependent. At the edge of depletion region, $\phi_1(x = X_d, y) = 0$. With these changes, the surface potential can be derived for the partially depleted devices similar to the fully depleted devices. It can be shown that for the partially depleted body

$$\phi_1\left(x=\frac{t_{si}}{2},y\right) = -2v_{kT}\cdot\ln\left(\cos\left\{\sqrt{\frac{q}{2K_{si}\varepsilon_0 v_{kT}}\frac{n_i^2}{N_b}\exp\left[\frac{-V_{ch}(y)}{v_{kT}}\right]}\cdot\frac{X_d}{2}\right\}\right) \tag{9.27}$$

$$V_{gs} = V_{fb} + \phi_s(y) + \frac{K_{si}\varepsilon_0}{C_{ox}}\sqrt{\frac{2qn_i}{K_{si}\varepsilon_0}\left(\begin{array}{c} v_{kT}\left\{\exp\left[\dfrac{\phi_s(y)}{v_{kT}}\right]-1\right\}\cdot\exp\left[\dfrac{-\phi_B-V_{ch}(y)}{v_{kT}}\right] \\ +\exp\left(\dfrac{\phi_B}{v_{kT}}\right)\cdot\phi_s(y) \end{array}\right)} \tag{9.28}$$

In order to obtain continuous expressions for terminal currents and charges, it is necessary to capture the transition between the fully depleted and partially depleted regimes in a smooth manner. Also, the solution of Equations 9.27 and 9.28 is computationally intensive due to the complex $\phi_2(x, y)$ term. To overcome these issues, a simplified expression is used for $\phi_2(x,y) = \phi_{pert}$ which is continuous between the partially depleted and fully depleted regimes. Here, ϕ_{pert} is used as a small perturbation term. Thus, using ϕ_{pert}, a surface potential in both the regimes is calculated through a single continuous equation. The transformation variable β is the argument of the cosine function in $\phi_1(t_{fin}/2, y)$ in Equation 9.13

$$\beta \equiv \frac{t_{fin}}{2} \sqrt{\frac{q}{2K_{si}\varepsilon_0 v_{kT}} \frac{n_i^2}{N_b}} \exp\left[\frac{\phi_0(y) - V_{ch}(y)}{v_{kT}}\right] \qquad (9.29)$$

and, from Equation 9.17, $\phi_{pert} \equiv \phi_2(t_{fin}/2, y)$ is given by

$$\phi_{pert} \equiv \phi_2\left(\frac{t_{fin}}{2}, y\right) = \frac{qN_b}{2K_{si}\varepsilon_0} \frac{t_{fin}^2}{4} \qquad (9.30)$$

Thus, through a change of variable, the unified surface potential ϕ_s equation can be written as

$$f(\beta) \equiv \ln(\beta) - \ln(\cos\beta) - \frac{V_{gs} - V_{fb} - V_{ch}}{2v_{kT}} + \ln\left(\frac{2}{t_{fin}}\sqrt{\frac{2K_{si}\varepsilon_0 v_{kT}N_b}{qn_i^2}}\right)$$

$$(9.31)$$

$$+ \frac{2K_{si}\varepsilon_0}{t_{fin}C_{ox}} \cdot \sqrt{\beta^2\left[\frac{\exp(\phi_{pert}/v_{kT})}{\cos^2\beta} - 1\right] + \frac{\phi_{pert}}{v_{kT}^2}\left[\phi_{pert} - 2v_{kT}\ln(\cos\beta)\right]} = 0$$

Equation 9.31 (implicit in β) is the basic surface potential equation (SPE) in Berkeley Short Channel IGFET Model (BSIM) CMG [50]. It is solved by first using an analytical approximation for the initial guess [61], followed by two Householder's cubic iterations (third-order Newton-Raphson iterations); together these make the model numerically robust and accurate. The surface potentials at the source end ϕ_{s0} and drain end ϕ_{sL} are calculated by setting $V_{ch}(y = 0) = V_s$ and $V_{ch}(y = L) = V_d$, respectively. For a lightly doped body, Equation 9.31 can be further simplified [62] to speed up the simulation.

From Equation 9.30: if $\phi_{pert} \approx 0$, then in Equation 9.31 we have $\exp(\phi_{pert}/v_{kT}) = 1$ and $(\phi_{pert}/v_{kT}^2)[\phi_{pert} - 2v_{kT}\ln(\cos\beta)] \approx 0$. Then

$$\left[\frac{\exp(\phi_{pert}/v_{kT})}{\cos^2\beta} - 1\right] = \frac{1}{\cos^2\beta} - 1 = \tan^2\beta$$

Therefore, we can simplify Equation 9.31 as

$$\ln(\beta) - \ln(\cos\beta) - \frac{V_{gs} - V_{fb} - V_{ch}}{2v_{kT}} + \ln\left(\frac{2}{t_{fin}}\sqrt{\frac{2K_{si}\varepsilon_0 v_{kT}N_b}{qn_i^2}}\right)$$

$$+ \frac{2K_{si}\varepsilon_0}{t_{fin}C_{ox}}\beta\tan\beta = 0 \qquad (9.32)$$

A separate surface potential expression is used for the cylindrical gate geometry [63].

9.3.1.2 Drain Current Model

The drain-to-source current I_{ds} for the long channel DG-FinFETs is obtained from the solution of drift-diffusion equation (Equation 4.63)

$$I_{ds}(y) = \mu(T)WQ_i(y)\frac{dV_{ch}}{dy} \tag{9.33}$$

where:
  $\mu(T)$ is the low-field and temperature-dependent mobility
  W is the total effective width
  Q_i is the inversion charge per unit area in the upper half part of the body

Equation 9.33 includes drift and diffusion transport mechanisms through the use of the quasi-Fermi potential. Integrating both sides of Equation 9.33, and considering the fact that under quasistatic operation I_{ds} is constant along the channel, we can express Equation 9.33 in its integral form:

$$I_{ds} = \left(\frac{W}{L}\right)\mu(T)\int_{Q_{is}}^{Q_{id}} Q_i \frac{dV_{ch}}{dQ_i}dQ_i \tag{9.34}$$

where:
  L is the effective channel length
  Q_{is} and Q_{id} are the inversion charge densities at the source and drain ends, respectively

From the relation $Q_s = (Q_i + Q_b)$, we get

$$\begin{aligned} Q_{is} &= C_{ox}\left(V_{gs} - V_{th} - \phi_{s0}\right) - Q_b \\ Q_{id} &= C_{ox}\left(V_{gs} - V_{th} - \phi_{sL}\right) - Q_b \end{aligned} \tag{9.35}$$

From Gauss's Law, we get the total charge in the fin, $Q_s = -K_{si}\varepsilon_0 E_{xs}$; then we can show from Equation 9.25

$$Q_s(y) = \sqrt{2qn_iK_{si}\varepsilon_0 \left[\begin{array}{l} v_{kT}\left(e^{\left[\phi_s(y)/v_{kT}\right]} - e^{\left[\phi_0(y)/v_{kT}\right]}\right).e^{\left\{\left[-\phi_B - V_{ch}(y)\right]/v_{kT}\right\}} \\ + e^{\left[\phi_B/v_{kT}\right]}.\left[\phi_s(y) - \phi_0(y)\right] \end{array} \right]} \tag{9.36}$$

Note that the second term in the square bracket is due to bulk charge. For lightly doped body, $Q_b \ll Q_i$; therefore, neglecting the bulk charge term in Equation 9.36, we can express inversion charge as

$$Q_i(y) \approx \sqrt{2qn_iK_{si}\varepsilon_0 \left[v_{kT}\left(e^{\left[\phi_s(y)/v_{kT}\right]} - e^{\left[\phi_0(y)/v_{kT}\right]}\right).e^{\left\{\left[-\phi_B - V_{ch}(y)\right]/v_{kT}\right\}} \right]} \tag{9.37}$$

Equation 9.37 can be further simplified as

$$Q_i(y) = \sqrt{2qn_iK_{si}\varepsilon_0v_{kT}}\, e^{\left[\phi_s(y)-\phi_B-V_{ch}(y)/2v_{kT}\right]}\sqrt{1-e^{\left[\phi_0(y)-\phi_s(y)/v_{kT}\right]}} \qquad (9.38)$$

In strong inversion $\phi_s(y) \gg \phi_0(y)$; therefore, $\sqrt{1-e^{\left[\phi_0(y)-\phi_s(y)/v_{kT}\right]}}$ approaches 1. In weak inversion, we can simplify this term assuming liner profile from $x = 0$ to $x = -t_{fin}/2$. If E_{avg} is the average electric field in the region between $x = -t_{fin}/2$ to the mid-potential at $x = 0$, then using Gauss's law, we can write

$$E_{avg} = -\frac{d\phi(y)}{dx} = \frac{Q_i}{K_{si}\varepsilon_0} \qquad (9.39)$$

If we assume that surface potential varies linearly from center potential $\phi_0(y)$ to the surface potential $\phi_s(y)$, then Equation 9.39 can be expressed as

$$-\frac{d\phi(y)}{dx} = \frac{\phi_s(y)-\phi_0(y)}{t_{fin}/2} = \frac{Q_i}{K_{si}\varepsilon_0} \qquad (9.40)$$

Thus, the inversion charge is given by

$$\phi_s(y)-\phi_0(y) = \frac{Q_i}{(2K_{si}\varepsilon_0)/t_{fin}} = \frac{Q_i}{2C_{si}} \qquad (9.41)$$

where $C_{si} = K_{si}\varepsilon_0/t_{fin}$; substituting Equation 9.41 in Equation 9.38 and performing Taylor's series expansion, the inversion charge for lightly doped DG-FETs is given by

$$Q_{i,LD}(y) \approx \sqrt{2qn_iK_{si}\varepsilon_0v_{kT}} \cdot \exp\left[\frac{\phi_s(y)-\phi_B-V_{ch}(y)}{2v_{kT}}\right]\cdot\sqrt{\frac{Q_i(y)}{Q_i(y)+2C_{si}v_{kT}}} \qquad (9.42)$$

Equation 9.42 is an implicit equation in Q_i and is solved iteratively to obtain drain current from Equation 9.33. Using $Q_s \approx Q_{i,LD}$ in Equation 9.19, we can compute V_{gs} versus inversion charge $Q_{i,LD} = -C_{ox}\left(V_{gs}-V_{fb}-\phi_s\right)$.

Similarly, the inversion charge density for heavily doped DG-FETs can be shown as

$$Q_{i,HD}(y) \approx \sqrt{2qn_iK_{si}\varepsilon_0v_{kT}} \cdot \exp\left[\frac{\phi_s(y)-\phi_B-V_{ch}(y)}{2v_{kT}}\right]\cdot\sqrt{\frac{Q_i(y)}{Q_i(y)+2Q_b}} \qquad (9.43)$$

From the similarities of charge expressions in Equations 9.42 and 9.43, a unified expression is used to calculate the inversion charge density for a wide range of devices as a function of Q_b and is given by

$$Q_i(y) = \sqrt{2qn_iK_{si}\varepsilon_0v_{kT}} \cdot \exp\left[\frac{\phi_s(y)-\phi_B-V_{ch}(y)}{2v_{kT}}\right]\sqrt{\frac{Q_i(y)}{Q_i(y)+Q_0}} \qquad (9.44)$$

where:

$Q_0 = 2Q_b + 5C_{si}v_{kT}$, with $C_{si} = K_{si}\varepsilon_0/t_{fin}$

Q_b is the fixed depletion charge and is given by $qN_b t_{fin}$

It is reported that the unified charge density model agrees very well with the inversion charge density calculated using an exact equation for a wide range of body doping concentration [60]. Then from Equation 9.44, the gradient in $V_{ch}(y)$, term dV_{ch}/dQ_i can be calculated as a function of Q_i using a simple but accurate implicit equation for Q_i [60]

$$\frac{dV_{ch}}{dy} = \frac{d\phi_s}{dy} + v_{kT}\frac{dQ_i}{dy}\left(\frac{2Q_b + 5C_{si}v_{kT}}{Q_i + 2Q_b + 5C_{si}v_{kT}} - \frac{2}{Q_i}\right) \quad (9.45)$$

Equation 9.34 can be integrated analytically using Equation 9.44 to calculate dV_{ch}/dQ_i to obtain the following basic equation for I_{ds}

$$I_{ds} = \left(\frac{W}{L}\right)\mu(T)\cdot\left[\frac{Q_{is}^2 - Q_{id}^2}{2C_{ox}} + 2v_{kT}\left(Q_{is} - Q_{id}\right) - v_{kT}Q_0\ln\left(\frac{Q_0 + Q_{is}}{Q_0 + Q_{id}}\right)\right] \quad (9.46)$$

Equation 9.46 describes the drain current model for symmetric DG-FETs. The model equation predicts the drain current in all operation regions: sub-threshold, linear, and saturation of both fully depleted and lightly depleted channel symmetric DG-FETs. Figure 9.6 shows the simulated *I–V* characteristics of a bulk FinFET device obtained by multigate drain current model with the measured data.

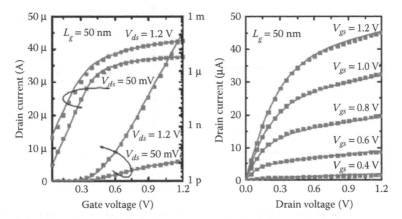

FIGURE 9.6
Drain current model used to compare the measured and simulated *I–V* characteristics of moderately doped symmetric bulk *n*-channel FinFET devices: (a) $I_{ds} - V_{gs}$ characteristics for different V_{ds}; (b) $I_{ds} - V_{ds}$ characteristics for different V_{gs}. Device data are $L = 50$ nm, $t_{fin} = 25$ nm, and TiN gate with equivalent $T_{ox} = 1.95$ nm; symbols are measured data and lines represent compact drain current model. (Data from M.V. Dunga et al., *IEEE Symposium on VLSI Technology*, pp. 60–61, 2007.)

A unique behavior of lightly doped DG-FETs with thin body is that the inversion charge is no longer confined to interface and the entire film is inverted. For any gate voltage, the electrostatic potential increases at the interfaces as well as in the volume of the film in all mode of device operation: the depletion, weak inversion, and the strong inversion. As a result, the potential shift or total band bending exceeds $2\phi_B$ in every region and in the entire film. This is referred to as the *volume inversion* [61, 63–65]. Due to volume inversion (1) the potential as well as the inversion carrier density is nearly independent of the position inside the body because of the negligible potential drop between the surface and the center of the body as shown in Figure 9.7a; (2) the potential as well as the inversion charge density is weakly dependent on the body thickness; any small increase in the gate voltage in the subthreshold region increases the potential throughout the entire body, causing inversion in the entire body; and (3) since the electronic potential is virtually independent of the body thickness, the total integrated charge inside the body is proportional to the body thickness. Thus, as a result of volume inversion, the subthreshold region drain current is also proportional to t_{fin} as shown in Figure 9.7.

9.3.2 Modeling Physical Effects of Real Device

This subsection briefly reviews some of the real-device effects for the modern multigate transistors, highlighting the key physical effects and implementations, and outlining the proper references for further details.

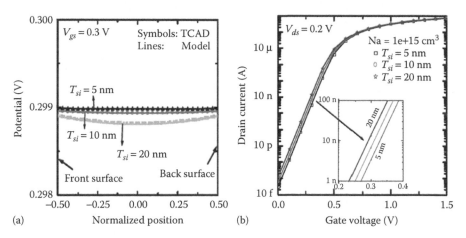

FIGURE 9.7
Drain current model showing volume inversion in lightly doped DG-nMOSFETs: (a) simulated potential profile in the body between the front and back surfaces in volume inversion; (b) subthreshold $I_{ds} - V_{gs}$ plots for different body thicknesses showing volume inversion (flat potential profile) simulated by the drain current model and numerical device simulation (TCAD); symbols represent TCAD and lines represent compact model; device data are $N_a = 1 \times 10^{15}$ cm^{-3} and $T_{ox} = 2$ nm; ($T_{si} \equiv t_{fin}$). (Data from M.V. Dunga et al., *IEEE Symposium on VLSI Technology*, pp. 60–61, 2007.)

9.3.2.1 Short Channel Effects

SCEs originate from 2D electrostatics where the drain significantly affects the potential barrier at the source due to its close proximity to source region. SCEs degrade the device performance through V_{th} roll-off and S degradation.

There are several approaches to model SCEs [66–70]. However, the approach assuming a parabolic potential function perpendicular to the silicon-insulator interface to solve the 2D Poisson's equation is shown to maintain a balance between the model accuracy and model computation time [68,69].

V_{th} *roll-off*: In order to model V_{th} roll-off in DG-FETs, 2D Poisson's equation is solved in the x direction into the body and in the y direction along the length of the channel, assuming that the inversion charge is negligible and the electric field E_x is independent of y whereas the electric field E_y is independent of x. Then assuming a parabolic potential distribution along the x direction, the minimum potential at the center of the channel $\phi_0(y)$ is determined [70]. Then the minimum potential $\phi_{c,min}$ [61] is expressed in terms of the terminal voltages V_{gs} and V_{ds}, L, and the characteristic field-penetration length λ, and is defined as

$$\lambda \equiv \sqrt{\frac{K_{si}\varepsilon_0}{2K_{ox}\varepsilon_0}\left(1 + \frac{K_{ox}\varepsilon_0 t_{si}}{4K_{si}\varepsilon_0 t_{ox}}\right)t_{fin}t_{ox}} \tag{9.47}$$

λ is known as the scale length that defines the extent of penetration of the electric field from the drain into the body as function of physical parameters T_{ox} and t_{fin} and, therefore, the amount of SCE in a transistor. The change in V_{th} is then defined as

$$\Delta V_{th}(L, \lambda, V_{ds}) \equiv \lim_{L \to \infty} \phi_{c,min}(L, \lambda, V_{gs}, V_{ds}) \tag{9.48}$$

The term $\Delta V_{th}(L, \lambda, V_{ds})$ is further enhanced with more parameters for simplicity of the parameter extraction procedure and to improve modeling accuracy [71]. In BSIM-CMG model, ΔV_{th} is subtracted from V_{fb} [72–74].

Figure 9.8 shows the dependence of ΔV_{th} on the gate oxide thickness and silicon body thickness. As the oxide thickness and body thickness decrease, the gate control on the body increases, thus suppressing SCE as expected [64].

Subthreshold slope degradation: The subthreshold swing, S, in a planar MOSFET is defined as (Equation 4.124)

$$S \equiv \left(\frac{d[\log(I_{ds})]}{dV_{gs}}\right)^{-1} \cong \ln(10)v_{kT}\left(1 + \frac{C_d}{C_{ox}} + \frac{C_{IT}}{C_{ox}} + \frac{C_{DSC}}{C_{ox}}\right) \tag{9.49}$$

where:
C_d is the depletion capacitance associated with the depletion region
C_{IT} is the capacitance due to interface states

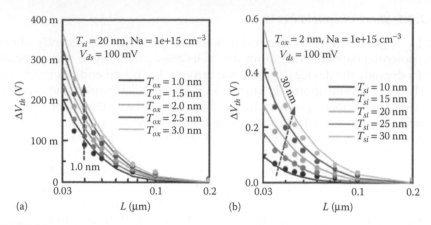

FIGURE 9.8
Drain current model used to simulate SCE in lightly doped DG-nMOSFETs through threshold voltage roll-off for different: (a) oxide thickness and (b) body thickness; symbols represent TCAD and lines represent compact model ($T_{si} \equiv t_{fin}$). (Data from M.V. Dunga et al., *IEEE Symposium on VLSI Technology*, pp. 60–61, 2007.)

C_{DSC} is the coupling capacitance between source/drain to channel, which has similar L, λ, and V_{ds} dependencies as ΔV_{th} discussed earlier

The degradation in subthreshold swing is then modeled through a modification in v_{kT} as

$$nv_{kT} \equiv \left(1 + \frac{C_d + C_{IT} + C_{DSC}}{C_{ox}} \right) v_{kT} \tag{9.50}$$

where nv_{kT} is substituted for v_{kT} in all bias-dependent calculations.

9.3.2.2 Quantum Mechanical Effects

Quantum mechanical confinement of inversion carriers is well known in bulk MOSFETs for a long time [13,75,76]. The large vertical electric field leads to strong band bending at the surface and the inversion carriers are confined to dimensions along the length and width of the transistor as shown in Figure 9.9a. This carrier confinement, also known as electrical confinement (EC), leads to splitting of energy bands into discrete sub-bands, which reflects as an increase in the threshold voltage of the transistor and a decrease in the gate capacitance, both of which act to reduce the current drive of the transistor [13,61].

In the case of DG-FETs, unlike bulk FETs, there is strong carrier confinement even at low electric fields, making the QME (quantum mechanical effect) even more complex [77]. The carriers are bounded by gate insulator on two sides, which is similar to carriers confined in a rectangular well [61,78–80].

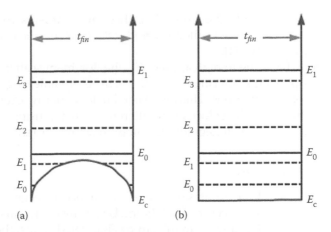

FIGURE 9.9
Energy-band diagrams showing the carrier confinement and associated quantization of electronic energy levels in DG-MOSFETs: (a) electrical confinement due to band bending at the top and bottom gate silicon/SiO$_2$ interface and (b) structural confinement due to ultrathin body.

This is referred to as structural confinement (SC) since it arises from the very physical structure of DG-FET as shown in Figure 9.9b. In order to capture the QME in its entirety it is necessary to model the effect of both EC and SC (Figure 9.9) on the performance of DG-FETs. Several groups have reported different analytical and numerical approaches to capture the QME in DG FETs [78–80].

The quantum mechanical confinement of the inversion carriers increases the device V_{th}, degrades the gate capacitance, and reduces the effective width of the device (see Figure 9.7a) due to a shift in the inversion charge centroid as discussed in Section 3.4.2.2 (Figure 3.19) away from the Si/SiO2 interface [13,61]. A shift in the bottom of the conduction/valence band due to the SC [61] is used to modify V_{ch} at the source and drain SPEs. In order to model EC, the bias-dependent charge centroid thickness Δz is used to modify T_{ox} (Equation 3.82) and calculate the reduction in the width of the device [79]. The simulation results are in an excellent agreement with those calculated from a self-consistent Schrödinger–Poisson approach [61].

9.3.2.3 Mobility Degradation

Similar to surface mobility degradation in bulk MOSFETs discussed in Section 5.3.1 (Figure 5.9b), the degradation of carrier mobility in FinFET also occurs due to four main scattering mechanisms: Coulomb scattering, acoustic phonon scattering, surface roughness scattering, and optical phonon scattering. The first three scattering mechanisms have vertical (transverse) field dependency and they are each dominant at different regions of device operation: Coulomb scattering at weak inversion, acoustic phonon scattering at mid-inversion, and surface roughness scattering at strong inversion (Figure 5.9b).

Similar to bulk MOSFETs (Section 5.3.1), these mechanisms together are modeled through a submodel called *low field mobility degradation* and used to get the *effective* mobility [71].

At high lateral field due to high applied V_{ds}, the dominant scattering mechanism is optical phonon scattering since the electrons are able to gain enough energy to emit optical phonons. This high lateral field scattering causes the carrier velocity saturation. The velocity saturation is calculated using a submodel called *current saturation* and it degrades the drain-to-source current directly [71].

9.3.2.4 Series Resistances

In thin body source-drain transistors, series resistance is large. In order to reduce the parasitic resistances in FinFETs and UTB transistors, raised source-drain regions are used in device architecture [Figure 9.4]. Thus, the parasitic source-drain resistance submodel includes a bias-dependent extension resistance R_{ext}, a spreading resistance R_{sp}, and a distributed contact resistance R_{con}.

The components of contact resistance include resistance ΔR_s of the raised source-drain bulk regions and silicon/silicide inter-face resistance ΔR_c. And, R_{con} is modeled as a lumped resistance using a distributed network.

The spreading resistance R_{sp} is due to *current crowding* as the current flows from the raised drain region into the drain extension; this results in an increase in the resistance by R_{sp}. The spreading resistance is, modeled in terms of the device and source-drain areas and a shape parameter [81].

The extension resistance R_{ext} contributes the most to the series resistance. The fringe field from the gate can cause surface accumulation at the interfaces of the extension region and the gate oxide/offset spacer; this modulates the resistivity of the region and makes R_{ext} bias-dependent. R_{ext} is modeled as a resistance network with two bias-independent resistances R_{ext1} and R_{ext2}, and a bias-dependent resistance R_{acc}. Since the exact extension doping profile is often unknown, analytical expressions with fitting parameters are used to obtain the values of these components of R_{ext} [81].

9.4 Independent Multiple-Gate FET Model

The model developed for common-gate FinFETs cannot be used for transistors with different gate dielectric thickness and independently biased gate terminals. In this section, we will derive a surface potential–based compact model targeted for UTB-SOI MOSFETs. The model could be used for computer analysis of emerging devices including graphene nanoribbon transistors [22,23,52]. Many of the real-device effects presented for a CMG model can be used with appropriate changes for independent gate operation. Thus, only a description of the core model is presented in the following section.

9.4.1 Electrostatics

In order to derive electrostatic potential of asymmetric independent DG-FETs, let us consider 2D cross-sectional view of the channel as shown in Figure 9.10. The asymmetric independent DG-FET includes different front- and back-gate dielectric thicknesses (T_{ox1} and T_{ox2}) and different gate-work functions (ϕ_{M1} and ϕ_{M2}). Since the threshold voltage of an independent DG-FET can be optimized by adjusting the back-gate bias (V_{bg}), there is no need for significant body doping, N_b. Therefore, we can develop surface potential-based model using a lightly doped body so that $Q_b \ll Q_i$.

Let us consider GCA, Boltzmann's distribution function, an undoped channel, and only the dominant mobile carriers in deriving the surface potential. Then Poisson's equation can be written as

$$\frac{d^2\phi(x,y)}{dx^2} = \frac{q}{K_{si}\varepsilon_0}\left\{ n_i \exp\left[\frac{\phi(x,y) - V_{ch}(y)}{v_{kT}} \right] \right\} \tag{9.51}$$

Again, using the identity (Equation 3.47) $(d/dx)(d\phi/dx)^2 = 2(d\phi/dx)\cdot(d^2\phi/dx^2)$ in Equation 9.51 and integrating the resultant expression along the x axis, we can show that

$$E_{s1}^2 - E_{s2}^2 = \frac{2qn_iv_{kT}}{K_{si}\varepsilon_0}\left\{ \exp\left[\frac{\phi_{s1} - V_{ch}(y)}{v_{kT}} \right] - \exp\left[\frac{\phi_{s2} - V_{ch}(y)}{v_{kT}} \right] \right\} \tag{9.52}$$

where:

E_{s1} and E_{s2} are the surface electric fields at the front and back gates, respectively

ϕ_{s1} and ϕ_{s2} are the front and back surface potentials, respectively, as shown in Figure 9.10

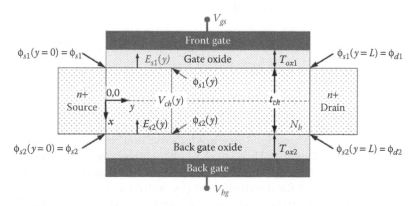

FIGURE 9.10
2D cross-sectional view of the channel region of a planar independent DG-FET; T_{ox1} and T_{ox2} are the front and back gate oxide thickness, respectively; t_{ch} and N_b are the substrate thickness and doping concentration, respectively.

Using Gauss's law at the front- and back-gate silicon surfaces, we can write

$$E_{s1} = \frac{1}{K_{si}\varepsilon_0} C_{ox1} \left(V_{fg} - V_{fb1} - \phi_{s1} \right)$$

$$E_{s2} = \frac{1}{K_{si}\varepsilon_0} C_{ox2} \left(V_{bg} - V_{fb2} - \phi_{s2} \right) \tag{9.53}$$

where:

V_{fg} and V_{bg} are the front- and back-gate voltages, respectively

V_{fb1} and V_{fb2} are the flat band voltages for the front and back gates, respectively

C_{ox1} and C_{ox2} are the front- and back-gate oxide capacitances given by $K_{ox}\varepsilon_0/T_{ox1}$ and $K_{ox}\varepsilon_0/T_{ox2}$, respectively, where T_{ox1} and T_{ox2} are the front and back oxide thicknesses, respectively, and K_{ox} is the dielectric constant of oxide

Substituting Equation 9.53 in Equation 9.52, we get an implicit equation in ϕ_{s1} and ϕ_{s2}.

Now, in order to solve the implicit Equation 9.52 with two interdependent unknowns, ϕ_{s1} and ϕ_{s2}, the back surface is approximated to be always in weak inversion. Using the equation for the potential of a capacitive divider node held between the two potentials ϕ_{s1} and V_{bg}, we can write

$$\phi_{s2} = \alpha_{si}\phi_{s1} + \alpha_{ox} \left(V_{bg} - V_{fb2} \right) \tag{9.54}$$

where

$$\alpha_{si} = \frac{C_{si}}{C_{si} + C_{ox2}} ; \alpha_{ox} = \frac{C_{ox2}}{C_{si} + C_{ox2}}$$

and, $C_{si} = K_{si}\varepsilon_0/t_{ch}$, with t_{ch} being the channel thickness.

Substituting Equation 9.54 in Equation 9.52, the implicit SPE for the IMG transistor basic model is obtained

$$f \equiv \left[\frac{C_{ox1} \left(V_{fg} - V_{fb1} - \phi_{s1} \right)}{K_{si}\varepsilon_0} \right]^2 - \left[\frac{V_{bg} - V_{fb2} - \phi_{s2}}{t_{ch} + \left(K_{si}/K_{ox} \right) T_{ox2}} \right]^2 - \frac{2qn_i v_{kT}}{K_{si}\varepsilon_0} \exp\left(\frac{\phi_{s1} - V_{ch}}{v_{kT}} \right)$$

$$+ \frac{2qn_i v_{kT}}{K_{si}\varepsilon_0} \exp\left[\frac{\alpha_{ch}\phi_{s1} + \alpha_{ox} \left(V_{bg} - V_{fb2} \right) - V_{ch}}{v_{kT}} \right] = 0 \tag{9.55}$$

Equation 9.55 is solved using Householder's method to obtain the front surface potential and electric field, ϕ_{s1} and E_{s1}, respectively, at the source end (by setting $V_{ch}(y = 0) = V_s$) [82]. The front surface potential and electric field, ϕ_{d1} and E_{d1}, are also found for the drain end (by setting $V_{ch}(y = L) = V_d$). The corresponding back-gate surface potentials ϕ_{s2} and ϕ_{d2} and electric fields E_{s2} and E_{d2} are then computed from Equations 9.54 and 9.53, respectively.

Finally, assuming lightly doped body, that is, $Q_b \ll Q_i$, so that $Q_s \cong Q_i$, we get the expression for the inversion charge density as

$$Q_i = K_{si}\varepsilon_0\left(E_{s1} - E_{s2}\right) \tag{9.56}$$

9.4.2 Drain Current Model

For long channel UTB-FET devices, the drain current is derived by solving drift-diffusion transport expression given by Equation 9.33. Integrating both sides of Equation 9.33 and considering the fact that under quasistatic operation I_{ds} is constant along the channel, it is possible to express Equation 9.33 in its integral form as

$$I_{ds} = \left(\frac{W}{L}\right)\mu(T)\int_0^L Q_i(y)\left[\frac{dV_{ch}(y)}{dy}\right]dy \tag{9.57}$$

Again, assuming that the back surface is weak, a simplified form of surface potential expression

$$E_{s1}^2 - E_{s2}^2 = \frac{2qn_i v_{kT}}{K_{si}\varepsilon_0}\left[\exp\left(\frac{\phi_{s1} - V_{ch}}{v_{kT}}\right)\right] \tag{9.58}$$

is used to compute the drain current by following the procedure described next:

1. Solving for F_{s1} in Equation 9.58 and using it in Equation 9.56, we can write

$$Q_i(y) = \sqrt{2qn_iK_{si}\varepsilon_0 v_t\left\{\exp\left[\frac{\phi_{s1} - V_{ch}(y)}{v_t}\right]\right\} + \left(K_{si}\varepsilon_0 E_{s2}\right)^2} - K_{si}\varepsilon_0 E_{s2} \tag{9.59}$$

2. Taking the derivatives of both sides of Equation 9.59 with respect to y, it is possible to write

$$Q_i(y)\frac{dV_{ch}(y)}{dy} = Q_i(y)\frac{d\phi_{s1}(y)}{dy} - \eta v_{kT}\frac{dQ_i(y)}{dy} \tag{9.60}$$

where

$$\eta = 2 - \frac{2\varepsilon_{si}E_{s2}(y)}{Q_i(y) + 2\varepsilon_{si}E_{s2}(y)} \tag{9.61}$$

Here, η varies from 1 to 2 going from subthreshold to strong inversion and is a function of y. To simplify the integral in Equation 9.57, η can be approximated to be independent of position, thus replacing $Q_i(y)$ and $E_{s2}(y)$ by their average values at the source and drain ends.

3. Evaluating the integral in Equation 9.57 using Equation 9.60 leads to the following basic equation for I_{ds} [49]

$$I_{ds} = \left(\frac{W}{L}\right)\mu(T)\cdot\left[\frac{Q_{is}+Q_{id}}{2} + v_{kT}\left(\phi_{s1,d}-\phi_{s1,s}\right) + \eta v_{kT}\left(Q_{is}-Q_{id}\right)\right] \qquad (9.62)$$

The model has been extensively verified for a wide range of reliability and scalability [83]

9.5 Dynamic Model

9.5.1 Common Multigate C–V Model

This section presents the dynamic model of the CMG DG-FETs for transient analysis of the devices in circuit CAD. The intrinsic capacitance model that describes the transient behavior of the transistors are derived from the terminal charges as described in Chapter 6.

For DG-FETs, the total charge in the body is given by the charges on the top- and bottom-gate electrodes. The total charge is computed by integrating the charge along the channel. Since the two gates are electrically interconnected, we have

$$Q_G = 2WC_{ox}\int_0^L\left[V_{gs}-V_{fb}-\phi(y)\right]\cdot dy \qquad (9.63)$$

where:
Q_G denotes the charge on the electrically interconnected gate

The inversion charge in the body is divided between the source and the drain terminals using Ward–Dutton charge partition approach discussed in Chapter 6 [84,85]. The charge on source terminal (Q_S) is given by

$$Q_S = -2WC_{ox}\int_0^L\left(1-\frac{y}{L}\right)\cdot\left[V_{gs}-V_{fb}-\phi(y)-\frac{Q_b}{C_{ox}}\right]\cdot dy \qquad (9.64)$$

Using charge conservation principle, the charge on the drain terminal (Q_d) can be expressed as

$$Q_D = -2WC_{ox}\int_0^L\frac{y}{L}\left[V_{gs}-V_{fb}-\phi(y)-\frac{Q_b}{C_{ox}}\right]\cdot dy \qquad (9.65)$$

The surface potential as a function of the position y along the length of the transistor, $\phi_s(y)$ is obtained using current continuity. Current continuity states that the current is conserved along the length of the transistor.

$$I_d(L) = I_d(y), \quad \text{where } 0 \le y \le L \tag{9.66}$$

The expression for the drain current in Equation 9.46 is very complex and is not practical for applying current continuity. For the purpose of determining $\phi_s(y)$, a simplified version of *I–V* model as shown below is used [61,72]

$$I_{ds}(y) = 2\mu \cdot \left(\frac{W}{y}\right) \left[g(Q_{is}) - g(Q_{id}) \right] \tag{9.67}$$

where the function $g(Q_i(y))$ follows from Equation 9.46 after neglecting the third term in the square bracket is defined as

$$g(Q_i) = \frac{Q_i^2}{2C_{ox}} + 2v_{kT}Q_i \tag{9.68}$$

The approximate Equations 9.67 and 9.68 retain good accuracy in the strong inversion regime but overestimate the drain current in the subthreshold regime. The advantage of using a mathematically simple analytical expression for terminal charges outweighs the resulting error in the accuracy of *C–V* model in the subthreshold regime. Using Equations 9.67 and 9.68, $\phi_s(y)$ can be expressed as

$$\frac{y}{l} \cdot (B - \phi_{s0} - \phi_{sL}) \cdot (\phi_{sL} - \phi_{s0}) = \left[B - \phi_{s0} - \phi_s(y) \right] \cdot (\phi_s(y) - \phi_{s0}) \tag{9.69}$$

where ϕ_{s0} and ϕ_{sL} represent the surface potential at the source and drain ends, respectively, and the parameter *B* is defined as

$$B = 2\left(V_{gs} - V_{fb} - \frac{Q_b}{C_{ox}} + 2v_{kT} \right) \tag{9.70}$$

The terminal charges are obtained by substituting $\phi_s(y)$ in Equations 9.63 through 9.65 and evaluating of the integrals [73] so that

$$Q_G = 2WLC_{ox} \left[V_g - V_{fb} - \frac{\phi_{s0} + \phi_{sL}}{2} + \frac{(\phi_{sL} - \phi_{s0})^2}{6(B - \phi_{sL} - \phi_{s0})} \right]$$

$$Q_D = -2WLC_{ox} \left[\begin{array}{l} \dfrac{V_g - V_{fb} - (Q_b/C_{ox})}{2} - \dfrac{\phi_{s0} + \phi_{sL}}{4} + \dfrac{(\phi_{sL} - \phi_{s0})^2}{60(B - \phi_{sL} - \phi_{s0})} \\[2ex] + \dfrac{(5B - 4\phi_{sL} - 6\phi_{s0}) \cdot (B - 2\phi_{sL}) \cdot (\phi_{s0} - \phi_{sL})}{60(B - \phi_{sL} - \phi_{s0})^2} \end{array} \right] \tag{9.71}$$

$$Q_S = -(Q_{fg} + Q_{bg} + Q_B + Q_D)$$

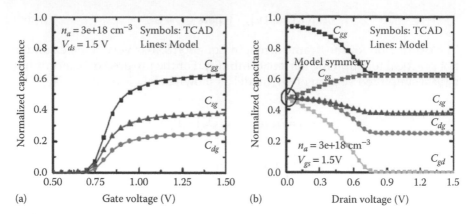

FIGURE 9.11
Dynamic model of symmetric DG-MOSFETs: modeling transcapacitances as a function of (a) gate voltage and (b) drain voltage; Model symmetry is seen at $V_{ds} = 0$ where $C_{dg(gd)} = C_{sg(gs)}$; n_a = body doping concentration; symbols represent TCAD and lines represent compact model. (Data from F. M.V. Dunga et al., *IEEE Symposium on VLSI Technology*, pp. 60–61, 2007.)

The expressions for terminal charges are continuous and are valid over subthreshold, linear, and saturation regimes of operation.

Equation 9.71 forms the *C–V* model for BSIM-CMG. The terminal charges are used as state variables in the circuit simulation. All the capacitances are derived from the terminal charges to ensure charge conservation. The capacitances are defined as

$$C_{ij} = \frac{\partial Q_i}{\partial V_j} \tag{9.72}$$

where:
  i and j denote the multigate FET terminals

Note that C_{ij} satisfies

$$\sum_i C_{ij} = \sum_j C_{ij} = 0 \tag{9.73}$$

due to charge conservation.

The capacitances from *C–V* model are plotted as a function of gate voltage and drain voltage in Figure 9.11a and b, respectively.

9.5.2 Independent Multigate C–V Model

We model the *C–V* using a charge-based approach [84,85] to ensure charge conservation. The charge associated with each terminal is modeled. The capacitive current flowing into each terminal is expressed as the time derivative of charge.

$$I_x = \frac{dQ_x}{dt} = \sum_y C_{xy} \frac{dV_y}{dt} \tag{9.74}$$

where $x, y = d, fg, bg, s$; each transcapacitance is defined as

$$C_{xy} = \frac{\partial Q_x}{\partial V_y} \tag{9.75}$$

The charge associated with the front gate fg can be calculated as

$$Q_{fg} = W \int_0^L C_{ox1} \left[V_{fg} - \Delta\Phi_1 - \phi_{s1}(y) \right] dy \tag{9.76}$$

where:

$\Delta\Phi_1$ is the work function of the front gate with reference to that of $n+$
source

In order to integrate Equation 9.76, the relation between front surface potential $\phi_{s1}(y)$ and position y is needed. This can be obtained by applying current continuity to Equation 9.62.

$$I_{ds}.y = \mu(T).W \left\{ \frac{Q_{is} + Q_i(y)}{2} \left[\phi_{s1}(y) - \phi_{s1,s} \right] + v_{kT} \left[Q_{is} - Q_i(y) \right] \right\} \tag{9.77}$$

Since $Q_i(y)$ is unknown, the capacitor divider approximation is used to relate the front surface potential ϕ_{s1} and charge Q_i:

$$Q_i(y) = C_{ox1} \left[V_{fg} - \Delta\Phi_1 - \phi_{s1}(y) \right] + \frac{C_{ox2} C_{si}}{C_{ox2} + C_{si}} \left[V_{bg} - \Delta\Phi_2 - \phi_{s2}(y) \right] \tag{9.78}$$

Combining Equations 9.77 and 9.78 and noting that $Q_{is} = Q_{is} \left[\phi_{s1}(y) = \phi_{s1,s} \right]$, we obtain the position dependence of surface potential as

$$y = \mu C_{ox1} \frac{W}{I_{ds}} \left[\phi_{s1}(y) - \phi_{s1,s} \right] \left\{ \begin{array}{l} V_{fg} - \Delta\Phi_1 - \dfrac{\phi_{s1,s} - \phi_{s1}(y)}{2} \\[2ex] + \gamma_c. \left[V_{bg} - \Delta\Phi_2 - \dfrac{\phi_{s1,s} - \phi_{s1}(y)}{2} \right] + v_{kT} \left(1 + \gamma_c \right) \end{array} \right\} \tag{9.79}$$

where:

$$\gamma_c = \frac{C_{ox2} \| C_{si}}{C_{ox1}}$$

Substituting Equation 9.79 in Equation 9.76 and performing integration, we get

$$Q_{fg} = C_{ox1}WL\left\{V_{fg} - \Delta\Phi_1 - \frac{\phi_{s1,s} + \phi_{s1,d}}{2} + \frac{B'\left(\phi_{s1,d} - \phi_{s1,s}\right)^2}{6\left[A' - B'\left(\phi_{s1,d} + \phi_{s1,s}\right)\right]}\right\} \quad (9.80)$$

where

$$A' = V_{fg} - \Delta\Phi_1 + \gamma_c \cdot \left(V_{bg} - \Delta\Phi_2\right) + v_{kT}\left(1 + \gamma_c\right) \quad (9.81)$$

and,

$$B' = \frac{1}{2}\left(1 + \gamma_c\right) \quad (9.82)$$

The charge associated with the back gate can be simply calculated by replacing $\phi_{s1,s(d)}$ with $\phi_{s2,s(d)}$, swapping $\left(V_{fg} - \Delta\Phi_1\right)$ and $\left(V_{bg} - \Delta\Phi_2\right)$, and swapping C_{ox1} and C_{ox2} in Equation 9.80, following an argument of symmetry.

The front- and back-gate charges are further partitioned into a source component and a drain component according to Ward–Dutton charge partition method [84,85]. The drain charge associated with the front gate is given by [79]

$$Q_{D1} = -W\int_0^L C_{ox1}\left[V_{fg} - \Delta\Phi_1 - \phi_{s1}(y)\right]\frac{y}{L}dy \quad (9.83)$$

After using Equation 9.79 and integrating, we obtain

$$Q_{D1} = -\frac{C_{ox1}WL}{2}\left\{\begin{array}{l} V_{fg} - \Delta\Phi_1 - \dfrac{\phi_{s1,s} + \phi_{s1,d}}{2} + \dfrac{B\left(\phi_{s1,d} - \phi_{s1,s}\right)^2}{30\left[A - B\left(\phi_{s1,s} + \phi_{s1,d}\right)\right]} \\[3mm] -\dfrac{\left(5A - 4B\phi_{s1,d} - 6B\phi_{s1,s}\right).\left(A - 2B\phi_{s1,d}\right)\left(\phi_{s1,d} - \phi_{s1,s}\right)}{30\left[A - B\left(\phi_{s1,s} + \phi_{s1,d}\right)\right]^2} \end{array}\right\} \quad (9.84)$$

Similarly, the drain charge, Q_{D2} associated with the back gate is obtained by replacing $\phi_{s1,s(d)}$ with $\phi_{s2,s(d)}$, swapping $\left(V_{fg} - \Delta\Phi_1\right)$ and $\left(V_{bg} - \Delta\Phi_2\right)$, and swapping C_{ox1} and C_{ox2} in Equation 9.84.

The total drain charge is the sum of Q_{D1} and Q_{D2}. Since Q_S, Q_D, Q_{fg}, and Q_{bg} must sum up to 0, the source charge can be calculated as

$$Q_S = -Q_{fg} - Q_{bg} - Q_D \quad (9.85)$$

Similar to Figure 9.11, the transcapacitances can be computed from the above terminal charges.

9.6 Summary

This chapter presented an overview of the present state-of-the-art surface potential–based compact models of thin-body CMG and IMG FET devices for circuit CAD. Each device model consists of a core model for large devices and real device submodels to analyze the physical and geometrical effects on these devices. The basic features of the model include capturing the important physics of thin-body multigate transistors such as the volume inversion and the dynamic V_{th} shift for body bias in ultrathin body transistors. The models are valid for digital as well as analog circuit analysis with the C–V models that simulate the transcapacitances. This chapter is intended to provide readers the present state-of-the-art modeling activities in thin-body FET devices. The detailed models and modeling methodologies including updates can be found in the literature [74].

Exercises

9.1 Complete the mathematical steps following the procedure described in Chapter 3 to derive Equation 9.13 for channel potential $\phi_1(x, y)$ at any point (x, y) in the channel of a typical symmetric DG-MOSFET device.

9.2 Complete the mathematical steps following the procedure described in Chapter 3 to derive Equation 9.17 for channel potential $\phi_2(x, y)$ at any point (x, y) in the channel of a typical symmetric DG-MOSFET device.

9.3 Use Equations from exercises 9.1 and 9.2 to derive:

    a. Vertical electrical field at any point y along the channel of the symmetric DG-MOSFET device

    b. Gate voltage for a fully depleted symmetrical DG-MOSFET structure

9.4 What is the volume inversion in DG-MOSFETs? Describe the effect of volume inversion on DG-MOSFET device performance.

9.5 Describe the difference between the electrical and structural Quantum Mechanical effects in DG-MOSFETs; qualitatively plot the centroid of inversion charge as a function of body thickness. Explain your results.

10

Beyond-CMOS Transistor Models: Tunnel FETs

10.1 Introduction

As the CMOS (complementary metal-oxide-semiconductor) technology approaches its ultimate scaling limit of MOSFET (metal-oxide-semiconductor field-effect transistors) device miniaturization, extensive global search for beyond-CMOS devices has been continued to *rebooting computing*. This new device technology must be *green* (i.e., energy efficient) and continue to increase packing density of devices as well as device functionalities in an IC (integrated circuit) chip in the same rate as the CMOS technology. A number of potential beyond-CMOS devices involving present as well as new state variables and communication frameworks have been reported [1–3]. Among the potential beyond-CMOS devices, the devices that compete directly with the MOSFETs in power, area, and speed in the commercial temperature range 0°C–75°C and can utilize the existing CMOS facility are of special interest to device technologists and IC manufacturers. These devices are aimed at supply voltages less than a 0.5 V with subthreshold swing (S) lower than that of MOSFETs.

The scaled MOSFET devices, discussed in Chapter 5, are limited by short channel effect (SCE) and S. As discussed in Chapter 9, the ultrathin-body (UTB) MOSFETs are adopted to surmount the challenges of SCEs. However, S in UTB-MOSFETs is still limited by the Boltzmann distribution of carriers to a minimum value of 60 mV per decade of channel current at room temperature. Therefore, the devices that can achieve switching mechanisms less than 60 mV per decade are highly desirable for beyond-CMOS *green IC technology*. The potential device structures with the desirable characteristics include tunneling [4–6], impact ionization [7–10], ferroelectric dielectrics [11], and mechanical gate [12–14] field-effect transistors (FETs). Among the emerging devices, the tunnel FET (TFET) is one of the potential candidates for beyond-CMOS technology that can be controlled at voltages well under a volt with steep S and does not have the delays associated with positive feedback that are intrinsic to impact ionization, ferroelectricity, and mechanical devices [15].

Therefore, in this chapter, an overview of the present state-of-the-art compact modeling activities on TFETs is presented. First of all, the basic features of TFET device structure are presented. Then the physics of TFET device operation is discussed. And, finally, the compact modeling activities on TFETs for circuit CAD is presented. TFET as a green transistor has the potential to provide an acceptable device performance as the supply voltage approaches to 0.1 V beyond-CMOS devices.

10.2 Basic Features of TFETs

The most commonly referred TFETs are gated *p-i-n* diodes or gated *p-n* diodes with an *intrinsic channel* as shown in Figure 10.1. In order to switch the device on, the *pn*-junction is reverse biased and a voltage (V_g) is applied to the gate to modulate the device characteristics. In order to be consistent with MOSFET device technology, the names of the TFET device terminals are chosen such that the biasing conditions for MOSFETs and TFETs are the same. Since a reverse bias with $V_g > 0$, $V(n+) > 0$, and $V(p+) = 0$ is needed across the *p-i-n* structure to trigger tunneling similar to the biasing condition of an nMOSFET with $V_g > 0$, $V_d > 0$, and $V_s = 0$, the *n+* region of a *p-i-n* TFET in Figure 10.1 is referred to as its drain and *p+* region as its source for an *n*-type TFET (nTFET). Similarly, for a *p*-type TFET (pTFET), *p+* region is referred to as the drain and *n+* region is the source to be consistent with biasing condition of a pMOSFET device discussed in Chapters 4 and 5.

Thus, Figure 10.1 shows an nTFET device structure, with a heavily doped *p+* source region and a heavily doped *n+* drain region. On the other hand, in a pTFET, the source is doped with *n+* and the drain is doped with *p+*. It is observed from Figure 10.1 that a *p-i-n* TFET device structure is similar to that of a conventional MOSFET except that the source is doped with the opposite dopant type with respect to the drain [4]. Thus, as shown in Figure 10.1,

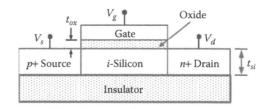

FIGURE 10.1
2D cross section of an ideal single gate *p–i–n* TFET device structure with a *p+* source, an intrinsic silicon (*i*-silicon) channel, and an *n+* drain regions on an insulating substrate; t_{ox} and t_{si} are the gate oxide thickness and body thickness, respectively; V_s, V_g, and V_d are the source, gate, and drain voltages, respectively.

a typical TFET device includes an ultrathin body on the top of a buried oxide layer, a gate electrode placed on the top of an ultrathin-gate dielectric, and a heavily doped source region with doping type opposite to a heavily doped drain region.

In principle, the same *p-i-n* TFET device structure shown in Figure 10.1 can be used for *n*-type or *p*-type operation by appropriate biasing conditions. In this respect, if a TFET is designed with symmetry between the *n*+ and *p*+ regions including similar doping levels, gate alignment, and geometries, the device shows *ambipolar behavior*, that is, the transfer characteristics resemble those of a pTFET when $V_{gs} < 0$ and $V_{ds} < 0$, and those of an nTFET when $V_{gs} > 0$ and $V_{ds} > 0$. Thus, in principle, the TFET is an ambipolar device showing *p*-type behavior with dominant hole conduction and *n*-type behavior with dominant electron conduction.

In another embodiment, a TFET can be used as a *fully depleted channel* device [16]. In the case of a fully depleted channel TFET, the metal gate work function of the gate is chosen to fully deplete the channel in the off-state. In the on-state, the Zener tunneling is enabled [17]. In order to achieve high current density, abrupt doping profiles are required with degenerately doped *n*+ and *p*+ regions [16].

One of the key challenges of TFET fabrication is that the gate must be self-aligned to the junction. If the gate is underlapped, that is, the junction is moved outside the gate edge, the field control is degraded along with the degradation of S. And, if the gate is overlapped, that is, the junction is under the gate metal, the field in the on-condition depletes carriers on the source side of the junction decreasing the tunneling injection. Thus, the gate must be placed with a high precision approaching that of the lateral potential variation length, which is typically less than 10 nm [18] in these heavily doped TFET structures.

Similar to MOSFETs, the gate control of the channel in TFETs can be improved by using double-gate (DG) structures. In order to increase the on-current (I_{on}), a degenerately doped pocket region can be used under the gate [19]. In addition to increasing I_{on}, the pocket also offers lower S by aligning the gate field with the internal tunnel junction field. TFET device structure is continuously evolving with the development of process technology to minimize access resistance, form abrupt degenerate junctions, self-align gate, and realize ultrathin channel.

A TFET-type device structure has been studied by Stuetzer [20] in 1952 predating Esaki's discovery of *pn*-interband tunneling [21]. In this study, the basic characteristics along with the ambipolar nature of the current–voltage (*I–V*) characteristics have been reported in the field gating of a lateral germanium *pn*-junction. This study also shows the dependence of the transistor characteristics on gate placement with respect to the *pn*-junction. In 1977, Quinn et al. [22] designed a surface-channel MOS tunnel junction by replacing the *n*+ source region of an nMOSFET device with a highly degenerate *p*+ source region to measure the sub-band splitting and transport properties of tunneling between a bulk source and a two-dimensional (2D) surface channel. This device structure is essentially a *lateral* TFET. The first known vertical TFET has been reported by Leburton et al. [23] in 1988 in the design of a high-speed transistor with the gate to control the *negative differential resistance* (NDR).

In 1992, Baba [24] independently proposed the lateral TFET device structure similar to that reported by Quinn to use the gate of the transistor in controlling NDR. This transistor has been referred to as the *surface tunnel transistor* (STT). In the 1990s, the STTs fabricated in different semiconductor materials such as gallium arsenide (GaAs), silicon-on-insulator (SOI), silicon (Si), and indium gallium arsenide (InGaAs) have been widely studied to show NDR at room temperature [25–33]. In this period, the focus of the STTs has been on field control of the forward-biased characteristic of the Esaki tunnel junction and in ways to utilize the NDR characteristics.

The interest on TFET as the potential device for beyond-CMOS technology has grown since the reported gating of the reverse Zener tunneling current of STTs to achieve better scaling due to the absence of punch-through by Reddick and Amartunga in 1995 [34]. Subsequently, the gating of the Zener side of the tunnel junction of a fabricated Si *vertical* TFET along with its potential for low off-current (I_{off}) relative to the MOSFET has been reported by Hansch et al. in 2000 [35]. In 2004, the device characteristics of a lateral SOI TFET have been reported by Aydin et al. [36], and low S in the TFET has been reported by Wang et al. [4], Bhuwalka et al. [5], and Appenzeller et al. [6]. Theoretically, it is shown that in a TFET, $S < 60$ mV per decade at room temperature [37,38]. However, less than 60-mV per decade S has been reported in only a few TFETs based on carbon nanotubes (CNTs) [6,39], Si [40–43], germanium (Ge) [44], and $p+Ge/n+Si$ [45] channels. TFET device structure is constantly evolving to outperform CMOS devices in comparable technology node [46].

10.3 Basic Theory of TFET Operation

10.3.1 Energy Band Diagram

The basic operating principle of a TFET can be understood from the energy band diagram shown in Figure 10.2 of the ideal device structure, shown in Figure 10.1. In Figure 10.2, the energy band diagram of a *p–i–n* TFET device under various biasing conditions is shown with reference to the structure in Figure 10.1. Figure 10.2a shows that in the off-state with zero bias ($V_{gs} = 0 = V_{ds}$), the majority carriers in the channel as well as in the drain regions see unsurmountable large potential barriers for tunneling and the only current flow through the device is due to the reverse-biased leakage current of the *p–i–n* structure. When a positive gate bias ($V_{gs} > 0$) is applied at the gate, the source channel junction is reverse biased, and therefore, the energy band of the channel region bends downward as shown in Figure 10.2b.

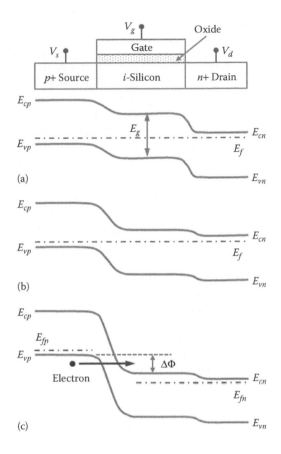

FIGURE 10.2
Energy band diagram taken laterally along the length of the *p-i-n* TFET structure: (a) off-state with $V_{gs} = V_{ds} = 0$; (b) gate modulation of the channel by $V_{gs} > 0$ and $V_{ds} = 0$; and (c) on-state with $V_{gs} > 0$ and $V_{ds} > 0$ leading to nFET-type behavior with the current flow set by the overlap of valence band electrons with the unfilled channel conduction band states. $\Delta\Phi$ is the window of tunneling; E_{cn} and E_{cp} represent the conduction band energies of the *n*-type and *p*-type semiconductors, respectively; E_{vn} and E_{vp} represent the valence band energies of the *n*-type and *p*-type semiconductors, respectively; E_f is the equilibrium Fermi level; E_{fn} and E_{fp} are the quasi-Fermi potentials of the *n*-type and *p*-type regions, respectively, under the applied bias; and E_g is the energy gap.

An additional positive drain bias ($V_{ds} > 0$) pulls down the Fermi levels in both the *n*-type drain and *i*-channel regions. If the downward shift of the bands is large enough to narrow the bandgap formed by the overlap of the conduction band and valence band at the source-channel junction, a tunneling path will be formed, allowing electrons to tunnel from the source to *i*-channel, as shown in Figure 10.2c. The tunneled electrons then move toward the *n*+ drain by drift-diffusion process, generating current flow in TFET devices. The gate modulation of the overlap region, defined as the *tunneling width* ($\Delta\Phi$), allows TFETs to achieve a lower *S* compared to the conventional MOSFETs.

10.3.2 Tunneling Mechanism

In a TFET the primary injection mechanism of charge carriers is *interband tunneling* [21] in which the charge carriers transfer from one energy band into another at a heavily doped *p+n+* junction in contrast to MOSFETs where the charge carriers are thermally injected over a barrier. Interband tunneling was first observed in 1957 by Esaki [21] while studying narrow forward-biased *p–n* junctions called tunnel diode. However, the interband tunneling concept was first used by Zener in 1934 to explain the dielectric breakdown at a high electric field [17] and is known as the *Zener tunneling*, which is also referred to as the *band-to-band tunneling* (BTBT).

The Zener or interband tunneling can be realized in a reverse-biased *p-i-n* structure as shown in Figure 10.2. In a TFET, the interband tunneling can be switched *on* and *off* abruptly by controlling the band bending in the channel region by applying V_{gs}. As shown in Figure 10.2a, for a *p-i-n* TFET structure at $V_{gs} = 0$, the tunneling barrier is large, and the device is in the off-state. A $V_{gs} > 0$ pulls the energy bands down and reduces the tunneling barrier. Due to reduced energy barrier, the carriers can tunnel from the valence band in the source to the conduction band in the channel and the tunneling current increases. For a *p+n+* tunnel junction, the tunneling current is determined by integrating the product of charge flux and the tunneling probability $T(E)$ from the energy states on the *p+* side to those on the *n+* side. And, $T(E)$ is calculated by applying Wentzel–Kramers–Brillouin (WKB) approximation of the *triangular* potential (Figure 10.3) at the tunnel junction [47–49] and is given by

$$T(E) \cong \exp\left(-\frac{4\sqrt{2m^* E_g^3}}{3q\hbar F} \right) \tag{10.1}$$

where:
   m^* is the effective mass
   E_g is the energy of the bandgap
   q is the electronic charge
   $\hbar$ is the reduced Plank's constant
   F is the maximum electric field at the tunneling junction

Equation 10.1 derived by WKB approximation works properly in direct bandgap semiconductors, such as indium arsenide (InAs), and has limited accuracy for silicon and Ge structures or when quantum effects and phonon-assisted tunneling become dominant [50]. However, it has been successfully applied for all TFET devices.

Equation 10.1 is a general expression for interband tunneling transmission and can be modified appropriately for tunneling mechanism in TFETs. In Figure 10.3b, it is shown that the height and the width of the triangular potential barrier are $\Delta\Phi + E_g$ and λ, respectively. The magnitude of F corresponds

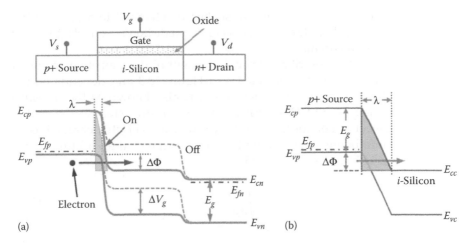

FIGURE 10.3
Interband tunneling mechanism in a TFET: (a) energy band diagram along the length of the
p-i-n TFET in the on-state (solid lines) and off-state (broken lines). In the off-state, no empty
states are available in the channel for tunneling from the source, so the off current is very
low; increasing V_g pulls the conduction band energy of the channel below the valence band
energy of the source so that interband tunneling can occur. This switches the device to the on-
state in which electrons in the energy window, $\Delta\Phi$, can tunnel from the source valence band
into the channel conduction band; (b) expanded schematic of the source-channel tunneling
region showing the WKB approximation of the triangular potential barrier; λ is the screening tunneling length; $\Delta\Phi$ is the window of tunneling; E_{cn} and E_{cp} represent the conduction
band energies of the n-type and p-type semiconductors, respectively; E_{vn} and E_{vp} represent the
valence band energies of n-type and p-type semiconductors, respectively; E_{fn} and E_{fp} are the
quasi-Fermi potentials of the n-type and p-type regions under the applied bias; and E_g is the
energy gap; and E_{cc} and E_{vc} are the conduction band and valence band energy of the channel,
respectively.

to the slope of the energy bands, so that $qF = (\Delta\Phi + E_g)/\lambda$. Therefore, the
tunneling probability for TFETs is given by [51]

$$T(E) \cong \exp\left[-\frac{4\lambda\sqrt{2m^*E_g^3}}{3\hbar\left(E_g + \Delta\Phi\right)}\right] \qquad (10.2)$$

where:
  $\Delta\Phi$ is the energy range over which the tunneling can take place
  λ is the *screening length* shown in Figure 10.3

There are four important conditions to trigger interband tunneling: available states to tunnel from, available states to tunnel to, a sufficiently narrow
energy barrier for tunneling to occur, and conservation of momentum [48].
For interband tunneling in an indirect band gap semiconductor such as silicon, crystal *phonons* are necessary to conserve momentum. Therefore, E_g in
the numerator of Equation 10.2 is replaced by $E_g - E_p$, where E_p is the phonon

energy and the effective mass m^* must then be changed to *reduce effective mass m_r^** in the tunneling direction for accurate prediction of tunneling current in the indirect semiconductors.

In Equation 10.2, λ describes the spatial extent of the transition region at the source-channel interface as shown in Figure 10.3 and depends on the biasing condition and device dimension. λ is also known as the *screening length, natural length,* and *Debye length* that physically refers to the spatial extent of the electric field or the length over which an electric charge has an influence before being screened out by the opposite charges around it [52]. For all silicon TFETs, λ is given by [51]

$$\lambda = \sqrt{\frac{\varepsilon_{si}}{a_g \varepsilon_{ox}} t_{ox} t_{si}} \tag{10.3}$$

where:

ε_{si} and t_{si} are the dielectric permittivity and thickness of silicon (or semiconductor material), respectively

ε_{ox} and t_{ox} are the dielectric permittivity and thickness of the gate dielectric

For a single-gate device, the parameter $a_g = 1$, whereas for a double gate, $a_g = 2$ [53]. Although, Equation 10.3 is derived to describe the conventional MOSFET behavior, it has been shown to be applicable for TFETs with appropriate use of the material parameters [51].

Using Equation 10.2 for $T(E)$, the drain current in a TFET device under high V_{gs} and V_{ds} is given by

$$I_{ds} = A \int_{E_c(C)}^{E_v(S)} \left[f_s(E) - f_d(E) \right].T(E) N_D N_S .dE \tag{10.4}$$

where:

$f_s(E)$ and $f_d(E)$ are the source- and drain-side Fermi-Dirac distributions (Equation 2.3)

N_S and N_D are the corresponding density of states

A is the area of the device

For a *p-i-n* TFET band structure (Figure 10.3b), the integral ranges from E_{cc} (channel conduction band) to E_{vp} (source valence band) represent the range of energies over which tunneling takes place. Note that Equation 10.4 is similar to the conventional tunnel diode equation [48]. This is justified for TFETs since the channel quasi-Fermi level is in equilibrium with the drain Fermi level at high V_{gs} and V_{ds}.

One of the challenges in TFETs is to achieve high on current I_{on} (at $V_{gs} = V_{ds}$) that depends on $T(E)$ as given in Equation 10.4. From Equation 10.1, we notice that $T(E)$ can be increased by increasing the electric field F (which is proportional to $(\Delta\Phi + E_g)/\lambda)$ along the channel. Higher F can be achieved by different ways: (1) thinner t_{ox} or higher dielectric constant (high-k) gate oxide; (2) low E_g channel materials such as silicon germanium (SiGe) or Ge [54,55] or a thin layer of SiGe material between the source and the channel [56]; and (3) light-m_r^* materials for source-channel junction. Low E_g channel materials increase I_{off} of TFETs and, therefore, require device optimization.

10.3.3 Device Characteristics

Let us consider a *p-i-n* structure for an nTFET device operation as shown in Figure 10.3. Figure 10.3a shows the off-state of the device with zero bias and on-state with $V_{gs} > 0$ and $V_{ds} > 0$. In the TFET off-state (broken line curve in Figure 10.3a), the conduction band edge of the channel is located above the valence band edge of the source, so interband tunneling is suppressed, leading to a very small off-state current (I_{off}) that is dictated by the reverse-biased *p-i-n* diode. The application of a $V_{gs} > 0$ pulls the energy bands down (solid line curve in Figure 10.3a). As soon as the channel conduction band is pulled below the source valence band, electrons from the source valence band can tunnel into the empty states of the channel conduction band. However, only the electrons within the *energy window* $\Delta\Phi$ can tunnel into the channel as shown in Figure 10.3a since the electrons from the high-energy ($E > kT$) tail of the Fermi distribution $f_s(E)$ are effectively cut off by the bandgap in the source as shown in Figure 10.4 and do not participate in the transport process [51].

To illustrate the interband tunneling from the degenerately doped p+ source of the *p-i-n* structure, only source-channel junction along with the Fermi

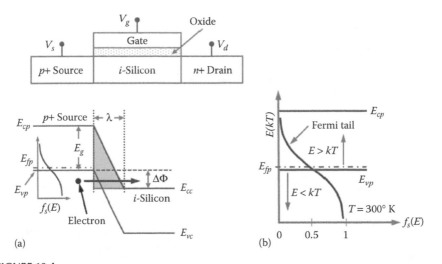

(a)

(b)

FIGURE 10.4
Energy band diagram of the source-channel *p-i-n* TFET device along the length of the device: (a) interband tunneling of carriers from the low energy ($E < kT$) regimes of the source Fermi distribution $f_s(E)$ and (b) expanded Fermi distribution of the source region at room temperature.

distribution function $f_s(E)$ of the source electrons is shown in Figure 10.4a, whereas Figure 10.4b shows the expanded $f_s(E)$ versus E distribution. Since the electrons with $E > kT$ of $f_s(E)$ are effectively filtered out from tunneling as shown in Figure 10.4, the current transport in TFETs is a *sub-kT* process. Thus, it can be considered that the electronic system is effectively *cooled down* acting as a conventional MOSFET at a lower temperature. Thus, in a TFET, primarily the cold carriers participate in the transport process, resulting in a subthreshold slope of less than 60 mV per decade. Note that in MOSFETs the subthreshold conduction is limited by Boltzmann distribution with higher-kT process (Chapters 4 and 5), thus limiting S to 60 mV per decade of current at room temperature.

The interband tunneling process in a TFET shown in Figure 10.4 is similar to a band-pass filter action wherein the high energy carriers are filtered out. This filtering function enables to achieve an S of below 60 mV per decade of current for TFETs. However, the channel conduction band E_{cc} can be pulled up or down by a small change in V_{gs}, that is, the tunneling width can be effectively changed by V_{gs} [38,49]. As a result, the value of S in a TFET is not a constant and depends on V_{gs} increasing with the increasing V_{gs}.

The above described physical mechanism of electron transport can be used to plot the transfer characteristics ($I_{ds} - V_{gs}$) of TFET devices. Again, let us consider a *p-i-n* TFET structure shown in Figure 10.5. As V_{gs} increases from $V_{gs} = 0$ to a certain trigger point, $V_{gs} = V_{off}$ at which the channel conduction band edge E_{cc} is pulled down to align with the source valence band edge, $E_{vp} \approx E_{fp}$; only the leakage current I_{off} of the *p-i-n* junction flows through the device as shown in Figure 10.5b. As V_{gs} increases above V_{off}, the overlap between E_{cc} and E_{vp} gradually increases triggering interband tunneling

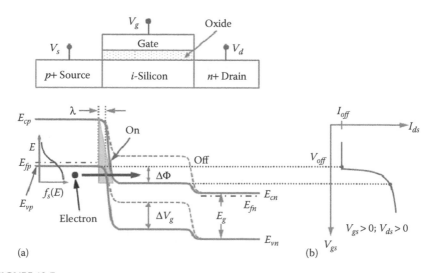

(a)   (b)

FIGURE 10.5
Current transport in a *p-i-n* TFET operation: (a) energy band diagram along the length of the *n*TFET in the on and off-states and (b) I_{ds} versus V_{gs} characteristics.

and the drain current I_{ds} increases. Since the *Fermi tail* of $f_s(E)$ is cut off from the tunneling window, only the electrons with $E < kT$ contribute to current transport, resulting in a sharp increase in I_{ds} with *steep slope* as shown in Figure 10.5b. On further increase of V_{gs} to the target supply voltage, the tunneling window reaches the corresponding final width $\Delta\Phi$ and the sub-kT electrons tunnel from the source to the channel. The tunneled electrons are then transported to the drain by supply voltage, V_{ds}, generating a steady flow of I_{ds} in the device as shown in Figure 10.5b.

In TFETs, I_{off} is low due to the filtering of electrons from the high energy Fermi tail. The value can be further reduced by widening the tunneling barrier at drain junction [4,16,57].

10.3.4 Subthreshold Swing

In Chapter 4, we have shown that S for a MOSFET device is defined by $S = \left(d \log I_{ds} / dV_{gs} \right)^{-1}$ in units of mV per (decade I_{ds}). For the TFETs, the tunnel current can be described approximately by [48]

$$I_{ds} \cong aV_{eff}F\exp\left(-\frac{b}{F}\right) \tag{10.5}$$

where:

a and b are the coefficients that depend on the material properties of the tunnel junction and the cross-sectional area of the device

V_{eff} and F are the effective reverse bias and electric field at the tunnel junction, respectively

The coefficients a and b are given by [16,38]

$$a = \frac{Aq^3\sqrt{2m_r^*}}{8\pi^2\hbar^2\sqrt{E_g}}$$

$$b = \left(-\frac{4\sqrt{2m_r^*E_g^3}}{3qh}\right) \tag{10.6}$$

The derivative of I_{ds} expression given in Equation 10.5 can be used to obtain a general expression for S of TFETs [38] as

$$S = \frac{dV_{gs}}{d\left(\log I_{ds}\right)} = \ln(10)\left[\frac{1}{V_{eff}}\frac{dV_{eff}}{dV_{gs}} + \frac{F+b}{F^2}\frac{dF}{dV_{gs}}\right]^{-1} \tag{10.7}$$

Equation 10.7 shows that S for TFETs depends on two terms that are not explicitly limited by kT/q unlike in MOSFETs. It is observed from Equation 10.7 that low S can be achieved by maximizing the two terms in the denominator. First of all, if we optimize the device in such a way so that $dV_{eff}/dV_{gs} \approx 1$, then the first term in the denominator of Equation 10.7 is inversely related to V_{gs}. Under this condition, S will decrease with decreasing V_{gs}. This can be achieved by TFET gate engineering so that V_{gs} directly controls the tunnel junction bias or band overlap $\Delta\Phi$, that is, gate has strong electrostatic control. The efficient gate electrostatics can be realized using a thin high-k gate dielectric and an ultrathin body channel region. Secondly, S can also be minimized by maximizing the second term in the denominator of Equation 10.7. This occurs when the gate is placed to align the applied field with the internal field of the tunnel junction so that the gate field adds to the internal field to increase the tunneling probability.

From the discussion of Equation 10.7, it is clear that S is a function of V_{gs} in sharp contrast to the conventional MOSFETs. This means that $\log(I_{ds}) - V_{gs}$ plot in the subthreshold region is not a straight line and S does not have a unique value as shown in Figure 10.5b. The value of S is lowest at the lowest V_{gs}, and increases as V_{gs} increases. Due to the changing values of S along $I_{ds} - V_{gs}$ curve it is useful to clearly define it for device characterization.

Several definitions have been used for TFET S [16,57,58]. The most commonly used method is to take the tangential inverse slope of $I_{ds} - V_{gs}$ curve at the steepest part of the characteristic called the *point swing* as shown in Figure 10.6. Bhuwalka et al. [58] and Boucart and Ionescu [57] have defined subthreshold region by an average swing as shown in Figure 10.6 and is given by

$$S_{avg} = \frac{V_{th} - V_{off}}{\log\left(I_{th} / I_{off}\right)} \tag{10.8}$$

where:
V_{th} is the threshold voltage
V_{off} is the voltage below V_{th} at which the drain current I_{off} is minimum

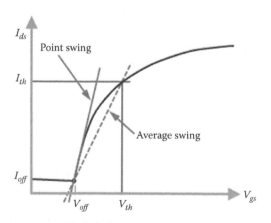

FIGURE 10.6
Subthreshold swing defined as a point swing is obtained at the steepest point of the $I_{ds} - V_{gs}$ curve and an average swing is defined as the average from turn-on to threshold voltage.

To use S_{avg}, it is important to use the appropriate value of V_{th} for accurate modeling of TFET device and circuit performance [59,60].

Another definition for S has been used to account for the voltage scaling attribute of low-S devices [16]. In this method, V_{th} is defined at $V_{th} = V_{dd}/2$; then the corresponding current $I_{th} = I_{ds}(V_{gs} = V_{dd}/2)$, where V_{dd} is the target supply voltage. In this definition, it is assumed that $V_{off} = 0$ so that $I_{off} = I_{ds}$ at $V_{gs} = 0$. Then the effective S is given by

$$S = \frac{V_{dd}}{2 \log \left(I_{th} - I_{off} \right)} \tag{10.9}$$

The basic DC performance of TFETs is characterized by specifying I_{on}, V_{dd}, and S. The TFET devices offer current gain, voltage gain, and input–output isolation, and have the basic attributes required for a complementary logic technology in a Boolean logic architecture. The current saturates with the saturation set by the source injection. Due to the ambipolarity, *the nTFET and pTFET devices can be designed to produce equal currents by using the same tunnel junction, that is, equal gate widths offer equal* I_{on} *and symmetric layouts are possible.* Since the Fermi tail is cut off by the bandgap, the S is not limited to 60 mV per decade and I_{off} can be significantly lower than that of MOSFETs.

The *scaling rule* for TFETs is different from that of the MOSFETs in which many parameters must be scaled simultaneously to keep the same electric field throughout the device [61]. In a TFET, the high electric fields exist only at the junctions. The current is determined by λ so that the device characteristics is independent of the length L of the intrinsic channel region for $L > L_{crit}$ (~20 nm for silicon TFETs) [61,62]. For $L > L_{crit}$ the *p-i-n* leakage becomes predominant. Thus, TFETs have a great potential to be devices for beyond-CMOS technology.

10.4 TFET Design Considerations

Typically, all-silicon TFET devices offer the lowest I_{off} and S, however, very low I_{on}; for example, for an nTFET $I_{off} < 100$ fA μm^{-1}, $S < 44$ mV per decade of I_{ds}, and $I_{on} < 0.1$ μA μm^{-1} [63]. Thus, the primary objective of TFET optimization is to achieve the highest possible I_{on} along with the lowest S over many orders of magnitude of I_{ds} and lowest possible I_{off}. To outperform CMOS transistors, the target parameters for TFETs are: I_{on} in the range of hundreds of mA; $S_{avg} \ll 60$ mV per decade for five decades of current at $T = 300$ K; $I_{on}/I_{off} > 1 \times 10^5$; and $V_{dd} < 0.5$ V. Since S decreases with V_{gs} [63], TFETs are targeted and optimized for low-voltage operation.

Equations 10.4 and 10.7 show that the tunneling current and S depend on the tunneling probability $T(E)$ of the source-channel junction. Therefore, in order to realize a high I_{on} and a steep slope, $T(E)$ of the source tunneling barrier should be close to unity for a small change in V_g. From Equation 10.2, it is found that $T(E)$ depends on λ, E_g, and m^*. Thus, high barrier transparency (i.e., $T(E) \sim 1$) can be achieved by minimizing E_g, m^*, and λ. Now, E_g and m^* depend on the materials of the TFET device structure, whereas λ depends on the device architecture including geometry, doping profiles, and gate capacitance [16,64,65]. Therefore, the performance of TFETs can be improved by *device architecture* and *energy band engineering*, that is, using low-E_g and low-m_r^* materials for forming the tunnel junction.

One way to improve TFET device performance by device architecture is to minimize λ. A small λ offers a strong modulation of the channel energy bands by the gate. This small value of λ, can be achieved by using a high-κ gate dielectric [57] with manufacturable ultrathin equivalent oxide thickness, ultrathin body [49,64], and degenerately doped abrupt doping profile of the tunnel junction [65,66]. Another technique to improve I_{on} by device architecture is to maximize the gate modulation of the tunneling barrier width by an appropriate alignment of the tunneling path with the direction of the electric field modulated by the gate. By overlapping the gate with the tunneling region, or designing a source region covered with an epitaxial *i*-channel layer under the top gate, I_{on} can be improved by a factor of more than 10 along with a low S_{avg} [19,67–69].

In addition to optimizing TFET device structure to improve device performance, the improvement in I_{on} and S can be achieved by band engineering, that is, use low-E_g and low-effective mass m_r^* materials to increase interband tunneling. This is achieved by heterostructure TFETs with different source materials with respect to the channel and drain, creating *staggered bandgap* structures. Device optimization should apply to both *n*- and *p*-type TFETs simultaneously, to offer a complementary TFET (CTFET) technology for logic circuits.

In heterostructure TFETs, a low E_g source material is used to reduce the width of the energy barrier at the source junction in the on-state, whereas a large E_g drain material is used to create the largest possible width of the energy barrier at the drain junction in the off-state to keep a low I_{off}. The device performance depends on the band's lineup with each other at the heterojunction [5,70,71]. The reported data show that a combination of steep S and high I_{on} can be achieved with moderate doping and a staggered band lineup [72,73].

In a reported theoretical study on staggered bandgap structures, the CTFET devices have been optimized by changing the source material from silicon to low E_g materials Ge and InAs for nTFETs and pTFETs, respectively [63]. The numerical simulation data on 50 nm silicon channel length of Ge-source nTFETs and InAs-source pTFETs show an improvement of I_{on} by a factor of 480 and 162, respectively, at $V_{ds} = V_{gs} = 1$ V over the identically designed all-silicon TFETs. For example, the simulated Ge-source nTFETs and InAs-source pTFETs show I_{on} of 244 µA µm^{-1} and 83 mA µm^{-1}, respectively, with much lower I_{on}/I_{off} than the comparable CMOS devices and $S \sim 60$ mV per decade of I_{ds} [63]. Thus, the heterostructure TFETs have a great potential to meet the target performance objectives of CTFET technology operating at $V_{dd} \ll 0.5$ V [70,71,74,75].

As discussed earlier, though all-silicon TFETs offer very low I_{on} than the conventional MOSFETs, they have shown lowest I_{off} with a small S. The current drivability in all-silicon TFETs can be improved by using high-k gate dielectric, abrupt doping profile at the tunnel junction, a thinner body, higher source doping, a double gate, a gate oxide aligned with the intrinsic region, and a shorter i-region (and gate length) [41,76–78]. A recent study on sub-60 nm all-silicon TFET devices shows $I_{on} \sim 100\ \mu A\ \mu m^{-1}$ [79].

In order to improve I_{on} by low m_r^* materials tunneling junction, III–V semiconductor-based TFETs are used in energy band engineering. The group III–V materials provide small tunneling mass as well as different band-edge alignments. Early experimental data on homojunction InGaAs p-i-n TFETs show higher I_{on} at a lower V_{gs} than all-silicon TFETs. [80,81]. Though the reported S is high, it is still above the thermal limit of MOSFETs [82]. The effective E_g can be further reduced by using III–V material-based hetrostructures with enhanced device performance compared to the homojunctions [70–72,83–85]. In this context, the tunneling barrier can be reduced by using InAs and GaAsSb for the source with AlGaSb and InGaAs for the channel.

The nanowire TFETs show a great potential for CTFET technology to mitigate the risk of lattice mismatch and defective material growth in InAs source and silicon channel pTFETs [70,86,87]. Experimental data on InAs-silicon Esaki tunnel diodes show high tunneling current and well-defined abrupt silicon-InAs heterojunction [87,88]. The vertical nanowire TFETs with *gate-all-around* device architecture offers an optimal geometry for minimizing λ and best electrostatic control [89]. The fabricated vertical n-i-p InAs–Si–Si nanowire heterojunction TFETs with InAs as a low E_g source [90] show a great promise for nanowire CTFETs.

CNT and graphene nanoribbons (GNR) are excellent choices for TFETs in terms of device architecture and energy band engineering due to the light m_r^* of their charge carriers, low and direct bandgap, and excellent electrostatic control of the gate over the ultrathin body channel. The ongoing theoretical and limited experimental studies show a great potential for carbon-based TFETs [6,51,91–93]. However, for the practical implementation of GNR-TFETs, a number of issues must be addressed including the influence of line edge roughness on the bandgap and transport properties and their effects on TFET device performance [94,95].

10.5 Compact TFET Models

From the discussions in Sections 10.3 and 10.4, it is found that CTFET technology is a viable candidate for beyond-CMOS technology due to its steep-slope complementary devices with $S < 60$ mV per decade at low V_{gs}, enabling supply voltage scaling nearing 0.1 V [67]. For concurrent development of

CTFET technology and TFET-based IC chips, there has been a tremendous interest in TFET-based exploratory novel circuit design [96,97]. However, due to the lack of available compact TFET models for circuit CAD, lookup tables or behavioral models are used for circuit analysis [98,99]. Recently, a number of current and capacitance models for TFET devices have been reported [100–107]. These reports include device models for multigate TFETs [100,104,106,107] as well as for heterojunction nanowire TFETs. Recently, Zhang et al. [103,107] reported a workable compact model and coded it with Verilog-A for public use. However, a general-purpose compact TFET model for circuit simulation is not yet available. Therefore, the underlying physical concepts to develop compact TFET models are presented in this section.

10.5.1 Threshold Voltage Model

Let us consider a *p-i-n* TFET device structure on an SOI substrate with heavily doped $p+$ source, intrinsic channel, and $n+$ drain regions as shown in Figure 10.7a. The energy band diagram at the source-channel tunnel junction of the device in the on-state is shown in Figure 10.7b. Since the $p+$ source region is degenerately doped, the Fermi level E_{fp} of the source-side tunneling junction is assumed at or below the valence band E_{vp} as shown in Figure 10.7b.

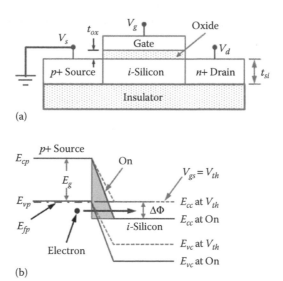

(a)

(b)

FIGURE 10.7
Threshold voltage modeling for TFETs: (a) 2D cross section of an ideal single gate *p–i–n* TFET device structure with a $p+$ source, an i-silicon channel, and an $n+$ drain regions on an SOI substrate; t_{ox} and t_{si} are the gate oxide thickness and body thickness, respectively; V_s, V_g, and V_d are the source, gate, and drain voltages, respectively and (b) energy band diagram of the source-channel junction along the length of the device at the threshold condition (broken lines) and on-state (solid lines).

Now, we define the threshold voltage V_{th} of TFETs as the gate voltage V_{gs} at which the interband tunneling sets in at the source-channel junction [106]. In other words, V_{th} is the value of the V_{gs} at which the Fermi energy level E_{fp} in the $p+$ source region aligns with the conduction band energy level E_{cc} in the channel as shown in Figure 10.7b. Considering the i-channel region as a conventional *long channel* MOSFET device, an inversion layer is formed at $V_{gs} = V_{th0}$, where V_{th0} is the long channel threshold voltage of the i-channel MOSFET is given by (Equation 4.13)

$$V_{th0} = V_{fb} + 2\phi_B + \gamma\sqrt{2\phi_B} \qquad (10.10)$$

where:
$\gamma = \sqrt{2K_{si}\varepsilon_0 q N_b}/C_{ox}$ and is defined in Equation 4.11
V_{fb} is the flat band voltage
C_{ox} is the oxide capacitance
N_b is the carrier concentration in the i-channel region
$\phi_B = v_{kT}\ln(N_b/n_i)$ is the bulk potential defined in Equation 3.35

For an nTFET device with $p+$ source (electrons are minority carriers), the injected number of minority carrier electrons from the source is insufficient to maintain the channel inversion layer. Therefore, the MOSFET device is in the off-state at $V_{gs} = V_{th0}$. If we further increase the V_{gs} beyond V_{th0}, the channel-side conduction band E_{cc} is pulled below the source side valence band E_{vp} and the tunneling window $\Delta\Phi$ is opened as shown in Figure 10.7b, causing current flow.

We know that the Fermi level of the i-channel region is at the center of the bandgap $(E_g/2)$, and for the degenerately doped $p+$ source region, we can assume $E_{vp} \approx E_{fs}$. Then at $V_{gs} = V_{th}$ when the channel E_{cc} is aligned with source $E_{vp} = E_{fp}$, the energy required to pull down E_{cc} from $V_{gs} = 0$ to $V_{gs} = V_{th}$ is about $E_g/2$. Therefore, the simplified expression for the threshold voltage of an nTFET can be written as

$$V_{thn} = V_{fb} + 2\phi_B + \gamma\sqrt{2\phi_B} + \frac{E_g}{2q} \qquad (10.11)$$

The same expression can be used for n-i-p structure with appropriate sign convention for the i-region pMOSFET biasing condition. An expression for V_{th} for short channel TFET devices has also been reported to model V_{th} roll-off [106]. However, experimental data show that the TFET device characteristics are independent of the length L of the intrinsic channel region for $L > L_{crit}$ (~20 nm for silicon TFETs) [61,62]. Therefore, Equation 10.11 is valid for threshold voltage modeling in most TFET devices. However, for $L < L_{crit}$ the p-i-n diode leakage current influences V_{th}, and therefore, appropriate channel length dependence in V_{th} must be used to account for the leakage currents in the short channel devices with $L < 20$ nm.

10.5.2 Drain Current Model

In order to develop a TFET I_{ds} model, we consider a TFET as a combination of an ideal *tunnel diode* in series with a *drain*-MOSFET device as shown in Figure 10.8. Therefore, the *interband tunneling* and *drift-diffusion* transport mechanisms must be considered in modeling I_{ds} for TFETs. Numerical simulation results show that the relatively large channel resistance due to the drift-diffusion causes additional potential drop and stronger lateral electric field in the channel, resulting in mobility (μ) and I_{on} degradation [107]. Thus, the modeling of channel transport in TFETs is important along with the interband tunneling. However, since the interband tunneling and channel transport mechanisms are coupled, the modeling of TFETs with two coupled transports increases the complexity of device modeling. Therefore, for the simplicity of compact TFET modeling, two separate I_{ds} models can be developed: (1) an ideal I_{ds} equation considering only the tunneling probability of the junction without the channel transport and (2) a second I_{ds} equation considering only the drift-diffusion transport in the channel-MOSFET. However, only the ideal current model can be accurately used for low-current drivability TFETs, whereas for high-current drivability devices, both the ideal and the channel transport I_{ds} equations must be used for accurate device analysis in circuit CAD.

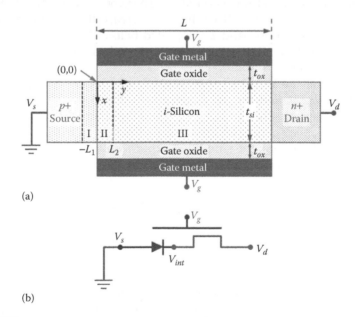

(a)

(b)

FIGURE 10.8
Schematics of a DG-nTFET for drain current modeling: (a) device structure with channel division in regions I, II, and III for current transports and (b) equivalent circuit representation of an ideal tunnel diode. L_1, L_2 are the length of the region I and II, respectively; V_{int} is the internal node; L is the channel length of the device; $V_{int} = V_{ds}$ for ideal current model and $V_{int} < V_{ds}$ for channel transport.

10.5.2.1 Ideal Drain Current Model

In the present state-of-the-art silicon TFETs with low-drive current, the channel transport is insignificant due to the comparatively large tunnel-junction resistance. Thus, for the simplicity of modeling TFETs, the channel transport can be neglected to develop an ideal TFET model by zero field approximation along the channel [105,107].

Let us consider a DG-nTFET shown in Figure 10.8a to illustrate the compact modeling techniques in TFETs. In order to derive I_{ds} model for DG-TFETs, the entire device is divided into three regions as shown in Figure 10.8a: the source junction region (I), channel junction region (II), and the channel transport region (III) [105,107]. With the zero field approximation, the quasi-Fermi potential at the boundary of regions II and III is equal to V_{ds}, and therefore, the channel transports can be neglected to develop an ideal I_{ds} model. Though a gradient of the electrostatic potential exists in region III, the quasi-Fermi potential at the *internal node* V_{int} is smaller than V_{ds}. In principle, V_{int} can be determined by ensuring current continuity between the quantum tunneling current and the drift-diffusion current.

In order to develop an ideal I_{ds} model without the channel transport, electrostatic potentials in TFETs are solved to derive the expression for tunneling current. The charges in region III are considered to model the exponential dependence of the output characteristics and the terminal capacitance properties of TFETs [105,107]. The potential solutions together with the zero electric field approximation in region III form the boundary conditions for region II.

In order to include the possible channel charge degenerations, the surface potential in region III is obtained by solving 1D Poisson's equation using Fermi-Dirac statistics [108, 109] given by

$$\frac{d^2u}{dx^2} = \left(\frac{1}{2} \frac{2q^2n_i}{\varepsilon_{si}kT} \right) \frac{F_{1/2}\left(u - v_{ch}(y) - E_g/2kT\right)}{F_{1/2}\left(-E_g/2kT\right)} \tag{10.12}$$

where:
  q is the electron charge
  u is the normalized potential by v_{kT}
  $v_{ch}(y)$ is the normalized quasi-Fermi channel potential at any point y along the channel
  n_i is the intrinsic carrier concentration
  E_g is the material bandgap
  ε_{si} is the dielectric constant of region III
  $F_{1/2}$ is the Fermi integral of the order 1/2

By integrating once, the vertical electric field at the surface is obtained as a function of the normalized surface potential u_s and center potential u_0, gate dielectric thickness t_{ox}, gate capacitance $C_{ox} = \varepsilon_{ox}/t_{ox}$, and $F_{3/2}$ the Fermi integral of the order 3/2

$$\left.\frac{du}{dx}\right|_{x=0} = \sqrt{\frac{2q^2 n_i}{\varepsilon_{si}kT}}\left[\frac{2F_{3/2}\left(u - v_{ch}(y) - E_g/2kT\right)\Big|_{u_0}^{u_s}}{3F_{1/2}\left(-E_g/2kT\right)}\right] \tag{10.13}$$

The gate control equation is derived using Equation 10.13 as

$$v_g - v_{fb} - u_s = \frac{\varepsilon_{si}}{C_{ox}}\left.\frac{du}{dx}\right|_{x=0} \tag{10.14}$$

The relationship between the surface potential (u_s) and center potential (u_0) in region III required to solve Equation 10.14 is obtained by numerical device simulation using surface potentials of DG-MOSFET [110] as the initial guess for u_0 [105,107]. With the surface potential ($\phi_{s3}(y)$) solution at any point y along the channel in region III, the expressions for the electrostatic potentials $\phi_{s1}(y)$ and $\phi_{s2}(y)$ at any point y in regions I and II, respectively, are derived. It is shown that $\phi_{s1}(y)$ is a function of the effective source doping N_{seff} and L_1; $\phi_{s2}(y)$ depends on V_{gs}, work function difference between the gate and p-doped source V_{fbs}, the work function difference between the gate and undoped channel V_{fbc}, built-in potential $V_{bi,s}$ of the source-channel junction, natural length of region II λ_{II}, and L_2; and $\phi_{s3}(y)$ depends on $V_{bi,s}$ and u_s [107].

The lenghts L_1 and L_2 are determined by matching the boundary conditions of regions I and II [105,107]. A tunneling distance W_T for any given energy level in the interband tunneling window (overlap between E_{cc} in region II and E_{vp} in region I) is found by deriving the classical turning points x_c in region II and x_v in region I from the derived surface potential profiles. Among all the tunneling paths, there exists a smallest tunneling distance $W_{T,min}$ with the largest tunneling probability, which will contribute to the peak generation rate of electrons at x_c and holes at x_v [111]. For interband tunneling in nTFETs at a certain potential level ϕ_I, W_T is found from the turning point x_c in region II (channel conduction band) and x_v in region I (source valence band). Then the potential level $\phi_{I,min}$ corresponding to the maximum generation rate is determined and is given by [107]

$$\phi_{I,min} = \left(V_{gs} - V_{fbs}\right) + \frac{qN_{seff}\lambda_{II}^2}{\varepsilon_{si}}$$

$$\overline{-\sqrt{\left(\frac{qN_{seff}\lambda_{II}^2}{\varepsilon_{si}}\right)^2 + \left(V_{gs} - V_{fbb} - \phi_{dg}\right)^2 + \frac{qN_{seff}\lambda_{II}^2}{\varepsilon_{si}\left[V_{gs} - V_{fbc} - \left(E_g/q\right)\right]}}} \tag{10.15}$$

and the minimum tunneling distance is [107]

$$W_{T,min} = L_2 - \lambda_{II}\cosh^{-1}\left(\frac{V_{gs} - V_{fbs} - \phi_{I,min}}{V_{gs} - V_{fbc} - \phi_{dg}}\right) - \sqrt{\frac{2\varepsilon_{si}}{qN_{seff}}\left(\phi_I - \frac{E_g}{q}\right)} + L_1 \tag{10.16}$$

where:

$\phi_{dg} = u_s v_{kT}$ is the surface potential of the channel region (III) DG-MOSFET

Now, from Kane's BTBT model [112], the expression for the tunneling probability is given by

$$G_T = A \frac{F^2}{\sqrt{E_g}} \exp\left(-B \frac{E_g^{3/2}}{|F|}\right) \tag{10.17}$$

where the maximum electric field across the tunneling junction is given by

$$|F| = \frac{E_g}{q} \frac{1}{W_{T,min}} \tag{10.18}$$

The peak electron generation rate is calculated by substituting Equation 10.18 into Equation 10.17 to obtain

$$G_{T,max} = A. \frac{E_g^{3/2}}{q^2} \frac{1}{W_{T,min}^2} \exp\left(-qB\sqrt{E_g} W_{T,min}\right) \tag{10.19}$$

where:

A and *B* are two parameters of the Kane's model [48,112]

Again, from Kane's model, we know that the generation rate of electrons decays exponentially with increasing tunneling distances. Therefore, the total tunneling current is obtained by integrating Equation 10.17 over the tunneling space $d\Omega_{tun}$ and is given by

$$I_{ds} = q \int G_T d\Omega_{tun} \tag{10.20}$$

To derive a closed-form solution of the above spatial integral of generation rates, a linearly changing tunnel distance is assumed [103]. Again, by assuming that the tunneling current is uniform across the channel thickness (a valid approximation for thin body DG-TFETs), the final I_{ds} expression can be shown as

$$I_{ds} = G_{T,max}. \left(\frac{2W.t_{si}}{B\sqrt{E_g}}\right) \tag{10.21}$$

where:

W is the channel width of TFETs

Equation 10.21 represents a simplified I_{ds} model for TFETs. However, it yields an unphysical nonzero current even at equilibrium states of TFETs.

Therefore, for physically acceptable simulation results, a correction factor is introduced as given by [107]

$$f_{fermi} = 2\left[\frac{1}{2} - \frac{1}{1+\exp\left(V_{ds}/f_n.v_{kT}\right)}\right]$$

(10.22)

where:

f_n is an empirical fitting parameter

Finally, the ideal I_{ds} model is given by

$$I_{ds} = G_{T,max}.\left(\frac{2W.t_{si}}{B\sqrt{E_g}}\right).f_{fermi}$$

(10.23)

Equation 10.23 is the ideal I_{ds} model for TFETs with $G_{T,max}$ and f_{fermi} given by Equations 10.19 and 10.22, respectively. The ideal I_{ds} is characterized by three model parameters, A, B, and f_n. When the potential $\phi_{L,min}$ given by Equation 10.15 is larger than the bandgap potential, the interband tunneling window is created and a tunneling current is observed. It can be shown that the bias-dependent S in TFETs is mainly determined by the V_{gs}-dependent tunneling distance given by Equation 10.16. The ideal I_{ds} model is valid in the operation regions with large V_{gs} due to the inclusion of channel charge in the surface potential model. Note that the I_{ds} expression in Equation 10.23 does not include channel length L. This is justified for $L > 20$ nm since the leakage current dominates in TFET devices with $L < 20$ nm as discussed earlier.

10.5.2.2 Modeling the Channel Transports Using Drain MOSFET

In contrast to low current drivability all-silicon TFETs (e.g., $\ll 100$ μA μm^{-1}), the III–V compound-based TFETs and heterojunction TFETs offer significantly high drive current (e.g., hundreds of μA μm^{-1}) [114]. In these devices, the resistance of the tunneling junction is comparable to the resistance of the channel region, and therefore, the drift-diffusion channel transport directly affects the device characteristics. Thus, for accurate modeling of high performance TFET devices, it is necessary to include the effect of channel transport in compact TFET modeling.

In order to model TFETs with coupled transports, a TFET device can be represented by an ideal TFET in series with a drift-diffusion MOSFET at the drain side of the device as shown in Figure 10.9. In this representation, a TFET includes two components that are coupled by the internal node with potential V_{int}. Thus, $V_{int} = V_d$ for the ideal TFET and $V_{int} = V_s$ for the drift-diffusion MOSFET of the DG-TFET device. At any applied biasing condition, V_{gs} is shared by both devices; however, the quasi-Fermi level V_{int} and/or electrostatic potential at the internal node is determined by setting the tunneling

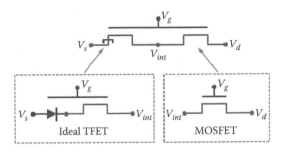

FIGURE 10.9
Schematic representation of a DG-TFET with coupled transports: The device is modeled by an ideal TFET in series with a MOSFET at the drain end of the device for modeling the channel transport; the internal node V_{int} is common to both TFET and the MOSFET and determined by current continuity at the node.

and drift-diffusion currents at the same value. From the discussion in the previous section we know that the tunneling current is a function of the surface potential in TFET region III and depends on its drain voltage ($V_d = V_{int}$) as shown in Figure 10.9. Similarly, the drift-diffusion current in a MOSFET depends on its source voltage ($V_s = V_{int}$). Thus, the current continuity at V_{int} can be achieved by iterations in a circuit CAD tool (e.g., SPICE).

The zero field approximation at V_{int} leading to $V_{int} = V_d$ simplifies the derivation of I_{ds} model as discussed in Section 10.5.2.1. However, $V_{int} = V_d$ assumption is physically invalid for modeling I_{ds} if the channel transport is considered. In this case, V_{int} is not uniquely defined since both the potential and the electric field are floating and a simple solution for Poisson's equation does not exist. Therefore, a *virtual node* method is used for modeling I_{ds} [107]. In this technique, a zero field is assumed to be still valid at V_{int} similar to the boundary condition used for the ideal TFET and a discontinuity in the lateral field is created keeping potential or quasi-Fermi level constant at the node. With this virtual node method, Equation 10.23 can be used as the component of the total I_{ds} due to the ideal TFET device shown in Figure 10.9 with its drain voltage as V_{di}. Again, V_{di} is a variable and is determined by current continuity at the internal node.

For any given biasing condition, the applied voltages V_g, V_s, and V_d and the pre-assumed quasi-Fermi potential at the internal node V_{di}, the potentials at the internal node ϕ_{int} and at the drain side ϕ_d are both derived from Equations 10.12 to 10.14 with the quasi-Fermi levels as V_{di} and V_d. Note that for the MOSFET element of the device ϕ_{int} and V_{di} are the surface potential and bias at the virtual source terminal, respectively. Then the electron charge densities at the source and drain ends of the MOSFET device are given by

$$q_{int} = C_{ox}\left(V_{gs} - \phi_{int}\right)$$
$$q_d = C_{ox}\left(V_{gs} - \phi_d\right)$$

$$(10.24)$$

Finally, the MOSFET current is simply calculated by summarizing the drift and diffusion current components and is given by

$$I_{ds,MOS} = 2\mu \frac{W}{L_{dg}} \left[v_{kT} \left(q_{int} - q_d \right) \right] + \frac{q_{int}^2 - q_d^2}{2C_{ox}} \tag{10.25}$$

where:

μ is the electron mobility

L_{dg} is the effective channel length of the MOSFET given by

$$L_{dg} = L_g - L_2 \tag{10.26}$$

L_{dg} depends on the external bias through L_2 [105,107].

Equations 10.23 and 10.25 constitute the total drain current in a DG-TFET to model both the interband tunneling and drift-diffusion transport. The values of quasi-Fermi level V_{di} and the potential at the internal node ϕ_{int} are determined iteratively to set the ideal TFET I_{ds} in Equation 10.23 equal to the drift-diffusion I_{ds} in Equation 10.25. However, a correction factor is needed for accurate modeling of channel transport in TFETs [107].

The complete set of model parameters including channel transport models in TFETs is given by $\{\mu, A, B, f_n\}$. A simple parameter extraction routine is used to extract the model parameters. The parameters, A and B, are optimized to fit the I–V characteristics in the subthreshold region of TFETs by setting a large value for μ. Then, μ is optimized to fit the I–V characteristics in the high V_{gs} region. The parameters A and μ determine the transition region of the transfer characteristics of TFET devices. Finally, A and B are reoptimized to fit the transconductance [107].

Though the drain-MOSFET method accurately models the effects of channel transport on TFET current and terminal charge characteristics, it requires additional iterative computations in circuit CAD and, therefore, is computationally inefficient compared to the ideal I_{ds} model described in Section 10.5.2.1. Since the drift-diffusion channel transport does not affect the terminal charge significantly, a computationally efficient simpler method can be used for modeling TFETs in circuit CAD as described in the following section [107].

10.5.2.3 Modeling the Channel Transports Using Source Resistance

The channel transport in TFETs can be modeled effectively by adding a source resistance (R_s) to the ideal TFET model as shown in Figure 10.10 [107]. R_s reduces both V_{gs} and V_{ds} simultaneously, due to the similarity in the exponential gate and drain control over the tunneling current and, therefore, effectively models the effects of channel transport in reducing the voltage drop over the tunnel junction. Since both V_{gs} and V_{ds} are reduced simultaneously, the saturation drain voltage, V_{dsat}, of TFETs does not change. R_s is, purely, a fitting parameter and is extracted along with the model parameters

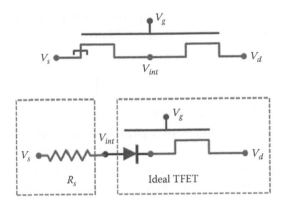

FIGURE 10.10
Schematic representation of a DG-TFET for coupled transports: The device is modeled by an ideal TFET in series with a source resistance, Rs; Rs is a fitting parameter to match the measured I–V characteristics of TFETs.

A, B, and f_n of the ideal I_{ds} model. Obviously, the accuracy of the method depends on the extraction of R_s.

The reported data show that the source-resistance method is computationally efficient compared to the drain-MOSFET method and provides acceptable simulation results for TFET devices [107].

10.6 Summary

This chapter introduces the emerging devices as the potential alternatives for beyond-CMOS devices. Among the emerging devices such as tunneling, impact ionization, ferroelectric dielectrics, and mechanical gate FETs, TFETs have recently been the subject of numerous investigations for a potential alternative to MOSFETs. Thus, in this chapter the fundamentals of TFETs are overviewed. First of all, the basic features of TFETs are discussed. Then the basic operating principle is presented to understand the basic mechanism of TFETs as the steep slope devices with S much lower than that of MOSFETs. Finally, the emerging compact TFET modeling techniques are presented.

Exercises

10.1 Consider an ideal all-silicon *n-i-p* TFET structure with *n+* source, *intrinsic*-silicon channel, and *p+* drain regions to explain the basic principle of pTFET operation.

a. Draw the energy band diagrams with reference to the ideal device structure for biasing conditions

    i. $V_{gs} = 0 = V_{ds}$

    ii. $V_{gs} < 0, V_{ds} = 0$

    iii. $V_{gs} < 0, V_{ds} < 0$

    Clearly label all relevant parameters;

b. Explain the operation of a pTFET device with reference to the band diagram for each of the biasing conditions in part (a).

10.2 Consider the ideal all-silicon *n-i-p* TFET structure described in exercise 10.1 to explain the interband tunneling mechanism in *p*TFET devices.

a. Draw the energy band diagrams in the off-state with $V_{gs} = 0 = V_{ds}$ and on-state with biasing condition $V_{gs} < 0$ and $V_{ds} < 0$ so that a tunneling window is created by overlapping bands at the source-channel junction. Clearly label all relevant parameters.

b. Explain the tunneling mechanism at the junction with reference to the band diagrams in part (a).

c. Sketch the transfer characteristics ($I_{ds} - V_{gs}$) of the device under the biasing conditions in part (a) with reference to the band diagrams in part (a) (similar to Figure 10.3); explain your plots.

10.3 Complete the mathematical steps to derive Equation 10.7 for subthreshold swing in TFET devices. Identify the critical parameters that can be used to optimize device performance for TFETs. Explain with examples how these parameters can be used to improve current drivability in TFETs.

10.4 Compare the carrier transport mechanisms in TFETs and MOSFETs. Explain why TFETs can offer lower S (<60 mV per decade of drain current at room temperature) than that of MOSFETs using carrier injection process and carrier statistics in each device.

10.5 Consider an all silicon *n-i-p* TFET device. Draw the energy band diagrams.

a. At the onset of threshold voltage.

b. In the on-state.

    Clearly label all relevant parameters and explain the device operation.

10.6 The carrier transport and drive current in TFETs are independent of gate length, L (i.e., length of the channel region). However, for $L < 20$ nm, the *p-i-n* diode leakage current dominates increasing the overall current flow in the device. Describe a simple technique to model this additional current in short channel TFETs for circuit CAD.

10.7 Write the complete coupled equations discussed in Section 10.5.2.2 that you would solve to model channel transport in TFETs. What are the compact model parameters and describe the methodology to extract these parameters to build compact TFET model for circuit CAD.

10.8 Are TFETs viable devices for beyond-CMOS technology? Explain your answer with examples.

11

Bipolar Junction Transistor Compact Models

11.1 Introduction

As described in Chapters 4 and 5, the *pn*-junctions are integral part of a MOSFET (metal-oxide-semiconductor field-effect transistor) device structure as the source and drain regions. Under the appropriate biasing condition of a MOSFET device, the source of the source-substrate *pn*-junction provides a steady supply of mobile carriers to form a conducting channel from the source to drain and the drain of the drain-substrate *pn*-junction collects the mobile carriers generating drain current. Two back-to-back *pn*-junctions form a bipolar junction transistor (BJT). BJTs are very often used in VLSI (very-large-scale-integrated) circuits. Therefore, a basic understanding of BJT modeling is necessary for engineers and researchers involved in device modeling. In this chapter, we present the basic but widely used BJT compact models for circuit CAD.

BJTs are active three-terminal devices and were the main active elements for ICs (integrated circuits) in the 1960s [1,2]. The areas of applications of BJTs include amplifiers, switches, high-power circuits, and high-speed logic circuits for high-speed computers. After the invention of bipolar transistors in 1947 [3], discrete BJTs were used to design circuits on printed circuit boards. In order to analyze the performance of BJTs, Ebers and Moll in 1954 reported a physics-based large signal BJT model, referred to as the *Ebers–Moll* or *EM* model [4]. The level 1 EM model, known as the EM1 model, is valid for the entire operating regime of BJTs from cutoff to active region. However, the application and accuracy of EM1 model are limited to evaluating the DC performance of the devices only due to several simplifying assumptions. In order to improve the modeling accuracy, EM1 model has been extended to EM2 and EM3 models for predicting the observed physical effects in BJTs including transient phenomena [5].

Though EM2 and EM3 models accurately predict most of the observed physical effects in BJTs, a more complete and unified physics-based BJT model was reported by Gummel and Poon in 1970 [6]. This model is known today as the *Spice Gummel–Poon* (SGP) model [7]. The SGP model uses an *integrated charge control* approach along with a very clear and standardized description of many observed effects in BJTs such as early effect [8], high current roll-off [9], and carrier transit time [10]. Due to its simple yet physical model formulation, SGP model was the most popular BJT model until mid-1990s.

With the continued scaling of modern transistors, some second-order effects that are not considered by SGP model, such as substrate network, self-heating effects, and avalanche effects, became more and more important for accurate modeling of BJT ICs. A number of advanced BJT models have been introduced to model emerging and second-order physical effects to provide more precise simulation results [11]. These models include Vertical Bipolar Inter Company model [12], Most Exquisite Transistor Model [13], and High Current Model [14]. However, the SGP BJT model continued to be used in circuit CAD because of its simplicity. Therefore, in this chapter only EM and SGP models are described to provide readers the basic idea of BJT modeling. In model derivations, the emphasis is placed on the understanding of the effect being modeled along with the explanation of the required parameters. Thus, in this chapter, we use a systematic methodology to derive SGP compact BJT model, starting from the basic EM compact BJT model that provides an extremely useful understanding of the basic BJT operation.

11.2 Basic Features of BJTs

A silicon BJT structure is a sandwich of alternating type of doped silicon layers. Depending on the sequence of layers, two types of BJTs are manufactured: *npn* and *pnp*. An *npn*-BJT is a sequence of *n-p-n* layers whereas a *pnp*-BJT is a sequence of *p-n-p* layers. The *npn*-BJTs are most widely used in ICs with BJT technologies. Again, the sequence of layers may be used vertically to fabricate *vertical* BJTs or laterally referred to as the *lateral* BJTs. Figure 11.1 shows the basic structure of a vertical *npn*-BJT.

As shown in Figure 11.1b, the basic structure includes a heavily doped *n*+ emitter (E), a lightly doped *n*-epitaxial layer, a *p*-type base (B), and a heavily doped *n*+ buried collector (C) on a *p*-type substrate. The *p*+ isolation regions are used to isolate the adjacent devices in an IC chip. Typically, the isolation regions are reverse-biased *pn*-junctions; however, in advanced BJTs, trench isolation is used to increase the packing density of IC chips. The intrinsic device consists of *n-p-n* vertical cross section as shown in Figure 11.1b. The one-dimensional (1D) doping profile along the cutline from the surface of the active device is shown in Figure 11.1c. Figure 11.2 shows a typical layout of an IC *npn*-BJT.

Figure 11.1c shows that the base region is nonuniformly doped. As a result, a built-in electric field is set up to establish an equilibrium between the mobile carriers attempt to *diffuse* away from the high concentration region and mobile carriers pulled by the electric field (*drift*) of the fixed ionized donors (N_d^+) or acceptors (N_a^-) left behind by mobile carriers. The built-in electric field is obtained by setting: *diffusion = drift* (Equations 2.45 and 2.46).

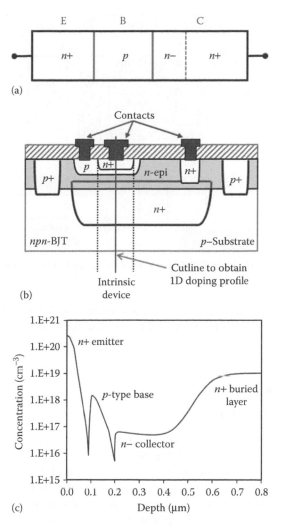

FIGURE 11.1
Basic feature of an *npn*-BJT: (a) an ideal structure showing an *n*+ emitter (E), a *p*-base (B), and
n-*n*+ collector (C) regions; (b) 2D-cross section of a typical vertical *npn*-BJT used in ICs and (c)
1D doping profile along the cutline through the intrinsic device.

The basic BJT structure in Figure 11.1b shows that the BJTs also include a
number of parasitic elements that must be accurately modeled in compact
BJT models. The basic BJT structure has *base resistance*, r_b, mainly from the
base contact to active area, a *collector resistance*, r_c (predominantly due to
n-epitaxial layer as shown in Figure 11.1c), and an *emitter resistance*, r_e (typically,
negligibly small). The *n*+ emitter and *p*-base junction includes an emitter-base
(EB) junction capacitance (C_{jE}). The *n*− collector region adjacent to the base
also includes a collector-base (CB) junction capacitance, C_{jC}. The advantages of
the *n*-layer include a reduction in C_{jC}, an improvement in the CB breakdown

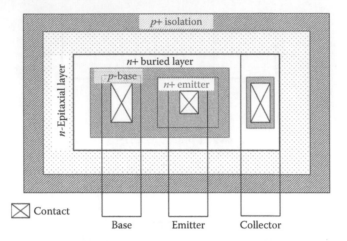

FIGURE 11.2
Typical layout of a vertical *npn*-BJT device shown in Figure 11.1(b) used for fabrication in an
IC chip.

voltage BV_{CB}, and a decrease in the base-width modulation by the collector
voltage at a cost of collector terminal series resistance R_C.

11.3 Basic Operation of BJTs

In order to describe the basic operation of BJTs, let us consider the structure
and biasing condition shown in Figure 11.3.

In a typical *npn*-BJT operation, an external potential, V_{BE} (≈ 0.7 V), is applied
across the EB-junction to forward bias it, as shown in Figure 11.3. *Electrons
are injected into the base by the emitter.* (*Also, holes are injected into the emitter
but their numbers are much lower because of the relative values of N_a and N_d.*) If the
effective base width $W_B \ll L_n$ (electron diffusion length) in the base, most of
the injected electrons get into the collector without recombining. A few do
recombine; the *holes* necessary for this are supplied as the base current, I_B. The
electrons reaching the collector are collected across the CB-junction depletion
region (X_{dCB}) under the reverse bias CB-junction, V_{BC}, and generates collector
current, I_C. The carrier transport process is shown in Figure 11.4a and the
circuit representation of an *npn*-BJT is shown in Figure 11.4b. In Figure 11.4,
I_E represents the emitter current. Conventionally, the current flowing into the
device terminal is defined as positive.

Since most of the injected electrons reach the collector, only a few holes are
injected into the emitter; therefore, $I_B \ll I_C$. As a result, the BJT device has a
substantial current gain (I_C/I_B). *Note that the built-in electric field across the base
also aids electron transport from E to C.*

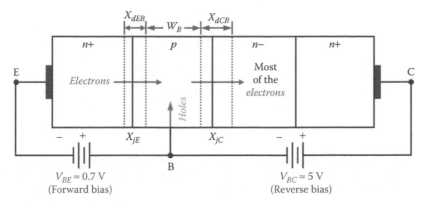

FIGURE 11.3
A typical biasing condition of an *npn*-BJT: the EB-junction is forward biased and CB-junction is reverse biased; W_B is the effective width of the neutral base region; X_{jE} and X_{jC} are the EB and CB metallurgical junctions, respectively; and X_{dEB} and X_{dCB} are the EB and CB depletion regions, respectively.

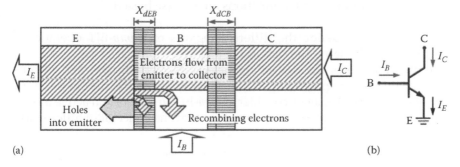

(a)   (b)

FIGURE 11.4
An *npn*-BJT operation: (a) carrier transport under the forward active mode of operation; the base-emitter junction is forward biased and base-collector junction is reverse biased and (b) circuit representation. I_C, I_B, and I_E are the collector, base, and emitter terminal currents, respectively.

11.4 Mode of Operations of BJTs

In order to develop compact BJT models for circuit CAD, let us define the operation regions of BJT devices. Depending on the biasing conditions, we can define four different modes of BJT operations as shown in Figure 11.5 for an *npn*-BJT device.

1. *Forward active* or *normal* mode: EB-junction is *forward* biased; and CB-junction is *reverse* biased. In this case, the collector current $I_C = \beta_F I_B$, where β_F is the forward current gain;

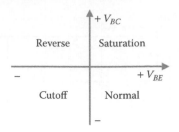

FIGURE 11.5
Four different regions of operation of an *npn*-BJT device depending on the biasing conditions: normal active, saturation, inverse active, and cutoff; where V_{BE} and V_{BC} are the EB-junction voltage and CB-junction voltage, respectively.

  2. *Reverse active* mode: EB-junction is *reverse* biased and CB-junction is *forward* biased. In this case, the reverse current at emitter terminal $I_E = \beta_R I_B$, where β_R is the reverse current gain ≈ 1;
  3. *Saturation*: both EB- and CB-junctions are forward biased;
  4. *Cutoff*: both EB- and CB-junctions are reverse biased.

Similarly, we can define the different regions of a *pnp*-BJT operation by appropriately changing the sign of V_{BE} and V_{CE}.
  Figure 11.6 shows the device characteristics of both *npn*- and *pnp*-BJTs showing different regions of operation. In order to analyze the device characteristics in Figure 11.6, let us consider an *npn*-BJT in the normal active mode of operation. From Figure 11.4b we can show that the collector–emitter voltage

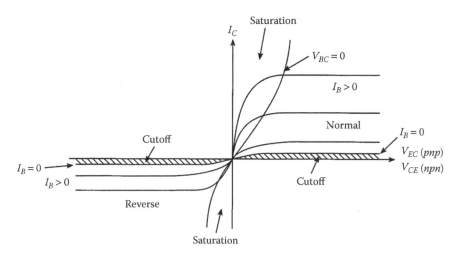

FIGURE 11.6
Collector current versus collector–emitter voltage with base current as the third parameter for both *npn*- and *pnp*-BJTs; plot shows all four regions of device operation, depending on the applied collector–emitter voltage.

$V_{CE} = V_{BE} - V_{BC}$. Therefore, at $V_{CE} = 0$, and $V_{BE} \geq \phi_{BE}$, $V_{BE} = V_{BC}$, where ϕ_{BE} is the built-in-potential of EB-junction (Equation 2.109). Under this condition, both the EB- and CB-junctions are forward biased, resulting in a decrease in the barrier height for electrons at both EB- and CB-junctions (Equation 2.100). Consequently, both emitter and collector junctions inject electrons into the base. Under this biasing condition, the electric field does not favor transport of electrons to the collector (or emitter) terminal and $I_C = 0$ and the device is in *saturation*. Thus, for $0 \leq V_{CE} < V_{BE}$, the *npn*-BJT operates in the saturation regime since both the EB- and CB-junctions are forward biased. As V_{CE} increases from $V_{CE} = 0$ due to the increasing collector supply voltage V_{CC}, V_{BC} gradually becomes less forward biased and the CB-junction barrier height gradually increases. Therefore, electron injection from the emitter to base dominates over that from the collector to base and I_C increases with the increase in V_{CE} as shown in Figure 11.6. At $V_{CE} = V_{BE}$ (or, $V_{BC} = 0$), the *npn*-BJT is at the onset of transition from the saturation region to the *normal active* mode of operation, and for $V_{CE} > V_{BE}$, the *npn*-BJT operates in the normal active *linear regime*. Therefore, the loci of the point $V_{CE} = V_{BE}$ on I_C–V_{CE} plot separates the saturation and linear regions of BJTs as shown in Figure 11.6.

Again, when $V_{CE} < 0$, both the EB- and CB-junctions are reverse biased. This increases the potential barrier height of electrons for both the *pn*-junctions and there is no electron injection from the emitter or collector to the base region of the transistor resulting in $I_C \approx 0$. Under this condition, the device operates in the *cutoff* region as shown in Figure 11.6. Similarly, we can explain the *pnp*-BJT characteristics. Note the difference between the MOSFET and BJT linear and saturation region of operations (Section 4.4.4.1).

11.5 Compact BJT Model

In order to develop a complete BJT model for circuit CAD, we first develop the basic DC model using the Ebers–Moll formulation and then include the different parasitic elements of the BJT structure and physical effects.

11.5.1 Basic DC Model: EM1

In order to derive a basic BJT current model, let us consider an *npn*-BJT device shown in Figure 11.7. As seen from Figure 11.7, a BJT structure can be considered as two back-to-back *pn*-junctions. For the simplicity of basic model formulation, we assume that all the parasitic elements such as series resistances and junction capacitances are negligibly small.

Now, let us assume that the *npn*-BJT is biased in the normal active mode of operation ($V_{BE} \geq \phi_{BE}$ and $V_{BC} < 0$). Then when the EB *pn*-junction is forward biased, a forward current I_F flows through the EB *pn*-junction and a current $\alpha_F I_F$ flows across the CB *pn*-junction, where α_F is the forward current gain

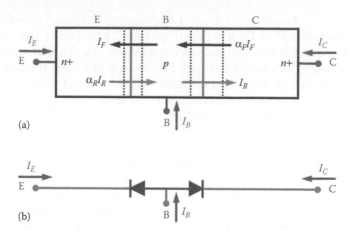

(a)

(b)

FIGURE 11.7
An ideal *npn*-BJT structure used to derive the basic Ebers–Moll model: (a) the transistor configuration with carrier injection due to the applied biases and (b) the *npn*-BJT structure represented by two back-to-back *pn*-junctions.

given by I_C/I_E for $V_{BE} > 0$. Similarly, for the reverse-biased CB *pn*-junction, a reverse current, I_R, flows through the CB *pn*-junction and a current, $\alpha_R I_R$, flows through the EB *pn*-junction, where α_R is the reverse current gain and is given by I_E/I_C for $V_{BC} < 0$. Then from Figure 11.7a, the terminal currents are given by

$$I_E = -I_F + \alpha_R I_R$$
$$I_C = \alpha_F I_F - I_R$$

(11.1)

The physical concept described in Figure 11.7 can be represented by an equivalent circuit shown in Figure 11.8.

Figure 11.8 represents the basic *npn*-BJT model for circuit CAD and is known as the basic Ebers–Moll or EM1 BJT model. From Figure 11.8 the terminal currents are given by

$$I_E = -I_F + \alpha_R I_R$$

(11.2)

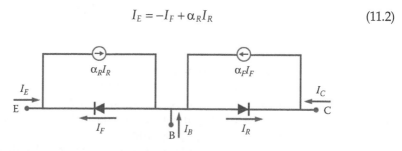

FIGURE 11.8
The basic Ebers–Moll model for an *npn*-BJT: the basic model is obtained by considering two back-to-back *pn*-junctions. Here, $\alpha_F I_F$ and $\alpha_R I_R$ are the current sources at the CB and EB *pn*-junctions, respectively.

$$I_C = \alpha_F I_F - I_R \tag{11.3}$$

$$I_B = I_F - \alpha_R I_R + I_R - \alpha_F I_F = (1 - \alpha_F) I_F + (1 - \alpha_R) I_R \tag{11.4}$$

From Equation 2.119, we can write the expression for the forward electron current flow through the EB *pn*-junction as

$$I_F = I_{nE} = I_{ES} \left[\exp\left(\frac{V_{BE}}{v_{kT}} \right) - 1 \right] \tag{11.5}$$

and, the reverse electron current flow through the CB *pn*-junction as

$$I_R = I_{nC} = I_{CS} \left[\exp\left(\frac{V_{BC}}{v_{kT}} \right) - 1 \right] \tag{11.6}$$

where:
   I_{ES} and I_{CS} are the saturation currents of the EB and CB *pn*-junctions, respectively
   V_{BE} and V_{BC} are the applied voltages at the EB and CB *pn*-junctions, respectively

Then from Equations 11.2 through 11.6, the terminal currents for an *npn*-BJT can be shown as

$$I_E = -I_{ES} \left[\exp\left(\frac{V_{BE}}{v_{kT}} \right) - 1 \right] + \alpha_R I_{CS} \left[\exp\left(\frac{V_{BC}}{v_{kT}} \right) - 1 \right] \tag{11.7}$$

And,

$$I_C = \alpha_F I_{ES} \left[\exp\left(\frac{V_{BE}}{v_{kT}} \right) - 1 \right] - I_{CS} \left[\exp\left(\frac{V_{BC}}{v_{kT}} \right) - 1 \right] \tag{11.8}$$

From the reciprocity property $\alpha_F I_{ES} = \alpha_R I_{CS} \equiv I_S$, where I_S is the reverse saturation current of an *npn*-BJT device, so we can express Equations 11.7 and 11.8 as

$$I_E = -\frac{I_S}{\alpha_F} \left[\exp\left(\frac{V_{BE}}{v_{kT}} \right) - 1 \right] + I_S \left[\exp\left(\frac{V_{BC}}{v_{kT}} \right) - 1 \right] \tag{11.9}$$

And,

$$I_C = I_S \left[\exp\left(\frac{V_{BE}}{v_{kT}} \right) - 1 \right] - \frac{I_S}{\alpha_R} \left[\exp\left(\frac{V_{BC}}{v_{kT}} \right) - 1 \right] \tag{11.10}$$

Again, using $\alpha_F = I_C/I_E$ and $\alpha_R = I_E/I_C$, we can show that the forward current gain $\beta_F = I_C/I_B$ and the reverse current gain $\beta_R = I_B/I_C$ are given by

$$\beta_F = \frac{\alpha_F}{(1-\alpha_F)}$$

$$\beta_R = \frac{\alpha_R}{(1-\alpha_R)}$$

(11.11)

From Equation 2.121, the temperature dependence of the saturation current I_S at any ambient temperature T with respect to reference temperature T_{NOM} is given by

$$I_S(T) = I_S(T_{NOM})\left(\frac{T}{T_{NOM}}\right)^3 \exp\left[\frac{E_g(T_{NOM})}{kT_{NOM}} - \frac{E_g(T)}{kT_{NOM}}\right]$$

(11.12)

where:
E_g is the energy gap of the silicon substrate

The BJT current model obtained in Equations 11.9 and 11.10 are known as the *injection version* of EM1 model.

To further simplify the model, we define the reference current source I_{CC} due to the forward injection at EB *pn*-junction by applied voltage V_{BE} and source current I_{EC} due to the reverse injection at CB *pn*-junction by applied bias V_{BC}. Then from Equation 2.119, we get

$$I_{CC} = I_S\left[\exp\left(\frac{V_{BE}}{v_{kT}}\right) - 1\right]$$

$$I_{EC} = I_S\left[\exp\left(\frac{V_{BC}}{v_{kT}}\right) - 1\right]$$

(11.13)

Using Equation 11.13, the model Equations 11.9 and 11.10 can be written as

$$I_E = \left(-\frac{1}{\alpha_F}\right)I_{CC} + I_{EC}$$

(11.14)

$$I_C = I_{CC} - \frac{1}{\alpha_R}I_{EC}$$

(11.15)

And, from Kirchhoff's current law, the base current $I_B = -(I_E + I_C)$ is given by

$$I_B = \left(\frac{1}{\alpha_F} - 1\right)I_{CC} + \left(\frac{1}{\alpha_R} - 1\right)I_{EC}$$

(11.16)

Equations 11.14 through 11.16 present the *npn*-BJT terminal currents with reference to source currents I_{CC} and I_{EC} given by Equation 11.13. This is referred

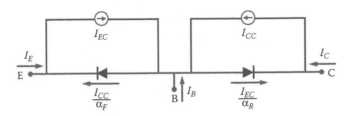

FIGURE 11.9
The transport version of the basic Ebers–Moll model for an *npn*-BJT; the model is derived with reference to source currents I_{CC} and I_{EC}.

to as the EM1 *transport model* and the corresponding equivalent circuit is shown in Figure 11.9.

We can further simplify the model by adding and subtracting I_{CC} on the right-hand side of Equation 11.14 to get

$$I_E = \left(1 - \frac{1}{\alpha_F}\right) I_{CC} - \left(I_{CC} - I_{EC}\right) = -\frac{I_{CC}}{\beta_F} - I_{CT} \tag{11.17}$$

Similarly, by adding and subtracting I_{EC} on the right-hand side of Equation 11.15, we get

$$I_C = \left(I_{CC} - I_{EC}\right) - \left(\frac{1}{\alpha_R} - 1\right) I_{EC} = I_{CT} - \frac{I_{EC}}{\beta_R} \tag{11.18}$$

where $I_{CT} = (I_{CC} - I_{EC})$; in Equations 11.17 and 11.18, we have used Equation 11.11 to replace α_F and α_R by β_F and β_R, respectively. Now, from Equation 11.13, the current source I_{CT} can be expressed as

$$I_{CT} = \left(I_{CC} - I_{EC}\right) = I_S \left\{ \left[\exp\left(\frac{V_{BE}}{v_{kT}}\right) - 1 \right] - \left[\exp\left(\frac{V_{BC}}{v_{kT}}\right) - 1 \right] \right\} \tag{11.19}$$

Note that I_{CT} models the current source in a BJT and describes the current flow through the device in the normal active model of operation. With reference to Equations 11.17 and 11.18, the equivalent circuit for the final version of EM1 transport model defined as the *nonlinear hybrid-π* model is shown in Figure 11.10

Thus, the terminal currents of the *transport version* of the basic EM1 model are given by

$$I_E = -\frac{I_{CC}}{\beta_F} - I_{CT}$$

$$I_C = I_{CT} - \frac{I_{EC}}{\beta_R} \tag{11.20}$$

$$I_B = \frac{I_{CC}}{\beta_F} + \frac{I_{EC}}{\beta_R}$$

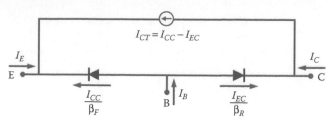

FIGURE 11.10
The nonlinear hybrid-π EM1 model for an *npn*-BJT; the model is derived with reference to the source current I_{CT}.

The currents I_{CT}, I_{CC}, and I_{EC} in the basic model (Equation 11.20) depend on I_S that depends on ambient temperature T_{NOM} and energy gap E_g, given by Equation 11.12. Thus, the EM1 transport model can be characterized by five model parameters: β_F, β_R, I_S, T_{NOM}, and E_g. The basic EM1 model can be used to analyze the performance of BJTs over the entire operating regimes at any temperature by five parameters only.

11.5.1.1 Linear Hybrid-π Small Signal Model

For small signal analysis, the total current (i_C) and total BE-junction voltage (v_{BE}) are denoted by their DC values and an incremental (small signal) quantity as

$$i_C = I_C + i_c$$

and   (11.21)

$$v_{BE} = V_{BE} + v_{be}$$

where:
  i_c and v_{be} are the small signal collector current and EB *pn*-junction voltage, respectively

The ratio of i_c and v_{be} at a given DC bias point (V_{BE}, I_C) is defined as the *transconductance* g_m and is given by

$$g_m = \left. \frac{i_c}{v_{be}} \right|_{(V_{BE}, I_C)} = \left. \frac{\partial i_C}{\partial v_{BE}} \right|_{(V_{BE}, I_C)}$$ (11.22)

The incremental change in i_C due to an increment in v_{BE} is represented by a voltage-controlled current source. From Equation 11.19, the collector current in the forward active region is given by

$$i_C \cong I_S \exp\frac{v_{BE}}{v_{kT}} \tag{11.23}$$

Now, using Equation 11.23 in Equation 11.22, we get

$$g_m = \frac{I_S}{v_{kT}}\exp\frac{v_{BE}}{v_{kT}} = \frac{I_C}{v_{kT}} \tag{11.24}$$

The transconductance connects the BE-junction terminal potential with the collector current and is the central element in the small signal model. To establish a relation of the current source in small signal model, let us write the exact expression for the collector current as a function of BE-junction voltage

$$i_C = I_C + i_c = I_S \exp\left[\frac{(V_{BE} + v_{be})}{v_{kT}}\right] = I_C \exp\left(\frac{v_{be}}{v_{kT}}\right) \tag{11.25}$$

Since $v_{be} \ll v_{kT}$, then by series expansion of Equation 11.25 and neglecting the higher order terms, we get

$$I_C + i_c = I_C \exp\left(\frac{v_{be}}{v_{kT}}\right) = I_C\left(1 + \frac{v_{be}}{v_{kT}} + \cdots\right) \tag{11.26}$$

From Equations 11.24 and 11.26, we get the expression for small signal current source as

$$i_c \cong g_m v_{be} \tag{11.27}$$

Input resistance: In order to find the small signal resistor that models the incremental base current due to v_{be}, we define the small signal forward current gain at the operating point $Q = (V_{BE}, I_C)$ as

$$\beta_F = \frac{\partial i_C}{\partial i_B}\bigg|_Q \tag{11.28}$$

The small signal current gain may vary with the operating point; however, for the first-order analysis we assume the small signal current gain (β_o) at Q is equal to the forward current gain given by Equation 11.28. Then the input resistance is defined by

$$\frac{1}{r_\pi} = \frac{\partial i_B}{\partial v_{BE}}\bigg|_Q \tag{11.29}$$

Applying the chain rule and substituting the small signal current gain from Equation 11.27, we find that

$$\frac{1}{r_\pi} = \left.\frac{\partial i_B}{\partial v_{BE}}\right|_Q = \left.\frac{\partial i_B}{\partial i_C}\right|_Q \left.\frac{\partial i_C}{\partial v_{BE}}\right|_Q = \frac{g_m}{\beta_F} \qquad (11.30)$$

where we identified the transconductance g_m from Equation 11.22. Substituting for g_m from Equation 11.24 at the operating point Q, we find the input resistance is

$$r_\pi = \frac{\beta_F v_{kT}}{I_C} = \frac{\beta_F}{g_m} \qquad (11.31)$$

From Equation 11.31, it is obvious that the input resistance of BJTs is inversely proportional to the DC collector current I_C and directly proportional to the small signal current gain β_F.

Output resistance: In the normal active mode of operation, the CB-junction is reverse biased. Therefore, the output resistance of the reverse-biased *pn*-junction is defined by

$$\frac{1}{r_o} = \left.\frac{\partial i_C}{\partial v_{CE}}\right|_Q \equiv g_o \qquad (11.32)$$

where:
g_o is the output conductance of the device

In EM1 BJT model, $r_o \equiv r_\mu$ is assumed to be extremely large, that is, open circuit. Therefore, the small signal equivalent circuit of the basic BJT model can be shown as in Figure 11.11

The basic EM1 model is fairly accurate only for modeling the DC characteristics of BJTs at any ambient temperature, T. However, the model cannot be used for transient analysis since in deriving the EM1 model we have neglected the effect of parasitic elements. In the next section, we will update

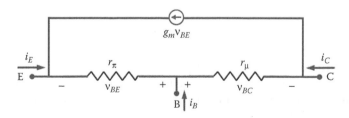

FIGURE 11.11
Small signal model of an *npn*-BJT derived from the basic transport version of the model shown in Figure 11.10.

the basic model by including the parasitic resistances and the capacitances to enable transient analysis and improve DC modeling accuracy. This updated EM1 model is often referred to as the EM2 BJT model and is described in the following section.

11.5.2 Enhancement of the Basic Model

11.5.2.1 Modeling Parasitic Circuit Elements

The basic BJT model is extended to improve DC modeling accuracy and enable transient simulation by including parasitic circuit elements in BJT modeling. The parasitic elements in BJTs include (1) the bulk-resistances r_e, r_b, and r_c of the neutral emitter, base, and collector regions, respectively; (2) EB-junction diffusion capacitance C_{DE} due to the diffusion of injected carriers from the emitter to the collector through the base and CB-junction diffusion capacitance C_{DC} due to the diffusion of injected carriers from the collector to the emitter through the base; and (3) junction capacitances C_{jE}, C_{jC}, and C_{sub} of the EB, CB, and collector-substrate pn-junctions, respectively. Different parasitic elements (except C_{DE} and C_{DC}) are shown in Figure 11.12a. The enhanced model is sometimes referred to as the EM2 model.

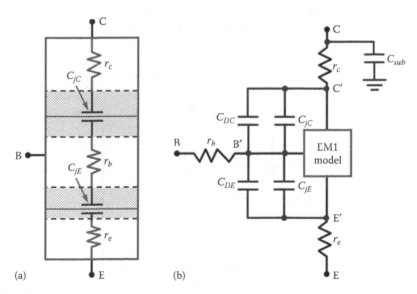

(a) (b)

FIGURE 11.12
The equivalent circuit of an enhanced vertical npn-BJT model: (a) parasitic elements in the ideal BJT structure and (b) addition of parasitic resistors and capacitors in the basic model to improve the DC modeling accuracy and transient modeling capability; E', B', and C' are the internal nodes of the emitter, base, and collector of the transistor, respectively; and C_{sub} is the collector-substrate pn-junction capacitance of the vertical npn-BJT structure.

With the consideration of the bulk resistances, the internal and terminal voltages are different due to the ohmic drop across the neutral bulk regions. Thus, including the ohmic bulk-resistors to improve DC characteristics and capacitors to model the charge storage effects in the basic model, we get the equivalent circuit of the model shown in Figure 11.12b.

Figure 11.12b shows the equivalent circuit of a vertical *npn*-BJT model to include its intrinsic parasitic resistive and capacitive elements in the basic EM1 model block. Replacing the EM1 model block in Figure 11.12b by the equivalent circuit of EM1 transport model shown in Figure 11.10, we get the revised compact *npn*-BJT model to account for the series resistance and charge storage effects in BJTs is shown in Figure 11.13.

From Equation 11.13, we can write the expressions for the currents flowing through the EB and CB *pn*-junctions in Figure 11.13 as

$$\frac{I_{CC}}{\beta_F} = \frac{I_S}{\beta_F}\left[\exp\left(\frac{V_{B'E'}}{v_{kT}}\right) - 1\right]$$

$$\frac{I_{EC}}{\beta_R} = \frac{I_S}{\beta_R}\left[\exp\left(\frac{V_{B'C'}}{v_{kT}}\right) - 1\right]$$

(11.33)

where:

$V_{B'E'} = V_{BE} - I_E r_e$ and $V_{B'C'} = V_{BC} - I_C r_c$ are the EB and CB internal voltages, respectively, as shown in Figures 11.13 and 11.14

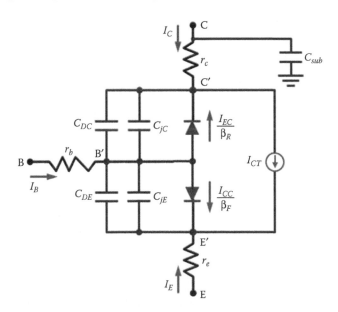

FIGURE 11.13

The equivalent circuit of the enhanced vertical *npn*-BJT model: parasitic resistances and capacitances are included to improve predictability of the simulation results.

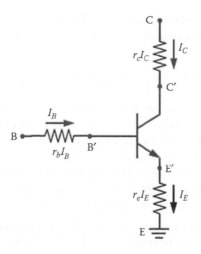

FIGURE 11.14
The parasitic series resistances on a typical *npn*-BJT device causing ohmic voltage drop at the respective neutral regions of the device.

Let us look at the impact of parasitic resistors shown in Figures 11.13 and 11.14 on the characteristics of a typical BJT device.

- *Effect of ohmic bulk collector resistor, r_c:* The collector series resistance r_c decreases the slope of I_C versus V_{CE} characteristics in the saturation region of BJT operation and improves the accuracy of modeling DC device characteristics as shown in Figure 11.15. Figure 11.15 shows (I_C, I_B) as a function of V_{CE} of an *npn*-BJT. As shown in Figure 11.15, r_c increases the transition voltage $V_{CE,sat}$ of BJTs. The typical value of r_c of modern IC BJTs is about 200 ohm.

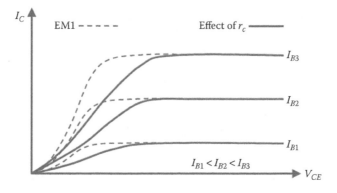

FIGURE 11.15
The effect of collector series resistance on the BJT device characteristics is to increase the saturation to linear region transition voltage $V_{CE,sat}$ of BJTs.

- *Effect of emitter series resistance,* r_e: The ohmic drop due to r_e reduces the EB-junction potential V_{BE} by a factor of $r_e I_E$ by the emitter current I_E so that

$$\Delta V_{BE} = I_E r_e = (I_C + I_B) r_e = I_B (1 + \beta_F) r_e \qquad (11.34)$$

From Equation 11.34 we find that r_e results in an equivalent base resistance of $(1 + \beta_F) r_e$. Since the emitter region is heavily doped ($\geq 1 \times 10^{19}$ cm^{-3}), r_e is negligibly small. However, due to the contact resistance at the emitter terminal and since $\beta_F \gg 1$, a typical value of r_e is about 5 ohm is obtained. Therefore, though the value of r_e is very small, it affects both I_C and I_B due to the voltage drop $\Delta V_{BE} = I_B (1 + \beta_F) r_e$ as shown in Figure 11.16.

- *Effect of base series resistance,* r_b: The base series resistance also reduces the EB-junction potential V_{BE} by a factor of $r_b I_B$ as shown in Figure 11.16. It effects the small signal and transient response of BJTs and difficult to measure accurately due to the dependence on r_e and operating point as shown in Figure 11.16.

- *Effect of junction capacitances*: The EB- and CB-junction capacitances per unit area C_{jE} and C_{jC}, respectively, model the incremental fixed charges stored in the EB- and CB-junction space-charge regions of BJTs due to the applied bias V_{BE} and V_{BC}, respectively. From Equation 2.139, we can write the expression for EB *pn*-junction capacitance in terms of internal node voltages as

$$C_{JE}(V_{BE}) = \frac{C_{jE0}}{\left[1 + (V_{B'E'}/\phi_{BE})\right]^{m_{jE}}} \qquad (11.35)$$

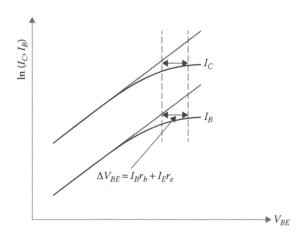

FIGURE 11.16
The saturation of I_C and I_B at higher values of V_{BE} due to the ohmic voltage drops at the base and emitter series resistances of BJT devices.

where:

C_{jE0} is the EB-junction capacitance per unit area at $V_{B'E'} = 0$
m_{jE} is the doping gradient coefficient
ϕ_{BE} is the EB-junction built-in potential that depends on the base doping concentration N_B and emitter doping concentration, N_E

We can show from Equation 2.109

$$\phi_{BE} = v_{kT} \ln\left(\frac{N_B N_E}{n_i^2}\right) \tag{11.36}$$

where:

$N_B = N_a$ (acceptor concentration)
$N_E = N_d$ (donor concentration) for *npn*-BJTs.

Similarly, the CB-junction capacitance due to V_{BC} is given by

$$C_{JC}(V_{BC}) = \frac{C_{jC0}}{\left[1 + \left(V_{B'C'}/\phi_{BC}\right)\right]^{m_{jC}}} \tag{11.37}$$

where:

C_{jC0} is the CB-junction capacitance per unit area at $V_{B'C'} = 0$
ϕ_{BC} is the CB built-in potential that depends on base doping concentration N_B and collector doping concentration, N_C

and is given by

$$\phi_{BC} = v_{kT} \ln\left(\frac{N_B N_C}{n_i^2}\right) \tag{11.38}$$

where:

$N_B = N_a$ (acceptor concentration)
$N_C = N_d$ (donor concentration) for *npn*-BJTs

• *Effect of diffusion capacitances*: The transition of injected minority carrier charge determines the speed of the transistor. The injected minority carriers from the emitter diffuse through the (1) EB-junction space-charge region, (2) neutral base region, and (3) CB-junction space-charge region. Thus, we consider three diffusion capacitances for forward injection and three for reverse injection.

Let us consider the capacitance effect due to the injected charge in the EB space-charge layer as shown in Figure 11.17. Let us define Q_E, Q_{BE}, Q_B, and Q_{BC} as the components of the total diffusion charge Q_{DE} in the emitter, EB-junction space-charge layer, neutral base, and CB-junction depletion regions, respectively. If τ_{Fdc} is the total

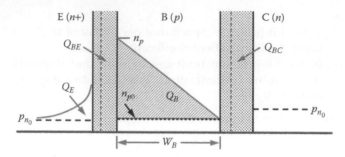

FIGURE 11.17
The components Q_E, Q_{BE}, Q_B, and Q_{BC} of the total diffusion charge Q_{DE} due to the forward injection of carriers at the EB-junction of an *npn*-BJT, resulting in the diffusion capacitance, C_{DE}; n_p is the injected electron concentration at the edge of the EB-depletion region inside the base; p_{no} and n_{po} are the equilibrium minority carrier concentrations in the n and p regions, respectively; and W_B is the width of the neutral base region.

forward transit time of the carriers to reach from the emitter to collector, then the total minority carrier charge due to forward current I_{CC} is given by

$$Q_{DE} = Q_E + Q_{BE} + Q_B + Q_{BC} = (\tau_E + \tau_{EB} + \tau_B + \tau_{CB})I_{CC} = \tau_{Fdc}I_{CC} \qquad (11.39)$$

where:
τ_{Fdc} is the total forward delay time consisting of emitter delay τ_E
EB space charge layer transit time τ_{EB}
base transit time τ_B
CB-space charge layer transit time τ_{CB}

From Equation 11.39, we can write the expression for the diffusion capacitance $C_{DE}(V_{BE})$ due to the applied bias V_{BE}

$$C_{DE}\left(V_{BE}\right) = \frac{Q_{DE}}{V_{B'E'}} = \frac{\tau_{Fdc}I_{CC}}{V_{B'E'}} \qquad (11.40)$$

The base transit time τ_B is the major contributor of total transistor delay time τ_{Fdc} given in Equation 11.39. Thus, τ_B is the most critical parameter to determine the speed of BJTs. In the absence of built-in electric fields in the base (i.e., constant N_a) with low-level injection, the injected *electron* concentration n_p varies linearly across the base from n_p to $n_{po} \approx 0$ as shown in Figure 11.17. Therefore, for low-level injection and uniformly doped base region, the total electron charge in the base is simply given by

$$Q_B = \frac{1}{2}qW_Bn_pA_E \qquad (11.41)$$

where:

A_E = emitter area

n_p is the injected electron concentration at the edge of EB-junction depletion region as shown in Figure 11.17

W_B is the width of the neutral base region

Then the minority carrier transit time across the base is given by

$$\tau_B = \frac{Q_B}{I_{CC}} \tag{11.42}$$

Since the built-in electric field within base is assumed to be negligible, from the Fick's first law of diffusion, we can show that the electron diffusion current (Equation 2.40) is

$$I_{CC} = qA_ED_n\frac{dn_p}{dx} \tag{11.43}$$

where:

D_n is the average electron diffusivity in the p-type base region of an npn-BJT

Assuming the equilibrium electron concentration, $n_{p0} \ll n_p$, we can express Equation 11.43 as

$$I_{CC} \cong qA_ED_n\frac{n_p}{W_B} \tag{11.44}$$

Now, substituting for Q_B and I_{CC} from Equations 11.41 and 11.44, respectively, in Equation 11.42, we get the expression for base transit time as

$$\tau_B = \frac{W_B^2}{2D_n} \tag{11.45}$$

To understand the importance of τ_B in determining the speed of BJTs, let us consider a vertical npn-BJT with $W_B = 1$ μm and lightly doped base so that $D_n \cong 38$ cm^2 sec^{-1}. Then from Equation 11.45, we find that the value of $\tau_B \cong 132$ psec for a uniformly doped base region. In reality, the base doping is graded, and therefore, an aiding electric field speeds up the carrier transit through the base. As a result, τ_B is further reduced. Also, in order to maintain the charge neutrality under high-level injection, the hole concentration in the base has a gradient similar to the electron gradient. This sets up an electron field, which also speeds up the electron transit through the base. Thus, τ_B is not the dominant frequency limitation in advanced IC BJTs.

Similarly, we can derive the expression for the reverse diffusion capacitance C_{DC} and transit time τ_{Rdc} with reference to Figure 11.18.

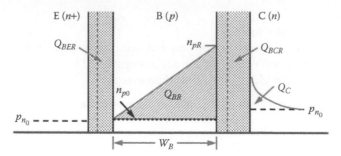

FIGURE 11.18
The components Q_C, Q_{BCR}, Q_{BR}, and Q_{BER} of the total diffusion charge Q_{DC} due to the reverse injection of carriers at the CB-junction of an *npn*-BJT resulting in the diffusion capacitance, C_{DC}; n_{pR} is the injected electron concentration at the edge of the CB-depletion region inside the base; p_{n_0} and n_{p_0} are the equilibrium minority carrier concentrations in the *n* and *p* regions, respectively; and W_B is the width of the neutral base region.

From Figure 11.18, the total minority carrier charge Q_{DC} due to the reverse current I_{EC} is given by

$$Q_{DC} = Q_C + Q_{BCR} + Q_{BR} + Q_{BER} = (\tau_C + \tau_{CB} + \tau_{BR} + \tau_{EB})I_{EC} = \tau_{Rdc}I_{EC} \quad (11.46)$$

where:

Q_C, Q_{BCR}, Q_{BR}, and Q_{BER} are the reverse-injected charge in the collector, CB-depletion, base, and EB-depletion regions, respectively

the total reverse delay time τ_{Rdc} consists of the collector delay τ_C, CB-junction reverse space-charge layer transit time τ_{CB}, reverse base transit time τ_{BR}, and reverse EB-junction space-charge layer transit time τ_{EB}

Therefore, the reverse diffusion capacitance due to CB applied bias V_{BC} is given by

$$C_{DC}(V_{BC}) = \frac{Q_{DC}}{V_{B'C'}} = \frac{\tau_{Rdc}I_{EC}}{V_{B'C'}} \quad (11.47)$$

11.5.2.2 Limitations of Basic Model

The enhanced BJT model includes parasitic elements in BJT structure to improve DC modeling accuracy and offers capability for transient analysis. However, the model is derived on the assumptions that (1) there is no recombination of minority carriers in the EB and CB *pn*-junction space-charge regions; (2) current gain β is independent of I_C; and (3) the neutral base width W_B is independent of applied bias V_{BC}, that is, no space-charge widening and *base-width modulation*. However, experimental data show β-degradation at the low values of EB-junction bias or low values of I_C, and β-roll-off at high

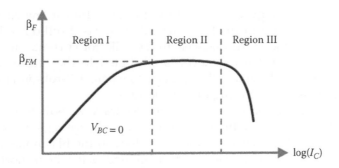

FIGURE 11.19
Forward current gain as a function of collector current of an *npn*-BJT showing three different $\beta - I_C$ regions: region I shows β-degradation in the low current level, region II shows the constant maximum current gain β_{FM}, and region III shows β roll-off at high current level.

current conditions as shown in Figure 11.19. Figure 11.19 shows $\beta_F - I_C$ plot at $V_{BC} = 0$. For the simplicity of discussions, we assume that the ohmic drop due to the parasitic resistances is negligibly small so that $V_{BE} \cong V_{B'E'}$ and $V_{BC} \cong V_{B'C'}$.

In order to understand the underlying physical mechanisms of $\beta_F - I_C$ plot shown in Figure 11.19, let us plot $\ln(I_C, I_B)$ as a function of V_{BE} at $V_{BC} = 0$ as shown in Figure 11.20. Figure 11.20 is often referred to as the *Gummel plot*.

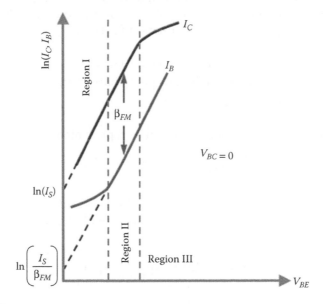

FIGURE 11.20
A typical Gummel plot of an *npn*-BJT: I_C and I_B versus V_{BE} at $V_{BC} = 0$ showing β-degradation at the low current level and β-roll-off at the high collector current level; β_{FM} is the maximum value of current gain.

It is observed from Figure 11.20 that the β-degradation at low current level is due to the increase in the base current, whereas β-roll-off at high current condition is due to the decrease in the collector current. The observed β-degradation at low current level is attributed to the recombination of injected carriers in the space-charge regions whereas β-roll-off at high current condition is due to the high-level injection.

Thus, in order to improve the simulation accuracy, we will develop models for minority carrier recombination in the space-charge regions, base-width modulation, and high-level injection to include in the BJT model described in Figure 11.13. First of all, we will develop models for the minority carrier recombination in the EB and CB *pn*-junction depletion regions and include in the model shown in Figure 11.13. Finally, we will include models for the base-width modulation and high-level injection and present a complete compact BJT model for circuit CAD.

11.5.3 Modeling Carrier Recombination in the Depletion Regions

In Region I of Figure 11.20, the increase in I_B is due to the minority carrier recombination in the EB and CB *pn*-junction depletion regions. For the simplicity of analysis, we assume a vertical *npn*-BJT in the normal active mode of operation and neglect the ohmic-bulk resistors (r_e, r_b, r_c) so that $V_{BE} = V_{B'E'}$ and $V_{BC} = V_{B'C'}$.

Let us consider the effect of the minority carrier recombination in the EB-depletion region only by setting $V_{BC} = 0$. In Region I, the decrease in β can be modeled by additional components of I_B from

- Carrier recombination at the surface, I_B(surface)
- Carrier recombination in the EB space-charge layer, I_B(EB-scl)
- EB surface channels, I_B(channel)

Thus, the overall excess base current can be represented by

$$\Delta I_B(\text{total}) = I_B(\text{surface}) + I_B(\text{EB-scl}) + I_B(\text{channel}) \tag{11.48}$$

In Equation 11.48, ΔI_B can be represented by an additional nonideal EB *pn*-junction in the model shown in Figure 11.13 with diode current given by

$$\Delta I_B = C_2 I_S(0)\left[\exp\left(\frac{V_{BE}}{n_E v_{kT}}\right) - 1\right] \tag{11.49}$$

where:
  n_E is the low-current forward emission coefficient (~2)
  C_2 models the various components of I_S in the low I_B regime

Here, $I_S(0)$ is the revised reverse saturation current of the transistor. Thus, combining Equations 11.20 and 11.49, the forward diode current can be modeled by

$$I_B = \frac{I_S(0)}{\beta_{FM}}\left[\exp\left(\frac{V_{BE}}{v_{kT}}\right) - 1\right] + C_2 I_S(0)\left[\exp\left(\frac{V_{BE}}{n_E v_{kT}}\right) - 1\right] \qquad (11.50)$$

Similarly, we can model the minority carrier recombination in the CB-depletion region only by another nonideal CB *pn*-junction by setting $V_{BE} = 0$ in the *inverse mode* of BJT operation with $V_{BC} > 0$. Two additional parameters used to model the components of I_B are low-current inverse emission coefficient n_C (~2) and component of I_S in the inverse region C_4. Thus, ΔI_B in the inverse region is through the nonideal CB-diode is given by

$$\Delta I_{BR} = C_4 I_S(0)\left[\exp\left(\frac{V_{BC}}{n_C v_{kT}}\right) - 1\right] \qquad (11.51)$$

Thus, the general expression for total I_B to model β-degradation in the low I_C region is given by

$$
\begin{aligned}
I_B &= \frac{I_S(0)}{\beta_{FM}}\left[\exp\left(\frac{V_{BE}}{v_{kT}}\right) - 1\right] + C_2 I_S(0)\left[\exp\left(\frac{V_{BE}}{n_E v_{kT}}\right) - 1\right] \\
&+ \frac{I_S(0)}{\beta_{RM}}\left[\exp\left(\frac{V_{BC}}{v_{kT}}\right) - 1\right] + C_4 I_S(0)\left[\exp\left(\frac{V_{BC}}{n_C v_{kT}}\right) - 1\right]
\end{aligned} \qquad (11.52)
$$

Note that in Equation 11.52, we have used maximum current gain (β_{FM}, β_{RM}) to model the EB and CB pn-junctions and included the excess I_B to account for the β-degradation in the low current level. Then the corresponding equivalent circuit for enhanced nonlinear hybrid *npn*-BJT model predicting β versus I_C at low current level can be represented by Figure 11.21.

Note that the series resistances (r_e, r_b, r_c) do not effect theoretical analysis and model equations. However, one should replace measured V_{BE} and V_{BC} with the internal voltages to account for the ohmic drops in the model equations.

In order to develop a complete compact BJT model for circuit CAD, we now include the effect of base-width modulation referred to as the *early effect* [8] and high-level injection to model β roll-off in the high current level shown in Figure 11.19 in the core model shown in Figure 11.21. In this effort, we will closely follow the *unified charge-control* model developed by Gummel and Poon [6].

11.5.4 Modeling Base-Width Modulation and High-Level Injection

The base-width modulation describes the change in the quasi-neutral base-region W_B due to the change in the reverse bias V_{BC} in the *normal active* mode

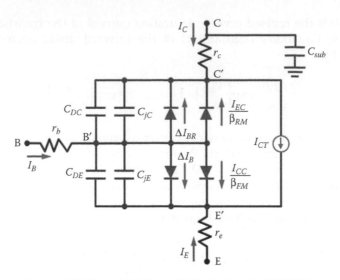

FIGURE 11.21
The equivalent circuit of the enhanced vertical *npn*-BJT model to model β-degradation in the low current level using two nonideal diodes to account for increased base current due to recombination in the depletion regions. $\Delta I_B = C_2 I_S(0) \left[\exp\left(V_{BE}/n_E v_{kT}\right) - 1 \right]$ is the excess forward base current due to the nonideal EB diodes and $\Delta I_{BR} = C_4 I_S(0) \left[\exp\left(V_{BC}/n_C v_{kT}\right) - 1 \right]$ is the excess reverse base current due to the nonideal CB diodes.

(or V_{BE} in the *inverse* mode). In the normal active mode of BJT operation, EB-junction is forward biased and CB-junction is reverse biased. As a result, the depletion width $X_d = f(V_{BC})$ and the neutral base width W_B as shown in Figure 11.22 change significantly with V_{BC}. This base-width modulation is originally reported by J. Early in 1952 and is called the *early effect* [8]. The concept is now used in MOSFET device characterization as discussed in Chapters 4 and 5. The early effect changes $I_C - V_{CE}$ characteristics of BJTs

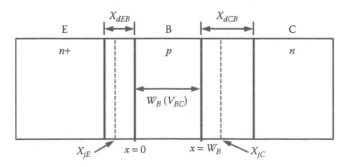

FIGURE 11.22
An idea *npn*-BJT device structure for modeling the early effect: X_{dEB} and x_{dCB} are the depletion widths of EB and CB *pn*-junction space-charge regions, respectively; $W(V_{BC})$ is the bias-dependent base width; $x = 0$ is the origin of *x*-axis at the edge of EB-junction depletion inside the base; and $x = W_B$ is at the edge of CB-junction depletion region inside the base.

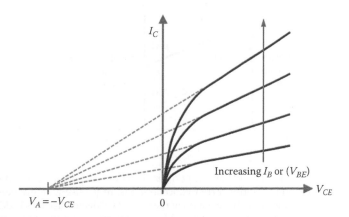

FIGURE 11.23

I_C versus V_{CE} characteristics of a typical *npn*-BJT for different values of I_B or V_{BE}; the increase in I_C is caused by base-width modulation at higher V_{BC}; V_A is defined at the intercept of $I_C - V_{CE}$ curve interpolated to $I_C = 0$.

significantly and, therefore, must be modeled for accurate simulation of BJTs in circuit CAD.

As the reverse bias V_{BC} across the BC-junction X_{jC} increases, BC-junction depletion-layer width X_{dCB} increases, resulting in a decrease in W_B. Due to the decrease in W_B, the injected minority carrier electron concentration gradient (dn/dx) increases. Then from Fick's first law of diffusion, I_C increases with V_{CE} as shown in Figure 11.23. The base-width modulation is modeled by two parameters called the forward early voltage (V_{AF}) and reverse early voltage (V_{AR}). Due to the early effect, the BJT device parameters I_S, β_F, and τ_F strongly depend on V_{BC} and V_A [5].

In order to derive the unified SGP charge control model [6] for base-width modulation and high-level injection, let us make the following simplifying assumptions:

Assumption 1: one-dimensional current equations hold.

Assumption 2: *npn*-BJT with EB-junction is forward biased and CB-junction is reverse biased.

Assumption 3: depletion approximation holds, that is, no mobile charge inside the depletion region.

Assumption 4: BJT gain is high, that is, $I_B \cong 0$.

Assumption 5: neglect ohmic-bulk resistors (r_e, r_b, r_c); that is, $V_{BE} = V_{B'E'}$ and $V_{BC} = V_{B'C'}$.

With the above simplifying assumptions, let us consider the 2D cross section of an *npn*-BJT shown in Figure 11.24.

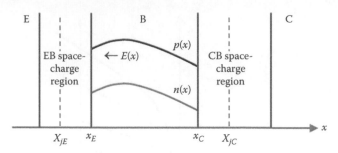

FIGURE 11.24
2D cross section of an ideal *npn*-BJT showing the *pn*-junctions and depletion regions: $p(x)$ and $n(x)$ are the majority and minority carrier concentrations, respectively, at any point x in the neutral base region; x_E and x_C are the position of the EB and CB depletion edges inside the base, respectively; and $E(x)$ is the built-in electric field from collector to emitter due to the nonuniform *p*-type base doping profile.

Using assumption 1, we can use 1D expression for the electron and hole current densities (Equations 2.76 and 2.77) as

$$J_n = q\mu_n n(x)E(x) + qD_n \frac{dn}{dx} \quad \text{(electrons)}$$

$$J_p = q\mu_p p(x)E(x) - qD_p \frac{dp}{dx} \quad \text{(holes)} \tag{11.53}$$

Now from assumption 4, for a high-gain *npn*-BJT, $I_B \cong 0$, that is, the hole current $\cong 0$; then from the hole current expression J_p in Equation 11.53, we get

$$0 = q\mu_p p(x)E(x) - qD_p \frac{dp}{dx} \tag{11.54}$$

After simplification of Equation 11.54 and using Equation 2.42, we can show that the built-in electric field $E(x)$ due to nonuniform base doping of *npn*-BJTs is given by (Equation 2.46)

$$E(x) = \frac{D_p}{\mu_p} \frac{1}{p(x)} \frac{dp}{dx} = v_{kT} \frac{1}{p(x)} \frac{dp}{dx} \tag{11.55}$$

The direction of the electric field $E(x)$ in Equation 11.55 aids the electron flow from the emitter to collector and retards the electron flow from the collector to emitter. Now, the flow of electrons from the emitter to collector is given by the electron current expression J_n in Equation 11.53. Then, substituting for $E(x)$ from Equation 11.55 to Equation 11.53 we get

$$J_n = q\mu_n n(x) \left[v_{kT} \frac{1}{p(x)} \frac{dp(x)}{dx} \right] + qD_n \frac{dn}{dx} \tag{11.56}$$

Using $\mu_n v_{kT} = D_n$ from Einstein's relation [Equation 2.42], we can express Equation 11.56 after simplification as

$$J_n = \frac{qD_n}{p(x)} \frac{d}{dx}\left(n(x)p(x)\right) \tag{11.57}$$

We integrate Equation 11.57 over the neutral base width W_B from $x = x_E$ to $x = x_C$ as shown in Figure 11.24 to get

$$\int_{x_E}^{x_C} J_n p(x)dx = q \int_{x_E}^{x_C} D_n d\left[n(x)p(x)\right] \tag{11.58}$$

Since the same collector current density J_n is flowing through the BJT, J_n is a constant. Then assuming D_n is a constant, we can show from Equation 11.58

$$J_n = \frac{qD_n\left[pn(x_C) - pn(x_E)\right]}{\int_{x_E}^{x_C} p(x)dx} \tag{11.59}$$

From Equation 11.59, we find that the electron current, that is, collector current density, depends on pn-products at the edges of the depletion regions of EB and CB pn-junctions inside the base and the integrated base doping in the denominator of Equation 11.59. Again, from pn-junction analysis (Equation 2.114), we can show that the pn-products at the edges of the collector and emitter depletion regions are

$$pn(x_C) = n_i^2 \exp\left(\frac{V_{BC}}{v_{kT}}\right)$$
$$pn(x_E) = n_i^2 \exp\left(\frac{V_{BE}}{v_{kT}}\right) \tag{11.60}$$

Now, substituting for pn-products from Equation 11.60 to Equation 11.59, we can show

$$J_n = \frac{qD_n n_i^2\left[\exp\left(V_{BC}/v_{kT}\right) - \exp\left(V_{BE}/v_{kT}\right)\right]}{\int_{x_E}^{x_C} p(x)dx} \tag{11.61}$$

If A_E is cross-sectional area of the emitter, then from Equation 11.61 we can show

$$I_n = \frac{qA_ED_nn_i^2\left[\exp(V_{BC}/v_{kT})-\exp(V_{BE}/v_{kT})\right]}{\displaystyle\int_{x_E}^{x_C}p(x)dx}$$

$$= -\frac{qA_ED_nn_i^2}{\displaystyle\int_{x_E}^{x_C}p(x)dx}\left\{\left[\exp\left(\frac{V_{BE}}{v_{kT}}\right)-1\right]-\left[\exp\left(\frac{V_{BC}}{v_{kT}}\right)-1\right]\right\}$$

(11.62)

where:
   I_n is the total DC current from the emitter to base in the positive *x*-direction due to the minority carrier electrons

Since at *low-level injection* $p(x) \cong N_a(x)$, in the neutral base region, $x_E \leq x \leq x_C$ as shown in Figure 11.24; then by replacing the injected $p(x)$ with the majority carrier concentration, we can write Equation 11.62 as

$$I_{n(low\text{-}level)} = -\frac{qA_ED_nn_i^2}{\displaystyle\int_{x_E}^{x_C}N_a(x)dx}\left\{\left[\exp\left(\frac{V_{BE}}{v_{kT}}\right)-1\right]-\left[\exp\left(\frac{V_{BC}}{v_{kT}}\right)-1\right]\right\}$$

(11.63)

We have shown that the current source for the basic BJT model (Equation 11.19) is given by

$$I_{CT} = (I_{CC}-I_{EC}) = I_S\left\{\left[\exp\left(\frac{V_{BE}}{v_{kT}}\right)-1\right]-\left[\exp\left(\frac{V_{BC}}{v_{kT}}\right)-1\right]\right\}$$

(11.64)

Therefore, comparing Equations 11.63 and 11.64, we can write for low-level injection

$$I_{CT(low\text{-}level)} = I_{SS}\left\{\left[\exp\left(\frac{V_{BE}}{v_{kT}}\right)-1\right]-\left[\exp\left(\frac{V_{BC}}{v_{kT}}\right)-1\right]\right\}$$

(11.65)

where:
   I_{SS} is the saturation leakage current at $V_{BE} = V_{BC} = 0$

and is given by

$$I_{SS} = -\frac{qA_ED_nn_i^2}{\displaystyle\int_{x_{E0}}^{x_{C0}}N_a(x)dx}$$

(11.66)

where:

X_{E0} and X_{C0} are the locations inside the neutral base region at the edge of the EB- and CB-junction depletion regions, respectively without the applied bias

Thus, I_{SS} defined in Equation 11.66 is a bias-independent fundamental constant. Since negative sign indicates the direction of collector current flowing out of the device terminal, we have omitted the negative sign in the above Equation 11.65. Now, Equation 11.63 can be expressed as

$$
I_{CT} = \frac{qA_E D_n n_i^2}{\displaystyle\int_{x_E}^{x_C} p(x)dx} \cdot \left(\frac{qA_E \displaystyle\int_{x_{E0}}^{x_{C0}} N_a(x)dx}{qA_E \displaystyle\int_{x_{E0}}^{x_{C0}} N_a(x)dx} \right) \times
$$

$$
\left\{ \left[\exp\left(\frac{V_{BE}}{v_{kT}}\right) - 1 \right] - \left[\exp\left(\frac{V_{BC}}{v_{kT}}\right) - 1 \right] \right\}
$$

$$
= \frac{qA_E D_n n_i^2}{\displaystyle\int_{x_{E0}}^{x_{C0}} N_a(x)dx} \cdot \left(\frac{qA_E \displaystyle\int_{x_{E0}}^{x_{C0}} N_a(x)dx}{qA_E \displaystyle\int_{x_E}^{x_C} p(x)dx} \right) \times
$$

$$
\left\{ \left[\exp\left(\frac{V_{BE}}{v_{kT}}\right) - 1 \right] - \left[\exp\left(\frac{V_{BC}}{v_{kT}}\right) - 1 \right] \right\}
$$

(11.67)

If we define Q_B and Q_{B0} as the neutral base charges with and without the applied biases, respectively, then we can show

$$
Q_{B0} = qA_E \int_{x_{E0}}^{x_{C0}} N_a(x)dx
$$

$$
Q_B = qA_E \int_{x_{E(V_{BE})}}^{x_C(V_{BC})} N_a(x)dx
$$

(11.68)

Now, using Equation 11.68 in Equation 11.67, we can show that the expression for current source for an *npn*-BJT is given by

$$
I_{CT} = I_{SS} \frac{Q_{B0}}{Q_B} \left\{ \left[\exp\left(\frac{V_{BE}}{v_{kT}}\right) - 1 \right] - \left[\exp\left(\frac{V_{BC}}{v_{kT}}\right) - 1 \right] \right\}
$$

$$
= \frac{I_{SS}}{q_b} \left\{ \left[\exp\left(\frac{V_{BE}}{v_{kT}}\right) - 1 \right] - \left[\exp\left(\frac{V_{BC}}{v_{kT}}\right) - 1 \right] \right\}
$$

(11.69)

In Equation 11.69, q_b is the normalized base charge and is defined as

$$q_b = \frac{Q_B}{Q_{B0}}$$
(11.70)

Equation 11.69 is the generalized expression for current source at all injection levels, where I_{SS} is a fundamental constant @ $V_{BE} = V_{BC} = 0$ and is given by Equation 11.66. The normalized majority carrier charge q_b in the neutral base region accurately models the base-width modulation. In the next section, we will express q_b in terms of bias-dependent measurable model parameters.

11.5.4.1 Components of Injected Base Charge

In order to evaluate q_b, we first determine the components of Q_B. For the simplicity of Q_B analysis, we assume that

- *npn*-BJT is in *saturation*, that is, $V_{BE} > 0$ and $V_{BC} > 0$, then
  - Minority carriers are injected into the base both from the *emitter* and from the *collector*
  - From the charge neutrality, the total increase in the majority carriers in the *base* = total increase in the minority carrier concentration
- *Superposition* of carriers in different regions holds, that is
  - Total *excess* majority carrier density = sum of the *excess* majority carrier density due to each junction separately
  - Excess majority carrier concentration in the base = excess carriers due to the forward voltage ($V_{BE} + V_{BC}$)
  - Depletion approximation holds

With the above assumptions, we define

- $p_F(x)$ as the majority carrier concentration at any point x in the base at $V_{BC} = 0$
- $p_R(x)$ as the majority carrier concentration at any point x in the base at $V_{BE} = 0$
- $N_a(x)$ as the base doping concentration at any point x

Then the excess majority carrier concentration in the base region is shown in Figure 11.25 and is given by

$$p'(x) = \left[p_F(x) - N_a(x) \right] + \left[p_R(x) - N_a(x) \right]$$
(11.71)

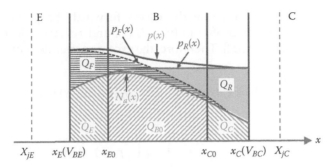

FIGURE 11.25
The components of base charge in an *npn*-BJT in saturation: Q_{B0} is the base charge in the neutral base region without applied biases; Q_E and Q_C are the increase in the base charge due to the EB and CB forward biases, respectively; Q_F and Q_R are the excess charge due to the high-level injection from the emitter and collector, respectively, to the base; $p_F(x)$ and $p_R(x)$ are the excess majority carrier concentration in the base region from emitter and collector, respectively

From Equation 11.68, the total majority carrier charge in the base consists of *equilibrium charge* due to N_a and *excess components* due to $p'(x)$ and is given by

$$Q_B = qA_E \int_{x_E(V_{BE})}^{x_C(V_{BC})} p(x)dx = \int_{x_E(V_{BE})}^{x_C(V_{BC})} qA_E N_a(x)dx + \int_{x_E(V_{BE})}^{x_C(V_{BC})} qA_E p'(x)(x)dx \quad (11.72)$$

The equilibrium base charge includes three components: Q_E due to the decrease in the depletion region by the forward-biased EB-junction space charge region, Q_{B0} due to the neutral base charge, and Q_C due to the decrease in the depletion charge by the forward-biased BC-junction depletion region as shown in Figure 11.25. Therefore, the total equilibrium component of the base charge under the saturation condition is given by

$$\int_{x_E(V_{BE})}^{x_C(V_{BC})} qA_E N_a(x)dx = Q_E + Q_{B0} + Q_C$$

$$= \int_{x_E(V_{BE})}^{x_{E0}} qA_E N_a(x)dx + \int_{x_{E0}}^{x_{C0}} qA_E N_a(x)dx + \int_{x_{C0}}^{x_C(V_{BC})} qA_E N_a(x)dx \quad (11.73)$$

Therefore, from Equations 11.72 and 11.73, the total base charge due to EB and CB-junction forward biases is given by

$$Q_B = \int_{x_E(V_{BE})}^{x_{E0}} qA_E N_a(x)dx + \int_{x_{E0}}^{x_{C0}} qA_E N_a(x)dx$$

$$+ \int_{x_{C0}}^{x_C(V_{BC})} qA_E N_a(x)dx + \int_{x_E(V_{BE})}^{x_C(V_{BC})} qA_E p'(x)dx \quad (11.74)$$

From Figure 11.25, we find that the excess forward charge is due to the forward injection carrier profile $[p_F(x) - N_a(x)]$ and reverse injection carrier density profile $[p_R(x) - N_a(x)]$. Therefore, the excess carrier density is given by

$$\int_{x_E(V_{BE})}^{x_C(V_{BC})} qA_E p'(x)dx = \int_{x_E(V_{BE})}^{x_C(V_{BC})} qA_E \left[p_F(x) - N_a(x) \right] dx$$

$$+ \int_{x_E(V_{BE})}^{x_C(V_{BC})} qA_E \left[p_R(x) - N_a(x) \right] dx \qquad (11.75)$$

$$= Q_F + Q_R$$

Therefore, the total base charge due to the applied EB- and CB-junction biases is given by

$$Q_B = Q_E + Q_{B0} + Q_C + Q_F + Q_R \qquad (11.76)$$

where:
Q_{B0} is the charge in the neutral base region at $V_{BE} = 0 = V_{BC}$
Q_E is the increase in Q_B under V_{BE} and is only a mathematical entity
Q_C is the increase in Q_B under V_{BC} and is only a mathematical entity
Q_F is the excess majority charge in the forward biased-device with $V_{BC} = 0$. It
     is only a mathematical entity and important under high level injection
Q_R is the excess majority charge in the forward biased-device with $V_{BE} = 0$. It
     is only a mathematical entity and important under *high-level injection*

It is clear from Figure 11.25 that $p_F(x) > N_a(x)$ and $p_R(x) > N_a(x)$, that is, Q_F and Q_R represent high-level injection. Then from Equation 11.76, we get the normalized components of base charge as

$$\frac{Q_B}{Q_{B0}} = \frac{Q_E}{Q_{B0}} + \frac{Q_{B0}}{Q_{B0}} + \frac{Q_C}{Q_{B0}} + \frac{Q_F}{Q_{B0}} + \frac{Q_R}{Q_{B0}} \qquad (11.77)$$

After simplification, we can show for the normalized base charge from Equation 11.77

$$q_b = 1 + q_e + q_c + q_f + q_r \qquad (11.78)$$

where:
$q_e = Q_E/Q_{B0}$, $q_c = Q_C/Q_{B0}$, $q_f = Q_F/Q_{B0}$, and $q_e = Q_R/Q_{B0}$ are the respective normalized components of base charge

In order to develop BJT compact model, each component of the normalized base charge is expressed in terms of measurable device model parameters. Next, we will evaluate each component of q_b given in Equation 11.78.

Evaluation of q_e: We defined Q_E as the increase in the majority carrier charge due to forward bias V_{BE}. Therefore, we can express

$$Q_E = \int_0^{V_{BE}} C_{jE}(V)dV \tag{11.79}$$

and,

$$q_e = \frac{1}{Q_{B0}} \int_0^{V_{BE}} C_{jE}(V)dV \tag{11.80}$$

For the simplicity of modeling, we consider an average value $\overline{C}_{jE}$ over the operating range of V_{BE}. Then from Equation 11.80, we get

$$q_e = \frac{\overline{C}_{jE}}{Q_{B0}} V_{BE} \equiv \frac{V_{BE}}{V_{AR}} \tag{11.81}$$

where:
V_{AR} is a model parameter that defines the effect of base-width modulation due to V_{BE} and the parameter V_{AR} is called the inverse early voltage

From Equation 11.81, we get

$$V_{AR} = \frac{Q_{B0}}{\overline{C}_{jE}} \tag{11.82}$$

However, for accurate modeling of q_e, C_{jE} must be integrated over the operating bias range so that

$$V_{AR} = \frac{Q_{B0}}{(1/V_{BE}) \int_0^{V_{BE}} C_{jE}(V)dV} \tag{11.83}$$

In Equation 11.81, V_{AR} models the base-width modulation due to the variation of BE-junction depletion layer under V_{BE} and is the inverse of the forward early voltage due to V_{BC} under the normal mode of BJT operation.

In Equation 11.82, a constant V_{AR} implies that C_{jE} is a constant, independent of V_{BE}. We observe from Figure 11.25 that $Q_E \ll Q_{B0}$, resulting in $q_e \ll 1$. Thus, q_e is not a dominant component of q_b. Therefore, using a constant C_{jE} to calculate q_e from Equation 11.81 is justified. However, a constant V_{AR} may cause a large error in q_e estimation, especially at $V_{BE} > 0$. The error in Equation 11.81 due to q_e for $V_{BE} > 0$ can be eliminated by integrating C_{jE} over the operating bias range and extracting V_{AR} from the slope of $\ln(I_C)$ versus V_{BE}/v_{kT} plot.

Now, in order to determine the effect of q_e on BJT device performance, we set: $q_c = q_r = q_f = 0$; then from Equation 11.78, we have $q_b = 1 + q_e$. Now, substituting for $q_b = (1 + q_e)$ in Equation 11.69, we get

$$I_C = I_{CT} = \frac{I_{SS}}{(1+q_e)}\left[\exp\left(\frac{V_{BE}}{v_{kT}}\right)-1\right] \tag{11.84}$$

We can calculate the slope of I_C versus V_{BE} plot by differentiating Equation 11.84 as

$$\frac{dI_C}{dV_{BE}} = \frac{I_C}{v_{kT}}\left[1-v_{kT}\frac{C_{jE}(V_{BE})}{q(1+q_e)Q_{B0}}\right]$$

or,

$$v_{kT}\frac{1}{I_C}\frac{dI_C}{dV_{BE}} = \frac{d(\ln(I_C))}{d(V_{BE}/v_{kT})} = \left[1-v_{kT}\frac{C_{jE}(V_{BE})}{q(1+q_e)Q_{B0}}\right] \tag{11.85}$$

The left-hand side of Equation 11.85 is the slope of $\ln(I_C)$ versus V_{BE}/v_{kT} plot and is given by

$$\frac{1}{n_E} = v_{kT}\frac{1}{I_C}\frac{dI_C}{dV_{BE}}\bigg|_{V_{BC}=0} = \left[1-v_{kT}\frac{C_{jE}(V_{BE})}{q(1+q_e)Q_{B0}}\right] \tag{11.86}$$

Thus,

$$n_E = \frac{1}{1-v_{kT}\left[C_{jE}(V_{BE})/q(1+q_e)Q_{B0}\right]} \tag{11.87}$$

Considering a constant average $\overline{C}_{jE} = Q_{B0}/V_{AR}$ from Equation 11.82, and $q_e = V_{BE}/V_{AR}$ from Equation 11.81, we can express Equation 11.87 as

$$n_E \cong \frac{1}{1-\left[v_{kT}/(V_{AR}+V_{BE})\right]} \tag{11.88}$$

The slope n_E is called the forward emission coefficient and is obtained from the $I_C - V_{BE}$ characteristics of BJTs at $V_{BC} = 0$. Since v_{kT}, V_{AR}, and V_{BE} are finite positive numbers, it is clear from Equation 11.88 that $n_E > 1$.

Evaluation of q_c: The parameter q_c models the base-width modulation due to the applied CB-junction voltage V_{BC} at the low current level during the forward active mode of a BJT operation and describes the forward early effect. Using the same procedure used for q_e, we can show

$$q_c = \frac{1}{Q_{B0}} \int_0^{V_{BC}} C_{jC}(V)dV \tag{11.89}$$

Again, considering a constant value of $\overline{C}_{jC}$ as the average value of BC-junction capacitance over the operating range of V_{BC}, we get from Equation 11.89

$$q_c = \frac{\overline{C}_{jC}}{Q_{B0}} V_{BC} \equiv \frac{V_{BC}}{V_{AF}} \tag{11.90}$$

where:
  V_{AF} is a model parameter that defines the effect of base-width modulation when BJTs operate in the forward active mode and the parameter V_{AF} is called the forward early voltage

Thus, from Equation 11.90, V_{AF} is defined as

$$V_{AF} = \frac{Q_{B0}}{\overline{C}_{jC}} \tag{11.91}$$

However, the accurate modeling of q_c for $V_{BC} > 0$ is achieved by integrating C_{jC} over the operating bias range so that

$$V_{AF} = \frac{Q_{B0}}{(1/V_{BC}) \int_0^{V_{BC}} C_{jC}(V)dV} \tag{11.92}$$

The parameter V_{AF} models the base-width modulation due to the variation in the CB-junction depletion layer with applied bias V_{BC}. In Equation 11.91, a constant V_{AF} implies that C_{jC} is a constant independent of V_{BC}. This constant C_{jC} is justified in the normal active model of BJT operation when CB-junction is reverse biased; that is, $V_{CB} < 1$. However, using a constant V_{AF} may cause a large error in estimating q_c, when CB-junction is forward biased; that is, the device is in the inverse region or saturation region. In these regions, a more accurate expression for q_c is required for accurate modeling of early voltage.

The effect of q_c on BJT device performance in the normal active region of operation is the finite output conductance g_o. In order to determine the effect of g_o accurately, we set $q_e = q_r = q_f = 0$, so that $q_b = 1 + q_c$. Then neglecting the bulk-ohmic resistances, we get from Equation 11.69 in the normal active region, we have

$$I_C \equiv I_{CT} = \frac{I_{SS}}{(1+q_c)}\left[\exp\left(\frac{V_{BE}}{v_{kT}}\right)-1\right]$$

$$= \frac{I_{SS}}{1+\left(V_{BC}/V_{AF}\right)}\left[\exp\left(\frac{V_{BE}}{v_{kT}}\right)-1\right]$$

(11.93)

From Equation 11.93, we can show

$$g_o = \frac{dI_C}{dV_{CE}}\bigg|_{V_{BE}=\text{constant}} \cong \frac{I_C(0)}{V_{AF}}$$

(11.94)

where:
  $I_C(0)$ is the collector current at $V_{CE} = 0$

Evaluation of q_f: The parameter q_f can be considered as the normalized excess carrier density in the base with EB-junction voltage V_{BE} only and models the high-level injection. From the charge neutrality condition, *total excess majority carriers = total excess minority carriers*. Therefore, for an *npn*-BJT in the normal active mode of operation with $|V_{BE}| > 0$ and $V_{BC} = 0$

$$Q_F = \int_{x_E}^{x_C} qA\left[p_F(x)-N_a(x)\right]dx = \int_{x_E}^{x_C} qA\left[n_F(x)-\frac{n_i^2}{N_a(x)}\right]dx$$

(11.95)

We have shown in Equation 11.42 that the minority carrier forward base transit time τ_B of a BJT and the base charge Q_B are given by $Q_B = \tau_B I_{CC}$. Therefore, from Equation 11.95, the forward base transit time can be expressed as

$$Q_F = \tau_{BF} I_{CC}$$

(11.96)

where:
  $I_{CC} = I_{CT}$ given by Equation 11.69 at $V_{BC} = 0$
  τ_{BF} is the forward base transit time

Therefore, the normalized forward injection charge is given by

$$q_f = \frac{Q_F}{Q_{B0}} = \frac{\tau_{BF}I_{CC}}{Q_{B0}} = \frac{\tau_{BF}}{Q_{B0}}\frac{I_{SS}}{q_b}\left[\exp\left(\frac{V_{BE}}{v_{kT}}\right)-1\right]$$

(11.97)

Evaluation of q_r: Similar to q_f, q_r can be considered as the normalized excess carrier density in the base due to CB-junction voltage V_{BC} only and models the high-level injection. Again, from the charge neutrality condition, *total excess majority carriers = total excess minority carriers*. Therefore, for an *npn*-BJT with $|V_{BC}| > 0$; $V_{BE} = 0$, we can show

$$Q_R = \int_{x_E}^{x_C} qA[p_R(x) - N_a(x)]dx = \int_{x_E}^{x_C} qA\left[n_R(x) - \frac{n_i^2}{N_a(x)}\right]dx \qquad (11.98)$$

Again, from Equation 11.47, we can show that the reverse base transit time for BJTs is given by

$$Q_R = \tau_{BR}I_{EC} \qquad (11.99)$$

Therefore, the normalized reverse injection charge is given by

$$q_r = \frac{Q_R}{Q_{B0}} = \frac{\tau_{BR}I_{EC}}{Q_{B0}} = \frac{\tau_{BR}}{Q_{B0}} \frac{I_{SS}}{q_b}\left[\exp\left(\frac{V_{BC}}{v_{kT}}\right) - 1\right] \qquad (11.100)$$

Equations 11.81, 11.90, 11.97, and 11.100 represent the components of the normalized base charge q_e, q_c, q_f, and q_r, respectively, in terms of measurable device parameters. We will substitute these components of base charges in Equation 11.78 to solve for q_b in the following section.

Evaluation of q_b: Substituting for q_e, q_c, q_f, and q_r from Equations 11.81, 11.90, 11.97, and 11.100, respectively, in Equation 11.78 we can show the expression for total normalized charge as

$$q_b = 1 + \frac{V_{BE}}{V_{AR}} + \frac{V_{BC}}{V_{AF}} + \frac{\tau_f}{Q_{B0}} \frac{I_{SS}}{q_b}\left[\exp\left(\frac{V_{BE}}{v_{kT}}\right) - 1\right] + \frac{\tau_r}{Q_{B0}} \frac{I_{SS}}{q_b}\left[\exp\left(\frac{V_{BC}}{v_{kT}}\right) - 1\right]$$

$$= \left(1 + \frac{V_{BE}}{V_{AR}} + \frac{V_{BC}}{V_{AF}}\right)$$

$$+ \frac{1}{q_b}\left\{\frac{\tau_f}{Q_{B0}}I_{SS}\left[\exp\left(\frac{V_{BE}}{v_{kT}}\right) - 1\right] + \frac{\tau_r}{Q_{B0}}I_{SS}\left[\exp\left(\frac{V_{BC}}{v_{kT}}\right) - 1\right]\right\} \qquad (11.101)$$

$$= q_1 + \frac{q_2}{q_b}$$

where we defined

$$q_1 = 1 + \frac{V_{BE}}{V_{AR}} + \frac{V_{BC}}{V_{AF}}$$

$$q_2 = \frac{\tau_f I_{ss}\left[\exp(V_{BE}/v_{kT}) - 1\right] + \tau_r I_{SS}\left[\exp(V_{BC}/v_{kT}) - 1\right]}{Q_{B0}} \qquad (11.102)$$

In Equations 11.101 and 11.102, τ_f is the effective forward base transit time including the mobile charge in the depletion region (without depletion approximation) and τ_r is the effective reverse base transit time including the mobile charge in the depletion region (without depletion approximation).

In Equation 11.101, q_1 models the base-width modulation and q_2 models the high-level injection. From Equation 11.101, we can show

$$q_b^2 - q_1 q_b - q_2 = 0 \qquad (11.103)$$

Equation 11.103 is a quadratic equation in q_b whose solution is given by

$$q_b = \frac{q_1}{2} \pm \frac{1}{2}\sqrt{q_1^2 + 4q_2} \qquad (11.104)$$

From Equation 11.78, we know $q_b > 0$; therefore, considering the positive solution only, we get from Equation 11.104

$$q_b = \frac{q_1}{2} + \sqrt{\left(\frac{q_1}{2}\right)^2 + q_2} \qquad (11.105)$$

Equation 11.105 offers a solution for I_C and defines the injection level. Let us consider the following cases:

$$\text{Case 1: } q_2 \ll \left(\frac{q_1}{2}\right)^2 \qquad (11.106)$$

Under this condition, we get from Equation 105, $q_b \cong q_1$. Then, setting $q_f = q_r = 0$, this condition represents the *low-level injection* and base-width modulation (q_1 in Equation 11.102)

$$\text{Case 2: } q_2 \gg \left(\frac{q_1}{2}\right)^2 \qquad (11.107)$$

Under this condition, we get from Equation 11.105, $q_b \cong \sqrt{q_2}$, and therefore represents the *high-level injection* in BJTs.

For the simplicity of modeling, we assume $V_{BC} = 0$ (i.e., $q_r = 0$). Then considering q_2 from Equation 11.102, we get from Equation 11.107 the expression for high-level injection in the forward active mode of *npn*-BJT operation as

$$q_b = \sqrt{q_2} \cong \sqrt{\frac{\tau_f I_{ss}}{Q_{B0}}\left[\exp\left(\frac{V_{BE}}{v_{kT}}\right)\right]} = \sqrt{\frac{\tau_f I_{ss}}{Q_{B0}}}\left[\exp\left(\frac{V_{BE}}{2v_{kT}}\right)\right] \qquad (11.108)$$

Then substituting for q_b in Equation 11.69, we get for high-level injection at $V_{BC} = 0$ and $V_{BE} \gg v_{kT}$

$$I_{C(high-level)} = I_{CT} \cong \frac{I_{SS}\left[\exp\left(V_{BE}/v_{kT}\right)-1\right]}{\sqrt{\tau_f I_{SS}/Q_{B0}}\,\exp\left(V_{BE}/2v_{kT}\right)} \cong \sqrt{\frac{Q_{B0}I_{SS}}{\tau_f}}\,\exp\left(\frac{V_{BE}}{2v_{kT}}\right) \quad (11.109)$$

Again, for *low-level injection*, if we assume $q_e = q_c = 0$, then from Equation 11.78 $q_b = 1$. Then from Equation 11.69, we get for low-level injection @ $V_{BC} = 0$ and $V_{BE} \gg v_{kT}$ as

$$I_{C(low-level)} \cong I_{SS}\exp\left(\frac{V_{BE}}{v_{kT}}\right) \quad (11.110)$$

At the transition point from the low-level injection to high-level injection, the collector current must be continuous and, therefore, equal. Let us assume that the intersection of high current and low current asymptote is given by (I_{KF}, V_{KF}) at $I_{C(high-level)} = I_{C(low-level)}$. Therefore, from Equations 11.109 and 11.110 at the transition point (I_{KF}, V_{KF}), we can show

$$I_{KF} = \sqrt{\frac{Q_{B0}I_{SS}}{\tau_f}}\,\exp\left(\frac{V_{KF}}{2v_{kT}}\right) \quad (11.111)$$

and

$$I_{KF} \cong I_{SS}\exp\left(\frac{V_{KF}}{v_{kT}}\right) \quad (11.112)$$

From the above equations, we can show that the collector current at the transition from the low-level to high-level injection, called as the forward *knee-current*, I_{KF} is given by

$$I_{KF} = \frac{Q_{B0}}{\tau_f} \quad (11.113)$$

Figure 11.26 shows the knee-point (I_{KF}, V_{KF}) in the $\ln(I_C)$ versus V_{BE} plot at $V_{BC} = 0$. It is observed from Figure 11.26 that the slope of $\ln(I_C) - V_{BE}/v_{kT}$ plot for high-level injection is, clearly, smaller (theoretically, about 1/2) than that due to low-level injection.

Similarly, we can show that in the *reverse mode* of BJT operation, the *inverse knee-current* I_{KR} is given by

$$I_{KR} = \frac{Q_{B0}}{\tau_r} \quad (11.114)$$

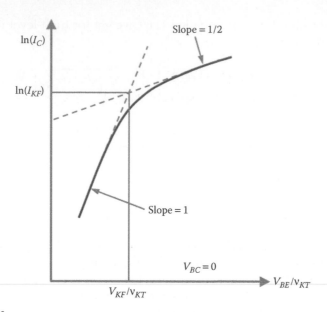

FIGURE 11.26
The forward $\ln(I_C)$ versus V_{BE}/v_{kT} plot of an *npn*-BJT: the plot shows the asymptotes of the low-level and high-level injections and the transition or knee-point (I_{KF}, V_{KF}) at the transition from the low-level to high-level injection.

Thus, the high-level injection in BJTs are modeled by forward and reverse knee-currents I_{KF} and I_{KR}, respectively, whereas the base-width modulation is modeled by the forward and reverse early voltages V_{AF} and V_{AR}, respectively.

The early effect and high-level injection model parameters (V_{AF}, V_{AR}, I_{SS}, I_{KF}, I_{KR}) are extracted from the following set of device characteristics.

- $\ln(I_C)$ versus V_{BE} plot at $V_{BC} = 0$ in the normal mode of BJT operation;
- $\ln(I_E)$ versus V_{BC} plot at $V_{BE} = 0$ in the inverse mode of BJT operation;
- I_C versus V_{CE} characteristics for different I_B (or V_{BE}) in the normal mode of BJT operation;
- I_E versus V_{EC} characteristics for different I_B (or V_{BC}) in the inverse mode of BJT operation.

11.5.5 Summary of Compact BJT Model

The complete set of BJT-model parameters consists of basic (EM1) dc parameters, EM2 model parameters for parasitic elements, space-charge layer recombination model parameters, and the unified integrated charge control model parameters describing the base-width modulation and high-level injection. The complete set of parameters is summarized below:

- Basic (EM1) DC model parameters: $\{\beta_{FM}, \beta_{RM}, T_{ref}, E_g, I_{SS}\}$
- Parasitic elements model parameters for: bulk-ohmic resistors, $\{r_c, r_e, r_b\}$ and charge storage elements: $\{C_{jE0}, \phi_{BE}, m_{jE}, C_{jC0}, \phi_{BC}, m_{jC}, \tau_f, \tau_r, C_{sub}\}$
- Space-charge layer recombination model parameters for modeling low-current β degradation: $\{C_2, n_E, C_4, n_C\}$
- Base-width modulation and high-level injection parameters: $\{V_{AF}, V_{AR}, I_{KF}, I_{KR}\}$

The equivalent circuit of the final SGP BJT model described above is represented by Figure 11.27 with redefined saturation current I_{SS} (Equation 11.66) and current source I_{CT} to model the early effect and high-level injection. The expression for I_{CT} in the SGP BJT equivalent circuit in Figure 11.27 is given by Equation 11.69 as

$$I_{CT} = \frac{I_{SS}}{q_b}\left\{\left[\exp\left(\frac{V_{BE}}{v_{kT}}\right) - 1\right] - \left[\exp\left(\frac{V_{BC}}{v_{kT}}\right) - 1\right]\right\} \qquad (11.115)$$

where the normalized base charge q_b is given by Equation 11.101. Again, the expressions for I_{EC} and I_{CC} are given by Equation 11.13. Then we can write the expressions for the terminal current in a BJT with reference to Figure 11.27.

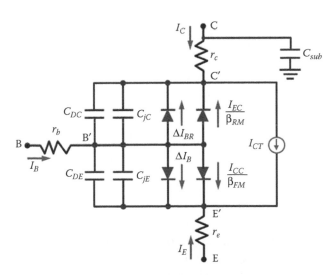

FIGURE 11.27
The equivalent circuit of SGP *npn*-BJT model: the current source I_{CT} is redefined to account for the base-width modulation and high-level injections.

With reference to Figure 11.27, the base current in SGP BJT model can be obtained from Equation 11.52 as

$$I_B = \frac{I_{SS}}{\beta_{FM}}\left[\exp\left(\frac{V_{BE}}{v_{kT}}\right) - 1\right] + C_2 I_{SS}\left[\exp\left(\frac{V_{BE}}{n_E v_{kT}}\right) - 1\right] + \frac{I_{SS}}{\beta_{RM}}\left[\exp\left(\frac{V_{BC}}{v_{kT}}\right) - 1\right]$$

$$+ C_4 I_{SS}\left[\exp\left(\frac{V_{BC}}{n_C v_{kT}}\right) - 1\right]$$

(11.116)

and the collector current is given by

$$I_C = \frac{I_{SS}}{q_b}\left[\exp\left(\frac{V_{BE}}{v_{kT}}\right) - \exp\left(\frac{V_{BC}}{v_{kT}}\right)\right] + \frac{I_{SS}}{\beta_{RM}}\left[\exp\left(\frac{V_{BC}}{v_{kT}}\right) - 1\right]$$

$$- C_4 I_{SS}\left[\exp\left(\frac{V_{BC}}{n_C v_{kT}}\right) - 1\right]$$

(11.117)

The corresponding model equations as implemented in SPICE are as follows:

$$I_B = \frac{IS_{eff}}{\beta_F}\left[\exp\left(\frac{V_{BE}}{n_F v_{kT}}\right) - 1\right] + ISE_{eff}\left[\exp\left(\frac{V_{BE}}{n_E v_{kT}}\right) - 1\right]$$

$$+ \frac{IS_{eff}}{\beta_R}\left[\exp\left(\frac{V_{BC}}{n_R v_{kT}}\right) - 1\right] + ISC_{eff}\left[\exp\left(\frac{V_{BC}}{n_C v_{kT}}\right) - 1\right]$$

(11.118)

and,

$$I_C = \frac{IS_{eff}}{q_b}\left[\exp\left(\frac{V_{BE}}{n_F v_{kT}}\right) - \exp\left(\frac{V_{BC}}{n_R v_{kT}}\right)\right] - \frac{IS_{eff}}{\beta_R}\left[\exp\left(\frac{V_{BC}}{n_R v_{kT}}\right) - 1\right]$$

$$- ISC_{eff}\left[\exp\left(\frac{V_{BC}}{n_C v_{kT}}\right) - 1\right]$$

(11.119)

Comparing derived Equations 11.116 and 11.117 with the SPICE implementation corresponding Equations 11.118 and 11.119, we find: $I_{SS} = IS_{eff}$; $C_2 I_{SS} = ISE_{eff}$; $C_4 I_{SS} = ISC_{eff}$; $\beta_{FM} = \beta_F$; and $\beta_{RM} = \beta_R$. In addition, the parameters n_F and n_R are included as the fitting parameters to improve the accuracy of data fitting with the model.

A complete set of SGP BJT compact model parameters is extracted from the following set of measurement data set for BJT parameter extraction including:

- Forward characteristics
    - Gummel plot (I_C, I_B) versus V_{BE} with $V_{BC} = 0$
    - I_C versus V_{CE}
    - Cut-off frequency, f_T versus I_C
- Reverse characteristics
    - Gummel plot (I_E, I_B) versus V_{EB} with $V_{BE} = 0$
    - I_E versus V_{CE}

The SGP model does not model the devices in the reverse mode as accurately as in the forward mode. Though BJTs are normally operated in the forward mode, both forward and reverse data are needed to extract series resistances.

In addition EB and CB *pn*-junction structures are used to extract capacitance model parameters $\{C_{j0}, \phi_{bi}, m_j\}$.

11.6 Summary

This chapter presented the basic BJT model for circuit simulation. A systematic methodology is presented to derive SGP BJT compact model starting from the basic EM model. An overview of the model parameter extraction is presented. The objective of this chapter is to expose readers to the basic understanding of BJT device modeling. The readers involved in BJT device engineering can extend the basic understanding from this study to more appropriate advanced BJT models.

Exercises

11.1 An *npn*-BJT is used as an open-collector *pn*-junction diode as shown in Figure E11.1. Then

a. Use the injection version of EM1 BJT model to derive an expression for the emitter current I_E as a function of V_{BE};

b. Use the expression derived in part (a) to calculate V_{BE} for $I_E = -1$ mA.

Given that: $\alpha_F = 0.98$; $\alpha_R = 0.49$; $I_{ES} = 10^{-16}$ A, and $T = 300$ K.

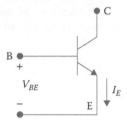

FIGURE E11.1

An open collector *npn*-BJT used as a two terminal EB *pn*-junction diode.

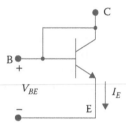

FIGURE E11.2

An *npn*-BJT is used as a two-terminal EB *pn*-junction diode with CB terminals shorted.

11.2 The *npn*-BJT in exercise 11.1 is used as a shorted base-collector diode as shown in Figure E11.2. Then use the parameters given in exercise 11.1 earlier to answer the following:

    a. Use the injection version of EM1 BJT model to derive an expression for the emitter current I_E as a function of V_{BE}

    b. Use the expression derived in part (a) to calculate V_{BE} for $I_E = -1\,\text{mA}$

11.3 The basic (EM1) *npn*-BJT compact model is discussed in Section 11.5.1. Following the same procedure, develop the EM1 type model equations for a lateral *pnp*-BJT as shown in Figure E11.3.

    a. Sketch the basic EM1 model; define and label all parameters.

    b. Write equations for the terminal currents; define and explain all parameters.

    c. If the *pnp*-BJT is used as a shorted base-collector diode, then from EM1 model equations in part (b) calculate the EB voltage at $I_E = 1\,\text{mA}$. Given that: $I_{ES} = 10^{-16}\,\text{A}$, $I_{CS} = 2.0 \times 10^{-16}\,\text{A}$, $\alpha_F = 0.98$, $\alpha_R = 0.49$, and $T = 300\,\text{K}$. Define and explain all parameters.

    d. Include the bulk-ohmic resistors and charge storage elements to your model in part (a) to generate and sketch the lateral *pnp*-BJT EM2 model. Define and label all parameters.

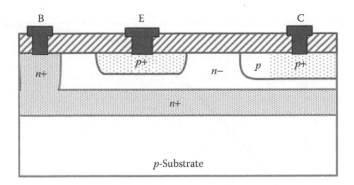

FIGURE E11.3
An ideal lateral *pnp*-BJT structure on a *p*-substrate: E, B, and C are the emitter, base, and collector terminals, respectively.

11.4 Consider an *npn*-BJT in the inverse active mode of operation in the high-level injection regime. For simplicity of modeling, you can assume uniformly doped base region.

a. Schematically show the components of base charge, Q_B, under the above operating condition. Label your plot and show the integration limits to compute different components of Q_B.

b. Write down the expression for the normalized base-charge responsible for base-width modulation in terms of junction biases and early voltages. Define all parameters and explain.

c. Write down the expression for SGP-model current source for base-width modulation in terms of early voltages only under the above specified operating condition. Define all parameters and explain.

11.5 The Gummel plot shown in Figure 11.20 for *npn*-BJT is obtained at $V_{BC} = 0$ and is used for DC compact model parameter extraction.

a. Describe graphically how I_C versus V_{BE} characteristics at $V_{BC} = 0$ can be obtained from I_C versus V_{CE} characteristics at different V_{BE} for an *npn*-BJT.

b. Mathematically describe the methodology to extract BJT saturation current I_{SS} for $I_C - V_{BE}$ plot at $V_{BC} = 0$.

Clearly define any assumptions you make.

11.6 The measured forward Gummel-plot of an *npn*-BJT is shown in Figure E11.4. Extract the following SGP-model parameters.

a. BJT saturation current, I_{SS}. Explain the extraction procedure.

b. Maximum forward current gain, β_{FM}. Explain the extraction procedure.

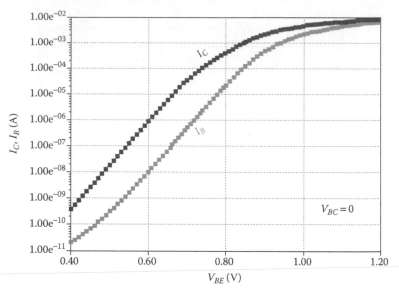

FIGURE E11.4
Measured forward Gummel plot of an *npn*-BJT used to extract SGP BJT device model parameters.

  c. Forward "Knee" current, I_{KF}. Explain the extraction procedure.
  d. The value of V_{BE} at $I_C = I_{KF}$. Explain the extraction procedure.
  e. If the extracted forward transit time $\tau_f = 1$ ns, calculate the zero-bias base charge, Q_{B0} of the device.
  f. Use the extracted value of Q_{B0} from part (e) to calculate the forward emission coefficient, n_E, that is, slope of $\ln(I_C)$ versus qV_{BE}/kT plot of the given characteristics at room temperature at the onset of high-level injection. Assume the width of the zero-bias neutral base region, $W_B = 0.2$ μm, $V_B = 10$ V, and area of the intrinsic BJT shown in Figure E11.4 is unity.

  State any assumptions you make.
  11.7 Consider a vertical *npn*-BJT operating in the normal active mode. The *p*-type base region is uniformly doped with concentration, $N_{aB} = 2.0 \times 10^{17}$ cm^{-3} and depth $= 1.0$ μm. The *n*-type emitter is formed by ion implantation with doping concentration, $N_{dE} = 2.0 \times 10^{20}$ cm^{-3} and depth $= 0.3$ μm. Neglect the space-charge recombination current for high-level injection to answer the following questions. Clearly define all parameters and explain.

  a. What is the minority-carrier density in the *p*-base at which the high-level injection is reached?

b. Calculate the base-emitter forward bias V_{KF} at which the *high-level injection* is reached.

c. If the effect of *high-level injection* on the current gain β starts when the injected minority-carrier density reaches 10% of the majority carrier density, calculate the value of V_{BE} at the onset of β roll-off at high collector current level, I_C.

d. Show the conditions obtained in part (b) and (c) on I_C vs. V_{BE} @ $V_{BC} = 0$ characteristic of the transistor.

Define and label all parameters in your plot(s). Explain.

11.8 The SGP-BJT model, presented in this chapter, cannot model the effect of parasitic substrate transistor on intrinsic devices. Consider the 2D cross section of dual-poly *npn*-BJT with a parasitic vertical *pnp*-BJT with the *p*-base as the emitter as shown in Figure E11.5. In this problem, you will modify the intrinsic vertical *npn*-BJT SGP model to include the effect of parasitic *pnp*-BJT. Clearly, state any assumptions you make, define all parameters, and label all terminal currents.

a. Draw the SGP equivalent network for the parasitic vertical *pnp*-BJT shown in Figure E11.5.

b. Use block diagrams to include the parasitic *pnp*-BJT network and the base-emitter and base-collector overlap oxide capacitances into the intrinsic *npn*-BJT model.

11.9 The small signal base-collector junction capacitance, C_{jC} versus V_{BC}, characteristics of an *npn*-BJT is shown in Figure E11.6. From the figure, extract the following diode model parameters. Clearly state any model you use and explain the procedure for each case.

a. CB-junction capacitance, C_{jC0} at $V_{BC} = 0$.

b. Built-in potential, ϕ_{BC}.

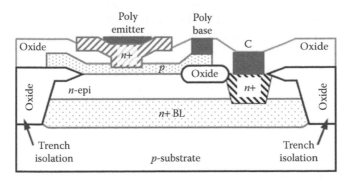

FIGURE E11.5
Trench isolated double polysilicon *npn*-BJT structure: poly emitter, poly base, and C represent emitter, base, and collector terminals, respectively.

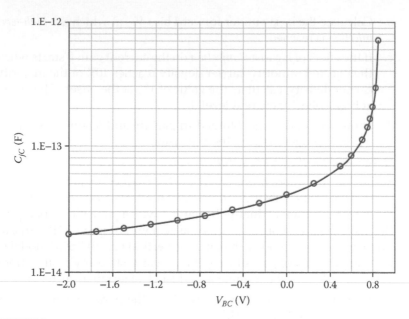

FIGURE E11.6

CB *pn*-junction C–V characteristics used to extract capacitance compact model parameters for *npn*-BJTs.

   c. Calculate the junction gradient factor, m_{jC}.

   d. If the integrated base charge, $Q_{B0} = 1.31 \times 10^{-12}$ C at $V_{BE} = V_{BC} = 0$, calculate the forward early voltage V_{AF} for the device operating in the normal active mode with CB-junction reverse biased at 2.0 V. Clearly state any assumptions you make.

11.10 A typical SGP *npn*-BJT compact model card of a typical bipolar technology is shown in Table E11.1. Consider an *npn*-BJT with emitter area, $A_E = 25$ μm² of this technology used in an integrated circuit to operate in the normal active mode at the biasing condition, $V_{BE} = 0.61$ V and $V_{CE} = 3$ V.

Use the relevant SGP model parameters from the given model card to answer the following questions. Define each parameter and explain your results. Assume that $V_{B'E'} = V_{BE}$ and $V_{B'C'} = V_{BC}$.

   a. Calculate the base charge Q_{B0} in the neutral base region at $V_{BE} = 0 = V_{BC}$.

   b. Calculate the normalized base charge q_1 that models the base-width modulation.

   c. Calculate the normalized base charge, q_2 that models the high-level injection.

TABLE E11.1

A Typical SGP Model Card of an *npn*-BJT for Circuit Simulation

.option gmin = 1.0000000–16		
.model NPN1 npn (	LEVEL = 1	
+ bf = 2.1139717E+01	br = 4.9802084E+0	brs = 0.0000000E+00
+ Gamma = 0.0000000E+00	ikf = 2.3796255E–04	ikr = 1.0738911E–04
+ irb = 0.0000000E+00	is = 3.7746394E–18	isc = 3.4311697E–15
+ ise = 4.5630707E–15	nc = 1.9324214E+00	ne = 1.7137365E+00
+ Nepi = 1.0000000E+00	nf = 1.0000000E+00	nkf = 5.0000000E–01
+ nr = 1.0000000E+00	rb = 6.7165161E+02	rbm = 1.0000000E–01
+ rc = 9.9999998E–03	re = 4.8533592E+01	vaf = 4.3480446E+01
+ var = 6.0068092E+00	vo = 0.0000000E+00	cjc = 1.3075999E–15
+ cje = 3.8835999E–15	cjs = 0.0000000E+00	fc = 5.0000000E–01
+ itf = 3.2536294E–03	mjc = 1.4907001E–01	mje = 1.7038001E–01
+ mjs = 5.0000000E–01	ptf = 0.0000000E+00	qco = 0.0000000E+00
+ tf = 7.7961665E–11	tr = 0.0000000E+00	vjc = 1.0000000E–01
+ vje = 1.7466000E–01	vjs = 1.0000000E+00	vtf = 2.2093861E+00
+ xcjc = 1.0000000E+00	xtf = 3.4448984E+00	tref = 2.7000000E+01
+ eg = 1.1100000E+00	xtb = 0.0000000E+00	xti = 3.0000000E+00
+ subs = 1)		

Note: The parameters in the model card: $is = I_{SS}$; $ise = C_2 I_{SS}$; and $i_{SC} = C_4 I_{SS}$.

    d.   Calculate the total normalized base charge, q_b. From your results, what is your conclusion on the injection level under the biasing condition?

    e.   Calculate the base current I_B at the operating point.

    f.   Calculate the collector current I_C at the operating point.

12

Compact Model Library for Circuit Simulation

12.1 Introduction

In Chapters 4 through 11, we have discussed compact model formulation for different VLSI (very-large-scale-integrated) devices. In this chapter a brief overview of the generation of compact model library for circuit simulation and compact model usage in circuit CAD (computer-aided design) is provided. A typical CMOS (complementary metal-oxide-semiconductor) technology and Berkeley Short Channel IGFET Model, version 4 (BSIM4) compact MOSFET (metal-oxide-semiconductor field-effect transistor) model are used to illustrate the methodology to build a compact model library for HSPICE (see Chapter 1) circuit CAD [1,2]. Note that the circuit CAD tools and the device models are continuously updated for improving the accuracy and simulation efficiency. So, the basic idea for compact model development presented in this chapter must be appropriately modified to the changing modeling and circuit CAD tools.

12.2 General Approach to Generate Compact Device Model

A generalized methodology to build a compact device model library is shown in Figure 12.1. As shown in Figure 12.1, the procedure involves data collection, data fitting to the target compact model, extraction of model parameters, generation of model library, and verification of model for accuracy and predictability. Each of these steps to generate a computationally robust compact model library depends on the device technology (e.g., MOSFETs), target model (e.g., BSIM4), design target (e.g., digital), and so on. In this chapter, we will use BSIM4 to illustrate the modeling methodology outlined in Figure 12.1 [2].

12.2.1 Data Collection

The first task in generating a model library is the data collection from the devices of the target technology representing the entire design space under the operating biasing conditions and ambient temperatures. Data collection

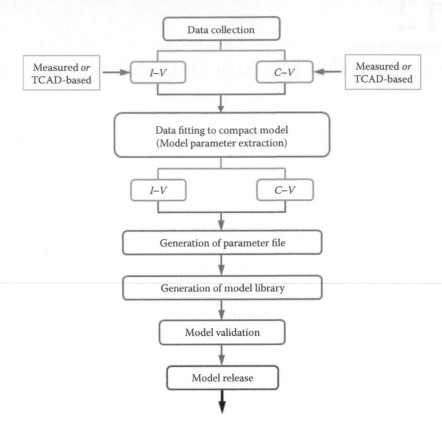

FIGURE 12.1
A generalized methodology to generate compact device model library for circuit CAD; each task in the flowchart depends on the target device technology, target compact model, and the target VLSI circuit for computer analysis.

includes the selection of an acceptable set of devices representing the entire IC (integrated circuit) design space and selection of device characteristics that account for the real-device effects to extract device model parameters to modeling physical, geometrical, and ambient temperature effects on the device performance in IC chips. The selection of devices and device characteristics for data acquisition is described in the following subsections.

12.2.1.1 Selection of Devices

In order to collect data for compact model parameter extraction, a set of devices are selected, representing the entire design space to properly characterize the detailed physics of the device operation formulated by the target compact model. These include devices to extract core model parameters

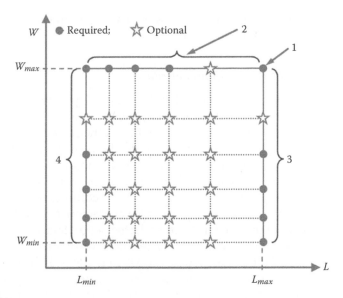

FIGURE 12.2
Typical device selection criteria for compact MOSFET models: device 1 is used to extract the core model parameters, group 2 devices are used to model the channel length dependence, group 3 devices are used to model the channel width dependence, and group 4 devices are used to model short devices; L_{min} and L_{max} represent the shortest and longest devices used in the circuit, respectively, and W_{min} and W_{max} represent the narrowest and widest devices used in the circuit, respectively; ● represents the devices required for modeling whereas ☆ represents the optional devices that can be used for model verification and further optimization of the model parameters.

and devices to extract real-device effects describing physical and geometrical effects. With reference to BSIM4 model, the device selection criteria for extracting the compact device model parameters of a target CMOS technology is shown in Figure 12.2 [2–4].

As shown in Figure 12.2, the set of selected devices must include the minimum and maximum geometries intended for IC chip design. If L_{min} and L_{max} represent the shortest and longest devices, respectively, used in the circuit and W_{min} and W_{max} represent the narrowest and widest devices, respectively, used in the circuit, then the set of devices must include $L_{min} \leq L \leq L_{max}$ and $W_{min} \leq W \leq W_{max}$. The accuracy of model library may be improved by fitting data from a large number of devices in the set. However, for efficient model generation, a minimum number of devices is selected as described in Figure 12.2 by required and optional devices. As shown in Figure 12.2, device 1 is used to extract the core model parameters independent of real-device effects, group 2 devices are used to model the channel length dependence (SCEs), group 3 devices are used to model the channel width dependence, and group 4 devices are used to model the width dependence of the shortest device. Note that in most cases, L_{min} represents the nominal device of the target technology node.

12.2.1.2 Selection of Device Characteristics

As discussed in Chapter 4 and 5 for MOSFETs and Chapter 11 for bipolar junctions transistors, the device operation is typically characterized by distinct regions, and compact models are developed to mathematically describe each region by separate equations or a single model. In most compact models, the device performance in each mode of device operation is described by a set of model parameters. Thus, a set of device characteristics such as current-voltage (I–V) and capacitance-voltage (C–V) is required to fit the model equations to each operating region of device characteristics and extract the corresponding model parameters of the selected set of devices.

Again, to illustrate the selection of device characteristics for compact modeling, let us use BSIM4 regional model. The MOSFET device is primarily characterized by subthreshold, linear, and saturation regions. Thus, to model the entire device characteristics for the selected set of devices of a target technology, the device data are obtained under the appropriate biasing conditions determined by the target supply voltage, V_{dd}, at the target operating range of the ambient temperature (T) of the devices in the IC chips. Table 12.1 shows a typical set of device characteristics required for compact modeling of the selected MOSFET devices of a specific technology under different gate-source, drain-source, and body-source voltages V_{gs}, V_{ds}, and V_{bs}, respectively.

A similar set of device data are obtained for PMOS (n-type body with $p+$ source-drain) devices by changing the sign of the operating applied voltages.

In addition, the source-drain pn-junction I–V and C–V characteristics are obtained for both NMOS (p-type body with $n+$ source-drain) and PMOS (n-type body with $p+$ source-drain) devices to extract the source-drain diode model parameters. The diode model is an integral part of MOSFET compact model, and therefore, diode characteristics are part of data collection for compact MOSFET modeling. For analog/RF modeling, the additional set of required device characteristics is obtained.

The *rev0* compact model of a target technology can be generated from the numerical device simulation data for the feasibility study and early IC design evaluation [5–7]. However, in order to generate the final compact device model of a technology for product design in CAD environment and release to process design kit (PDK), the measurement data are collected from a single die, referred to as the *golden die* of a specific silicon wafer, referred to as the *golden wafer*. The golden die of the golden wafer provides the target device performance of the target technology node [6].

In order to develop a *statistical model library*, the required set of data shown in Table 12.1 is collected from different silicon die, wafers, and different wafer lots over a period of time. Then from the statistical distribution of data

TABLE 12.1

Selection of NMOS Device Characteristics for the Basic BSIM4 Compact Model Parameter Extraction: All Characteristics are Obtained in the Ambient Temperature Range $-55°C < T < 125°C$

Device Data	V_{ds} (V)	V_{gs} (V)	V_{bs} (V)	Objective
$I_{ds} - V_{gs}$	Constant: ~50 mV	Ramp: 0 to $1.1 \cdot V_{dd}$ step ~ 50 mV	Constant: 0 to $-V_{dd}$, step ~ $-1.1 \cdot V_{dd}/4$	Subthreshold region and linear region parameter extraction
$I_{ds} - V_{gs}$	Constant: $1.1 \cdot V_{dd}$	Ramp: 0 to $1.1 \cdot V_{dd}$ step ~ 50 mV	Constant: 0 to $-V_{dd}$, step ~ $-1.1 \cdot V_{dd}/4$	Saturation region parameter extraction
$I_{ds} - V_{ds}$	Ramp: 0 to $1.1 \cdot V_{dd}$, step ~50 mV	Constant: 0 to $1.1 \cdot V_{dd}$ step ~ $1.1 \cdot V_{dd}/4$	Constant: 0, $-V_{dd}/2$, $-V_{dd}$	High field effect parameter extraction, e.g., output resistance and early voltage parameters
$C_{gg} - V_{gs}$	0	Ramp: $-1.1 \cdot V_{dd}$ to $+1.1 \cdot V_{dd}$	0	Intrinsic capacitance model parameter extraction
$C_{gs} - V_{gd}$	Ramp: 0 to $1.1 \cdot V_{dd}$	Constant: 0 to $1.1 \cdot V_{dd}$ step ~ $1.1 \cdot V_{dd}/4$	0	Intrinsic capacitance model parameter extraction
$C_{gd} - V_{gd}$	Ramp: 0 to $1.1 \cdot V_{dd}$	Constant: 0 to $1.1 \cdot V_{dd}$ step ~ $1.1 \cdot V_{dd}/4$	0	Intrinsic capacitance model parameter extraction

set, process variability–induced device model parameters are obtained as discussed in Chapter 8 [6–9].

12.2.2 Data Fitting to Extract Compact Model Parameters

After the required data acquisition for modeling, the data set is formatted to the required format of the parameter extraction tool used for compact model parameter extraction [3,4]. The detailed parameter extraction routine is described in each tool [3,4]. A brief outline for fitting the data to the device compact model is shown in the flowchart in Figure 12.3.

Figure 12.4a–f shows the typical measured and fitted nMOSFET device characteristics of an advanced CMOS technology obtained by BSIMProPlus [3]. The fitted data are within the acceptable range of tolerance to build the model card of the representative CMOS technology.

In addition to fitting the basic device characteristics, the first and second derivatives of I–V curves are fitted to extract the model parameters related to g_m, R_{out}, and so on for both NMOS and PMOS devices.

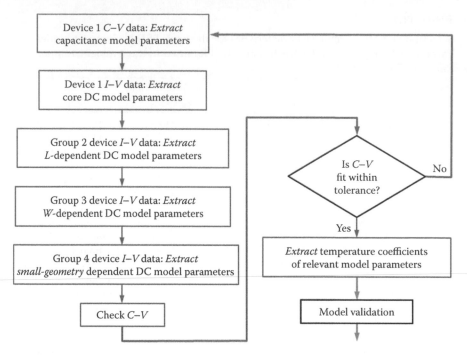

FIGURE 12.3
A general outline to extract compact MOSFET model parameters of a target VLSI technology; the regional compact MOSFET model like BSIM4 is used to illustrate the parameter extraction procedure.

12.2.3 Generation of Parameter Files

After the parameter optimization to fit the measured device characteristics with the simulation data obtained by extracted model parameters, the parameters for both NMOS and PMOS devices are saved into the respective parameter files. A typical parameter file includes device information such as W, L, device type, and the list of optimized model parameters of the compact model used.

The parameter files are verified to check the accuracy of fitting. This can be achieved graphically using the parameter extraction tool by (1) comparing the simulated device characteristics obtained by the parameter file to the measured device characteristics of a different set of device dimensions other than that used for parameter extraction, (2) comparing the simulated device characteristics with the measured device characteristics at temperatures other than that used for parameter extraction, (3) comparing the simulated device characteristics with the measured device characteristics at different bias point other than that used for extraction, (4) checking for discontinuities in the first- or second-order derivative of current (g_m, R_{out}, etc.), and (5) using external circuit CAD tool (e.g., HSPICE) to check for convergence issues. The verified parameter files are then used to generate the final model library.

12.2.4 Generation of Compact Model Library

The parameter files for both NMOS and PMOS devices obtained by fitting device characteristics with the target model (e.g., BSIM4) are assembled together to form the model card. A typical industry standard compact model library consists of a set of model cards that include logic with performance options, analog/RF, and SRAM as well as interconnect models. In this chapter, we describe the methodology to generate a simple compact MOSFET model that includes the real-device effects and process variability for device analysis in circuit CAD. To illustrate the methodology to generate compact model library, examples of a MOSFET and an SRAM model cards are presented in section, Sample Model Cards at the end of this chapter.

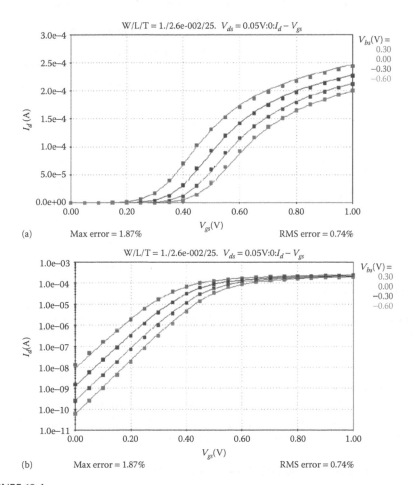

FIGURE 12.4
Measured data of an nMOSFET device with $W = 1\,\mu m$ and $L = 26\,nm$ fitted with BSIM4 model: (a) $I_d - V_{gs}$ characteristics at low V_{ds} to extract linear region model parameters; (b) $\log(I_d) - V_{gs}$ characteristics at low V_{ds} to extract subthreshold model parameters. (*Continued*)

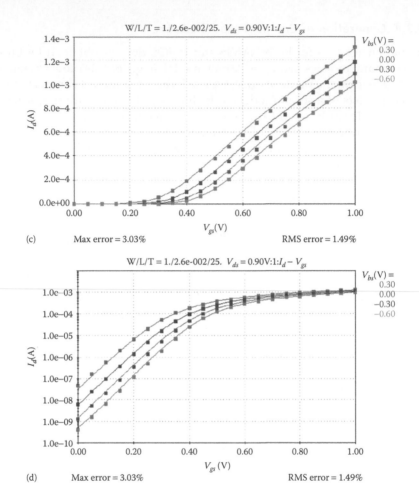

FIGURE 12.4 (Continued)

Measured data of an nMOSFET device with $W = 1$ μm and $L = 26$ nm fitted with BSIM4 model: (c) $I_d - V_{gs}$ characteristics at $V_{ds} = V_{dd}$ to extract saturation region model parameters; (d) $\log(I_d) - V_{gs}$ characteristics at low $V_{ds} = V_{dd}$ to extract off-state leakage current model parameters. *(Continued)*

A typical model card includes separate subsections describing (1) general information of the contents and usage of the model card, (2) process corners to model the device and circuit performance variability due to process variability, and finally (3) the properly parameterized model parameters to simulate different performance corners as discussed next.

12.2.4.1 Modeling Systematic Process Variability

Modeling process variation is critical in advanced ICs to design variability-aware VLSI circuits and IC chips [6–9]. The detailed modeling of process variability is described in Chapter 8 including the selection of process

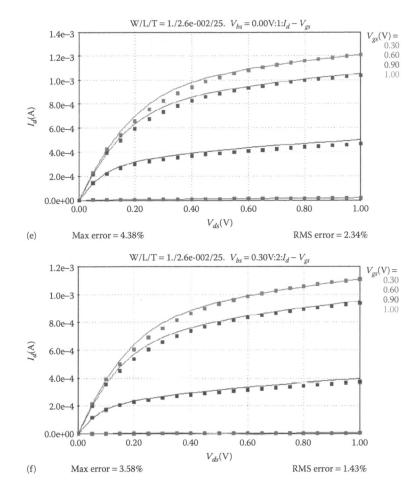

(e)   Max error = 4.38%   RMS error = 2.34%

(f)   Max error = 3.58%   RMS error = 1.43%

FIGURE 12.4 (Continued)
Measured data of an nMOSFET device with $W = 1$ μm and $L = 26$ nm fitted with BSIM4 model: (e) $I_d - V_{ds}$ characteristics for different V_{gs} and $V_{bs} = 0$ to extract saturation region model parameters including early voltage and output resistance model parameters; and (f) $I_d - V_{ds}$ characteristics for different V_{gs} and V_{bs} to extract body bias–dependent saturation region model parameters.

variability-sensitive model parameters for corner modeling. Table 12.2 shows a typical corner file for nMOSFETs with selected parameters to illustrate the corner file generation technique.

In Table 12.2, TT defines the *typical values* of the extracted model parameters at room temperature for NMOS and PMOS devices; SS, FF, SF, and FS define the parameters for slow NMOS and slow PMOS, fast NMOS and fast PMOS, slow NMOS and fast PMOS, and fast NMOS and slow PMOS, respectively, as described in Chapter 8 [6,7]. It is to be noted that SS, FF, SF, and FS represent the worst case speed (WS), worst case power (WP), worst case zero (WZ), and worst case one (WO), respectively [6,7]. Each instance parameter (delta) defines the variance of the corresponding model parameter. In Table 12.2,

TABLE 12.2

Example Corner Parameters in the Compact Model Library: Only a Selected List of nMOSFET Model Parameters is Shown to Illustrate the Corner Modeling Technique in BSIM4 Model Library

Compact Model Parameter	Extracted Typical Value (TT)	Instance Parameter	TT Value	SS (WS) delta (%)	SS (WS) Value (delta)	FF (WP) delta (%)	FF (WP) Value (delta)	SF (WZ) Value (delta)	FS (WO) Value (delta)	Statistical (delta)
TOXE/TOXM	1.667E-09	toxen_svt	1.667E-09	2.6	1.710E-09	-2.6	1.6237E-09	1.667E-09	1.667E-09	4.2948E-11
TOXP	1.770E-09	toxpn_svt	1.770E-09	2.6	1.816E-09	-2.6	1.724E-09	1.770E-09	1.770E-09	4.5613E-11
XL	1.000E-08	dxln_svt	0	*	2.000E-09	*	-2.000E-09	1.000E-09	-1.0000E-09	2.0000E-09
XW	0	dxwn_svt	0	*	-5.000E-09	*	5.000E-09	-5.000E-09	5.000E-09	-5.0000E-09
VTH0	0.3827	dvthn_svt	0	*	-0.0079	*	0.0021	-0.0053	0.00138	-7.9330E-03
RDSW	1.597E+02	drdswn_svt	0	3.0	4.7916	-3.0	-4.7916	0	0	4.7916E+00
K1	0.5658	dk1n_svt	0	2.0	0.0113	-2.0	-0.0113	0	0	1.1317E-02
CJS/CJD	3.201E-03	cjn_svt	3.201E-03	3.0	3.297E-03	-3.0	3.105E-03	3.297E-03	3.105E-03	9.6030E-05
CJSWS/CJSWD	1.000E-14	cjswn_svt	1.000E-14	3.0	1.030E-14	-3.0	9.700E-15	1.030E-14	9.700E-15	3.0000E-16
CJSWGS/CJSWGD	1.938E-10	cjswgn_svt	1.938E-10	3.0	1.996E-10	-3.0	1.880E-10	1.997E-10	1.880E-10	5.8149E-12
CGSO/CGDO	6.206E-11	cgon_svt	6.206E-11	-3.0	6.020E-11	3.0	6.392E-11	6.020E-11	6.392E-11	-1.8619E-12
CGDL/CGSL	5.228E-11	cgln_svt	5.228E-11	-3.0	5.071E-11	3.0	5.385E-11	5.07E-11	5.385E-11	-1.5684E-12

delta is the 3σ variation of the mean value of the corresponding parameter; where σ is the variance of the statistical distribution of the model parameter.

Let us consider a typical parameter file of a standard V_{th} (*svt*) device of a CMOS technology. A typical parameter is then parameterized in the model card as: TT ± delta. Thus, the model parameter VTH0 is parameterized for SS and FF corners as

$$VTH0 = VTH0 \pm dvthn\_svt \tag{12.1}$$

where:

dvthn_svt defines the 3σ value of V_{th} variation of *svt* devices and is defined in the header file of a typical compact model card for each corner used in the model card

It has different values for different corners as determined by the variance shown in Table 12.2.

For *statistical compact modeling*, each instance parameter is obtained from the statistical distribution of the corresponding parameter, whereas for a *fixed corner* model, a historical standard percentage of variation of the selected parameters is used to define process corners [6,7]. For Monte Carlo (MC) and statistical corner modeling, the σ values obtained from the statistical distribution of each variability sensitive model parameters are used, as shown in Table 12.2. The detailed statistical modeling methodology and variability-aware circuit design is discussed in Chapter 8. In HSPICE circuit CAD tool, the variation for MC analysis is defined by

$$dvthn\_svt = globalmcflag \cdot (3\sigma) \cdot agauss(0,1,3) \tag{12.2}$$

where:

globalmcflag is a *switch* or *flag* that is used to turn *on* or *off* the model for circuit simulation

agauss(0,1,3) defines the probability distribution function of V_{th} variation

Equations 12.1 and 12.2 are used in the model card shown at the end of this chapter.

12.2.4.2 Modeling Mismatch

The mismatch modeling and parameter selection techniques are described in Chapter 8. Considering only threshold voltage mismatch, primarily due to random discrete doping, the mismatch in threshold voltage is given by (Equation 8.20)

$$\sigma V_{TH0,mismatch} = \frac{1}{\sqrt{2}} \sigma_{\Delta V_{TH0}} = \frac{1}{\sqrt{2}} \frac{A_{vt}}{\sqrt{WL}} \tag{12.3}$$

where:

A_{vt} is the mismatch coefficient

$\sqrt{WL}$ is the area of the device (Chapter 8)

Let us define *sigvtp_svtp* as the mismatch in V_{th} of *svt* PMOS devices, *cvtp* is the mismatch coefficient, and $1/\sqrt{WL}$ is the *geometrical factor* (*geo_fac*); then the V_{th} mismatch is modeled by

$$sigvtp\_svtp = (cvtp\_svtp) \cdot (geo\_fac) \cdot mismatchflag \qquad (12.4)$$

where

cvtp_svtp = 'avtp/1.414214';

lef = '1-5.6E-09'; wef = 'w/n_fingers+0E-09'; and *geo_fac* = '1/sqrt(multi*n_fingers*lef*wef)' with

lef and wef are used to define the effective channel length and width (w), respectively, n_fingers represents the number of gate geometries in the layout of a transistor, and multi is a number.

Again, the *mismatchflag* is used to turn on and off the model in circuit analysis.

In corner simulation, only geometry dependent fixed mismatch can be modeled by using a flag to appropriately call the model if desired without MC simulation use

$$'sigvtp\_vtp = (cvtp\_svtp) \cdot (geo\_fac) \cdot globalsigmavtflag \qquad (12.5)$$

For MC statistical model the mismatch is modeled by one-sigma variation of the parameter in space

$$'sigvtp\_svtp = (cvtp\_svtp) \cdot (geo\_fac) \cdot agaus(0,1,1) \cdot mismatchflag \quad (12.6)$$

where the mismatch parameter, *avtp*, for *p*-channel devices is extracted from the sigma(delta_vt) versus $1/\sqrt{WL}$ plots. The expressions given in Equations 12.4 through 12.6 are used in generating compact model cards in the examples shown at the end of this chapter.

12.2.4.3 Generate Model Card

Finally, use the above formulations to develop a compact model card for circuit CAD. In reality, the corner files, header files, and other files can be kept separate in the model library and a simple script file can be used to call the relevant models for circuit analysis. In this chapter all the relevant files are integrated into a single comprehensive compact model card. To illustrate the basic procedure, a compact MOSFET model card *ex1mod0p1.l* and an SRAM model card *sram127hp.l* are presented at the end of this chapter. Note that due to continuous updates of compact model formulations and CAD environment, appropriate changes in the parameters of the model library may be needed to use the model for circuit CAD. Again, the model cards represent BSIM4 model for HSPICE circuit CAD tool. The model parameters are defined in BSIM4 user's manual [2–4].

The model card *ex1mod0p1.l* is a statistical model that can be used for corner analysis of a VLSI circuit as well as MC statistical analysis of the circuit by selecting appropriate switches or flags. The different flags are described in the header section of the model card. The model card can be set up for corner simulation by turning off all the flags for MC simulation. Different flags used for MC simulation and their functions are summarized in Table 12.3. Some of the parameters are repeated in the following section (Figure 12.5).

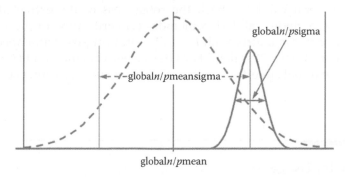

FIGURE 12.5
The parameters of the systematic and random Gaussian distribution functions used in generating the compact model card using HSPICE circuit CAD.

TABLE 12.3

Definition of Different Parameters and Their Functions as Used in the Model Card

Parameter	Description
globalnmean	Sets the mean of the systematic variation agauss for NMOS in MC simulation; e.g., globalnmean = 2 for a mean of 2-sigma (default = 0)
globalpmean	Sets the mean of the systematic variation agauss for PMOS in MC simulation; e.g., globalpmean = 2 for a mean of 2-sigma (default = 0)
globalnsigma	Sets the width of the local systematic variation agauss for NMOS (default: 1)
globalpsigma	Sets the width of the local systematic variation agauss for PMOS (default: 1)
globalnmeansigma	Sets the spread of the distribution of die means for NMOS in total MC variation (default: 2.8284) (die-to-die)
globalpmeansigma	Sets the spread of the distribution of die means for PMOS in total MC variation (default: 2.8284) (die-to-die)
globalsigmavtflag = 1	Adds a fixed 1-sigma VTH variation offset to a fixed systematic corner (default)
globalmismatchflag = 1	Enables mismatch (random variation) in MC simulations
globalmismatchnsigma	Set the width of the mismatch agauss (random variation) for NMOS (default: 3)
globalmismatchpsigma	Set the width of the mismatch agauss (random variation) for PMOS (default: 3)
mismatchflag = 1	Enables mismatch per device (instance parameter)
sigmavt	Sets the point on the distribution for VTH for a fixed corner per device (instance parameter); this includes ONLY RDD and LER

12.2.5 Model Validation

After the generation of the model library, a number of simulation experiments are performed to verify the accuracy and predictability of the model prior to release to production for circuit CAD. Different model cards (e.g., logic and SRAM) have different requirements for model validation and there are several ways to check the robustness of the extracted model. In general, the model validation includes (1) verification of the simulated device performance matrix such I_{on}, I_{off}, V_{th}, and ring-oscillator speed to the target specifications; (2) scalability of device performance; and (3) compatibility with external circuit simulation tools to ensure convergence of circuit simulation.

12.3 Model Usage

The usage of the model cards or library is described in the example model library. The model card similar to *ex1mod0p1.l* can be used for corner simulation using systematic ±3σ extreme corners (SS, FF) and (FS, SF) intermediate corners defined in Table 12.2 as well as MC statistical analysis. In order to select appropriate simulation setup (e.g., corners or MC), the appropriate input command must be used to select the flags in HSPICE netlist. The appropriate switches (flags) to turn on or off the target method of simulation using *ex1mod0p1.l* are shown in Table 12.4.

Thus, for circuit analysis using the model card in *ex1mod0p1.l*, the following commands are used in HSPICE circuit input file or netlist.

TABLE 12.4

Assignment of Different Flags in the HSPICE Netlist to Set up Circuit Analysis Using the Compact Model in Example1

Variation	Corner	global-mismatch-flag	global-sigmavt-flag	global-mcflag	global [n/p]-mean	global [n/p] meansigma	global [n/p]sigma
Systematic	Fixed corners	0	0	0	*	*	*
Total variation	Fixed corners	0	1	0	*	*	*
Total variation	MC	1	0	1	Systematic mean	Die-to-die spread	Within die spread
Variation around mean	MC	1	0	1	Systematic mean	0	Within die spread
Variation around a corner	MC	1	0	0	*	*	*

```
.lib 'lib' TT (or SS, FF, FS, and SF)
.param globalmismatchflag = 0     $ disable MC mismatch
.param globalsigmavtflag = 1      $ enable fixed V_th mismatch
                                    offset
.param globalmcflag = 0   $ disable MC systematic variation
```

Figure 12.6a shows the simulated corner points along with the simulated TT values of *n*MOSFET on current, IONN versus *p*MOSFET on current, IONP for an advanced CMOS technology.

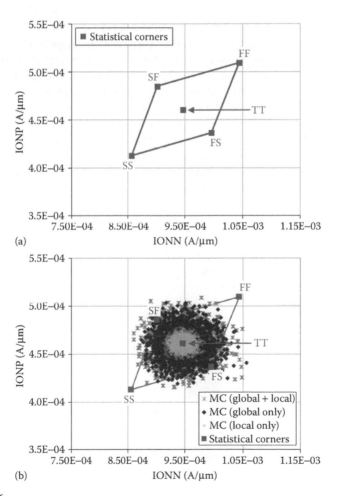

(a)

(b)

FIGURE 12.6
Simulated IONN versus IONP of a typical CMOS technology obtained by HSPICE circuit CAD using a model card similar to *ex1mod0p1.l*: (a) corner simulation and (b) MC analysis with simulated corner values superimposed.

For MC simulation using ±3σ distribution for systematic variation and ±1σ variation in mismatch, we can use the following commands in the HSPICE netlist

```
.lib 'lib' MC
.param globalmismatchflag = 1    $ enable MC mismatch
.param globalsigmavtflag = 0     $ disable fixed V_th mismatch
                                   offset
.param globalmcflag = 1    $ enable MC systematic variation
```

Figure 12.6b shows the MC simulation results of IONN versus IONP for the same CMOS technology. The corner values are shown for comparison only.

Figure 12.7a and b show the simulated distribution of IONN versus IONP obtained by MC (local and global) analysis around TT and MC mismatch simulation around SS and FF corners, respectively. Again, the corner values are shown for comparison only. The simulated mismatch distribution in Figure 12.7a shows that some of the worst-case (SS) speed die are pulled toward the TT values. This offers realistic prediction of actual speed since all transistors are not pinned to the worst-case device value. Similarly, the simulated mismatch distribution in Figure 12.7b shows that some of the worst-power (FF) die are pulled toward the TT values, thus offering a better estimate of average power of the FF devices.

Table 12.5 presents a qualitative evaluation of simulation outcome using different options for modeling process variability in advanced CMOS technology.

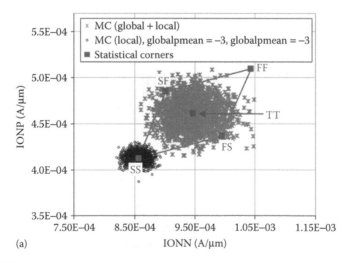

(a)

FIGURE 12.7
Simulated IONN versus IONP of a typical CMOS technology obtained by HSPICE circuit CAD using a model card similar to *ex1mod0p1.l*: (a) total MC distribution around the TT values and MC mismatch distribution at SS corner.   *(Continued)*

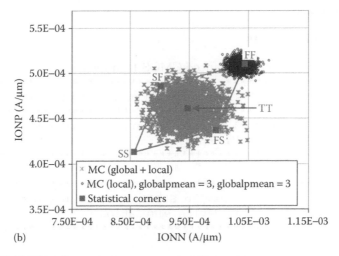

(b)

FIGURE 12.7 (Continued)
Simulated IONN versus IONP of a typical CMOS technology obtained by HSPICE circuit CAD using a model card similar to *ex1mod0p1.l*: (b) total MC distribution around the TT values and MC mismatch distribution at FF corner.

TABLE 12.5

Comparison of Nanoscale MOSFET Circuit Simulation Results using Different Methods for Process Variability Modeling

Simulation Method	Simulation Results	Model Set	Model Extraction	Simulation Time
Worst-case fixed corner	Pass/Fail: pessimistic	Discrete; artificial	Easy	Fast to moderate
Statistical corner	Pass/Fail: realistic	Discrete wafer data	Complex	Moderate
Monte Carlo	Yield: mismatch	Distribution	Complex	Long

12.4 Summary

A brief overview of generating statistical compact MOSFET model library and its usage is described. The generation of compact model library involves selection of device structures and device characteristics to capture the real-device effects in describing the device performance in IC chips. The data acquisition and data fitting to generate the model parameters and parameterization of the process variability-sensitive model parameters are critical for statistical compact model. It is shown that a comprehensive simple statistical model library can be developed and used by implementing different switches or flags to turn on or off the relevant models for circuit CAD. The example model files are provided to expose readers to the statistical models and model development from a minimum set of devices and device characteristics.

Sample Model Cards

1. Compact Model of an Advanced CMOS Technology

```
**************************************************************
* Example1: 20 nm CMOS technology node MOS Hspice model library
**************************************************************
* Version:    ex1mod0p1.l   *** January 21, 2015 ***
* MODEL:      BSIM4 ( V4.5 )
**************************************************************
* Model Information
**************************************************************
* This Rev0 version of sample SPICE model is based on two-
* dimensional numerical device simulation
**************************************************************
*.LIB INTERNAL
*-------------------------------------------------------------
* History of model updates
*-------------------------------------------------------------
* Jan 20, 2015: model created from TCAD-based data - only
* selected parameters are shown in the model card to illustrate
* the basic procedure for the generation of compact model
* library
*-------------------------------------------------------------
*.ENDL INTERNAL
**************************************************************
* Begin header
**************************************************************
* Usage:
*
* Hspice Version: 2007.03, 2008.03, 2009.03
*
* Library includes standard-Vth (svt) corner libraries:
* .lib 'ex1mod0p1.l' TT
* .lib 'ex1mod0p1.l' FF
* .lib 'ex1mod0p1.l' SS
* .lib 'ex1mod0p1.l' SF
* .lib 'ex1mod0p1.l' FS
* .lib 'ex1mod0p1.l' MC
*
* .options scale=1
* XXX = svt
* Transistor sub-circuits
*  p4_XXX : PMOS XXX vt Ldrawn = 0.026-u, Wdrawn = 1-u
*      x0 d g s b
*      +p4_svt w=w l=l          $$ REQUIRED
*
```

```
* ------------------------- optional -------------------------
*          + z=z                     $$ default = 0.1n: s/d length from
* channel
*          + ad=ad as=as pd=pd ps=ps  $$ default = function of w, l, z
*          + n_fingers=#fingers       $$ default=1
*          + sigmavt=(instance sigvt) $$ default=0
*          + mismatchflag=(0|1)       $$ default=0
*
* n4_XXX : NMOS XXX vt Ldrawn = 0.026u, Wdrawn = 1u
*       x0 d g s b
*       +n4_svt w=w l=1              $$ REQUIRED
*
* -------------------------- optional --------------------------
*          + z=z                     $$ default = 0.1n : s/d length from
* channel
*          + ad=ad as=as pd=pd ps=ps  $$ default = function of w, l, z
*          + n_fingers=#fingers       $$ default=1
*          + sigmavt=(instance sigvt) $$ default=0
*          + mismatchflag=(0|1)       $$ default=0
*
* For statistical modeling using Monte Carlo simulation, use:
* .lib 'ex1mod0p1.l' MC
*
* Use flags to simulate different sources of process
* variability, For example, to simulate local variation, set
* the following flags
*
* .param globalsigmavtflag=0
* .param globalmcflag=0
* .param globalmismatchflag=1
*
* MC simulations       globalsigmavtflag  globalmcflag  globalmismatchflag
* --------------       -----------------  ------------  ------------------
* MC (Global + Local)       0                 1                 1
* MC (Global)               0                 1                 0
* MC (Local)                0                 0                 1
*
* Example of user-defined (mismatch parameter) "avt" values in
* Hspice netlist:
*          .param avtp=1.0e-9
*          .param avtn=1.0e-9
*
* Use the following command to include the appropriate library
* for typical N and typical P (TT) models
* .lib 'ex1mod0p1.l' TT
*
* Temperature Range : 25C
* Vds Range         : 0 ~ 0.9 V
```

```
* Vbsn Range           : 0.3 V (forward) ~ -0.5 V (reverse)
* Vbsp Range           :-0.3 V (forward) ~ 0.5 V (reverse)
*
***************************************************************
* end header
***************************************************************
.LIB TT
.lib 'ex1mod0p1.l' MOD_GLOBAL
.lib 'ex1mod0p1.l' TT_svt
.ENDL TT
*
.LIB SS
.lib 'ex1mod0p1.l' MOD_GLOBAL
.lib 'ex1mod0p1.l' SS_svt
.ENDL SS
*
.LIB FF
.lib 'ex1mod0p1.l' MOD_GLOBAL
.lib 'ex1mod0p1.l' FF_svt
.ENDL FF
*
.LIB SF
.lib 'ex1mod0p1.l' MOD_GLOBAL
.lib 'ex1mod0p1.l' SF_svt
.ENDL SF
*
.LIB FS
.lib 'ex1mod0p1.l' MOD_GLOBAL
.lib 'ex1mod0p1.l' FS_svt
.ENDL FS
*
.LIB MC
.lib 'ex1mod0p1.l' MOD_GLOBAL
.lib 'ex1mod0p1.l' MC_svt
.ENDL MC

.LIB MOD_GLOBAL
***************************************************************
* global model parameters
***************************************************************
.param globalmcflag=0

.param
+globalnmean=0
+globalpmean=0
+globalnsigma=1
+globalpsigma=1
+globalnmeansigma=2.8284
+globalpmeansigma=2.8284
```

```
.param globalsigmavtflag=1
.param globalmismatchflag=0
.param globalmismatchpsigma=3
.param globalmismatchnsigma=3

.param wmin_p_svt=0.99e-8
.param wmax_p_svt=10.01e-6
.param wmin_n_svt=0.99e-8
.param wmax_n_svt=10.01e-6

.ENDL MOD_GLOBAL
****************************************************************
* VT TYPE = SVT (standard VT)
****************************************************************

**************** SVT Library of Typical Case ****************
.LIB TT_svt

.param sdvtncorn = 0
.param sdvtpcorn = 0

** Need these defined since the parameters created for MC are
*  in the sub-circuit
** Sub-circuit will therefore, over-ride 'tox, dxl, dxw, dvth,
*  cj, cjsw, cjswn, cgo, cgl' in this LIB.
** g[np]sigma is always 0 when using fixed corners
.param gnmean = 0
.param gpmean = 0
.param gnsigma = 0
.param gpsigma = 0
*
.param
+toxn_svt = 1.0800E-09    dxln_svt = 0
+dxwn_svt = 0             dvthn_svt = 0
+cjn_svt =  3.2010E-03    cjswn_svt = 1.0000E-14
+cjswgn_svt = 1.9383E-10  cgon_svt = 1.5000E-11
+cgln_svt = 1.0000E-11    drdswn_svt = 0
.param
+toxp_svt = 1.1398E-09    dxlp_svt = 0
+dxwp_svt = 0             dvthp_svt = 0
+cjp_svt = 3.9056E-03     cjswp_svt = 1.0000E-14
+cjswgp_svt = 2.5218E-10  cgop_svt = 3.3205E-12
+cglp_svt =  6.4850E-12   drdswp_svt = 0
*
.lib 'ex1mod0p1.l' SUBCKTS_SVT
.ENDL TT_svt

********** SVT Library of SNSP Corner Case with RDD **********
```

```
.LIB SS_svt

.param sdvtncorn = 1
.param sdvtpcorn = 1


** Need these defined since the parameters created for MC are
*   in the sub-circuit
** Sub-circuit will therefore, over-ride 'tox, dxl, dxw, dvth,
*   cj, cjsw, cjswn,cgo, cgl' in this LIB.
** g[np]sigma is always 0 when using fixed corners
.param gnmean = 1
.param gpmean = 1
.param gnsigma = 0
.param gpsigma = 0
*
.param
+toxn_svt = 1.1081E-09      dxln_svt = 1.2000E-09
+dxwn_svt = -2.7000E-09     dvthn_svt = 1.5058E-02
+cjn_svt = 3.4251E-03       cjswn_svt = 1.0700E-14
+cjswgn_svt = 2.0740E-10    cgon_svt = 1.3950E-11
+cgln_svt = 9.3000E-12      drdswn_svt = 4.9500E+00
.param
+toxp_svt = 1.1683E-09      dxlp_svt = 1.2000E-09
+dxwp_svt = -2.7000E-09     dvthp_svt = -2.3075E-02
+cjp_svt = 4.1799E-03       cjswp_svt = 1.0700E-14
+cjswgp_svt = 2.6983E-10    cgop_svt = 3.0881E-12
+cglp_svt = 6.0311E-12      drdswp_svt = 5.3496E+00
*
.lib 'ex1mod0p1.1' SUBCKTS_SVT
.ENDL SS_svt


********* SVT Library of FNFP Corner Case with RDD **********
.LIB FF_svt

.param sdvtncorn = -1
.param sdvtpcorn = -1


** Need these defined since the parameters created for MC are
*   in the sub-circuit
** Sub-circuit will therefore, over-ride 'tox, dxl, dxw, dvth,
*   cj, cjsw, cjswn, cgo, cgl' in this LIB.
** g[np]sigma is always 0 when using fixed corners
.param gnmean = -1
.param gpmean = -1
.param gnsigma = 0
.param gpsigma = 0
```

```
*
.param
+toxn_svt = 1.0519E-09     dxln_svt = -1.2000E-09
+dxwn_svt = 2.7000E-09     dvthn_svt = -1.5058E-02
+cjn_svt = 2.9769E-03      cjswn_svt = 9.3000E-15
+cjswgn_svt = 1.8026E-10   cgon_svt = 1.6050E-11
+cgln_svt = 1.0700E-11     drdswn_svt = -4.9500E+00
.param
+toxp_svt = 1.1113E-09     dxlp_svt = -1.2000E-09
+dxwp_svt = 2.7000E-09     dvthp_svt = 2.3075E-02
+cjp_svt = 3.6313E-03      cjswp_svt = 9.3000E-15
+cjswgp_svt = 2.3453E-10   cgop_svt = 3.5529E-12
+cglp_svt = 6.9390E-12     drdswp_svt = -5.3496E+00
*
.lib 'ex1mod0p1.l' SUBCKTS_SVT
.ENDL FF_svt


********** SVT Library of SNFP Corner Case with RDD **********
.LIB SF_svt

.param sdvtncorn = 0
.param sdvtpcorn = -0

** Need these defined since the parameters created for MC are
*  in the sub-circuit
** Sub-circuit will therefore, over-ride 'tox, dxl, dxw, dvth,
*  cj, cjsw, cjswn, cgo, cgl' in this LIB.
** g[np]sigma is always 0 when using fixed corners
.param gnmean = 1
.param gpmean = -1
.param gnsigma = 0
.param gpsigma = 0
*
.param
+toxn_svt = 1.0800E-09     dxln_svt = 0
+dxwn_svt = -2.7000E-09    dvthn_svt = 1.0040E-02
+cjn_svt = 3.4251E-03      cjswn_svt = 1.0700e-14
+cjswgn_svt = 2.0740E-10   cgon_svt = 1.3950E-11
+cgln_svt =  9.3000E-12    drdswn_svtc = 0
.param
+toxp_svt = 1.1398e-09     dxlp_svt = 0
+dxwp_svt = -2.7000E-09    dvthp_svt = 1.5380E-02
+cjp_svt = 3.6313E-03      cjswp_svt = 9.3000E-15
+cjswgp_svt = 2.3453E-10   cgop_svt = 3.5529E-12
+cglp_svt = 6.9390E-1      drdswp_svt =  0
*
.lib 'ex1mod0p1.l' SUBCKTS_SVT
.ENDL SF_svt
```

```
********** SVT Library of FNSP Corner Case with RDD **********
.LIB FS_svt

.param sdvtncorn = -0
.param sdvtpcorn = 0

** Need these defined since the parameters created for MC are
*   in the sub- circuit
** Sub-circuit will therefore, over-ride 'tox, dxl, dxw, dvth,
*   cj, cjsw, cjswn, cgo, cgl' in this LIB.
** g[np]sigma is always 0 when using fixed corners
.param gnmean = -1
.param gpmean = 1
.param gnsigma = 0
.param gpsigma = 0
*
.param
+toxn_svt = 1.0800E-09      dxln_svt = 0
+dxwn_svt = 2.7000E-09      dvthn_svt = -1.0040E-02
+cjn_svt = 2.9769E-03       cjswn_svt = 9.3000E-15
+cjswgn_svt = 1.8026E-10    cgon_svt = 1.6050e-11
+cgln_svt = 1.0700E-11      drdswn_svt = 0
.param
+toxp_svt = 1.1398e-09      dxlp_svt = 0
+dxwp_svt = -2.7000e-09     dvthp_svt = -1.5380E-02
+cjp_svt = 4.1799E-03       cjswp_svt = 1.0700e-14
+cjswgp_svt = 2.6983E-10    cgop_svt = 3.0881E-12
+cglp_svt = 6.0311E-12      drdswp_svt = 0
*
.lib 'ex1mod0p1.l' SUBCKTS_SVT
.ENDL FS_svt


**************************************************************
*                  SVT MC MODEL LIBRARY                      *
**************************************************************
.lib MC_svt

** create 1 gnmean agauss and 1 gpmean agauss
.param grandom0n=agauss(0,1,3)
.param grandom0p=agauss(0,1,3)


.param Agmn='grandom0n*(globalmcflag==1)'
.param Agmp='grandom0p*(globalmcflag==1)'


** globalp/nmean=n shifts mean of systematic agauss by n*sigma
** global[pn]meansigma variation of mean to be added to g[np]sigma
```

```
** defaults are g[np]sigma=1, g[np]msigma=2.83
** sampling should be the same for each iteration (i.e. not
   instance based)
.param gnmean='(globalmcflag==1)*((Agmn*globalnmeansigma) +
globalnmean)/3'
.param gpmean='(globalmcflag==1)*((Agmp*globalpmeansigma) +
globalpmean)/3'

** globalp/nsigma=m multiplies stdev of systematic agauss by m
** this scale factor is passed to subcircuit to vary systematic
*  variation by instance
.param gnsigma=globalnsigma*(globalmcflag==1)/3
.param gpsigma=globalpsigma*(globalmcflag==1)/3

.param sdvtpcorn = 0
.param sdvtncorn = 0


.LIB 'ex1mod0p1.l' SUBCKTS_SVT
.endl MC_svt


.LIB SUBCKTS_SVT

**************************************************************
* Subcircuit references
**************************************************************
*
* weff = w/nf
* wnflag = 1 size bin w/nf
* wnflag = 0 size bin w
*
* fixed nf=1 and use instance mult factor to capture folding
* setting instance of these subckts with parameter nf!=1 have
* no effect area/peri default calculation assumes nf=1


.subckt p4_svt D G S B
.param w=0 l=0 z=0.100e-6 multi=1
.param pd='2*(w+z)' ad='w*z'
.param ps='2*(w+z)' as='w*z'
.param pw='2*((2*z)+l+w)' aw='((2*z)+l)*w'
.param lw='(2*z)+l'
.param n_fingers=1
*
.param mismatchflag=0
.param avtp=1.12e-09
** must make default sigmavt=0, i.e., no instance parameter
** sdvtp/ncorn in TT,SS,... takes care of corners sigvt when
*  globalsigmavtflag=1
```

```
.param sigmavt=0
.param delvto=0

** systematic **
** instance based sampling of systematic variation
** g[pn]mean varying in clock step for each instance
** defined in calling library MC_svt using global[np]mean and
*  global[np]meansigma
** existing global[np]sigma varies by instance  and is defined by
*  global[np]sigma
.param random0p=agauss(0,1,3)
.param random1p=agauss(0,1,3)
.param random2p=agauss(0,1,3)
.param random3p=agauss(0,1,3)
.param random4p=agauss(0,1,3)
.param random5p=agauss(0,1,3)
.param random6p=agauss(0,1,3)
.param random7p=agauss(0,1,3)
.param random8p=agauss(0,1,3)
.param random9p=agauss(0,1,3)
.param A0p='random0p*(globalmcflag==1)'
.param A1p='random1p*(globalmcflag==1)'
.param A2p='random2p*(globalmcflag==1)'
.param A3p='random3p*(globalmcflag==1)'
.param A4p='random4p*(globalmcflag==1)'
.param A5p='random5p*(globalmcflag==1)'
.param A6p='random6p*(globalmcflag==1)'
.param A7p='random7p*(globalmcflag==1)'
.param A8p='random8p*(globalmcflag==1)'
.param A9p='random9p*(globalmcflag==1)'
.param
+toxp_svt = '1.1398E-009 + ( 2.8414E-011*((A0p*gpsigma) + gpmean))'
+dxlp_svt = '0.0000E+000 + ( 1.2000E-009*((A1p*gpsigma) + gpmean))'
+dxwp_svt = '0.0000E+00 + (-2.7000E-009*((A2p*gpsigma) + gpmean))'
+dvthp_svt = '0.0000E+00 + (-2.3075E-002*((A3p*gpsigma) + gpmean))'
+cjp_svt = '3.9056E-003 + ( 2.7427E-004*((A4p*gpsigma) + gpmean))'
+cjswp_svt = '1.0000E-014 + ( 7.0000E-016*((A5p*gpsigma) + gpmean))'
+cjswgp_svt = '2.5218E-01 + ( 1.7653E-011*((A6p*gpsigma) + gpmean))'
+cgop_svt = '3.3205E-012 + (-2.3244E-013*((A7p*gpsigma) + gpmean))'
+cglp_svt = '6.4850E-012 + (-4.5395E-013*((A8p*gpsigma) + gpmean))'
+drdswp_svt = '0.0000E+00 + ( 5.3496E-000*((A9p*gpsigma) + gpmean))'

*** mismatch ***
.param AM0=agauss(0,1,1)

.param
+cvtp_svt='avtp/1.414214'
+lef='l-8.3256E-11' wef='w/n_fingers+0E-09'
+geo_fac='1/sqrt(multi*n_fingers*lef*wef)'
```

```
+sigvtp_svt='AM0*cvtp_svt*geo_fac*globalmismatchpsigma/3 *
((globalmismatchflag==1)||(mismatchflag==1))'

** fixed sigmavt offset **
.param
+sdvtp_svt='-cvtp_svt*geo_fac*((sigmavt*(globalsigmavtflag==0)) +
(sdvtpcorn*(globalsigmavtflag==1)))'

** no width effect
mp_svt D G S B pch_svt w=1e-6 l=1 pd=pd ad=ad ps=ps as=as
m='w/1.0e-6' nf=n_fingers wnflag=1 delvto=delvto
** width effect
*mp_svt D G S B pch_svt w=w l=1 pd=pd ad=ad ps=ps as=as
nf=n_fingers wnflag=1 delvto=delvto

**************************************************************
* BSIM4.5.0 model card for p-type devices
**************************************************************
.model  pch_svt.1 pmos  ( level = 54
**************************************************************
*                MODEL FLAG PARAMETERS
**************************************************************
+lmin = 2.59e-008     lmax = 2.51e-007    wmin = 'wmin_p_svt'
+wmax = 'wmax_p_svt'  version = 4.5       binunit = 1
+paramchk = 1         mobmod = 1          capmod = 2
+igcmod = 2           igbmod = 1          geomod = 0
+diomod = 1           rdsmod = 0          rbodymod= 0
+rgeomod = 0          rgatemod= 0         permod = 1
+acnqsmod= 0          trnqsmod= 0
**************************************************************
*                GENERAL MODEL PARAMETERS
**************************************************************
+tnom = 25            toxe = 'toxp_svt'   toxp = 1.012e-009
+toxm = 'toxp_svt'    dtox = 2.5e-010     epsrox = 3.9
+toxref = 1.2e-009    wmlt = 1            wint = 0
+lint = 4.1628e-011   ll = 0              wl = 0
+lln = 1              wln = 1             lw = 0
+ww = 0               lwn = 1             wwn = 1
+lwl = 0              wwl = 0             xl = '0+dxlp_svt'
+xw = '0+dxwp_svt'    dlc = 2.6887e-009   dwc = 0
+xpart = 0
**************************************************************
*                DC PARAMETERS
**************************************************************
+vth0 = '-0.46149+dvthp_svt+sdvtp_svt+sigvtp_svt'  k1 = 0.35256
+lk1 = -0.0024567     k2 = 0.017398       k3 = 0.27044
+k3b = 0.07434        w0 = 1e-007         dvt0 = 0.52272
+dvt1 = 0.50091       dvt2 = -0.021065    dvt0w = 0.013
+dvt1w = 5984800      dvt2w = 0.05        dsub = 4.1202
```

```
+minv = 0.69601          voffl = -8.0716e-011  dvtp0 = 1e-013
+dvtp1 = 1e-013          lpe0 = 1.5509e-011    lpeb = 4.606e-008
+vbm = -3                xj = 5.81e-008        ngate = 1.5e+022
+ndep = 1.7e+017         nsd = 1e+020          phin = 0
+cdsc = 0.0036589        cdscb = 0.001341      cdscd = 0.00027994
 +cit = -3.2858e-018     voff = -0.11885       nfactor = 4.1034
+eta0 = 1.6875           etab = -19.429        u0 = 0.0056074
+lu0 = 0.0026234         ua = 1.1889e-009      lua = 1.448e-011
+ub = 2.8685e-019        uc = -0.11407         eu = 1
+vsat = 127540           a0 = 3.7101           ags = 2.4363
+lags = -0.25867         a1 = 0                a2 = 1
+b1 = 0                  keta = 0.057369       lketa = 0.0049156
+dwg = 0                 dwb = 0               pclm = 0.14801
+pdiblc1 = 0.028077      pdiblc2 = 0.00016557  pdiblcb = 0.029513
+drout = 0.51504         pvag = 0              delta = 0.002695
+ldelta = 0.0004167      pscbe1 = 7.884e+008   pscbe2 = 1.9903e-006
+fprout = 0              pdits = 0             pditsd = 0
+pditsl = 0              rsh = 0
+rdsw = '178.32+drdswp_svt'                    rsw = 100
+rdw = 100               rdswmin = 0           rdwmin = 0
+rswmin = 0              prwg = 0.0031883      prwb = 0.25478
+wr = 1                  alpha0 = 7e-011       alpha1 = 7.2e-011
+beta0 = 18.96           agidl = 1.4811e-010   bgidl = 2250600
+cgidl = 433.08          egidl = 0.021835      aigbacc = 0.43
+bigbacc = 0.054         cigbacc = 0.075       nigbacc = 1
+aigbinv = 0.35          bigbinv = 0.03        cigbinv = 0.006
+eigbinv = 1.1           nigbinv = 3           aigc = 0.43
+bigc = 0.054            cigc = 0.075          aigsd = 0.43
+bigsd = 0.054           cigsd = 0.075         nigc = 1
+poxedge = 1             pigcd = 1             ntox = 1
*************************************************************
*                  CAPACITANCE PARAMETERS
*************************************************************
+cgso = 'cgop_svt'       cgdo = 'cgop_svt'     cgbo = 1.7739e-009
+cgdl = 'cglp_svt'       cgsl = 'cglp_svt'     clc = 1.2714e-011
+cle = 1                 ckappas = 0.12        ckappad = 0.12
+vfbcv = -0.5008         acde = 0.414          moin = 3.2553
+noff = 2.8698           voffcv = -0.01272     lvoffcv = 0.0024
*************************************************************
*                  TEMPERATURE PARAMETERS
*************************************************************
+kt1 = -0.47607          kt1l = -1.025e-010    kt2 = -0.050313
+ute = -1.75             ua1 = 4.187e-011      ub1 = -2.882e-019
+uc1 = -6.5038e-010      prt = 0               at = 71599
*************************************************************
*                  NOISE PARAMETERS
*************************************************************
+fnoimod = 1             tnoimod = 0           em = 4.1e+007
+ef = 1                  noia = 6.25e+041      noib = 3.125e+026
+noic = 8.75e+009        ntnoi = 1
```

```
*************************************************************
*                    DIODE PARAMETERS
*************************************************************
+jss = 7.6065e-005      jsws = 6.8173e-014     jswgs = 0
+njs = 1.2059           ijthsfwd= 0.017908     ijthsrev= 0.1
+bvs = 10               xjbvs = 1              pbs = 1.6835
+cjs = 'cjp_svt'        mjs = 0.72601          pbsws = 1
+cjsws = 'cjswp_svt'    mjsws = 0.5            pbswgs = 0.76056
+cjswgs = 'cjswgp_svt'  mjswgs = 0.424         cjd = 'cjp_svt'
+cjswd = 'cjswp_svt'    cjswgd = 'cjswgp_svt'  tpb = 0.0020847
+tcj = 0.00098          tpbsw = 0              tcjsw = 0
+pbswg = 6e-005         tcjswg = 0.00047385    xtis = 3
*************************************************************
*              LAYOUT RELATED PARAMETERS
*************************************************************
+dmcg = 0               dmdg = 0              dmcgt = 0
+xgw = 0                xgl = 0
*************************************************************
*                    RF PARAMETERS
*************************************************************
+rshg = 0.1             gbmin = 1e-012        rbpb = 50
+rbpd = 50              rbps = 50             rbdb = 50
+rbsb = 50              ngcon = 1             xrcrg1 = 12
+xrcrg2 = 1
*************************************************************
*                  STRESS PARAMETERS
*************************************************************
+saref = 1e-006         sbref = 1e-006        wlod = 0
+kvth0 = 0              lkvth0 = 0            wkvth0 = 0
+pkvth0 = 0             llodvth = 0           wlodvth = 0
+stk2 = 0               lodk2 = 1             lodeta0 = 1
+ku0 = 0                lku0 = 0              wku0 = 0
+pku0 = 0              llodku0 = 0            wlodku0 = 0
+kvsat = 0             steta0 = 0             tku0 = 0        )
.ends
*
.subckt n4_svt D G S B
.param w=0 l=0 z=0.100e-6 multi=1
.param pd='2*(w+z)'ad='w*z'
.param ps='2*(w+z)'as='w*z'
.param pw='2*((2*z)+l+w)'aw='((2*z)+l)*w'
.param lw='(2*z)+l'
.param n_fingers=1

.param mismatchflag=0
.param avtn=1.12e-09

** must make default sigmavt=0, i.e. no instance parameter
** sdvtp/ncorn in TT,SS,... takes care of corners sigvt when
*   globalsigmavtflag=1
```

```
.param sigmavt=0
.param delvto=0

** systematic **
** instance based sampling of systematic variation
** g[pn]mean varying in clock step for each instance
** defined in calling library MC_svt using globap[np]mean and
*   global[np]meansigma
** existing global[np]sigma varys by instance  and is defined by
*   global[np]sigma
.param random0n=agauss(0,1,3)
.param random1n=agauss(0,1,3)
.param random2n=agauss(0,1,3)
.param random3n=agauss(0,1,3)
.param random4n=agauss(0,1,3)
.param random5n=agauss(0,1,3)
.param random6n=agauss(0,1,3)
.param random7n=agauss(0,1,3)
.param random8n=agauss(0,1,3)
.param random9n=agauss(0,1,3)

.param A0n='random0n*(globalmcflag==1)'
.param A1n='random1n*(globalmcflag==1)'
.param A2n='random2n*(globalmcflag==1)'
.param A3n='random3n*(globalmcflag==1)'
.param A4n='random4n*(globalmcflag==1)'
.param A5n='random5n*(globalmcflag==1)'
.param A6n='random6n*(globalmcflag==1)'
.param A7n='random7n*(globalmcflag==1)'
.param A8n='random8n*(globalmcflag==1)'
.param A9n='random9n*(globalmcflag==1)'

.param
+toxn_svt = '1.0800E-009 + ( 2.7831E-011*((A0n*gnsigma) + gnmean))'
+dxln_svt = '0.0000E+000 + ( 1.2000E-009*((A1n*gnsigma) + gnmean))'
+dxwn_svt = '0.0000E+000 + (-2.7000E-009*((A2n*gnsigma) + gnmean))'
+dvthn_svt = '0.0000E+000 + ( 1.5058E-002*((A3n*gnsigma) + gnmean))'
+cjn_svt = '3.2010E-003 + ( 2.2407E-004*((A4n*gnsigma) + gnmean))'
+cjswn_svt = '1.0000E-014 + ( 7.0000E-016*((A5n*gnsigma) + gnmean))'
+cjswgn_svt = '1.9383E-010 + ( 1.3568E-011*((A6n*gnsigma) + gnmean))'
+cgon_svt = '1.5000E-011 + (-1.0500E-012*((A7n*gnsigma) + gnmean))'
+cgln_svt = '1.0000E-011 + (-7.0000E-013*((A8n*gnsigma) + gnmean))'
+drdswn_svt = '0.0000E+00 + ( 4.9500E-000*((A9n*gnsigma) + gnmean))'

*** mismatch ***
.param AM0=agauss(0,1,1)

.param
+cvtn_svt='avtn/1.414214'
```

```
+lef='l-7.5802E-09' wef='w/n_fingers+0E-09'
+geo_fac ='1/sqrt(multi*n_fingers*lef*wef)'
+sigvtn_svt='AM0*cvtn_svt*geo_fac*globalmismatchnsigma/3 *
((globalmismatchflag==1)||(mismatchflag==1))'

** fixed sigmavt offset **
.param
+sdvtn_svt='cvtn_svt*geo_fac*((sigmavt*(globalsigmavtflag==0)) +
(sdvtncorn*(globalsigmavtflag==1)))'

** no width effect
mn_svt D G S B nch_svt w=1e-6 l=1 pd=pd ad=ad ps=ps as=as
m='w/1.0e-6' nf=n_fingers wnflag=1  delvto=delvto
** width effect
*mn_svt D G S B nch_svt w=w l=1 pd=pd ad=ad ps=ps as=as
nf=n_fingers wnflag=1 delvto=delvto

****************************************************************
* BSIM4.5.0 model card for n-type devices
****************************************************************
.model  nch_svt.1 nmos  ( level = 54
*
****************************************************************
*                MODEL FLAG PARAMETERS
****************************************************************
+lmin = 2.59e-008     lmax = 2.51e-007    wmin = 'wmin_n_svt'
+wmax = 'wmax_n_svt'  version = 4.5       binunit = 1
+paramchk= 1          mobmod = 1          capmod = 2
+igcmod = 2           igbmod = 1          geomod = 0
+diomod = 1           rdsmod = 0          rbodymod= 0
+rgeomod = 0          rgatemod= 0         permod = 1
+acnqsmod= 0          trnqsmod= 0
****************************************************************
*                GENERAL MODEL PARAMETERS
****************************************************************
+tnom = 25            toxe = 'toxn_svt'   toxp = 1.05e-009
+toxm = 'toxn_svt'    dtox = 2.3e-010     epsrox = 3.9
+toxref = 9.043737e-010  wmlt = 1         wint = 0
+lint = 4.1539e-009   ll = 2.455641e-024  wl = 0
+lln = 1.000054       wln = 1             lw = 0
+ww = 0               lwn = 1             wwn = 1
+lwl = 0              wwl = 0             xl = '0+dxln_svt'
+xw = '0+dxwn_svt'    dlc = 1.1811e-009   dwc = 1e-009
+xpart = 0
****************************************************************
*                DC PARAMETERS
****************************************************************
+vth0 = '0.30116+dvthn_svt+sdvtn_svt+sigvtn_svt' k1 = 0.40102
+lk1 = -0.0018       k2 = 0.036607        k3 = 80
```

```
+k3b = 11.372          w0 = 1.4976e-008      dvt0 = 0.030258
+dvt1 = 0.27353        dvt2 = -0.1292        dvt0w = 0
+dvt1w = 10000000      dvt2w = 0.01          dsub = 0.079604
+minv = 0.57893        voffl = -3.2236e-015  dvtp0 = 0
+dvtp1 = 0             lc = 5e-009           lambda = 0
+vtl = 200000          lpe0 = 0              lpeb = 9.7128e-009
+vbm = -3              xj = 5.81e-008        ngate = 1.1557e+021
+ndep = 1.7e+017       nsd = 1e+020          phin = 0
+cdsc = 0.00038013     cdscb = 0.00029951    cdscd = 0
+cit = 0.00029514      voff = -0.091517      nfactor = 4.6201
+eta0 = 0.00011433     etab = -0.00018655    u0 = 0.053874
+ua = 1.3724e-009      ub = 2.4451e-021      uc = -0.3327
+eu = 1.67             vsat = 81402          a0 = 4.034
+ags = 0.51            a1 = 1e-005           a2 = 1
+b0 = 6.0098e-015      b1 = 0                keta = 0.010603
+lketa = 0.0030012     dwg = 0               dwb = 0
+pclm = 0.26825        lpclm = 0.0057        pdiblc1 = 0.030607
+pdiblc2 = 2.582e-005  pdiblcb = 0.01        drout = 1.1625
+pvag = 0              delta = 0.0045611     ldelta = 0.0002262
+pscbe1 = 1.1753e+009  pscbe2 = 1e-005       fprout = 0.01
+pdits = 0             pditsd = 0            pditsl = 0
+rsh = 0               rdsw = '159.72+drdswn_svt'
+rsw = 100             rdw = 100             rdswmin = 0
+rdwmin = 0            rswmin = 0            prwg = 0.15293
+prwb = 0.080695       wr = 1                alpha0 = 0
+alpha1 = 0            beta0 = 30            agidl = 2.6078e-010
+bgidl = 4984400       cgidl = 1974.8        egidl = 0.057969
+aigbacc = 0.43        bigbacc = 0.054       cigbacc = 0.075
+nigbacc = 1           aigbinv = 0.35        bigbinv = 0.03
+cigbinv = 0.006       eigbinv = 1.1         nigbinv = 3
+aigc = 0.43           bigc = 0.054          cigc = 0.075
+aigsd = 0.43          bigsd = 0.054         cigsd = 0.075
+nigc = 1              poxedge = 1           pigcd = 1
+ntox = 1
*****************************************************************
*                    CAPACITANCE PARAMETERS
*****************************************************************
+cgso = 'cgon_svt'     cgdo = 'cgon_svt'     cgbo = 2.191e-009
+cgdl = 'cgln_svt'     cgsl = 'cgln_svt'     clc = 2.905e-010
+cle = 1               ckappas = 0.6         ckappad = 0.6
+vfbcv = -1.1698       acde = 0.37365        lacde = -0.00145
+moin = 3.8485         lmoin = 0.022         noff = 2.118
+lnoff = 0.0114        voffcv = 0.048569     lvoffcv = 0.00012324
*****************************************************************
*                    TEMPERATURE PARAMETERS
*****************************************************************
+kt1 = -0.38457        kt1l = -9.0624e-012   kt2 = -0.029313
+ute = -2.895          ua1 = 1.4986e-009     ub1 = -3.2473e-018
```

```
+uc1 = -3.5e-011    prt = 67.655            at = 36837
+lat = 16049
*******************************************************************
*                   NOISE PARAMETERS
*******************************************************************
+fnoimod = 1        tnoimod = 0            em = 4.1e+007
+ef = 1             noia = 6.25e+041       noib = 3.125e+026
+noic = 8.75e+009   ntnoi = 1
*******************************************************************
*                   DIODE PARAMETERS
*******************************************************************
+jss = 8.2539e-005  jsws = 7.9765e-012     jswgs = 8e-012
+njs = 1.2512       ijthsfwd= 4.5539e-005  ijthsrev= 0.1818
+bvs = 10           xjbvs = 1.08           pbs = 0.91838
+cjs = 'cjn_svt'    mjs = 0.3616           pbsws = 1
+cjsws = 'cjswn_svt'  mjsws = 0.5          pbswgs = 0.73407
+cjswgs = 'cjswgn_svt'  mjswgs = 0.3464    cjd = 'cjn_svt'
+cjswd = 'cjswn_svt'  cjswgd = 'cjswgn_svt'  tpb = 0.0019412
+tcj = 0.00075514991  tpbsw = 0            tcjsw = 0
+tpbswg = 1.5577e-017  tcjswg = 1.0211e-017  xtis = 3
*******************************************************************
*                   LAYOUT RELATED PARAMETERS
*******************************************************************
+dmcg = 0           dmdg = 0               dmcgt = 0
+xgw = 0            xgl = 0
*******************************************************************
*                   RF PARAMETERS
*******************************************************************
+rshg = 0.1         gbmin = 1e-012         rbpb = 50
+rbpd = 50          rbps = 50              rbdb = 50
+rbsb = 50          ngcon = 1              xrcrg1 = 12
+xrcrg2 = 1
*******************************************************************
*                   STRESS PARAMETERS
*******************************************************************
+saref = 1e-006     sbref = 1e-006         wlod = 0
+kvth0 = 0          lkvth0 = 0             wkvth0 = 0
+pkvth0 = 0         llodvth = 0            wlodvth = 0
+stk2 = 0           lodk2 = 1              lodeta0 = 1
+ku0 = 0            lku0 = 0               wku0 = 0
+pku0 = 0           llodku0 = 0            wlodku0 = 0
+kvsat = 0          steta0 = 0             tku0 = 0          )
.ends
*
.ENDL SUBCKTS_SVT
*******************************************************************
```

2. Sample SRAM Compact Model of an Advanced CMOS Technology
```
*****************************************************************
* Example2: 20-nm CMOS technology node SRAM Hspice model library
*****************************************************************
* Version:    sram127hp.l *** January 20, 2015 ***
* MODEL:    BSIM4 ( V4.5)
*****************************************************************
* Model Information
*****************************************************************
* This Rev0 version of sample SPICE model is based on
  two-dimensional
* numerical device simulation
*****************************************************************
.LIB INTERNAL
*-------------------------------------------------------------
* History of model updates
*-------------------------------------------------------------
* Jan 20, 2015: model created from TCAD-based model
*               : bit cell127: wxn_pd = 76.5; wxn_pg = 63.9;
                  wxp_pu = 50.4; *  L = 35.1 (nm)
*-------------------------------------------------------------
.ENDL INTERNAL
*****************************************************************
* Begin header
*****************************************************************
* Usage:
*
* Hspice Version: 2007.03, 2008.03, 2009.03
*
* .lib 'sram127hp.l' TT
* .lib 'sram127hp.l' FF
* .lib 'sram127hp.l' SS
* .lib 'sram127hp.l' SF
* .lib 'sram127hp.l' FS
* .lib 'sram127hp.l' MC
*
* .options scale=1
*
* Transistor sub-circuits
* p4_pu_svt : PMOS PULL UP Ldrawn = 0.0351um,
* Wdrawn = 0.0504u
*     x0 d g s b
*     +p4_pu_svt w = w  l = l    $$ REQUIRED
*
*-------------------------optional-----------------------------
*     + z=z                      $$ default = 0.1n: s/d length from
* channel
*     + ad=ad as=as pd=pd ps=ps  $$ default = function of w, l, z
*     + n_fingers=#fingers       $$ default=1
```

```
*      + sigmavt=(instance sigvt)    $$ default=0
*      + mismatchflag=(0|1)          $$ default=0
*
* n4_pd_svt : NMOS PULL DOWN  Ldrawn = 0.0351u,
* Wdrawn = 0.0765u
*      x0 d g s b
*      +n4_pd_svt w=w  l=l            $$ REQUIRED
*
*-------------------------optional----------------------------
*      + z=z                         $$ default = 0.1n: s/d length from
* channel
*      + ad=ad as=as pd=pd ps=ps     $$ default = function of w, l, z
*      + n_fingers=#fingers          $$ default=1
*      + sigmavt=(instance sigvt)    $$ default=0
*      + mismatchflag=(0|1)          $$ default=0
*
* n4_pg_svt : NMOS PASS GATE Ldrawn = 0.0351u,
* Wdrawn = 0.0639u
*      x0 d g s b
*      + n4_pg_svt w=w  l=l  $$ REQUIRED
*
*-------------------------optional----------------------------
*      + z=z                         $$ default = 0.1n: s/d length from
* channel
*      | ad=ad as=as pd=pd ps=ps     $$ default = function of w, l, z
*      + n_fingers=#fingers          $$ default=1
*      + sigmavt=(instance sigvt)    $$ default=0
*      + mismatchflag=(0|1)          $$ default=0
*
* For statistical modeling using Monte Carlo simulation, use:
* .lib 'sram127hp.l' MC
*
* Example of user-defined (mismatch parameter) "avt" values in Hspice
* netlist:
*         :.param avtp=1.0e-9
*         :.param avtn=1.0e-9
*
***********************************************************************
* end header
***********************************************************************
.LIB TT
.lib 'sram127hp.l' MOD_GLOBAL
.lib 'sram127hp.l' TT_svt
.ENDL TT
*
.LIB SS
.lib 'sram127hp.l' MOD_GLOBAL
.lib 'sram127hp.l' SS_svt
.ENDL SS
```

```
*
.LIB FF
.lib 'sram127hp.l' MOD_GLOBAL
.lib 'sram127hp.l' FF_svt
.ENDL FF
*
.LIB SF
.lib 'sram127hp.l' MOD_GLOBAL
.lib 'sram127hp.l' SF_svt
.ENDL SF
*
.LIB FS
.lib 'sram127hp.l' MOD_GLOBAL
.lib 'sram127hp.l' FS_svt
.ENDL FS
*
.LIB MC
.lib 'sram127hp.l' MOD_GLOBAL
.lib 'sram127hp.l' MC_svt
.ENDL MC

.LIB MOD_GLOBAL
***************************************************************
* global model parameters
***************************************************************
.param globalmcflag=0

.param
+globalnmean=0
+globalpmean=0
+globalnsigma=1
+globalpsigma=1
+globalnmeansigma=2.8284
+globalpmeansigma=2.8284

.param globalsigmavtflag=1
.param globalmismatchflag=0
.param globalmismatchpsigma=3
.param globalmismatchnsigma=3

.param wmin_n_pd_svt=0.999e-6
.param wmax_n_pd_svt=1.01e-6
.param wmin_n_pg_svt=0.999e-6
.param wmax_n_pg_svt=1.01e-6
.param wmin_p_pu_svt=0.999e-6
.param wmax_p_pu_svt=1.01e-6

.ENDL MOD_GLOBAL
***************************************************************
* SVT SRAM - Bitcell127
```

```
****************** Library of Typical Case ******************
.LIB TT_svt

.param sdvtncorn = 0
.param sdvtpcorn = 0

** Need these defined since the parameters created for MC are
*   in the sub-circuit
** Sub-circuit will therefore, over-ride 'tox, dxl, dxw, dvth,
*   cj, cjsw, cjswn,cgo, cgl' in this LIB.
** g[np]sigma is always 0 when using fixed corners
.param gnmean = 0
.param gpmean = 0
.param gnsigma = 0
.param gpsigma = 0
*
*device: pull down
.param
+toxn_pd_svt = 1.6e-09        dxln_pd_svt = 0.0
+dxwn_pd_svt = 0             dvthn_pd_svt = 0
+cjn_pd_svt = 0.0030468587   cjswn_pd_svt = 1.000E-14
+cjswgn_pd_svt = 3.0727e-010  cgon_pd_svt = 9.8e-12
+cgln_pd_svt = 4.05e-11
*
*device: pull up
.param
+toxp_pu_svt = 1.6e-09        dxlp_pu_svt = 0.0
+dxwp_pu_svt = 0             dvthp_pu_svt = 0
+cjp_pu_svt = 0.0033668797   cjswp_pu_svt = 1.000e-014
+cjswgp_pu_svt = 3.10333e-010  cgop_pu_svt = 1.0505e-012
+cglp_pu_svt = 5.3856e-011
*
*device: pass gate
.param
+toxn_pg_svt = 1.6e-09        dxln_pg_svt = 0.0
+dxwn_pg_svt = 0             dvthn_pg_svt = 0
+cjn_pg_svt = 0.0030468587   cjswn_pg_svt = 1.000E-14
+cjswgn_pg_svt = 3.0727e-010  cgon_pg_svt = 9.8e-12
+cgln_pg_svt = 4.05e-11
*
.LIB 'sram127hp.l' SUBCKTS_SVT
.ENDL TT_svt

*********** Library of SNSP Corner Case with RDD ***********
.LIB SS_svt

.param sdvtncorn = 1
.param sdvtpcorn = 1

** Need these defined since the parameters created for MC are
```

```
*  in the sub-circuit
** Sub-circuit will therefore, over-ride `tox, dxl, dxw, dvth,
*  cj, cjsw, cjswn, cgo, cgl' in this LIB.
** g[np]sigma is always 0 when using fixed corners
.param gnmean = 1
.param gpmean = 1
.param gnsigma = 0
.param gpsigma = 0
*
```

*device: pull down
```
.param
+toxn_pd_svt = 1.6412E-09     dxln_pd_svt = 1.2E-09
+dxwn_pd_svt = -2.70E-09      dvthn_pd_svt = 1.6768E-02
+cjn_pd_svt = 0.003260        cjswn_pd_svt = 1.070E-14
+cjswgn_pd_svt = 3.2878e-010  cgon_pd_svt = 9.114e-12
+cgln_pd_svt = 3.7665e-011
*
```

*device: pull down
```
.param
+toxp_pu_svt = 1.6399E-09     dxlp_pu_svt = 1.2E-09
+dxwp_pu_svt = -2.70E-09      dvthp_pu_svt = -1.8080E-02
+cjp_pu_svt = 0.003603        cjswp_pu_svt = 1.070E-14
+cjswgp_pu_svt = 3.3206e-010  cgop_pu_svt = 9.7696E-13
+cglp_pu_svt = 5.0086E-11
*
```

*device: pass gate
```
.param
+toxn_pg_svt = 1.6412E-09     dxln_pg_svt = 1.2E-09
+dxwn_pg_svt = -2.70E-09      dvthn_pg_svt = 1.6768E-02
+cjn_pg_svt = 0.003260        cjswn_pg_svt = 1.070E-14
+cjswgn_pg_svt = 3.2878e-010  cgon_pg_svt = 9.114e-12
+cgln_pg_svt = 3.7665e-011
*
.LIB `sram127hp.l' SUBCKTS_SVT
.ENDL SS_svt

************* Library of FNFP Corner Case with RDD **************
.LIB FF_svt

.param sdvtncorn = -1
.param sdvtpcorn = -1

** Need these defined since the parameters created for MC are
*  in the sub-circuit
** Sub-circuit will therefore, over-ride `tox, dxl, dxw, dvth,
```

```
*   cj, cjsw, cjswn, cgo, cgl' in this LIB.
** g[np]sigma is always 0 when using fixed corners
.param gnmean = -1
.param gpmean = -1
.param gnsigma = 0
.param gpsigma = 0
*
*device: pull down
.param
+toxn_pd_svt = 1.5588E-09     dxln_pd_svt = -1.2E-09
+dxwn_pd_svt = 2.70E-09       dvthn_pd_svt = -1.6768E-02
+cjn_pd_svt = 0.002834        cjswn_pd_svt = 9.3e-015
+cjswgn_pd_svt = 2.8576e-010  cgon_pd_svt = 1.0486E-11
+cgln_pd_svt = 4.3335E-11
*
*device: pull up
.param
+toxp_pu_svt = 1.5601E-09     dxlp_pu_svt = -1.2E-09
+dxwp_pu_svt = 2.70E-09       dvthp_pu_svt = 1.8080E-02
+cjp_pu_svt = 0.003130        cjswp_pu_svt = 9.3e-015
+cjswgp_pu_svt = 2.8861E-010  cgop_pu_svt = 1.124E-12
+cglp_pu_svt = 5.7626E-11
*
*device: pass gate
.param
+toxn_pg_svt = 1.5588E-09     dxln_pg_svt = -1.2E-09
+dxwn_pg_svt = 2.70E-09       dvthn_pg_svt = -1.6768E-02
+cjn_pg_svt = 0.002834        cjswn_pg_svt = 9.3e-015
+cjswgn_pg_svt = 2.8576e-010  cgon_pg_svt = 1.0486E-11
+cgln_pg_svt = 4.3335E-11
*
.LIB 'sram127hp.l' SUBCKTS_SVT
.ENDL FF_svt


*********** Library of SNFP Corner Case with RDD ************
.LIB SF_svt

.param sdvtncorn = 1
.param sdvtpcorn = -1

** Need these defined since the parameters created for MC are
*   in the sub-circuit
** Sub-circuit will therefore, over-ride 'tox, dxl, dxw, dvth,
*   cj, cjsw, cjswn, cgo, cgl' in this LIB.
** g[np]sigma is always 0 when using fixed corners
```

```
.param gnmean = 1
.param gpmean = -1
.param gnsigma = 0
.param gpsigma = 0
*
*device: pull down
.param
+toxn_pd_svt = 1.6E-09        dxln_pd_svt = 0.0
+dxwn_pd_svt = -2.70E-09      dvthn_pd_svt = 1.1179E-02
+cjn_pd_svt = 0.003260        cjswn_pd_svt = 1.070E-14
+cjswgn_pd_svt = 3.2878e-010  cgon_pd_svt = 9.114E-12
+cgln_pd_svt = 3.767e-11
*
*device: pull up
.param
+toxp_pu_svt = 1.6e-09        dxlp_pu_svt = 0.0
+dxwp_pu_svt = 2.70E-09       dvthp_pu_svt = 1.2053E-02
+cjp_pu_svt = 0.003130        cjswp_pu_svt = 9.3E-15
+cjswgp_pu_svt = 2.8861E-010  cgop_pu_svt = 1.124E-12
+cglp_pu_svt = 5.7626E-11
*
*device: pass gate
.param
+toxn_pg_svt = 1.6E-09        dxln_pg_svt = 0.0
+dxwn_pg_svt = -2.70E-09      dvthn_pg_svt = 1.1179E-02
+cjn_pg_svt = 0.003260        cjswn_pg_svt = 1.070E-14
+cjswgn_pg_svt = 3.2878e-010  cgon_pg_svt = 9.114E-12
+cgln_pg_svt = 3.767e-11
*
.LIB 'sram127hp.l' SUBCKTS_SVT
.ENDL SF_svt

************ Library of FNSP Corner Case with RDD ************
.LIB FS_svt

.param sdvtncorn = -1
.param sdvtpcorn = 1

** Need these defined since the parameters created for MC are
*   in the sub-circuit
** Sub-circuit will therefore, over-ride 'tox, dxl, dxw, dvth,
*   cj, cjsw, cjswn, cgo, cgl' in this LIB.
** g[np]sigma is always 0 when using fixed corners
.param gnmean = -1
.param gpmean = 1
.param gnsigma = 0
.param gpsigma = 0
*
*device: pull down
```

```
.param
+toxn_pd_svt = 1.6E-09          dxln_pd_svt = 0.0
+dxwn_pd_svt = 2.70E-09         dvthn_pd_svt = -1.1179E-02
+cjn_pd_svt = 0.002834          cjswn_pd_svt = 9.3E-15
+cjswgn_pd_svt = 2.8576e-010    cgon_pd_svt = 1.049E-11
+cgln_pd_svt = 4.3335E-11
*
*device: pull up
.param
+toxp_pu_svt = 1.6e-09          dxlp_pu_svt = 0.0
+dxwp_pu_svt = -2.70E-09        dvthp_pu_svt = -1.2053E-02
+cjp_pu_svt = 0.003603          cjswp_pu_svt = 1.070E-14
+cjswgp_pu_svt = 3.3206e-010    cgop_pu_svt = 9.7696E-13
+cglp_pu_svt = 5.0086E-11
*
*device: pass gate
.param
+toxn_pg_svt = 1.6E-09          dxln_pg_svt = 0.0
+dxwn_pg_svt = 2.70E-09         dvthn_pg_svt = -1.1179E-02
+cjn_pg_svt = 0.002834          cjswn_pg_svt = 9.3E-15
+cjswgn_pg_svt = 2.8576e-010    cgon_pg_svt = 1.049E-11
+cgln_pg_svt = 4.3335E-11
*
.LIB 'sram127hp.l' SUBCKTS_SVT
.ENDL FS_svt
*****************************************************************
*                    SRAM MC MODEL LIBRARY                      *
*****************************************************************
.lib MC_svt

** create 1 gnmean agauss and 1 gpmean agauss
.param grandom0n=agauss(0,1,3)
.param grandom0p=agauss(0,1,3)

.param Agmn='grandom0n*(globalmcflag==1)'
.param Agmp='grandom0p*(globalmcflag==1)'

** globalp/nmean=n shifts mean of systematic agauss by n*sigma
** global[pn]meansigma variation of mean to be added to g[np]sigma
** defaults are g[np]sigma=1, g[np]msigma=2.83
** sampling should be the same for each iteration (i.e., not
*  instance based)
.param gnmean='(globalmcflag==1)*((Agmn*globalnmeansigma) +
globalnmean)/3'
.param gpmean='(globalmcflag==1)*((Agmp*globalpmeansigma) +
globalpmean)/3'

** globalp/nsigma=m multiplies stdev of systematic agauss by m
** this scale factor is passed to subckt to vary systematic
```

```
*   variation by instance
.param gnsigma=globalnsigma*(globalmcflag==1)/3
.param gpsigma=globalpsigma*(globalmcflag==1)/3

.param sdvtpcorn = 0
.param sdvtncorn = 0

.LIB 'sram127hp.l' SUBCKTS_SVT
.endl MC_svt

.LIB SUBCKTS_SVT

*************************************************************
* Subcircuit references
*************************************************************
*
* weff = w/nf
* wnflag = 1  size bin w/nf
* wnflag = 0  size bin w
*
* fixed nf=1 and use instance mult factor to capture folding
* setting instance of these subckts with parameter nf!=1 have
  no effect
* area/peri default calculation assumes nf=1

* DEVICE: PULL UP
.subckt p4_pu_svt D G S B
.param w=0  l=0  z=0.100e-6  multi=1
.param pd='2*(w+z)'  ad='w*z'
.param ps='2*(w+z)'  as='w*z'
.param pw='2*((2*z)+l+w)'  aw='((2*z)+l)*w'
.param lw='(2*z)+l'
.param n_fingers=1

.param mismatchflag=0
.param avtp=2.8e-09

** must make default sigmavt=0, ie, no instance parameter
** sdvtp/ncorn in TT,SS,... takes care of corners sigvt when
** globalsigmavtflag=1
.param sigmavt=0
.param delvto=0

** systematic **
** instance based sampling of systematic variation
** g[pn]mean varying in lock step for each instance
** defined in calling library MC_svt using globap[np]mean and
** global[np]meansigma
** existing global[np]sigma varies by instance and is defined by
```

```
*   global[np]sigma
.param random0p=agauss(0,1,3)
.param random1p=agauss(0,1,3)
.param random2p=agauss(0,1,3)
.param random3p=agauss(0,1,3)
.param random4p=agauss(0,1,3)
.param random5p=agauss(0,1,3)
.param random6p=agauss(0,1,3)
.param random7p=agauss(0,1,3)
.param random8p=agauss(0,1,3)
.param A0p='random0p*(globalmcflag==1)'
.param A1p='random1p*(globalmcflag==1)'
.param A2p='random2p*(globalmcflag==1)'
.param A3p='random3p*(globalmcflag==1)'
.param A4p='random4p*(globalmcflag==1)'
.param A5p='random5p*(globalmcflag==1)'
.param A6p='random6p*(globalmcflag==1)'
.param A7p='random7p*(globalmcflag==1)'
.param A8p='random8p*(globalmcflag==1)'

.param
+toxp_pu_svt = '1.600E-09 + (3.9886E-011*((A0p*gpsigma) + gpmean))'
+dxlp_pu_svt = '0.000E+00 + (1.2000E-009*((A1p*gpsigma) + gpmean))'
+dxwp_pu_svt = '0.00E+00 + (-2.700E-009*((A2p*gpsigma) + gpmean))'
+dvthp_pu_svt = '0.00E+00 + (-1.8080E-02*((A3p*gpsigma) + gpmean))'
+cjp_pu_svt = '3.3669E-003 + (2.3644E-04*((A4p*gpsigma) + gpmean))'
+cjswp_pu_svt = '1.00E-014 + (7.000E-016*((A5p*gpsigma) + gpmean))'
+cjswgp_pu_svt = '3.103E-10 + (2.172e-11*((A6p*gpsigma) + gpmean))'
+cgop_pu_svt = '1.0505E-12 + (-7.354E-14*((A7p*gpsigma) + gpmean))'
+cglp_pu_svt = '5.3856E-11 + (-3.770E-12*((A8p*gpsigma) + gpmean))'

*** mismatch ***
.param AM0=agauss(0,1,1)

.param
+cvtp_svt='avtp/1.414214'
+lef='l-5.6E-09' wef='w/n_fingers+0E-09'
+geo_fac='1/sqrt(multi*n_fingers*lef*wef)'
+sigvtp_pu_svt='AM0*cvtp_svt*geo_fac*globalmismatchpsigma/3 *
((globalmismatchflag==1)||(mismatchflag==1))'

** fixed sigmavt offset **
.param
+sdvtp_pu_svt='-cvtp_svt*geo_fac*((sigmavt*(globalsigmavtflag=
=0)) + (sdvtpcorn*(globalsigmavtflag==1)))'

** no width effect
mp_pu_svt D G S B pch_svt w=1e-6 l=l pd=pd ad=ad ps=ps as=as
m='w/1.0e-6' nf=n_fingers wnflag=1 delvto=delvto
```

```
** width effect
*mp_pu_svt D G S B pch_svt w=w l=l pd=pd ad=ad ps=ps as=as
nf=n_fingers wnflag=1 delvto=delvto

***********************************************************
* BSIM4.5.0 model card for p-type devices
***********************************************************
*
.model pch_svt.1 pmos ( level = 54
***********************************************************
*                 MODEL FLAG PARAMETERS
***********************************************************
+lmin = 3e-008      lmax = 3.999e-008    wmin = 'wmin_p_pu_svt'
+wmax = 'wmax_p_pu_svt' version = 4.5    binunit = 1
+paramchk= 1        mobmod = 0           capmod = 2
+igcmod = 2         igbmod = 1           geomod = 0
+diomod = 1         rdsmod = 0           rbodymod= 0
+rgeomod = 0        rgatemod= 0          permod = 1
+acnqsmod= 0        trnqsmod= 0
***********************************************************
*                 GENERAL MODEL PARAMETERS
***********************************************************
+tnom = 27          toxe = 'toxp_pu_svt' toxp = 1.35e-009
+toxm = 'toxp_pu_svt' dtox = 2.5e-010    epsrox = 3.9
+toxref = 1.2e-009  wmlt = 1             wint = 0
+lint = 2.8175e-009 ll = 0               wl = 0
 +lln = 1           wln = 1              lw = 0
+ww = 0             lwn = 1              wwn = 1
+lwl = 0            wwl = 0              xl = '0+dxlp_pu_svt'
+xw = '0+dxwp_pu_svt'                    dlc = 4.0311e-009
+dwc = 2.2731e-009                       xpart = 0
***********************************************************
*                 DC PARAMETERS
***********************************************************
+vth0 = '-0.452+dvthp_pu_svt+sdvtp_pu_svt+sigvtp_pu_svt'
+k1 = 0.3           k2 = -0.06           k3 = 0.27044
+k3b = 0.07434      w0 = 1e-007          dvt0 = 0.01345
+dvt1 = 0.070243    dvt2 = -0.038        dvt0w = 0.013
+dvt1w = 5984800    dvt2w = 0.05         dsub = 1.7101
+minv = 0           voffl = -8.0716e-011 dvtp0 = 1e-013
+dvtp1 = 1e-013     lpe0 = 1.5509e-011   lpeb = 4.606e-008
+vbm = -3           xj = 5.81e-008       ngate = 0
+ndep = 1.7e+017    nsd = 1e+020         phin = 0
+cdsc = 2.2788e-007 cdscb = 6.9999e-006  cdscd = 2.6e-005
+cit = 0.0006652    voff = 0.01155       lvoff = -9e-005
+nfactor = 1        lnfactor= 0.205      eta0 = 1.45
+etab = -0.59359    u0 = 0.0054227       ua = 2.3103e-011
+ub = 5.0326e-020   uc = -2.2601e-010    eu = 1
+vsat = 342000      lvsat = -3620        a0 = 1.7533
+ags = 1.28         a1 = 0               a2 = 1
```

```
+b0 = 6.0098e-015    b1 = 0              keta = 0.17583
+lketa = -0.0019227  dwg = 0            dwb = 0
+pclm = 0.2423       pdiblc1 = 0.0012528 pdiblc2 = 0.0006
+pdiblcb = 0.029513  drout = 0.59157     pvag = 0
+delta = 0.047796    pscbe1 = 7.884e+008 pscbe2 = 1.9903e-06
+fprout = 0          pdits = 0          pditsd = 0
+pditsl = 0          rsh = 0            rdsw = 251.13
+rsw = 100           rdw = 100          rdswmin = 0
+rdwmin = 0          rswmin = 0         prwg = 0.068211
+prwb = -0.60085     wr = 1             alpha0 = 7e-011
+alpha1 = 7.2e-011   beta0 = 18.96      agidl = 2.0464e-009
+bgidl = 2250600     cgidl = 3149.6     egidl = 0.016014
+aigbacc = 0.43      bigbacc = 0.054    cigbacc = 0.075
+nigbacc = 1         aigbinv = 0.35     bigbinv = 0.03
+cigbinv = 0.006     eigbinv = 1.1      nigbinv = 3
+aigc = 0.43         bigc = 0.054       cigc = 0.075
+aigsd = 0.43        bigsd = 0.054      cigsd = 0.075
+nigc = 1            poxedge = 1        pigcd = 1
+ntox = 1
*************************************************************
*              CAPACITANCE PARAMETERS
*************************************************************
+cgso = 'cgop_pu_svt'  cgdo = 'cgop_pu_svt'
+cgbo = 1.8e-012       cgdl = 'cglp_pu_svt'
+cgsl = 'cglp_pu_svt'  clc = 1.2714e-011      cle = 1
+ckappas = 0.12        ckappad = 0.12         vfbcv = 0.5008
+acde = 0.45177        moin = 14.182          noff = 3.1539
+voffcv = 0.12863
*************************************************************
*              TEMPERATURE PARAMETERS
*************************************************************
+kt1 = -0.47607       kt1l = -1.025e-010  kt2 = -0.050313
+ute = -1.75          ua1 = 4.187e-011    ub1 = -2.882e-019
+uc1 = -6.5038e-010   prt = 0             at = 71599
+lat = -200
*************************************************************
*              NOISE PARAMETERS
*************************************************************
+fnoimod = 1          tnoimod = 0         em = 4.1e+007
+ef = 1               noia = 6.25e+041    noib = 3.125e+026
+noic = 8.75e+009     ntnoi = 1
*************************************************************
*              DIODE PARAMETERS
*************************************************************
+nigc = 1             poxedge = 1         pigcd = 1
+jss = 2.628e-005     jsws = 6.6059e-014  jswgs = 0
+njs = 0.97789        ijthsfwd= 0.1       ijthsrev= 0.1
+bvs = 10             xjbvs = 1           pbs = 0.9353
+cjs = 'cjp_pu_svt'   mjs = 0.44398       pbsws = 1
+cjsws = 'cjswp_pu_svt'                   mjsws = 0.5
```

```
+pbswgs = 0.76458    cjswgs = 'cjswgp_pu_svt'
+mjswgs = 0.44846    cjd = 'cjp_pu_svt'
+cjswd = 'cjswp_pu_svt'              cjswgd = 'cjswgp_pu_svt'
+tpb = 0.0020847    tcj = 0.00098       tpbsw = 0
+tcjsw = 0          tpbswg = 6e-005     tcjswg = 0.00047385
+xtis = 3
**************************************************************
*              LAYOUT RELATED PARAMETERS
**************************************************************
+dmcg = 0            dmdg = 0            dmcgt = 0
+xgw = 0             xgl = 0
**************************************************************
*                   RF PARAMETERS
**************************************************************
+rshg = 0.1          gbmin = 1e-012      rbpb = 50
+rbpd = 50           rbps = 50           rbdb = 50
+rbsb = 50           ngcon = 1
+xrcrg1 = 12         xrcrg2 = 1
**************************************************************
*                 STRESS PARAMETERS
**************************************************************
+saref = 1e-006      sbref = 1e-006      wlod = 0
+kvth0 = 0           lkvth0 = 0          wkvth0 = 0
+pkvth0 = 0          llodvth = 0         wlodvth = 0
 +stk2 = 0           lodk2 = 1           lodeta0 = 1
+ku0 = 0             lku0 = 0            wku0 = 0
+pku0 = 0            llodku0 = 0         wlodku0 = 0
+kvsat = 0           steta0 = 0          tku0 = 0        )
.ends

* DEVICE: PULL DOWN
.subckt n4_pd_svt D G S B
.param w=0 l=0 z=0.100e-6 multi=1
.param pd='2*(w+z)' ad='w*z'
.param ps='2*(w+z)' as='w*z'
.param pw='2*((2*z)+l+w)' aw='((2*z)+l)*w'
.param lw='(2*z)+l'
.param n_fingers=1

.param mismatchflag=0
.param avtn=2.8e-09

** must make default sigmavt=0, ie, no instance parameter
** sdvtp/ncorn in TT,SS,... takes care of corners sigvt when
*   globalsigmavtflag=1
.param sigmavt=0
.param delvto=0

** systematic **
** instance based sampling of systematic variation
```

```
** g[pn]mean varying in lock step for each instance
** defined in calling library MC_svt using globap[np]mean and
*  global[np]meansigma
** existing global[np]sigma varys by instance  and is defined by
*  global[np]sigma
.param random0n=agauss(0,1,3)
.param random1n=agauss(0,1,3)
.param random2n=agauss(0,1,3)
.param random3n=agauss(0,1,3)
.param random4n=agauss(0,1,3)
.param random5n=agauss(0,1,3)
.param random6n=agauss(0,1,3)
.param random7n=agauss(0,1,3)
.param random8n=agauss(0,1,3)

.param A0n='random0n*(globalmcflag==1)'
.param A1n='random1n*(globalmcflag==1)'
.param A2n='random2n*(globalmcflag==1)'
.param A3n='random3n*(globalmcflag==1)'
.param A4n='random4n*(globalmcflag==1)'
.param A5n='random5n*(globalmcflag==1)'
.param A6n='random6n*(globalmcflag==1)'
.param A7n='random7n*(globalmcflag==1)'
.param A8n='random8n*(globalmcflag==1)'

.param
+toxn_pd_svt = '1.600E-009 + (4.1231E-011*((A0n*gnsigma) + gnmean))'
+dxln_pd_svt = '0.000E+00 + (1.2000E-009*((A1n*gnsigma) + gnmean))'
+dxwn_pd_svt = '0.000E+00 + (-2.700E-009*((A2n*gnsigma) + gnmean))'
+dvthn_pd_svt = '0.000E+00 + (1.6768E-002*((A3n*gnsigma) + gnmean))'
+cjn_pd_svt = '3.0469E-003 + (2.1328E-004*((A4n*gnsigma) + gnmean))'
+cjswn_pd_svt = '1.000E-014 + (7.000E-016*((A5n*gnsigma) + gnmean))'
+cjswgn_pd_svt = '3.073E-10 + (2.1509e-11*((A6n*gnsigma) + gnmean))'
+cgon_pd_svt = '9.800E-012 + (-6.860E-013*((A7n*gnsigma) + gnmean))'
+cgln_pd_svt = '4.050E-011 + (-2.835E-012*((A8n*gnsigma) + gnmean))'

*** mismatch ***
.param AM0=agauss(0,1,1)

.param
+cvtn_svt='avtn/1.414214'
+lef='l-2.0E-09' wef='w/n_fingers+0E-09'
+geo_fac ='1/sqrt(multi*n_fingers*lef*wef)'
+sigvtn_pd_svt='AM0*cvtn_svt*geo_fac*globalmismatchnsigma/3 *
((globalmismatchflag==1)||(mismatchflag==1))'

** fixed sigmavt offset **
.param
```

```
+sdvtn_pd_svt='cvtn_svt*geo_fac*((sigmavt*(globalsigmavtflag==0)) +
(sdvtncorn*(globalsigmavtflag==1)))'

** no width effect
mn_pd_svt D G S B nch_svt w=1e-6 l=1 pd=pd ad=ad ps=ps as=as
m='w/1.0e-6' nf=n_fingers wnflag=1  delvto=delvto
** width effect
*mn_pd_svt D G S B nch_svt w=w l=1 pd=pd ad=ad ps=ps as=as
nf=n_fingers wnflag=1 delvto=delvto

****************************************************************
* BSIM4.5.0 model card for n-type devices
****************************************************************
.model  nch_svt.1 nmos  ( level = 54
****************************************************************
*                 MODEL FLAG PARAMETERS
****************************************************************
+lmin = 30.0e-009    lmax = 39.99e-009    wmin = 'wmin_n_pd_svt'
+wmax = 'wmax_n_pd_svt'        version = 4.5 binunit = 1
+paramchk= 1         mobmod = 0           capmod = 2
+igcmod = 2          igbmod = 1           geomod = 0
+diomod = 1          rdsmod = 0           rbodymod= 0
+rgeomod = 0         rgatemod= 0          permod = 1
+acnqsmod= 0         trnqsmod= 0
****************************************************************
*                 GENERAL MODEL PARAMETERS
****************************************************************
+tnom = 27             toxe = 'toxn_pd_svt' toxp = 1.1822e-009
+toxm = 'toxn_pd_svt+ 1.6e-10'            dtox = 2.3e-010
+epsrox = 3.9          toxref = 1.1822e-009 wmlt = 1
+wint = 0              lint = 1.288e-009    ll = 0
+wl = 0                lln = 1              wln = 1
+lw = 0                w = 0                lwn = 1
+wwn = 1               lwl = 0              wwl = 0
+xl = '0+dxln_pd_svt'                       xw = '0+dxwn_pd_svt'
+dlc = 4.877e-009      dwc = 0              xpart = 0
****************************************************************
*                 DC PARAMETERS
****************************************************************
+vth0 = '0.4192+dvthn_pd_svt+sdvtn_pd_svt+sigvtn_pd_svt'
+k1 = 0.4991          k2 = -0.0050218      k3 = 80
+k3b = 11.372         w0 = 1.4976e-008     dvt0 = 35.565
+dvt1 = 1.8           dvt2 = 0.064         dvt0w = 1.577
+dvt1w = 10000000     dvt2w = 0.01         dsub = 0.079604
+minv = 0             voffl = -3.2236e-015 dvtp0 = 0
+dvtp1 = 0            lc = 5e-009          lambda = 0
+vtl = 200000         lpe0 = 3.0161e-008   lpeb = 9.9848e-009
+vbm = -3             xj = 5.81e-008       ngate = 0
+ndep = 1.7e+017      nsd = 1e+020         phin = 0
+cdsc = 7.3987e-005 cdscb = -0.058894      cdscd = 0.066319
```

```
+cit = 0.0023514      voff = -0.013951     nfactor = 3.1059
+lnfactor= 0.12517    eta0 = 0.0014643     etab = 8.2903e-005
+u0 = 0.12346         ua = 1e-009          ub = 7.5477e-018
+uc = 1.0139e-009     eu = 1.67            vsat = 430000
+lvsat = -6065.7      a0 = 1               ags = 30.591
+a1 = 0               a2 = 1               b0 = 6.0098e-015
+b1 = 0               keta = -0.51811      lketa = 0.015916
+dwg = 0              dwb = 0              pclm = 0.00312
+pdiblc1 = 0.16186    pdiblc2 = 0          pdiblcb = 0.01
+drout = 2.7295       pvag = 0.12445       delta = 0.11364
+pscbe1 = 6.9889e+08  pscbe2 = 1e-005      fprout = 0.01
+pdits = 0            pditsd = 0           pditsl = 0
+rsh = 0              rdsw = 265.78        rsw = 100
+rdw = 100            rdswmin = 0          rdwmin = 0
+rswmin = 0           prwg = 0.15168       prwb = 0.10181
+wr = 1               alpha0 = 0           alpha1 = 0
+beta0 = 30           agidl = 1.1566e-011  bgidl = 4984400
+cgidl = 267.18       egidl = 0.057969     aigbacc = 0.43
+bigbacc = 0.054      cigbacc = 0.075      nigbacc = 1
+aigbinv = 0.35       bigbinv = 0.03       cigbinv = 0.006
+eigbinv = 1.1        nigbinv = 3          aigc = 0.43
+bigc = 0.054         cigc = 0.075         aigsd = 0.43
+bigsd = 0.054        cigsd = 0.075        nigc = 1
+poxedge = 1          pigcd = 1            ntox = 1
*************************************************************
*               CAPACITANCE PARAMETERS
*************************************************************
+cgso = 'cgon_pd_svt' cgdo = 'cgon_pd_svt'  cgbo = 1.2078e-09
+cgdl = 'cgln_pd_svt' cgsl = 'cgln_pd_svt'  clc = 2.9050e-010
+cle = 1              ckappas = 0.13608    ckappad = 0.13608
+vfbcv = -1.016       acde = 0.57228       moin = 8.6897
+noff = 2.7073        voffcv = 0.08287     lvoffcv = 0.001368
*************************************************************
*               TEMPERATURE PARAMETERS
*************************************************************
+kt1 = -0.43841       kt1l = 2.175e-009    kt2 = -0.023067
+ute = -1.5           ua1 = 2.0242e-008    ub1 = -1.4227e-017
+uc1 = -7.6509e-010   prt = 41.946         at = 153850
+lat = 1000
*************************************************************
*               NOISE PARAMETERS
*************************************************************
+fnoimod = 1          tnoimod = 0          em = 4.1e+007
+ef = 1               noia = 6.25e+041     noib = 3.125e+026
+noic = 8.75e+009     ntnoi = 1
*************************************************************
*               DIODE PARAMETERS
*************************************************************
+jss = 0.0001         jsws = 9e-012        jswgs = 0
+njs = 1              ijthsfwd= 0.1        ijthsrev= 0.1
```

```
+bvs = 10              xjbvs = 1            pbs = 0.71899
+cjs = 0.00346         mjs = 0.3515         pbsws = 1
+cjsws = 1e-014        mjsws = 0.5          pbswgs = 0.6134
+cjswgs = 5.0727e-10   mjswgs = 0.41349     cjd = 0.00346
+cjswd = 1e-014        cjswgd = 5.0727e-010 tpb = 0.0015686
+tcj = 0.00076331      tpbsw = 0            tcjsw = 0
+tpbswg = 0            tcjswg = 0           xtis = 3
*****************************************************************
*                 LAYOUT RELATED PARAMETERS
*****************************************************************
+dmcg = 0              dmdg = 0             dmcgt = 0
+xgw = 0               xgl = 0
*****************************************************************
*                      RF PARAMETERS
*****************************************************************
+rshg = 0.1            gbmin = 1e-012       rbpb = 50
+rbpd = 50             rbps = 50            rbdb = 50
+rbsb = 50             ngcon = 1            xrcrg1 = 12
+xrcrg2 = 1
*****************************************************************
*                    STRESS PARAMETERS
*****************************************************************
+saref = 1e-006        sbref = 1e-006       wlod = 0
+kvth0 = 0             lkvth0 = 0           wkvth0 = 0
+pkvth0 = 0            llodvth = 0          wlodvth = 0
+stk2 = 0              lodk2 = 1            lodeta0 = 1
+ku0 = 0               lku0 = 0             wku0 = 0
+pku0 = 0              llodku0 = 0          wlodku0 = 0
+kvsat = 0             steta0 = 0           tku0 = 0        )
.ends

* DEVICE: PASS GATE
.subckt n4_pg_svt D G S B
.param w=0  l=0  z=0.100e-6  multi=1
.param pd='2*(w+z)' ad='w*z'
.param ps='2*(w+z)' as='w*z'
.param pw='2*((2*z)+l+w)' aw='((2*z)+l)*w'
.param lw='(2*z)+l'
.param n_fingers=1

.param mismatchflag=0
.param avtn=2.8e-09

** must make default sigmavt=0, ie, no instance parameter
** sdvtp/ncorn in TT,SS,... takes care of corners sigvt when
** globalsigmavtflag=1
.param sigmavt=0
.param delvto=0

** systematic **
```

```
** instance based sampling of systematic variation
** g[pn]mean varying in lock step for each instance
** defined in calling library MC_svt using globap[np]mean and
*  global[np]meansigma
** existing global[np]sigma varys by instance  and is defined by
*  global[np]sigma
.param random0n=agauss(0,1,3)
.param random1n=agauss(0,1,3)
.param random2n=agauss(0,1,3)
.param random3n=agauss(0,1,3)
.param random4n=agauss(0,1,3)
.param random5n=agauss(0,1,3)
.param random6n=agauss(0,1,3)
.param random7n=agauss(0,1,3)
.param random8n=agauss(0,1,3)
.param A0n='random0n*(globalmcflag==1)'
.param A1n='random1n*(globalmcflag==1)'
.param A2n='random2n*(globalmcflag==1)'
.param A3n='random3n*(globalmcflag==1)'
.param A4n='random4n*(globalmcflag==1)'
.param A5n='random5n*(globalmcflag==1)'
.param A6n='random6n*(globalmcflag==1)'
.param A7n='random7n*(globalmcflag==1)'
.param A8n='random8n*(globalmcflag==1)'

.param
+toxn_pg_svt = '1.600E-09 + (4.1231E-011*((A0n*gnsigma) + gnmean))'
+dxln_pg_svt = '0.000E+00 + (1.2000E-09*((A1n*gnsigma) + gnmean))'
+dxwn_pg_svt = '0.000E+00 + (-2.700E-09*((A2n*gnsigma) + gnmean))'
+dvthn_pg_svt = '0.000E+00 + (1.6768E-02*((A3n*gnsigma) + gnmean))'
+cjn_pg_svt = '3.0469E-03 + (2.1328E-004*((A4n*gnsigma) + gnmean))'
+cjswn_pg_svt = '1.000E-014 + (7.000E-16*((A5n*gnsigma) + gnmean))'
+cjswgn_pg_svt = '3.073E-10 + (2.1509e-11*((A6n*gnsigma) + gnmean))'
+cgon_pg_svt = '9.800E-012 + (-6.86E-013*((A7n*gnsigma) + gnmean))'
+cgln_pg_svt = '4.050E-011 + (-2.835E-012*((A8n*gnsigma) + gnmean))'

*** mismatch ***
.param AM0=agauss(0,1,1)

.param
+cvtn_svt='avtn/1.414214'
+lef='l-2.0E-09' wef='w/n_fingers+0E-09'
+geo_fac ='1/sqrt(multi*n_fingers*lef*wef)'
+sigvtn_pg_svt='AM0*cvtn_svt*geo_fac*globalmismatchnsigma/3 *
((globalmismatchflag==1)||(mismatchflag==1))'

** fixed sigmavt offset **
.param
+sdvtn_pg_svt='cvtn_svt*geo_fac*((sigmavt*(globalsigmavtflag==0)) +
(sdvtncorn*(globalsigmavtflag==1)))'
```

```
** no width effect
mn_pg_svt D G S B nch_svt w=1e-6 l=1 pd=pd ad=ad ps=ps as=as
m='w/1.0e-6' nf=n_fingers wnflag=1  delvto=delvto
** width effect
*mn_pg_svt D G S B nch_svt w=w l=1 pd=pd ad=ad ps=ps as=as
nf=n_fingers wnflag=1 delvto=delvto

***************************************************************
* BSIM4.5.0 model card for n-type devices
***************************************************************
.model  nch_svt.2 nmos  ( level = 54
***************************************************************
*                   MODEL FLAG PARAMETERS
***************************************************************
+lmin = 40.0e-009   lmax = 45.10e-009   wmin = 'wmin_n_pg_svt'
+wmax = 'wmax_n_pg_svt' version = 4.5   binunit = 1
+paramchk= 1         mobmod = 0         capmod = 2
+igcmod = 2          igbmod = 1         geomod = 0
+diomod = 1          rdsmod = 0         rbodymod= 0
+rgeomod = 0         rgatemod= 0        permod = 1
+acnqsmod= 0         trnqsmod= 0
***************************************************************
*                   GENERAL MODEL PARAMETERS
***************************************************************
+tnom = 27           toxe = 'toxn_pg_svt' toxp = 1.1822e-009
+toxm = 'toxn_pg_svt+ 1.6e-10'           dtox = 2.3e-010
+epsrox = 3.9        toxref = 1.1822e-009 wmlt = 1
+wint = 0            lint = 1.288e-009    ll = 0
+wl = 0              lln = 1              wln = 1
+lw = 0              ww = 0               lwn = 1
+wwn = 1             lwl = 0              wwl = 0
+xl = '0+dxln_pg_svt'                     xw = '0+dxwn_pg_svt'
+dlc = 4.877e-009    dwc = 0              xpart = 0
***************************************************************
*                   DC PARAMETERS
***************************************************************
+vth0 = '0.4192+dvthn_pg_svt+sdvtn_pg_svt+sigvtn_pg_svt'
+k1 = 0.4991         k2 = -0.0050218      k3 = 80
+k3b = 11.372        w0 = 1.4976e-008     dvt0 = 33.065
+dvt1 = 1.5462       dvt2 = 0.0613        dvt0w = 1.577
+dvt1w = 10000000    dvt2w = 0.01         dsub = 0.079604
+minv = 0            voffl = -3.2236e-015 dvtp0 = 0
 +dvtp1 = 0          lc = 5e-009          lambda = 0
+vtl = 200000        lpe0 = 3.0161e-008   lpeb = 9.9848e-009
+vbm = -3            xj = 5.81e-008       ngate = 0
+ndep = 1.7e+017     nsd = 1e+020         phin = 0
+cdsc = 7.3987e-05   cdscb = -0.058894    cdscd = 0.066319
+cit = -0.00015514   voff = 0.0060067     nfactor = 3.1059
+lnfactor= 0.22042   eta0 = 0.001923      etab = 8.2903e-005
+u0 = 0.12346        ua = 1e-009          ub = 7.5477e-018
```

```
+uc = 1.0139e-009    eu = 1.67          vsat = 602800
+lvsat = -6065.7     a0 = 1             ags = 30.591
+a1 = 0              a2 = 1             b0 = 6.0098e-015
+b1 = 0              keta = -0.42485    lketa = 0.015916
+dwg = 0             dwb = 0            pclm = 0.00312
+pdiblc1 = 0.16186   pdiblc2 = 0        pdiblcb = 0.01
+drout = 2.7295      pvag = 0.12445     delta = 0.11364
+pscbe1 = 6.9889e+08 pscbe2 = 1e-005    fprout = 0.01
+pdits = 0           pditsd = 0         pditsl = 0
+rsh = 0             rdsw = 265.78      rsw = 100
+rdw = 100           rdswmin = 0        rdwmin = 0
+rswmin = 0          prwg = 0.15168     prwb = 0.10181
+wr = 1              alpha0 = 0         alpha1 = 0
+beta0 = 30          agidl = 3.5057e-010  bgidl  = 4984400
+cgidl = 267.18      egidl = 0.057969   aigbacc = 0.43
+bigbacc = 0.054     cigbacc = 0.075    nigbacc = 1
+aigbinv = 0.35      bigbinv = 0.03     cigbinv = 0.006
+eigbinv = 1.1       nigbinv = 3        aigc = 0.43
+bigc = 0.054        cigc = 0.075       aigsd = 0.43
+bigsd = 0.054       cigsd = 0.075      nigc = 1
+poxedge = 1         pigcd = 1          ntox = 1
***********************************************************
*               CAPACITANCE PARAMETERS
***********************************************************
+cgso = 'cgon_pg_svt'   cgdo = 'cgon_pg_svt'   cgbo = 1.2078e-09
+cgdl = 'cgln_pg_svt'   cgsl = 'cgln_pg_svt'   clc = 2.9050e-010
+cle = 1             ckappas = 0.13608  ckappad = 0.13608
+vfbcv = -1.016      acde = 0.57228     moin = 8.6897
+noff = 2.7073       voffcv = 0.08287   lvoffcv = 0.001368
***********************************************************
*               TEMPERATURE PARAMETERS
***********************************************************
+kt1 = -0.43841      kt1l = 2.175e-009  kt2 = -0.023067
+ute = -1.5          ua1 = 2.0242e-008  ub1 = -1.4227e-017
+uc1 = -7.6509e-010 prt = 41.946       at = 153850
+lat = 1000
***********************************************************
*               NOISE PARAMETERS
***********************************************************
+fnoimod = 1         tnoimod = 0        em = 4.1e+007
+ef = 1              noia = 6.25e+041   noib = 3.125e+026
+noic = 8.75e+009    ntnoi = 1
***********************************************************
*               DIODE PARAMETERS
***********************************************************
+jss = 0.0001        jsws = 9e-012      jswgs = 0
+njs = 1             ijthsfwd= 0.1      ijthsrev= 0.1
+bvs = 10            xjbvs = 1          pbs = 0.71899
+cjs = 0.00346       mjs = 0.3515       pbsws = 1
```

```
+cjsws = 1e-014      mjsws = 0.5          pbswgs = 0.6134
+cjswgs = 5.073e-10 mjswgs = 0.41349     cjd = 0.00346
+cjswd = 1e-014      cjswgd = 5.0727e-010 tpb = 0.0015686
+tcj = 0.00076331    tpbsw = 0            tcjsw = 0
+tpbswg = 0          tcjswg = 0           xtis = 3
*******************************************************************
*                LAYOUT RELATED PARAMETERS
*******************************************************************
+dmcg = 0            dmdg = 0             dmcgt = 0
+xgw = 0             xgl = 0
*******************************************************************
*                     RF PARAMETERS
*******************************************************************
+rshg = 0.1          gbmin = 1e-012       rbpb = 50
+rbpd = 50           rbps = 50            rbdb = 50
+rbsb = 50           ngcon = 1            xrcrg1 = 12
+xrcrg2 = 1
*******************************************************************
*                     STRESS PARAMETERS
*******************************************************************
+saref = 1e-006      sbref = 1e-006       wlod = 0
+kvth0 = 0           lkvth0 = 0           wkvth0 = 0
+pkvth0 = 0          llodvth = 0          wlodvth = 0
+stk2 = 0            lodk2 = 1            lodeta0 = 1
+ku0 = 0             lku0 = 0             wku0 = 0
+pku0 = 0            llodku0 = 0          wlodku0 = 0
+kvsat = 0           steta0 = 0           tku0 = 0        )
*
.ends
.ENDL SUBCKTS_SVT
```

References

Chapter 1

1. M.S. Lundstrom and D.A. Antonidis, "Compact models and the physics of nanoscale FETs," *IEEE Transactions on Electron Devices*, vol. 61, no. 2, pp. 225–233, 2014.
2. C.C. McAndrew, "Practical modeling for circuit simulation," *IEEE Journal of Solid-State Circuits*, vol. 33, no. 3, pp. 439–448, 1998.
3. J.S.C. Kilby, "Miniaturized electronics circuits," US Patent 3138743, June 23, 1964.
4. Y.S. Chauhan, S. Venugopalan, M.-A. Chalkiadaki, M.A.U. Karim, N. Paydavosi, J.P. Duarte, C.C. Enz, A.M. Niknejad, and C. Hu, "BSIM6: Analog and RF compact model for bulk MOSFET," *IEEE Transactions on Electron Devices*, vol. 61, no. 2, pp. 234–244, 2014.
5. S.K. Saha, "Modeling process variability in scaled CMOS technology," *IEEE Design & Test of Computers*, vol. 27, no. 2, pp. 8–16, 2010.
6. S.K. Saha, N.D. Arora, M.J. Deen, and M. Miura-Mattausch, "Advanced compact models and 45-nm modeling challenges," *IEEE Transactions on Electron Devices*, vol. 53, no. 9, pp. 1957–1960, 2006.
7. S.K. Saha, "Introduction to technology computer aided design," in *Technology Computer Aided Design: Simulation for VLSI MOSFET*, C.K. Sarkar (ed.). CRC Press, Boca Raton, FL, 2013.
8. S.K. Saha, "Managing technology CAD for competitive advantage: An efficient approach for integrated circuit fabrication technology development," *IEEE Transactions on Engineering and Management*, vol. 46, no. 2, pp. 221–229, 1999.
9. N. Arora, *MOSFET Models for VLSI Circuit Simulation: Theory and Practice*, Springer-Verlag, Wien, Austria, 1993.
10. S.K. Saha, M.J. Deen, and H. Masuda, "Compact interconnect models for gigascale integration," *IEEE Transactions on Electron Devices*, vol. 56, no. 9, pp. 1784–1786, 2009.
11. W. Shockley, "Semiconductor amplifier," US Patent 2502488, April 4, 1950.
12. J. Bardeen and W.H. Brattain, "Three-electrode circuit element utilizing semiconductor materials," US Patent 2524035, October 3, 1950.
13. J.J. Ebers and J.L. Moll, "Large-signal behavior of junction transistors," *Proceedings of IRE*, vol. 42, no. 12, pp. 1761–1772, 1954.
14. H.K. Gummel and C.C. Poon, "An integral charge control model for bipolar transistors," *Bell System Technical Journal*, vol. 49, no. 5, pp. 827–852, 1970.
15. L. Nagel and D. Pederson, "Simulation program with integrated circuit emphasis," University of California, Berkeley, CA, Electronics Research Laboratory Memorandum No. UCB/ERL M352, 1973.

16. C.C. Enz, F. Krummenacher, and E.A. Vittoz, "An analytical MOS transistor model valid in all regions of operation and dedicated to low voltage and low-current applications," *Journal of Analog Integrated Circuit and Signal Processing*, vol. 8, no. 1, pp. 83–114, 1995.

17. M. Chan, K.Y. Hui, C. Hu, and P.K. Ko, "A robust and physical BSIM3 non-quasi-static transient and AC small signal model for circuit simulation," *IEEE Transactions on Electron Devices*, vol. 45, no. 4, pp. 834–841, 1998.

18. M. Miura-Mattausch, H. Ueno, M. Tanaka, H.J. Mattausch, S. Kamashiro, T. Yamaguchi, K.Yamashita, and N. Nakayama, "HiSIM: A MOSFET model for circuit simulation connecting device performance with technology," *IEEE International Electron Devices Meeting Technical Digest*, pp. 109–112, 2002.

19. G. Gildenblat, X. Li, W. Wu, H. Wang, A. Jha, R. van Langevelde, G.D.J. Smit, A.J. Scholten, and D.B.M. Klaassen, "PSP: An advanced surface-potential-based MOSFET model for circuit simulation," *IEEE Transactions on Electron Devices*, vol. 53, no. 9, pp. 1979–1993, 2006.

20. Y. Tsividis and C. McAndrew, *Operation and Modeling of the MOS Transistor*, Oxford University Press, Oxford, 2010.

21. H.K.J. Ihantola and J.L. Moll, "Design theory of a surface field-effect transistor," *Solid-State Electronics*, vol. 7, no. 6, pp. 423–430, 1964.

22. H.C. Pao and C.T. Sah, "Effects of diffusion current on characteristics of metal-oxide (insulator)-semiconductor transistors," *Solid-State Electronics*, vol. 9, no. 10, pp. 927–937, 1966.

23. J.R. Brews, "A charge sheet model of the MOSFET," *Solid-State Electronics*, vol. 21, no. 2, pp. 345–355, 1978.

24. B.J. Sheu, D.L. Scharfetter, P.-K. Ko, and M.-C. Jeng, "BSIM: Berkeley short-channel IGFET model for MOS transistors," *IEEE Journal of Solid-State Circuits*, vol. 22, no. 4, pp. 558–556, 1987.

25. C.T. Sah, "Evolution of MOS transistor—Form conception to VLSI," *Proceedings of IEEE*, vol. 76, no. 10, pp. 1280–1326, 1988.

26. C.T. Sah, "Characteristics of the metal-oxide-semiconductor transistors," *IEEE Transactions on Electron Devices*, vol. 11, no. 7, pp. 324–345, 1964.

27. L.W. Nagel, "SPICE2—A computer program to simulate semiconductor circuits," University of California, Berkeley, CA, Electronic Research Laboratory Memo ERL-M250, 1975.

28. K.J. Kuhn, "Considerations for ultimate CMOS scaling," *IEEE Transactions on Electron Devices*, vol. 59, no. 7, pp. 1813–1828, 2012.

29. S. Saha, "Device characteristics of sub-20-nm silicon nanotransistors," in *Proceedings of the SPIE Conference on Design and Process Integration for Microelectronic Manufacturing*, vol. 5042, pp. 172–179, 2003.

30. S.K. Saha, "Transistors having optimized source-drain structures and methods for making the same," US Patent 6344405, February 5, 2002.

31. S. Saha, "Design considerations for 25 nm MOSFET devices," *Solid-State Electronics*, vol. 45, no. 10, pp. 1851–1857, 2001.

32. S. Saha, "Scaling considerations for high performance 25 nm metal-oxide-semiconductor field-effect transistors," *Journal of Vacuum Science & Technology B*, vol. 19, no. 6, pp. 2240–2246, 2001.

33. H. Iwai, "CMOS technology—Year 2010 and beyond," *IEEE Journal of Solid-State Circuits*, vol. 34, no. 3, pp. 357–366, 1999.

34. H. Schichmann and D. Hodges, "Modeling and simulation of insulated-gate field-effect transistor switching circuits," *IEEE Journal of Solid-State Circuits*, vol. 3, no. 3, pp. 285–289, 1968.
35. A. Vladimirescu and S. Liu, "The simulation of MOS integrated circuits using SPICE2," University of California, Berkeley, CA, Electronics Research Laboratory Memorandum No. UCB/ERL M80/7, 1980.
36. R.H. Dennard, F.H. Gaensslen, H.N. Yu, V.L. Rideout, E. Bassous, and A.R. LeBlanc, "Design of ion-implanted MOSFETs with very small physical dimensions," *IEEE Journal of Solid-State Circuits*, vol. 9, no. 5, pp. 256–268, 1974.
37. F. van de Wiele, "A long channel MOSFET model," *Solid-State Electronics*, vol. 22, no. 12, pp. 991–997, 1979.
38. C.C. McAndrew and J.J. Victory, "Accuracy of approximations in MOSFET charge models," *IEEE Transactions on Electron Devices*, vol. 49, no. 1, pp. 72–81, 2002.
39. S. Liu, "A unified CAD model for MOSFETs," University of California, Berkeley, CA, Electronics Research Laboratory Memorandum No. UCB/ERL M81/31, 1981.
40. M.C. Jeng, "Design and modeling of deep-submicrometer MOSFETs," University of California, Berkeley, CA, Electronics Research Laboratory Memorandum No. UCB/ERL M90/90, 1990.
41. *HSPICE User's Manual*, Meta Software, Campbell, CA, 1993.
42. Synopsys Inc., HSPICE User Guide: Simulation and Analysis, B-2008.09, Synopsys Inc., September 2008.
43. J.H. Huang, Z.H. Liu, M.-C. Jeng, K. Hui, M. Chan, P.K. Ko, and C. Hu, "BSIM3 manual—Version 2.0," University of California, Berkeley, CA, 1994.
44. Y. Cheng, M. Chan, K. Hui, M.-C. Jeng, Z. Liu, J. Huang, K. Chen, J. Chen, R. Tu, P.K. Ko, and C. Hu, "BSIM3 version 3.0 user's manual," University of California, Berkeley, CA, 1995.
45. Y. Cheng, M. Chan, K. Hui, M.-C. Jeng, Z. Liu, J. Huang, K. Chen, J. Chen, P.K. Ko, and C. Hu "BSIM3 version 3.1 user's manual," University of California, Berkeley, CA, Electronics Research Laboratory Memorandum No. UCB/ERL M97/2, 1997.
46. Y. Cheng, M.-C. Jeng, Z. Liu, J. Huang, M. Chan, K. Chen, P.K. Ko, and C. Hu, "A physical and scalable I–V in BSIMv3 for analog/digital simulation," *IEEE Transactions on Electron Devices*, vol. 44, no. 2, pp. 277–287, 1997.
47. Y. Cheng, T. Sugii, K. Chen, and C. Hu, "Modeling of small size MOSFETs with reverse short channel and narrow width effects for circuit simulation," *Solid-State Electronics*, vol. 41, no. 9, pp. 1227–1231, 1997.
48. Si2. n.d. "Si2 (CMC) homepage." https://www.si2.org/cmc_index.php, accessed May 26, 2015.
49. C. Hu et al., "Description of the code changes in BSIM3v3.1 versus BSIM3v3.0," 1997. http://www-device.eecs.berkeley.edu/bsim/?page=BSIM3_Arc, accessed May 26, 2015.
50. W. Liu, X. Jin, and C. Hu, "BSIM3v3.2 Model enhancements and improvements relative to BSIM3v3.1," University of California, Berkeley, CA, 1998.
51. Y. Cheng and C. Hu, *MOSFET Modeling and BSIM3 User's Guide*, Kluwer Academic Publishers, London, 1999.
52. H. deGraaff and F. Klaassen, *Compact Transistor Modeling for Circuit Design*, Springer-Verlag: Wein, Austria, 1990.
53. R.M.D.A. Velghe, D.B.M. Klaassen, and F.M. Klaassen, "Compact MOS modeling for analog circuit simulation," in *IEEE International Electron Devices Meeting Technical Digest*, pp. 485–488, 1993.

54. R.M.D.A. Velghe, D.B.M. Klaassen, and F.M. Klaassen, "MOS 9 model," Unclassified report NL-UR 003/94, Philips Electronics N.V., Amsterdam, the Netherlands, 1994.

55. N. Paydavosi, T.H. Morsged, D.D. Lu, W. Yang, M.V. Dunga, X. Xi, J. He, W. Liu, K. cao, X. Jin, J.J. Ou, M. Chan, A.M. Niknejad, and C. Hu, "BSIM4v4.8.0 MOSFET Model User's Manual," University of California, Berkeley, CA, 2013.

56. W. Liu, *MOSFET models for SPICE simulation, including BSIM3 and BSIM4*, Wiley, New York, 2001.

57. X. Xi et al., *BSIM4.3.0 MOSFET Model—User's Manual*, University of California, Berkeley, CA, 2003.

58. M. Bagheri and Y. Tsividis, "A small signal dc-to-high-frequency nonquasi-static model for the four-terminal MOSFET valid in all regions of operation," *IEEE Transactions on Electron Devices*, vol. 32, no. 12, pp. 2383–2391, 1985.

59. P.P. Guebels and F. Van de Wiele, "A charge sheet model for small geometry MOSFET's," in *IEEE International Electron Devices Meeting Technical Digest*, pp. 211–214, 1981.

60. A.M. Ostrowsky, *Solutions of Equations and Systems of Equations*, Academic Press, New York, 1973.

61. C. Turchetti and G. Masetti, "A CAD-oriented analytical MOSFET model for high-accuracy applications," *IEEE Transactions on Computer-Aided Design of ICAS*, vol. 3, no. 2, pp. 117–122, 1984.

62. N.D. Arora, R. Rios, C.L. Huang, and K. Raol, "PCIM: A physically based continuous short-channel IGFET model for circuit simulation," *IEEE Transactions on Electron Devices*, vol. 41, no. 6, pp. 988–997, 1994.

63. H.-J. Park, P.-K. Ko, and C. Hu, "A charge sheet capacitance model of short channel MOSFETs for SPICE," *IEEE Transactions on Computer-Aided Design*, vol. 10, no. 3, pp. 376–389, 1991.

64. J.W. Sleight and R. Rios, "A continuous compact MOSFET model for SOI with automatic transitions between fully and partially depleted device behavior," in *IEEE International Electron Devices Meeting Technical Digest*, pp. 143–246, 1996.

65. R. Rios, N.D. Arora, C.-L. Huang, N. Khalil, J. Faricelli, and L. Gruber, "A physical compact MOSFET model including quantum mechanical effects for statistical circuit design applications," in *IEEE International Electron Devices Meeting Technical Digest*, pp. 937–940, 1995.

66. M. Miura-Mattausch and H. Jacobs, "Analytical model for circuit simulation with quarter micron metal-oxide semiconductor field-effect transistors: Subthreshold characteristics," *Japanese Journal of Applied Physics*, vol. 29, no. 12 A, pp. L2279–L2282, 1990.

67. M. Miura-Mattausch, "Analytical MOSFET model for quarter micron technologies," *IEEE Transactions on Computer-Aided Design*, vol. 13, no. 5, pp. 610–615, 1994.

68. M. Miura-Mattausch, U. Feldmann, A. Rahm, M. Bollu, and D. Savignac, "Unified complete MOSFET model for analysis of digital and analog circuits," *IEEE Transactions on Computer-Aided Design*, vol. 15, no. 1, pp. 1–7, 1996.

69. H. Uneo, S. Jinbou, H. Kawano, K. Morikawa, N. Nakayama, A. Miura-Mattausch, and H.J. Mattausch, "Drift-diffusion-based modeling of the non-quasi-static small-signal response for RF-MOSFET applications," in *International Conference on Simulation of Semiconductor Processes and Devices*, pp. 71–74, 2002.

70. J. Watts, C. McAndrew, C. Enz, C. Galup-Montoro, G. Gildenblat, C. Hu, R. van Langevelde, M. Miura-Mattausch, R. Rios, and C.-T. Sah, "Advanced

compact models for MOSFETs," in *Technical Proceedings, International Workshop on Compact Modeling,* NSTI Nanotech, pp. 3–11, 2005.

71. M. Miura-Mattausch et al., "HiSIM2: An advanced MOSFET model valid for RF circuit simulation," *IEEE Transactions on Electron Devices,* vol. 53, no. 9, pp. 1994–2007, 2006.

72. R. van Langevelde, A.J. Scholten, and D.B.M. Klaassen, "Physical background of MOS11," Level 1101 Nat. Lab. Unclassified Report 2003/00239, Koninklijke Philips Electronics, Amsterdam, the Netherlands, April 2003. http://www.nxp.com/wcm_documents/models/mos-models/model-11/nl_tn2003_00239.pdf, accessed May 26, 2015.

73. R. van Langevelde, J. Paasschens, A. Scholten, R. Havens, L. Tiemeijer, and D. Klaassen, "New compact model for induced gate current noise," in *IEEE International Electron Devices Meeting Technical Digest,* pp. 867–870, 2003.

74. R. van Langevelde, A.J. Scholten, R. Duffy, F.N. Cubaynes, M.J. Knitel, and D.B.M. Klaassen, "Gate current: Modeling, ΔL extraction and impact on RF performance," in *IEEE International Electron Devices Meeting Technical Digest,* 2001, pp. 289–292.

75. R. van Langevelde and F.M. Klaassen, "Effect of gate-field dependent mobility degradation on distortion analysis in MOSFETs," *IEEE Transactions on Electron Devices,* vol. 44, no. 11, pp. 2044–2052, 1997.

76. R. van Langevelde and F.M. Klaassen, "Accurate drain conductance modeling for distortion analysis in MOSFETs," in *IEEE International Electron Devices Meeting Technical Digest,* pp. 313–316, 1997.

77. R. van Langevelde, L. Tiemeijer, R. Havens, M. Knitel, R. Roes, P. Woerlee, and D. Klaassen, "RF-distortion in deep-submicron CMOS technologies," in *IEEE International Electron Devices Meeting Technical Digest,* pp. 807–810, 2000.

78. D.B.M. Klaassen, R.V. Langevelde, and A.J. Scholten, "Compact CMOS modeling for advanced analogue and RF applications," *IEICE Transactions on Electronics,* vol. E87-C, no. 6, pp. 854–866, 2004.

79. K. Joardar, K.K. Gullapalli, C.C. McAndrew, M.E. Burnham, and A. Wild, "An improved MOSFET model for circuit simulation," *IEEE Transactions on Electron Devices,* vol. 45, no. 1, pp. 134–148, 1998.

80. R. van Langevelde, A.J. Scholten, R.J. Havens, L.F. Tiemeijer, and D.B.M. Klaassen, "Advanced compact MOS modelling," in *Proceedings of the European Solid-State Device Research Conference,* pp. 81–88, 2001.

81. A.J. Scholten, R. Duffy, R. van Langevelde, and D.B.M. Klaassen, "Compact MOS modelling of pocket-implanted MOSFETs," in *Proceedings of the European Solid-State Device Research Conference,* pp. 311–314, 2001.

82. R. van Langevelde and F.M. Klaassen, "An explicit surface-potential based MOSFET model for circuit simulation," *Solid-State Electronics,* vol. 44, no. 11, pp. 409–418, 2000.

83. M. Miura-Mattausch and U. Weinert, "Unified MOSFET model for all channel lengths down to quarter micron," *IEICE Transactions on Electronics,* vol. E75-C, no. 2, pp. 172–180, 1992.

84. C.-L. Huang and G.S. Gildenblat, "Correction factor in the split C–V method for mobility measurements," *Solid-State Electronics,* vol. 36, no, 4, pp. 611–615, 1993.

85. G. Gildenblat, H. Wang, T.-L. Chen, X. Gu, and X. Cai, "SP: An advanced surface-potential-based compact MOSFET model," *IEEE Journal of Solid-State Circuits,* vol. 39, no. 9, pp. 1394–1406, 2004.

86. W. Wu, T.-L. Chen, G. Gildenblat, and C.C. McAndrew, "Physics-based mathematical conditioning of the MOSFET surface potential equation," *IEEE Transactions on Electron Devices*, vol. 51, no. 7, pp. 1196–1199, 2004.

87. T.-L. Chen and G. Gildenblat, "Analytical approximation for the MOSFET surface potential," *Solid-State Electronics*, vol. 45, no. 2, pp. 335–339, 2001.

88. X. Gu, T.-L. Chen, G. Gildenblat, G.O. Workman, S. Veeraraghavan, S. Shapira, and K. Stiles, "A surface potential-based compact model of n-MOSFET gate-tunneling current," *IEEE Transactions on Electron Devices*, vol. 51, no. 1, pp. 127–135, 2004.

89. J. Victory, Z. Yan, G. Gildenblat, C. McAndrew, and J. Zheng, "A physically based, scalable MOS varactor model and extraction methodology for RF applications," *IEEE Transactions on Electron Devices*, vol. 52, no. 7, pp. 1343–1353, 2005.

90. H. Wang and G. Gildenblat, "A robust large signal non-quasi-static MOSFET model for circuit simulation," in *Proceedings of the IEEE Custom Integrated Circuits Conference*, pp. 5–8, 2004.

91. T. Nguyen and J. Plummer, "Physical mechanisms responsible for short channel effects in MOS devices," in *IEEE International Electron Devices Meeting Technical Digest*, pp. 596–599, 1981.

92. C. Turchetti, G. Masetti, and Y. Tsividis, "On the small-signal behavior of the MOS transistor in quasi-static operation," *Solid-State Electronics*, vol. 26, no. 10, pp. 941–949, 1983.

93. H. Wang, T.-L. Chen, and G. Gildenblat, "Quasi-static and nonquasi-static compact MOSFET models based on symmetric linearization of the bulk and inversion charges," *IEEE Transactions on Electron Devices*, vol. 50, no. 11, pp. 2262–2272, 2003.

94. P. Bendix, P. Rakers, P. Wagh, L. Lemaitre, W. Grabinski, C.C. McAndrew, X. Gu, and G. Gildenblat, "RF distortion analysis with compact MOSFET models," in *Proceedings of the IEEE Custom Integrated Circuits Conference*, pp. 9–12, 2004.

95. T. Grotjohn and B. Hofflinger, "A parametric short-channel MOS transistor model for subthreshold and strong inversion current," *IEEE Journal of Solid-State Circuits*, vol. 19, no. 1, pp. 100–112, 1984.

96. G. Gildenblat, "One-flux theory of a nonabsorbing barrier," *Journal of Applied Physics*, vol. 91, no. 12, pp. 9883–9886, 2002.

97. H. Wang and G. Gildenblat, "Scattering matrix based compact MOSFET model," in *IEEE International Electron Devices Meeting Technical Digest*, pp. 125–128, 2002.

98. M. Lundstrom and Z. Ren, "Essential physics of carrier transport in nanoscale MOSFETs," *IEEE Transactions on Electron Devices*, vol. 49, no. 1, pp. 133–141, 2002.

99. M.A. Maher and C.A. Mead, "A physical charge-based model for MOS transistors," in *Advanced Research in VLSI*, P. Losleben (ed.). MIT Press, Cambridge, MA, 1987.

100. Y.H. Byun, K. Lee, and M. Shur, "Unified charge control model and subthreshold current in heterostructure field-effect transistors," *IEEE Electron Devices Letters*, vol. 11, no. 1, pp. 50–53, 1990.

101. C.K. Park, C.-Y. Lee, K. Lee, B.-J. Moon, Y.H. Byun, and M. Shur, "A unified current-voltage model for long-channel nMOSFETs," *IEEE Transactions on Electron Devices*, vol. 38, no. 2, pp. 399–406, 1991.

102. A.I.A. Cunha, M.C. Schneider, and C. Galup-Montoro, "An explicit physical model for the long-channel MOS transistor including small-signal parameters," *Solid-State Electronics*, vol. 38, no. 11, pp. 1945–1952, 1995.

103. H.K. Gummel and K. Singhal, "Inversion charge modeling," *IEEE Transactions on Electron Devices*, vol. 48, no. 8, pp. 1585–1593, 2001.
104. J. He, X. Xi, M. Chan, A. Niknejad, and C. Hu, "An advanced surface-potential-plus MOSFET model," in *Technical Proceedings, Workshop on Compact Modeling*, NSTI Nanotech, pp. 307–310, 2005.
105. C. Galup-Montoro, M.C. Schneider, and I.J.B. Loss, "Series-parallel association of FET's for high gain and high frequency applications," *IEEE Journal of Solid-State Circuits*, vol. 29, no. 9, pp. 1094–1101, 1994.
106. O.C. Gouveia Filho, A.I.A. Cunha, M.C. Schneider, and C.G. Montoro, "The ACM model for circuit simulation and equations for smash," September 1997. http://www.dolphin.fr/medal/downloads/notes/acm_report.pdf, accessed May 26, 2015.
107. A.T. Behr, M.C. Schneider, F.S. Noceti, and C.G. Montoro, "Nonlinearities of capacitors realized by MOSFET gates," in *Proceedings of the IEEE International Conference on Circuits and Systems* (ISCAS'92), pp. 1284–1287, 1992.
108. A.I.A. Cunha, M.C. Schneider, and C. Galup-Montoro, "An MOS transistor model for analog circuit design," *IEEE Journal of Solid-State Circuits*, vol. 33, no. 10, pp. 1510–1519, 1998.
109. C. Galup-Montoro, M.C. Schneider, and A.I.A. Cunha, "Current-based MOSFET model for integrated circuit design," in *Low-Voltage/Low-Power Integrated Circuits and Systems: Low Voltage Mixed-Signal Circuits*, E. Sanchez-Sinencio, and A.G. Andreou (eds.). pp. 7–55, IEEE Press, New York, 1999.
110. A.I.A. Cunha, M.C. Schneider, and C. Galup-Montoro, "Derivation of the unified charge control model and parameter extraction procedure," *Solid-State Electronics*, vol. 43, no. 3, pp. 481–485, 1999.
111. O.C. Gouveia Filho, A.I.A. Cunha, M.C. Schneider, and C. Galup-Montoro, "Advanced compact model for short-channel MOS transistors," *Proceedings of the IEEE Custom Integrated Circuits Conference*, pp. 209–212, IEEE, 2000.
112. A. Arnaud and C. Galup-Montoro, "A compact model for flicker noise in MOS transistors for analog circuit design," *IEEE Transactions on Electron Devices*, vol. 50, no. 9, pp. 1815–1818, 2003.
113. C. Galup-Montoro, M.C. Schneider, A. Arnaud, and H. Klimach, "Self-consistent models of DC, AC, noise and mismatch for the MOSFET," *Proceedings of the Nanotech Workshop on Compact Modeling*, vol. 2, pp. 494–499, 2004.
114. E. Vittoz, and J. Fellrath, "CMOS analog integrated circuits based on weak inversion operation," *IEEE Journal of Solid-State Circuits*, vol. 12, no. 3, pp. 224–231, 1977.
115. J.D. Chatelaine, *Dispositifs a Semiconducteur*, vol. 7, 1st edition, Dunod, Paris, France, 1979.
116. A.I.A. Cunha, O.C. Gouveia-Filho, M.C. Schneider, and C. Galup-Montoro, "A current based model for the MOS transistor," *IEEE International Symposium on Circuits and Systems*, vol. 3, pp. 1608–1611, 1997.
117. M. Bucher, C. Lallement, C. Enz, F. Théodoloz, F. Krummenacher, "Scalable GM/I based MOSFET model," in *International Semiconductor Device Research Symposium*, pp. 615–618, 1997.
118. J.-M. Sallese and A.-S. Porret, "A novel approach to charge-based non-quasistatic model of the MOS transistor valid in all modes of operation," *Solid-State Electronics*, vol. 44, no. 6, pp. 887–894, 2000.
119. C. Lallement, J.-M. Sallese, M. Bucher, W. Grabinski, and P. Fazan, "Accounting for quantum effects and polysilicon depletion from weak to strong inversion in

a charge-based design-oriented MOSFET model," *IEEE Transactions on Electron Devices*, vol. 50, no. 2, pp. 406–417, 2003.

120. C. Lallement, M. Bucher, and C. Enz, "Modelling and characterization of non-uniform substrate doping," *Solid-State Electronics*, vol. 41, no. 12, pp. 1857–1861, 1997.

121. E. Vittoz, C. Enz, and F. Krummenacher, "A basic property of MOS transistors and its circuit implications," in *Technical Proceedings, International Workshop on Compact Modeling*, NSTI Nanotech, pp. 246–249, 2003.

122. J.-M. Sallese, "Advancements in DC and RF MOSFET Modeling with the EPFL-EKV Charge Based Model," Special session: MOS Transistor: Compact Modeling and Standardization Aspects, in *8th International Conference MIXDES 2001*, 21–23, 2001.

123. M. Bucher, J.-M. Sallese, C. Lallement, W. Grabinski, C.C. Enz, and F. Krummenacher, "Extended charges modeling for deep submicron CMOS," in *International Semiconductor Device Research Symposium*, pp. 397–400, 1999.

124. A.-S. Porret, J.-M. Sallese, and C. Enz, "A compact non-quasi-static extension of a charge-based MOS model," *IEEE Transactions on Electron Devices*, vol. 48, no. 8, pp. 1647–1654, 2001.

125. C. Enz, "An MOS transistor model for RF IC design valid in all regions of operation," *IEEE Transactions on Microwave Theory and Techniques*, vol. 50, no. 1, pp. 342–359, 2002.

126. A.-S. Porret and C.C. Enz, "Non-quasi-static (NQS) thermal noise modelling of the MOS transistor," *IEE Proceedings Circuits, Devices and Systems*, vol. 151, no. 2, pp. 155–166, 2004.

127. A.S. Roy and C. Enz, "Compact modeling of thermal noise in the MOS transistor," *IEEE Transactions on Electron Devices*, vol. 52, no. 4, pp. 611–614, 2005.

128. J.-M. Sallese, M. Bucher, F. Krummenacher, and P. Fazan, "Inversion charge linearization in MOSFET modeling and rigorous derivation of the EKV compact model," *Solid-State Electronics*, vol. 47, no. 4, pp. 677–683, 2003.

129. M. Bucher, "The EKV compact MOS transistor model version 3: Accounting for deep-submicron aspects," in *NANOTECH Workshop on Compact Modeling*, San Juan, PR, April 23–25, 2002.

130. M. Bucher, A. Bazigos, F. Krummenacher, J.-M. Sallese, C. Enz, and W. Grabinski, "Advances in MOSFET charges modeling: EKV3.0 MOSFET model," in *NANOTECH Workshop on Compact Modeling*, Anaheim, CA, May 10–12, 2005.

131. C.C. Enz and E.A. Vittoz, *Charge-Based MOS Transistor Modeling: The EKV Model for Low-Power and RF IC Design*, Wiley, New York, 2006.

132. J.M. Sallese, F. Krummenacher, F. Prégaldiny, Ch. Lallement, A. Roy, and C. Enz, "A design oriented charge-based current model for symmetric DG MOSFET and its correlation with the EKV formalism," *Solid-State Electronics*, vol. 49, no. 3, pp. 485–489, 2005.

133. J. He, X. Xi, H. Wan, M. Dunga, M. Chan, and A.M. Niknejad, "BSIM5: An advanced charge-based MOSFET model for nanoscale VLSI circuit simulation," *Solid-State Electronics*, vol. 51, no. 3, pp. 433–444, 2007.

134. C.-H. Lin, X. Xi, J. He, R.Q. Williams, M.B. Ketchen, W.E. Haensch, M. Dunga, S. Balasubramanian, A.M. Niknejad, M. Chan, and C. Hu, "Compact modeling of FinFETs featuring independent-gate operation mode," in *Proceedings of the IEEE VLSI-TSA Symposium*, pp. 120–121, 2005.

135. Y.S. Chauhan, M.A. Karim, S. Venugopalan, S. Khandelwal, P. Thakur, N. Paydavosi, A.B. Sachid, A. Niknejad, and C. Hu, "BSIM6: Symmetric bulk

MOSFET model," in *Technical Proceedings Workshop on Compact Modeling*, NSTI Nanotech, pp. 724–729, 2012.

136. H. Agarwal, S. Khandelwal, J.P. Duarte, Y.S. Chauhan, A. Niknejad, and C. Hu, "BSIM6.1.0 MOSFET Compact Model Technical Manual," University of California, Berkeley, CA, 2014.

137. S.K. Saha, "Modelling the effectiveness of computer-aided development projects in the semiconductor industry," *International Journal of Engineering Management and Economics*, vol. 1, no. 2/3, pp. 162–178, 2010.

Chapter 2

1. A.S. Grove, *Physics and Technology of Semiconductor Devices*, John Wiley & Sons, New York, 1967.
2. R.A. Smith, *Semiconductors*, 2nd edition, Cambridge University Press, London, 1978.
3. B.G. Streetman and S.K. Banerjee, *Solid State Electronic Devices*, 7th edition, Prentice Hall, Englewood Cliffs, NJ, 2014.
4. R.M. Warner, Jr. and B.L. Grung, *Transistors—Fundamentals for the Integrated Circuit Engineer*, John Wiley & Sons, New York, 1983.
5. S.M. Sze and K.K. Ng, *Physics of Semiconductor Devices*, John Wiley & Sons, New York, 2007.
6. R.S. Muller and T.I. Kamins, *Device Electronics for Integrated Circuits*, John Wiley & Sons, New York, 1986.
7. R.F. Pierret, *Advanced Semiconductor Fundamentals*, Addison-Wesley, Reading, MA, 1987.
8. E.S. Yang, *Microelectronic Devices*, McGraw-Hill, New York, 1988.
9. S. Wang, *Fundamentals of Semiconductor Theory and Devices*, Prentice Hall, Englewood Cliffs, NJ, 1989.
10. M. Zambuto, *Semiconductor Devices*, McGraw-Hill, New York, 1989.
11. M. Shur, *Physics of Semiconductor Devices*, Prentice Hall, Englewood Cliffs, NJ, 1990.
12. G.W. Neudeck, *The PN Junction Diode*, Addison-Wesley, Reading, MA, 1989.
13. D.J. Roulston, *Bipolar Semiconductor Devices*, McGraw-Hill, New York, 1990.
14. P. Balk, P.G. Burkhardt, and L.V. Gregor, "Orientation dependence of built-in surface charge on thermally oxidized silicon," *IEEE Proceedings*, vol. 53, no. 12, pp. 2133–2134, 1965.
15. M. Aoki, K. Yano, T. Masuhara, S. Ikeda, and S. Meguro, "Optimum crystallographic orientation of submicron CMOS devices," in *IEEE International Electron Devices Meeting Technical Digest*, pp. 577–580, 1985.
16. M.A. Green, "Intrinsic concentration, effective density of states, and effective mass in silicon," *Journal of Applied Physics*, vol. 67, no. 6, pp. 2944–2954, 1990.
17. N. Arora, *MOSFET Models for VLSI Circuit Simulation: Theory and Practice*, Springer-Verlag, Wien, Austria, 1993.
18. L. Nagel and D. Pederson, "Simulation program with integrated circuit emphasis," University of California, Berkeley, CA, Electronics Research Laboratory Memorandum No. UCB/ERL M352, 1973.

19. Y.P. Varshni, "Temperature dependence of energy gap in semiconductors," *Physica*, vol. 34, no. 1, pp. 149–154, 1967.
20. S.K. Ghandhi, *The Theory and Practice of Microelectronics*, Wiley, New York, 1984.
21. N.D. Arora, J.R. Hauser, and D.J. Roulston, "Electron and hole mobilities in silicon as a function of concentration and temperature," *IEEE Transactions on Electronic Devices*, vol. 29, no. 2, pp. 292–295, 1982.
22. W. Shockley and W.T. Read, "Statistics of the recombination of holes and electrons," *Physical Review*, vol. 87, no. 5, pp. 835–842, 1952.
23. R.N. Hall, "Electron-hole recombination in germanium," *Physical Review*, vol. 87, no. 2, pp. 387–387, 1952.
24. S. Saha, C.S. Yeh, and B. Gadepally, "Impact ionization rate of electrons for accurate simulation of substrate current in submicron devices," *Solid-State Electronics*, vol. 36, no. 10, pp. 1429–1432, 1993.
25. S. Saha, "Extraction of substrate current model parameters from device simulation," *Solid-State Electronics*, vol. 37, no. 10, pp. 1786–1788, 1994.
26. G.W. Neudeck, *The PN Junction Diode*, vol. II, 2nd edition, Modular Series on Solid-State Devices, Addision-Wesley, Reading, MA, 1987.
27. C. Zener, "A theory of electrical breakdown of solid dielectrics," *Proceedings of the Royal Society of London*, vol. A145, no. 8555, pp. 523–529, 1934.
28. E.O. Kane, "Zener tunneling in semiconductors," *Journal of Physics and Chemistry of Solids*, vol. 12, no. 2, 181–188, 1960.

Chapter 3

1. E.H. Nicollian and J.R. Brews, *MOS (Metal Oxide Semiconductor) Physics and Technology*, John Wiley & Sons, New York, 1982.
2. N.D. Arora, *MOSFET Models for VLSI Circuit Simulation: Theory and Practice*, Springer-Verlag, Wien, Austria, 1993.
3. R.S. Muller and T.I. Kamins with M. Chan, *Device Electronics for Integrated Circuits*, 3rd edition, John Wiley & Sons, New York, 2003.
4. P. Ranade, H. Takeuchi, T.-J. King, and C. Hu, "Work function engineering of molybdenum gate electrodes by nitrogen implantation," *Electrochemical and Solid-State Letters*, vol. 4, no. 11, pp. G85–G87, 2001.
5. S.M. Sze, *Semiconductor Physics and Technology*, John Wiley & Sons, New York, 1982.
6. T.P. Chow and A.J. Steckl, "Refractory metal silicides: Thin film properties and processing technology," *IEEE Transactions on Electronic Devices*, vol. 30, no. 11, pp. 1480–1497, 1983.
7. Y. Tsividis, *Operation and Modeling of the MOS Transistors*, 2nd edition, Oxford University Press, New York, 1999.
8. N. Lifshitz, "Dependence of the work function difference between the poly-silicon gate and silicon substrate on the doping level in polysilicon," *IEEE Transactions on Electronic Devices*, vol. 32, no. 3, pp. 617–621, 1985.

9. T. Kamins, *Polycrystalline Silicon for IC Application*, 2nd edition, Kluwer Academic Publisher, Boston, MA, 1998.

10. B.E. Deal, "Standardized terminology for oxide charge associated with thermally oxidized silicon," *IEEE Transactions on Electronic Devices*, vol. 27, no. 3, pp. 606–608, 1980.

11. D.K. Schroder, *Semiconductor Materials and Device Characterization*, John Wiley & Sons, New York, 1990.

12. B.G. Streetman and S. Banerjee, *Solid State Electronic Devices*, 7th edition, Prentice Hall, Englewood Cliffs, NJ, 2014.

13. M.J. van Dort, P.H. Woerlee, and A.J. Walker, "A simple model for quantisation effects in heavily-doped silicon MOSFETs at inversion conditions," *Solid-State Electronics*, vol. 37, no. 3, pp. 411–414, 1994.

14. R.A. Chapman, C.C. Wei, D.A. Bell, S. Aur, G.A. Brown, and R.A. Haken, "0.5 micron CMOS for high performance at 3.3 V," in *IEEE International Electron Devices Meeting Technical Digest*, pp. 52–55, 1988.

15. C.Y. Wong, J.Y.-C. Sun, Y.T. Aur, C.S. Oh, R. Angelucci, and B. Davari, "Doping of N+ and P+ poly_si in a dual-gate CMOS process," in *IEEE International Electron Devices Meeting Technical Digest*, pp. 238–241, 1988.

16. C.Y. Lu, J.M. Sung, H.C. Kirch, S.J. Hellenius, T.E. Smith, and L. Manchanda, "Anomalous C-V characteristics of implanted poly MOS structure in n+/p+ dual-gate CMOS technology," *IEEE Electron Device Letters*, vol. 10, no. 5, pp.192–194, 1989.

17. P. Habas and S. Selberherr, "On the effect of nondegenerate doping of polysilicon gate in thin oxide MOS devices–analytical modeling," *Solid-State Electronics*, vol. 33, no. 12, pp. 1539–1544, 1990.

18. S. Saha, G. Srinivasan, G.A. Rezvani, and M. Farr, "Effects of inversion layer quantization and polysilicon gate depletion on tunneling current of ultra-thin SiO_2 gate material," *Materials Research Society Symposium Proceedings*, vol. 567, pp. 275–282, 1999.

19. G. Srinivasan, S.K. Saha, and G.A. Rezvani, "Estimation of quantum mechanical and polysilicon depletion effects for ultra-thin SiO_2 gate dielectric," in *Proceedings of the SPIE Conference on Microelectronic Device Technology*, vol. 3881, pp. 168–174, 1999.

Chapter 4

1. J.E. Lilenfield, "Methods and apparatus for controlling electric currents," US Patent 1745175, January 28, 1930.

2. H.C. Pao and C.T. Sah, "Effects of diffusion current on characteristics of metal-oxide (insulator)-semiconductor transistors," *Solid-State Electronics*, vol. 9, no. 10, pp. 927–937, 1966.

3. R.F. Pierret and J.A. Shields, "Simplified long-channel MOSFET theory," *Solid-State Electronics*, vol. 26, no. 2, pp. 143–147, 1983.

4. A. Naussbaum, R. Sinha, and D. Dokos, "The theory of the long-channel MOSFET," *Solid-State Electronics*, vol. 27, no. 1, pp. 97–106, 1984.

5. J.R. Brews, "A charge sheet model of the MOSFET," *Solid-State Electronics*, vol. 21, no. 2, pp. 345–355, 1978.

6. S.K. Saha, "Modeling process variability in scaled CMOS technology," *IEEE Design & Test of Computers*, vol. 27, no. 2, pp. 8–16, 2010.

7. S. Saha, "Device characteristics of sub-20-nm silicon nanotransistors," in *Proceedings of the SPIE Conference on Design and Process Integration for Microelectronic Manufacturing*, vol. 5042, pp. 172–179, 2003.

8. S.K. Saha, "Transistors having optimized source-drain structures and methods for making the same," US Patent 6344405, February 5, 2002.

9. S. Saha, "Scaling considerations for high performance 25 nm metal-oxide-semiconductor field-effect transistors," *Journal of Vacuum Science & Technology B*, vol. 19, no. 6, pp. 2240–2246, 2001.

10. S. Saha, "Design considerations for 25 nm MOSFET devices," *Solid-State Electronics*, vol. 45, no. 10, pp. 1851–1857, 2001.

11. S.K. Saha, "Drain profile engineering for MOSFET devices with channel lengths below 100 nm," in *Proceedings of the SPIE Conference on Microelectronic Device Technology*, vol. 3881, pp. 195–204, 1999.

12. Y. Cheng and C. Hu, *MOSFET Modeling and BSIM3 User's Guide*, Kluwer Academic Publishers, Boston, MA, 1999.

13. N.D. Arora, *MOSFET Models for VLSI Circuit Simulation: Theory and Practice*, Springer-Verlag, Wien, Austria, 1993.

14. *Taurus MEDICI Manuals*, Synopsys, Inc., Mountain View, CA, 2007.

15. T. Binder, K. Dragosits, T. Grasser, R. Klima, M. Knaipp, H. Kosins, R. Mlekus et al., *MINIMOS-NT User's Guide: Institut für Mikroelektronik*, Technical University, Vienna, Austria, August 2013.

16. *Sentaurus TCAD Manuals*, Synopsys, Inc., Mountain View, CA, 2012.

17. *Atlas User's Manual: Device Simulation Software*, Silvaco, Inc., Santa Clara, CA, 2013.

18. J.R. Brews, "Physics of MOS transistor," in *Silicon Integrated Circuits*, Part A, D. Kahng (ed.). Applied Solid State Science Series, Academic Press, New York, 1981.

19. P.P. Guebels and F. Van de Wiele, "A small geometry MOSFET models for CAD applications, *Solid-State Electronics*," vol. 26, no. 4, pp. 267–273, 1983.

20. G. Baccarani, M. Rudan, and G. Spadini, "Analytical i.g.f.e.t model including drift and diffusion currents," *IEEE Journal on Solid-State and Electron Devices*, vol. 2, no. 2, pp. 62–68, 1987.

21. F. Van de Wiele, "A long channel MOSFET model," *Solid-State Electronics*, vol. 22, no. 12, pp. 991–997, 1979.

22. C. Turchetti and G. Masetti, "A CAD-oriented analytical MOSFET model for high–accuracy applications," *IEEE Transactions on Computer-Aided Design*, vol. 3, no. 2, pp. 117–122, 1984.

23. A.M. Ostrowsky, *Solutions of Equations and Systems of Equations*, Academic Press, New York, 1973.

24. Y.P. Tsividis, "Moderate inversion in MOS devices," *Solid-State Electronics*, vol. 26, no. 11, pp. 1099–1104, 1982.

25. Y.P. Tsividis, *Operation and Modeling of the MOS Transistor*, 2nd edition, Oxford University Press, New York, 1999.

26. M. Miura-Mattausch, H.J. Mattausch, and T. Ezaki, *The Physics and Modeling of MOSFETS: Surface-Potential Model HiSIM*, World Scientific, Singapore, 2008.

27. C.T. Sah, "Characteristics of the metal-oxide-semiconductor transistors," *IEEE Transactions on Electron Devices*, vol. 11, no. 7, pp. 324–345, 1964.
28. H. Shichman and D.A. Hodges, "Modeling and simulation of insulated-gate field-effect transistor switching circuits," *IEEE Journal Solid-State Circuits*, vol. 3, no. 3, pp. 285–289, 1968.
29. H.K.J. Ihantola and J.L. Moll, "Design theory of a surface field-effect transistor," *Solid-State Electronics*, vol. 7, no. 6, pp. 423–430, 1964.
30. B.J. Sheu, D.L. Scharfetter, P.K. Ko, and M.C. Jeng, "BSIM: Berkeley short-channel IGFET models for MOS transistors," *IEEE Journal on Solid-State Circuits*, vol. 22, no. 4, pp. 558–565, 1987.
31. G.T. Wright, "Physical and CAD model for the implanted-channel VLSI MOSFET," *IEEE Transactions on Electron Devices*, vol. 34, no. 4, pp. 823–833, 1987.
32. R.M. Swanson and J.D. Meindel, "Ion implanted complementary MOS transistors in low-voltage circuits," *IEEE Journal on Solid-State Circuits*, vol. 7, no. 2, pp. 146–143, 1972.

Chapter 5

1. K.J. Kuhn, "Considerations for ultimate CMOS scaling," *IEEE Transactions on Electron Devices*, vol. 59, no. 7, pp. 1813–1828, 2012.
2. H. Iwai, "CMOS technology—Year 2010 and beyond," *IEEE Journal of Solid-State Circuits*, vol. 34, no. 3, pp. 357–366, 1999.
3. S. Saha, "Scaling considerations for high performance 25 nm metal-oxide-semiconductor field-effect transistors," *Journal of Vacuum Science & Technology B*, vol. 19, no. 6, pp. 2240–2246, 2001.
4. S. Saha, "Design considerations for 25 nm MOSFET devices," *Solid-State Electronics*, vol. 45, no. 10, pp. 1851–1857, 2001.
5. C. Duvvury, "A guide to short channel effects in MOSFETS," *IEEE Circuit and Systems Magazine*, vol. 2, no.6, pp. 6–10, 1986.
6. C.Y. Lu and J.M. Sung, "Reverse short channel effects on threshold voltage in submicron salicide devices," *IEEE Electron Device Letters*, vol. l0, p. 446, 1989.
7. E.H. Li et al., "The narrow channel effect in MOSFET with semi-recessed oxide structures," *IEEE Transactions on Electron Devices*, vol. 37, no. 3, p. 692, 1990.
8. L.A. Akers, "The inverse narrow width effect," *IEEE Electron Device Letters*, vol. 7, no. 7, p. 419, 1986.
9. W. Fichtner and H.W. Potzl, "MOS modeling by analytical approximations-subthreshold current and subthreshold voltage," *International Journal of Electronics*, vol. 46, no. 1, pp. 33–35, 1979.
10. R.R. Troutman, "VLSI limitations from drain-induced barrier lowering," *IEEE Transactions on Electron Devices*, vol. 26, no, 4, pp. 461–469, 1979.
11. C.G. Sodini, P.K. Ko, and J.L. Moll, "The effect of high fields on MOS device and circuit simulation," *IEEE Transactions on Electron Devices*, vol. 31, no. 10, pp. 1386–1393, October 1984.

12. M.S. Liang, J.Y. Choi, P.-K. Ko, and C. Hu, "Inversion layer capacitance and mobility of very thin oxide MOSFETs," *IEEE Transactions on Electron Devices,* vol. 33, no. 3, pp. 409–413, 1986.

13. C. Hu, "Hot Carrier Effects," in *Advanced MOS Device and Physics,* N.G. Einspruch, and G. Gildenblat, eds., vol. 18, VLSI Electronics Microstructure Science, Academic Press, New York, vol. 18, pp. 119–139, 1989.

14. I.C. Chen, C.W. Teng, D.J. Coleman, and A. Nishimura, "Interface-trap enhanced gate-induced leakage current in MOSFET," *IEEE Electron Device Letters,* vol. 10, no. 5, pp. 216–218, 1989.

15. F. Assderaghi, P.K. Ko, and C. Hu, "Observation of velocity overshoot in silicon in version layers," *IEEE Electron Device Letters,* vol. 14, no. 10, p. 484–486, 1993.

16. Y. Cheng and T.A. Fjeldly, "Unified physical I–V model including self-heating effect for fully depleted SOI/MOSFET's," *IEEE Transactions on Electron Devices,* vol. 43, no. 8, pp. 1291–1296, 1996.

17. Y.-C. King, H. Fujioka, S. Kamohara, W.C. Lee, and C. Hu, "AC charge centroid model for quantization of inversion layer in NMOSFET," in *International Symposium on VLSI Technology, Systems, and Applications, Proceedings of Technical Papers,* Taipei, Taiwan, pp. 245–249, June 1997.

18. M.J. van Dort, P.H. Woerlee, and A.J. Walker, "A simple model for quantisation effects in heavily-doped silicon MOSFETs at inversion conditions," *Solid-State Electronics,* vol. 37, no. 3, pp. 411–414, 1994.

19. S. Saha, "Effects of inversion layer quantization on channel profile engineering for nMOSFETs with 0.1 μm channel lengths," *Solid-State Electronics,* vol. 42, no. 11, pp. 1985–1991, 1998.

20. K. Chen et al., "Polysilicon gate depletion effect on IC performance," *Solid-State Electronics,* vol. 38, no. 11, pp. 1975–1977, 1995.

21. S.K. Saha, "Modeling process variability in scaled CMOS technology," *IEEE Design & Test of Computers,* vol. 27, no. 2, pp. 8–16, 2010.

22. S.K. Saha, "Compact MOSFET modeling for process variability-aware VLSI circuit design," *IEEE Access,* vol. 2, pp. 104–115, 2014.

23. S.K. Saha, "Method for forming channel-region doping profile for semiconductor device," US Patent 6,323,520, November 27, 2001.

24. S.K. Saha, "Transistors having optimized source-drain structures and methods for making the same," US Patent 6,344,405, February 5, 2002.

25. S.K. Saha, "Drain profile engineering for MOSFET devices with channel lengths below 100 nm," *Proceedings of the SPIE Conference on Microelectronic Device Technology,* vol. 3881, pp. 195–204, 1999.

26. S. Saha, "Device characteristics of sub-20-nm silicon nanotransistors," *Proceedings of the SPIE Conference on Design and Process Integration for Microelectronic Manufacturing,* vol. 5042, pp. 172–179, 2003.

27. J.H. Huang, Z.H. Liu, M.C. Jeng, P. Ko, and C. Hu, "A physical model for MOSFET output resistance," *International Electron Devices Meeting. Technical Digest,* pp. 569–572, IEEE, 1992.

28. N. Paydavosi, T.H. Morshed, D.D. Lu, W. Yang, M.V. Dunga, X. Xi, J. He et al., *BSIM4v4.8.0 MOSFET Model User's Manual,* University of California, Berkeley, CA, 2013.

29. Y. Cheng and C. Hu, *MOSFET modeling and BSIM3 user's Guide,* Kluwer Academic Publishers, Boston, MA, 1999.

30. K.M. Cao, W. Liu, X. Jin, K. Vasanth, K. Green, J. Krick, T. Vrotsos, and C. Hu, "Modeling of pocket implanted MOSFETs for anomalous analog behavior," in *IEEE International Electron Devices Meeting Technical Digest*, pp. 171–174, 1999.

31. S. Mudanai, W.-K. Shih, R. Rios, X. Xi, J.-H. Rhew, K. Kuhn, and P. Packan, "Analytical modeling of output conductance in long-channel MOSFETs," *IEEE Transactions on Electron Devices*, vol. 53, no. 9, pp. 2091–2097, 2006.

32. G. Merckel, "A simple model of the threshold voltage of short and narrow channel MOSFETs," *Solid-State Electronics*, vol. 23, no. 12, pp. 1207–1213, 1980.

33. L.D. Yau, "A simple theory to predict the threshold voltage of short-channel IGFET's," *Solid-State Electronics*, vol. 17, no. 10, pp. 1059–1063, 1974.

34. Z.-H. Liu, C. Hu, J.-H. Huang, T.-Y. Chan, M.-C. Jeng, P.K. Ko, and Y.C. Cheng, "Threshold voltage model for deep-submicron MOSFETs," *IEEE Transactions on Electron Devices*, vol. 40, no. 1, pp. 86–98, 1993.

35. N.D. Arora and G.Sh. Gildenblat, "A semi-empirical model of the MOSFET inversion layer mobility for low temperature operation," *IEEE Transactions on Electron Devices*, vol. 34, no. 1, pp. 89–93, 1987.

36. S.C. Sun and J.D. Plummer, "Electron mobility in inversion and accumulation layers on thermally grown oxidized silicon surfaces," *IEEE Transactions on Electron Devices*, vol. 27, no. 8, pp. 1497–1508, 1980.

37. D.T. Amm, H. Mingam, P. Delpech, and T.T. D'ouville, "Surface mobility in n+ and p+ doped polysilicon gate PMOS transistors," *IEEE Transactions on Electron Devices*, vol. 36, no. 5, pp. 963–968, 1989.

38. M.S. Liang, J.Y. Choi, P.-K. Ko, and C. Hu, "Inversion layer capacitance and mobility of very thin oxide MOSFETs," *IEEE Transactions on Electron Devices*, vol. 33, no. 3, pp. 409–413, 1986.

39. K. Lee, J.Y. Choi, S.P. Sim, and C.K. Kim, "Physical understanding of low field carrier mobility in silicon inversion layer," *IEEE Transactions on Electron Devices*, vol. 38, no. 8, pp. 1905–1912, 1991.

40. K. Chen, H.C. Wann, J. Dunster, P.K. Ko, and C. Hu, "MOSFET carrier mobility model based on gate oxide thickness, threshold and gate voltages," *Solid-State Electronics*, vol. 39, no. 10, pp. 1515–1518, 1996.

41. Y. Cheng, M.-C. Jeng, Z. Liu, J. Huang, M. Chan, K. Chen, P.K. Ko, and C. Hu, "A physical and scalable BSIM3v3 I-V model for analog/digital circuit simulation," *IEEE Transactions on Electron Devices*, vol. 44, no. 2, pp. 227–287, 1997.

42. S. Saha, C.S. Yeh, and B. Gadepally, "Impact ionization rate of electrons for accurate simulation of substrate current in submicron devices," *Solid-State Electronics*, vol. 36, no. 10, pp. 1429–1432, 1993.

43. S. Saha, "Hot-carrier reliability in sub-0.1 μm nMOSFET devices," in *Materials Research Society Symposia Proceedings*, vol. 428, p. 379–384, 1996.

44. J.W. Slotboom, G. Streutker, G.J.T. Davids, and P.B. Hartong, "Surface impact ionization in silicon devices," in *IEEE International Electron Devices Meeting Technical Digest*, pp. 494–497, 1987.

45. N.D. Arora, *MOSFET Models for VLSI Circuit Simulation: Theory and Practice*, Springer-Verlag, Wien, Austria, 1993.

46. T.Y. Chan, P.K. Ko, and C. Hu, "A simple method to characterize substrate current in MOSFET's," *IEEE Electron Device Letters*, vol. 5, no. 12, pp. 505–507, 1984.

47. S. Saha, "Extraction of substrate current model parameters from device simulation,"*Solid-State Electronics*, vol. 37, no. 10, pp. 1786–1788, 1994.

48. S.-H. Lo, D.A. Buchanan, Y. Taur, and W. Wang, "Quantum-mechanical modeling of electron tunneling current from the inversion layer of ultra-thin-oxide nMOS-FET's," *IEEE Electron Device Letters*, vol. 18, no. 5, pp. 209–211, 1997.

49. W.-C. Lee and C. Hu, "Modeling gate and substrate currents due to conduction- and valence-band electron and hole tunneling," in *Symposium on VLSI Technology*, pp. 198–199, 2000.

50. N. Yang, W.K. Hension, J.R. Hauser, and J.J. Wortman, "Modeling study of ultrathin gate oxides using direct tunneling current and CV measurements in MOS devices," *IEEE Transactions on Electron Devices*, vol. 46, no. 7, pp. 1464–1471, 1999.

51. W.-C. Lee, and C. Hu, "Modeling CMOS tunneling currents through ultrathin gate oxide due to conduction- and valence-band electron and hole tunneling," *IEEE Transactions on Electron Devices*, vol. 48, no. 7, pp. 1366–1373, July 2001.

52. J.B. Jacobs and D. Antoniadis, "Channel profile engineering for MOSFET's with 100 nm channel lengths," *IEEE Transactions on Electron Devices*, vol. 42, no. 5, pp. 870–875, 1995.

53. S.A. Parke, J.E. Moon, H.C. Wann, P.K. Ko, and C. Hu, "Design for suppression of gate-induced drain leakage in LDD MOSFET's using a quasi-two dimensional analytical model," *IEEE Transactions on Electron Devices*, vol. 39, no. 7, pp. 1694–1703, 1992.

54. N. Lindert, M. Yoshida, C. Wann, and C. Hu, "Comparison of GIDL in p+-poly PMOS and n+-poly PMOS devices," *IEEE Electron Device Letters*, vol. 17, no. 6, pp. 285–287, 1996.

55. T.Y. Chan, J. Chen, P.K. Ko, and C. Hu, "The impact of gate-induced drain-leakage current on MOSFET scaling," in *IEEE International Electron Devices Meeting Technical Digest*, pp. 718–721, 1987.

56. H.J. Wann, P.K. Ko, and C. Hu, "Gate-induced band-to-band tunneling leakage current in LDD MOSFET's," in *IEEE International Electron Devices Meeting Technical Digest*, pp. 147–150, 1992.

Chapter 6

1. Y. Tsividis and C. McAndrew, *Operation and Modeling of the MOS Transistor*, Oxford University Press, Oxford, 2011.

2. J. Meyer, "MOS models and circuit simulation," *RCA Review*, vol. 32, no. 1, pp. 42–63, 1971.

3. N. Arora, *MOSFET Models for VLSI Circuit Simulation: Theory and Practice*, Springer-Verlag, Wien, Austria, 1993.

4. D.E. Ward and R.W. Dutton, "A charge-oriented model for MOS transistor capacitances," *IEEE Journal of Solid-State Circuits*, vol. 13, no. 5, pp. 703–707, 1978.

5. S.Y. Oh, D.E. Ward, and R.W. Dutton, "Transient analysis of MOS transistors," *IEEE Transactions on Electron Devices*, vol. 27, no. 2, pp. 1571–1578, 1980.

6. M.F. Sevat, "On the channel charge division in MOSFET modeling," in *IEEE International Conference on Computer-Aided Design, Digest of Technical Papers,* pp. 208–210, 1987.
7. G.W. Taylor, W. Fichtner, and J.G. Simmons, "A description of MOS intermodal capacitances for transient simulation," *IEEE Transactions on Computer-Aided Design,* vol. 1, no. 4, pp. 150–156, 1982.
8. P. Yang, B.D. Epler, and P. Chatterjee, "An investigation of the charge conversation problem for MOSFET circuit simulation," *IEEE Journal of Solid-State Circuits,* vol. 18, no. 1, pp. 128–138, 1983.
9. K.Y. Tong, "AC model for MOS transistors from transient-current computations," *IEE Proceedings on Solid-State and Electron Devices,* vol. 130, no. 1, pp. 33–36, 1983.
10. L.W. Nagel, *SPICE2—A Computer Program to Simulate Semiconductor Circuits,* Memo ERL-M250, Electronic Research Laboratory, University of California, Berkeley, CA, May 1975.
11. M.A. Cirit, "The Meyer model revisited: Why is charge not conserved?," *IEEE Transactions on Computer-Aided Design,* vol. 8, no. 10, pp. 1033–1037, 1989.
12. K.A. Sakallah, Y.-T. Yen, and S.S. Greenberg, "The Meyer model revisited: Explaining and correcting the charge non-conservation problem," in *IEEE International Conference on Computer-Aided Design, Digest of Technical Papers,* pp. 204–207, IEEE, 1987.
13. T. Smedes and F.M. Klaassen, "Effects of the lightly doped drain configuration on capacitance characteristics of submicron MOSFETs," in *IEEE International Electron Devices Meeting Technical Digest,* pp. 197–200, 1990.
14. K.A. Sakallah, Y.-T. Yen, and S.S. Greenberg, "A first order charge conserving model MOS capacitor model," *IEEE Transactions on Computer-Aided Design,* vol. 9, no. 1, pp. 99–108, 1990.
15. B.J. Sheu, D.L. Scharfetter, C. Hu, and D.O. Pederson, "A compact IGFET charge model," *IEEE Transactions on Circuits and Systems,* vol. 31, no. 8, pp. 745–748, 1984.
16. D.E. Ward, "Charge-base modeling of capacitances in MOS transistors," Stanford Electronics Laboratory Technical Report, G201–11, Stanford University, Stanford, CA, 1981.
17. J.G. Fossum, H. Jeong, and S. Veeraraghavan, "Significance of the channel-charge partition in the transient MOSFET model," *IEEE Transactions on Electron Devices,* vol. 33, no. 10, pp. 1621–1623, 1986.
18. P. Yang, "Capacitance modeling for MOSFETs," in *Circuit Analysis, Simulation and Design,* A.E. Ruehli, ed., Elsevier, Amsterdam, the Netherlands, pp. 107–130, 1986.
19. B.J. Sheu, W.-J. Hsu, and P.K. Ko, "An MOS transistor charge model for VLSI design," *IEEE Transactions on Computer-Aided Design,* vol. 7, no. 4, pp. 520–527, 1988.
20. R. Shrivastava and K. Fitzpatrick, "A simple model for the overlap capacitance of a VLSI MOS device," *IEEE Transactions on Electron Devices,* vol. 29, no. 12, pp. 1870–1875, 1982.
21. N. Paydavosi, T.H. Morshed, D.D. Lu, W. Yang, M.V. Dunga, X. Xi, J. He, W. Liu, K. cao, X. Jin, J.J. Ou, M. Chan, A.M. Niknejad, and C. Hu, "BSIM4v4.8.0 MOSFET Model User's Manual," University of California, Berkeley, CA, 2013.
22. Y. Cheng and C. Hu, *MOSFET Modeling and BSIM3 User's Guide,* Kluwer Academic Publishers, Boston, MA, 1999.

Chapter 7

1. A. van der Ziel, *Noise in Solid State Devices and Circuit*, Wiley-Interscience, New York, 1986.
2. A. Ambrozy, *Electronic Noise*, McGraw-Hill, New York, 1982.
3. P.R. Gray and R.G. Meyer, *Analysis and Design of Analog Integrated Circuits*, Wiley, New York, 1993.
4. Y. Tsividis and C. McAndrew, *Operation and Modeling of the MOS Transistor*, Oxford University Press, Oxford, 2010.
5. M. Haartman and M. Ostling, *Low Frequency Noise in Advanced MOSA Devices*, Springer, New York, 2007.
6. M.J. Deen, C.-H. Chen, S. Asgaran, G.A. Rezvani, J. Tao, and Y. Kiyota, "High-frequency noise of modern MOSFETs: Compact modeling and measurement issues," *IEEE Transactions on Electron Devices*, vol. 53, no. 9, pp. 2062–2081, 2006.
7. P. Jindal, "Compact noise models for MOSFETs," *IEEE Transactions on Electron Devices*, vol. 53, no. 9, pp. 2051–2061, 2006.
8. A. Van der Ziel, *Noise: Source. Characterization, and Measurements*, Prentice Hall, Englewood Cliffs, NJ, 1970.
9. A.G. Jordan and N.A. Jordan, "Theory of the noise in metal oxide semiconductor devices," *IEEE Transactions on Electron Devices*, vol. 12, no. 3, pp. 148–156, 1965.
10. B. Wang, R. Hellums, and C.G. Sodini, "MOSFET thermal noise modeling for analog integrated circuits," *IEEE Journal of Solid-State Circuits*, vol. 29, no. 7, pp. 833–835, 1994.
11. D.P. Triantis, A.N. Birbas, and D. Kondis, "Thermal noise modeling for short channel MOSFETs," *IEEE Transactions on Electron Devices*, vol. 43, no. 11, pp. 1950–1955, 1996.
12. L.W. Nagel, *SPICE2—A Computer Program to Simulate Semiconductor Circuits*, Memo ERL-M250, Electronic Research Laboratory, University of California, Berkeley, CA, 1975.
13. P. Jindal and A. Van der Ziel, "Carrier fluctuation noise in a MOSFET channel due to traps in the oxide," *Solid-State Electronics*, vol. 216, pp. 901–903, 1978.
14. F.M. Klaassen, "Characterization of low 1/f noise in MOS transistors," *IEEE Transactions on Electron Devices*, vol. 18, no. 10, pp. 887–891, 1971.
15. B.J. Gross and C.G. Sodini, "1/f noise in MOSFETs with ultrathin gate dielectric," in *IEEE International Electron Devices Meeting Technical Digest*, pp. 881–884, 1992.
16. T.G.M. Kleinpenning, "On 1/f trapping noise in MOST's," *IEEE Transactions on Electron Devices*, vol. 37, no. 9, pp. 2084–2089, 1990.
17. L.K.J. Vandamme, X. Li, and D. Rigaud, "1/f noise in MOS devices, mobility or number fluctuations?," *IEEE Transactions on Electron Devices*, vol. 41, no. 11, pp. 1936–1945, 1994.
18. F.N. Hooge, "1/f noise sources," *IEEE Transactions on Electron Devices*, vol. 41, no. 11, pp. 1926–1935, 1994.
19. O. Jantsch, "Flicker (1/f) noise generated by a random walk of electrons in interface," *IEEE Transactions on Electron Devices*, vol. 34, no. 5, pp. 1100–1113, 1987.
20. H. Mikeoshiba, "1/f noise in n-channel silicon-gate MOS transistors," *IEEE Transactions on Electron Devices*, vol. 29, no. 6, pp. 965–970, 1982.

21. K.K. Hung, P.K. Ko, C. Hu, and Y.C. Cheng, "A unified model for the flicker noise in metal-oxide-semiconductor field-effect transistors," *IEEE Transactions on Electron Devices*, vol. 37, no. 3, pp. 654–665, 1990.

22. Z. Celik-Butler and T.Y. Hsiang, "Spectral dependence of $1/f\gamma$ noise on gate bias in n-MOSFETs," *Solid-State Electronics*, vol. 30, no. 4, pp. 419–423, 1987.

23. C. Surya, and T.Y. Hsiang, "Theory and experiments on the $1/f'\gamma$ noise in p-channel metal-oxide-semiconductor field-effect transistors at low drain bias," *Physics Review*, vol. B33, no. 7, pp. 4898–4905, 1986.

24. L.K.J. Vandamme, "Model for l/f noise in MOS transistor based in linear region," *Solid-State Electronics*, vol. 23, no. 4, pp. 317–323, 1980.

25. R.P. Jindal and A. van der Ziel, "Phonon fluctuation model for flicker noise in elemental semiconductor," *Journal of Applied Physics*, vol. 52, no. 4, pp. 2884–2888, 1981.

26. H.S. Park, A. van der Ziel, and S.T. Liu, "Comparison of two l/f noise models in MOSFETs," *Solid-State Electronics*, vol. 25, no. 3, pp. 213–217, 1982.

27. S. Liu and L.W. Nagel, "Small signal MOSFET models for analog circuit design," *IEEE Journal of Solid-State Circuits*, vol. 17, no. 6, pp. 983–998, 1982.

28. Y. Cheng et al., *BSIM3 version 3 User's Manual*, Memorandum No. UCBIERL M97/2, University of California, Berkeley, CA, 1997.

29. K.K. Hung, P.K. Ko, C. Hu, and Y.C. Cheng, "A physics-based MOSFET noise model for circuit simulation," *IEEE Transactions on Electron Devices*, vol. 37, no. 5, pp. 1323–1333, 1990.

30. G. Reimbold, "Modified l/f trapping noise theory and experiments in MOS transistors biased from weak to strong inversion influence of interface state," *IEEE Transactions on Electron Devices*, vol. 31, no. 9, pp. 1190–1198, 1984.

31. S.C. Sun, and J.D. Plummer, "Electron mobility in inversion and accumulation layers on thermally oxidized silicon surfaces," *IEEE Transactions on Electron Devices*, vol. 27, no. 8, pp. 1497–1508, 1980.

32. N. Paydavosi, T.H. Morshed, D.D. Lu, W. Yang, M.V. Dunga, X. Xi, J. He, W. Liu, K. cao, X. Jin, J.J. Ou, M. Chan, A.M. Niknejad, and C. Hu, "BSIM4v4.8.0 MOSFET Model User's Manual," University of California, Berkeley, CA, 2013.

33. J.J. Paulous and D.A. Antoniadis, "Limitations of quasi-static capacitance models for the MOS transistors," *IEEE Electron Device Letters*, vol. 4, no. 7, pp. 221–224, 1983.

34. S.Y. Oh, D.E. Ward, and R.W. Dutton, "Transient analysis of MOS transistors," *IEEE Journal of Solid-State Circuits*, vol. 15, no. 4, pp. 636–643, 1980.

35. M. Chan, K.Y. Hui, C. Hui, and P.K. Ko, "A robust and physical BSIM3 non-quasi-static transient and AC small signal model for circuit simulation," *IEEE Transactions on Electron Devices*, vol. 45, no. 4, pp. 834–841, 1998.

36. P. Yang and P.K. Chatterjee, "SPICE modeling for small geometry MOSFET circuits," *IEEE Transactions on Computer-Aided Design*, vol. 1, no. 4, pp. 169–182, 1982.

37. Y.P. Tsividis and G. Masetti, "Problems in the precision modeling of the MOS transistor for analog applications," *IEEE Transactions on Computer-Aided Design*, vol. 3, no. 1, pp. 72–79, 1984.

38. T.L. Quarles, *SPICE 3 Implementation Guide*, Memorandum No. UCB/ERL M89/44, April 1989.

39. R. Singh, A. Juge, R. Joly, and G. Morin, "An investigation into the nonquasi-static effects in MOS devices with an wafer S-parameter techniques," in *Proceedings of the IEEE International Conference on Microelectron Test Structures,* Barcelona, Spain, 1993.

40. Y. Cheng, M. Schroter, C. Enz, M. Matloubian, and D. Pehlke, "RF modeling issues of deep-submicron MOSFETs for circuit design," in *5th International Conference on Solid-State and Integrated-Circuit Technology,* pp. 416–419, 1998.

41. M. Bagheri and Y. Tsividis, "A small signal dc-to-high frequency non-quasistatic model for the four-terminal MOSFET valid in all regions of operation," *IEEE Transactions on Electron Devices,* vol. 32, no. 11, pp. 2383–2391, 1985.

42. H.J. Park, P.K. Ko, and C. Hu, "A charge-conserving non-quasi-static MOSFET model for SPICE transient analysis," in *IEEE International Electron Devices Meeting Technical Digest,* pp. 652–655, 1987.

43. C. Turchetti, P. Mancini, and G. Masetti, "A CAD-oriented non-quasi-static approach for the transient analysis of MOS IC's," *IEEE Journal of Solid-State Circuits,* vol. 21, no. 5, pp. 827–836, 1986.

44. Y.P. Tsividis, *Operation and Modeling of the MOS Transistor.* New York: McGraw-Hill, 1987.

45. W.C. Elmore, "The transient response of damped linear networks with particular regard to wideband amplifiers," *Journal of Applied Physics,* vol. 19, no. 1, pp. 55–63, January 1948.

46. J.G. Fossum, H. Jeong, and S. Veeraraghavan, "Significance of the channel-charge partition in the transient MOSFET model," *IEEE Transactions on Electron Devices,* vol. 33, no. 10, pp. 1621–1623, October 1986.

47. M.F. Sevat, "On the channel charge division in MOSFET modeling," in *IEEE International Conference on Computer-Aided Design, Digest of Technical Papers,* pp. 208–210, 1987.

48. B.J. Sheu and P.K. Ko, "Measurement and modeling of short-channel MOS transistor gate capacitances," *IEEE Journal of Solid-State Circuits,* vol. 22, no. 3, pp. 464–472, 1987.

49. W. Liu, C. Bowen, and M.-C. Chang, "A CAD-compatible non-quasi-static MOSFET model," in *IEEE International Electron Devices Meeting Technical Digest,* pp. 151–154, 1996.

50. A.A. Abidi, A. Rofougaran, G. Chang, J. Rael, J. Chang, M. Rofougaran, and P. Chang, "The future of CMOS wireless transceivers," in *ISSCC Technical Digest,* pp. 118–119, 1997.

51. W. Liu, R. Gharpurey, M.C. Chang, U. Erdogan, R. Aggarwal, and J.P. Mattia, "RF MOSFET modeling accounting for distributed sub-state and channel resistance with emphasis on the BSIM3v3 SPICE model," in *IEEE International Electron Devices Meeting Technical Digest,* pp. 309–312, 1997.

52. M.C. Ho, F. Brauchler, and J.Y. Yang, "Scalable RF Si MOSFET distributed lumped element model based on BSIM3v3," *Electronics Letters,* vol. 33, no. 23, pp. 1992–1993, 1997.

53. D.R. Pehlke, M. Schroter, A. Burnstein, M. Matloubian, and M.F. Chang, "High frequency application of MOS compact model and their development for scalable RF model libraries," in *Proceedings of the IEEE Custom Integrated Circuits Conference,* pp. 219–222, 1998.

54. J.-J. Ou, J. Xiaodong, I. Ma, C. Hu, and P.R. Gray, "CMOS RF modeling for GHz communication IC's," in *Symposium on VLSI Technology,* pp. 94–95, 1998.

55. S.H.M. Jen, C.C. Enz, D.R. Pehlke, M. Schroter, and B.J. Sheu, "Accurate modeling and parameter extraction for MOS transistor valid up to 10-GHz," *IEEE Transactions on Electron Devices*, vol. 46, no. 11, pp. 2217–2227, 1999.

56. R. Goyal, *High-Frequency Analog Integrated Circuit Design*, John Wiley & Sons, New York, 1994.

57. W. Liu and M.-C. Chang, "Transistor transient studies including transcapacitive current and distributive gate resistance for inverter circuits," *IEEE Transactions on Circuit and Systems I: Fundamental Theory and Applications*, vol. 45, no. 4, pp. 416–422, 1998.

58. X. Jin, J.-J. Ou, C.-H. Chen, W. Liu, M.J. Deen, P.R. Gray, and C. Hu, "An effective gate resistance model for CMOS RF and noise modeling," in *IEEE International Electron Devices Meeting Technical Digest*, pp. 961–964, 1998.

59. C. Enz and Y. Cheng, "MOS transistor modeling issues for RF circuit design," *Workshop of Advances in Analog Circuit Design*, Nice, France, March 1999.

Chapter 8

1. S.K. Saha, "Modeling process variability in scaled CMOS technology," *IEEE Design & Test of Computers*, vol. 27, no. 2, pp. 8–16, 2010.

2. R.H. Dennard, F.H. Gaensslen, H.N. Yu, V.L. Rideout, E. Bassous, and A.R. LeBlanc, "Design of ion-implanted MOSFETs with very small physical dimensions," *IEEE Journal of Solid-State Circuits*, vol. SC-9, no. 5, pp. 256–268, 1974.

3. H. Iwai, "CMOS technology—Year 2010 and beyond," *IEEE Journal of Solid-State Circuits*, vol. 34, no. 3, pp. 357–366, 1999.

4. K.J. Kuhn, "Considerations for ultimate CMOS scaling," *IEEE Transactions on Electron Devices*, vol. 59, no. 7, pp. 1813–1828, 2012.

5. S. Saha, "Scaling considerations for high performance 25 nm metal-oxide-semiconductor field-effect transistors," *Journal of Vacuum Science & Technology B*, vol. 19, no. 6, pp. 2240–2246, 2001.

6. S. Saha, "Design considerations for 25 nm MOSFET devices," *Solid-State Electronics*, vol. 45, no. 10, pp. 1851–1857, 2001.

7. S.K. Saha, "Transistors having optimized source-drain structures and methods for making the same," US Patent 6,344,405, February 5, 2002.

8. S. Saha, "Device characteristics of sub-20 nm silicon nanotransistors," in *Proceedings of the SPIE Conference on Design and Process Integration for Microelectronic Manufacturing*, vol. 5042, pp. 172–179, 2003.

9. S.K. Saha, "Compact MOSFET modeling for process variability-aware VLSI circuit design," *IEEE Access*, vol. 2, pp. 104–115, 2014.

10. K.J. Kuhn, M.D. Giles, D. Becher, P. Kolar, A. Kornfeld, R. Kotlyar, S.T. Ma, A. Maheshwari, and S. Mudanai, "Process technology variation," *IEEE Transactions on Electron Devices*, vol. 58, no. 8, pp. 2197–2208, 2011.

11. C.M. Mezzomo, A. Bajolet, A. Cathignol, R. Di Frenza, and G. Ghibaudo, "Characterization and modeling of transistor variability in advanced CMOS technologies," *IEEE Transactions on Electron Devices*, vol. 58, no. 8, pp. 2235–2248, 2011.

12. K. Bernstein, D.J. Frank, A.E. Gattiker, W. Haensch, B.L. Ji, S.R. Nassif, E.J. Nowak, D.J. Pearson, and N.J. Rohrer, "High-performance CMOS variability in the 65-nm regime and beyond," *IBM Journal of Research and Development* vol. 50, no. 4/5, pp. 433–449, 2006.

13. S.K. Springer, S. Lee, N. Lu, E.J. Nowak, J.O. Plouchart, J.S. Watts, R.Q. Williams, and N. Zamder, "Modeling of variation in submicrometer CMOS ULSI technologies," *IEEE Transactions on Electron Devices*, vol. 53, no. 9, pp. 2168–2178, 2006.

14. K. Singhal and V. Visvanathan, "Statistical models for worst case files and electrical data," *IEEE Transactions on Semiconductor Manufacturing*, vol. 12, no. 4, pp. 470–484, 1999.

15. S.M. Kang and A. Dharchoudhury, "Worst-case analysis and optimization of VLSI circuit performances," *IEEE Transactions on Computer-Aided Design*, vol. 14, pp. 481–492, 1995.

16. I.N. Hajj, T.N. Trick, T.-K. Yu, and S.M. Kang, "Statistical performance modeling and parametric yield estimation of MOS VLSI," *IEEE Transactions on Computer-Aided Design*, vol. CAD-6, pp. 1013–1022, 1987.

17. J.C. Chen, C. Hu, C.-P. Wan, P. Bendix, and A. Kapoor, "E-T based statistical modeling and compact statistical circuit simulation methodologies," in *IEEE International Electron Devices Meeting Technical Digest*, pp. 635–638, 1996.

18. C.-H. Lin, M.V. Dunga, D.D. Lu, A.M. Niknejad, and C. Hu, "Performance-aware corner model for design for manufacturing," *IEEE Transactions on Electron Devices*, vol. 56, no. 4, pp. 595–600, 2009.

19. G. Rappitsch, E. Seebacher, M. Kocher, and E. Stadlober, "SPICE Modeling of process variation using location depth corner models," *IEEE Transactions on Semiconductor Manufacturing*, vol. 17, no. 2, pp. 201–213, 2004.

20. C.C. McAndrew, "Statistical modeling for circuit simulation," in *Proceedings of the International Symposium on Quality Electronic Design*, pp. 496–501, 2003.

21. C.C. McAndrew, I. Stevanovic, X. Li, and G. Gildenblat, "Extensions to backward propagation of variance for statistical modeling," *IEEE Design & Test of Computers*, vol. 27, no. 2, pp. 36–43, 2010.

22. International Technology Roadmap for Semiconductors, ITRS 2012 updates, http://www.itrs.net/reports.html, access date: June 9, 2015.

23. P. Stolk, F. Widdershoven, and D. Klaassen, "Modeling statistical dopant fluctuations in MOS transistors," *IEEE Transactions on Electron Devices*, vol. 45, no. 9, pp. 1960–1971, 1998.

24. T. Mizuno, J.-I. Okamura, and A. Toriumi, "Experimental study of threshold voltage fluctuation due to statistical variation of channel dopant number in MOSFETs," *IEEE Transactions on Electron Devices*, vol. 41, no. 11, pp. 2216–2221, 1994.

25. A. Asenov, "Random-dopant-induced threshold voltage lowering and fluctuations in sub-0.1-μm MOSFETs: A 3-D 'atomistic' simulation study," *IEEE Transactions on Electron Devices*, vol. 45, no. 12, pp. 2505–2513, 1998.

26. A. Asenov, S. Kaya, and A.R. Brown, "Intrinsic parameter fluctuations in decananometer MOSFETs introduced by line edge roughness," *IEEE Transactions on Electron Devices*, vol. 50, no. 5, pp. 1254–1260, 2003.

27. C.H. Diaz, H.-J. Tao, Y.-C. Ku, A. Yen, and K. Young, "An experimentally validated analytical model for gate line edge roughness (LER) effects on technology scaling," *IEEE Electron Device Letters*, vol. 22, no. 6, pp. 287–289, 2001.

28. M. Koh, W. Mizubayashi, K. Iwamoto, H. Murakami, T. Ono, M. Tsuno, T. Mihara, K. Shibahara, S. Miyazaki, and M. Hirose, "Limit of gate oxide thickness scaling in MOSFETs due to apparent threshold voltage fluctuation introduced by tunneling leakage current," *IEEE Transactions on Electron Devices*, vol. 48, no. 2, pp. 259–264, 2001.

29. A. Asenov, S. Kaya, and J.H. Davies, "Intrinsic threshold voltage fluctuations in decananometer MOSFETs due to local oxide thickness variations," *IEEE Transactions on Electron Devices*, vol. 49, no. 1, pp. 112–119, 2002.

30. A.R. Brown, G. Roy, and A. Asenov, "Poly-Si-gate-related variability in decananometer MOSFETs with conventional architecture," *IEEE Transactions on Electron Devices*, vol. 54, no. 11, pp. 3056–3063, 2007.

31. Y. Zhang, J. Li, M. Grubbs, M. Deal, B. Magyari-Kope, B.M. Clemens, and Y. Nishi, "Physical model of the impact of metal grain work function variability on emerging dual metal gate MOSFETs and its implication for SRAM reliability," in *IEEE International Electron Devices Meeting Technical Digest*, pp. 57–60, 2009.

32. V.S. Kaushik, B.J. O'Sullivan, G. Pourtois, N. Van Hoornick, A. Delabie, S. Van Elshocht, W. Deweerd, T. Schram, L. Pantisano, E. Rohr, L.-A. Ragnarsson, S. De Gendt, and M. Heyns, "Estimation of fixed charge densities in hafnium–silicate gate dielectrics," *IEEE Transactions on Electron Devices*, vol. 53, no. 10, pp. 2627–2633, 2006.

33. H.C. Wen, H.R. Harris, C.D. Young, H. Luan, H.N. Alshareef, K. Choi, D.-L. Kwong, P. Majhi, G. Bersuker, and H. Lee, "On oxygen deficiency and fast transient charge-trapping effects in high-k dielectrics," *IEEE Electron Device Letters*, vol. 27, no. 12, pp. 984–987, 2006.

34. L. Capodieci, "From optical proximity correction to lithography-driven physical design (1996–2006): 10 years of resolution enhancement technology and the road-map enablers for the next decade," in *Proceedings of the SPIE*, vol. 6154, p. 615401, 2006.

35. S. Nag, A. Chatterjee, K. Taylor, I. Ali, S. O'Brien, S. Aur, J.D. Luttmer, and I.C. Chen, "Comparative evaluation of gap-fill dielectrics in shallow trench isolation for sub-0.25-µm technologies," in *IEEE International Electron Devices Meeting Technical Digest*, pp. 841–845, 1996.

36. J. Steigerwald, "Chemical mechanical polish: The enabling technology," in *IEEE International Electron Devices Meeting Technical Digest*, pp. 37–40, 2008.

37. Y.L. Tsang, S. Chattopadhyay, S. Uppal, E. Escobedo-Cousin, H.K. Ramakrishnan, S.H. Olsen, and A.G. O'Neill, "Modeling of the threshold voltage in strained Si/Si1−xGexSi1−yGex CMOS architectures," *IEEE Transactions on Electron Devices*, vol. 54, no. 11, pp. 3040–3048, 2007.

38. O. Weber et al., "High immunity to threshold voltage variability in undoped ultrathin FDSOI MOSFETs and its physical understanding," in *IEEE International Electron Devices Meeting Technical Digest*, pp. 245–248, 2008.

39. L.-T. Pang, K. Qian, C.J. Spanos, and B. Nikolic, "Measurement and analysis of variability in 45-nm strained-Si CMOS technology," *IEEE Journal of Solid-State Circuits*, vol. 44, no. 8, pp. 2233–2243, 2009.

40. T. Tanaka, T. Usuki, T. Futatsugi, Y. Momiyama, and T. Sugii, "Vth fluctuation induced by statistical variation of pocket dopant profile," in *IEEE International Electron Devices Meeting Technical Digest*, pp. 271–274, 2000.

41. I. Ahsan et al., "RTA-driven intradie variations in stage delay and parametric sensitivities for 65-nm technology," in *Symposium on VLSI Technology*, pp. 170–171, 2006.

42. M.J.M. Pelgrom, A.C.J. Duinmaijer, and A.P.G. Welbers, "Matching properties of MOS transistors," *IEEE Journal of Solid-State Circuits*, vol. 24, no. 5, pp. 1433–1439, 1989.

43. M. Pelgrom, H. Tuinhout, and M. Vertregt, "Modeling of MOS Matching," in *Compact Modeling: Principles, Techniques and Applications*, G. Gildenblat, (ed.) Springer, London, 2010.

44. J. Mazurier, O. Weber, F. Andrieu, A. Toffoli, O. Rozeau, T. Poiroux, F. Allain, P. Perreau, C. Fenouillet-Beranger, O. Thomas, M. Belleville, and O. Faynot, "On the variability in planar FDSOI technology: From MOSFETs to SRAM cells," *IEEE Transactions on Electron Devices*, vol. 58, no. 8, pp. 2326–2336, 2011.

45. N. Paydavosi, T.H. Morshed, D.D. Lu, W. Yang, M.V. Dunga, X. Xi, J. He, W. Liu, K. cao, X. Jin, J.J. Ou, M. Chan, A.M. Niknejad, and C. Hu, "BSIM4v4.8.0 MOSFET Model User's Manual," University of California, Berkeley, CA, 2013.

46. N.D. Arora, *MOSFET Models for VLSI Circuit Simulation: Theory and Practice*, Springer-Verlag, Wien, Austria, 1993.

47. J.A. Croon, W. Sansen, and H.E. Maes, "Matching properties of deep sub-micron MOS transistors," Springer, Amsterdam, the Netherlands, 2005.

48. K. Fujita, Y. Tori, M. Hori, J. Oh, L. Shifren, P. Ranade, M. Nakagawa, K. Okabe, T. Miyake, K. Ohkoshi, M. Kuramae, T. Mori, T. Tsuruta, S. Thompson, and T. Ema, "Advanced channel engineering achieving aggressive reduction of VT variation for ultra-low-power applications," in *IEEE International Electron Devices Meeting Technical Digest*, pp. 749–752, 2011.

49. S. Saha, "Effects of inversion layer quantization on channel profile engineering for nMOSFETs with 0.1 μm channel lengths," *Solid-State Electronics*, vol. 42, no. 11, pp. 1985–1991, 1998.

50. S.K. Saha, "Method for forming channel-region doping profile for semiconductor device," US Patent 6,323,520, November 2001.

51. S. Saha, C.S. Yeh, and B. Gadepally, "Impact ionization rate of electrons for accurate simulation of substrate current in submicron devices," *Solid-State Electronics*, vol. 36, no. 10, pp. 1429–1432, 1993.

52. S. Saha, "Extraction of substrate current model parameters from device simulation," *Solid-State Electronics*, vol. 37, no. 10, pp. 1786–1788, 1994.

53. S.K. Saha, "Introduction to technology computer aided design," in *Technology Computer Aided Design: Simulation for VLSI MOSFET*: C.K. Sarkar, (ed.). CRC Press, Boca Raton, FL, 2013.

54. S.K. Saha, "Modelling the effectiveness of computer-aided development projects in the semiconductor industry," *International Journal of Engineering Management and Economics*, vol. 1, no. 2/3, pp. 162–178, 2010.

55. S.K. Saha, "Managing technology CAD for competitive advantage: An efficient approach for integrated circuit fabrication technology development," *IEEE Transactions on Engineering and Management*, vol. 46, no. 2, pp. 221–229, 1999.

56. Synopsys Inc., HSPICE User Guide: Simulation and Analysis, B-2008.09, Synopsys Inc., September 2008.

57. M. Kanno, A. Shibuya, M. Matsumura, K. Tamura, H. Tsuno, S. Mori, Y. Fukuzaki, T. Gocho, H. Ansai, and N. Nagashima, "Empirical characteristics and extraction of overall variations for 65-nm MOSFETs and beyond," in Symposium on VLSI Technology, pp. 88–89, 2007.

58. J.D. Plummer, M.D. Deal, and P.B. Griffin, *Silicon VLSI Technology: Fundamentals, Practice and Modeling*, Prentice Hall, Upper Saddle River, NJ, 2000.
59. S.K. Saha, "Transistors structure and method with an epitaxial layer over multiple halo implants," US Patent Application 14/229,102, 2014.
60. J.-P. Colinge, (ed.), *FinFETs and Other Multi-Gate Transistors*. Springer, New York, 2008.

Chapter 9

1. D. Hisamoto, W.-C. Lee, J. Kedzierski, H. Takeuchi, K. Asano, C. Kuo, E. Anderson, T.-J. King, J. Bokor, and C. Hu, "FinFET—A self-aligned double-gate MOSFET scalable to 20 nm," *IEEE Transactions on Electron Devices*, vol. 47, no. 12, pp. 2320–2325, 2000.
2. H.-S.P. Wong, D.J. Frank, P.M. Solomon, and Y. Taur, "Device design considerations for double gate, ground plane, and single-gate ultra-thin SOI MOSFETs at the 25 nm channel length generation," in *IEEE International Electron Devices Meeting Technical Digest*, pp. 407–410, 1998.
3. G. Bertrand, S. Deleonibus, B. Previtali, G. Guegan, X. Jehl, M. Sanquer, and F. Balestra, "Toward the limits of conventional MOSFETs: Case of sub 30 nm NMOS devices," *Solid-State Electronics*, vol. 48, no. 4, pp. 505–509, 2004.
4. D.J. Frank, R.H. Dennard, E. Nowak, P.M. Solomon, Y. Taur, and H.-S. PhilipWong, "Device scaling limits of Si MOSFETs and their application dependencies," *Proceedings of the IEEE*, vol. 89, no. 3, pp. 259–288, 2001.
5. Y. Taur, D.A. Buchanan, W. Chen, D.J. Frank, K.E. Ismail, S.-H. Lo, G.A. Sai-Halasz, R.G. Viswanathan, H.-J.C. Wann, S.J. Wind, and H.-S. Wong, "CMOS scaling into the nanometer regime," *Proceedings of the IEEE*, vol. 85, no. 4, pp. 486–504, 1997.
6. K.J. Kuhn, "Moore's law past 32nm: Future challenges in device scaling," in *13th International Workshop on Computer Electronics*, pp. 1–6, 2009.
7. C. Hu, *Modern Semiconductor Devices for Integrated Circuits*, 1st edition, Pearson Education, Upper Saddle River, NJ, 2010.
8. S.K. Saha, "Modeling process variability in scaled CMOS technology," *IEEE Design & Test of Computers*, vol. 27, no. 2, pp. 8–16, 2010.
9. K.J. Kuhn, M.D. Giles, D. Becher, P. Kolar, A. Kornfeld, R. Kotlyar, S.T. Ma, A. Maheshwari, and S. Mudanai, "Process technology variation," *IEEE Transactions on Electron Devices*, vol. 58, no. 8, pp. 2197–2208, 2011.
10. L.D. Yau, "A simple theory to predict the threshold voltage of short-channel IGFET's," *Solid-State Electronics*, vol. 17, no. 10, pp. 1059–1063, 1974.
11. Z.H. Liu, C. Hu, J.-H. Huang, T.-Y. Chan, J. Min-Chie, P.K. Ko, and Y.C. Cheng, "Threshold voltage model for deep-submicrometer MOSFETs," *IEEE Transactions*, vol. 40, no. 1, pp. 86–95, 1993.
12. R.R. Troutman, "VLSI limitations from drain-induced barrier lowering," *IEEE Journal of Solid-State Circuits*, vol. 14, no. 2, pp. 383–391, 1979.
13. S. Saha, "Effects of inversion layer quantization on channel profile engineering for nMOSFETs with 0.1 μm channel lengths," *Solid-State Electronics*, vol. 42, no. 11, pp. 1985–1991, 1998.

14. S.K. Saha, "Method for forming channel-region doping profile for semiconductor device," US Patent 6,323,520, November 27, 2001.
15. S.K. Saha, "Drain profile engineering for MOSFET devices with channel lengths below 100 nm," in *Proceedings of the SPIE Conference on Microelectronic Device Technology*, vol. 3881, pp. 195–204, 1999.
16. S.K. Saha, "Transistors having optimized source-drain structures and methods for making the same," US Patent 6,344,405, February 5, 2002.
17. S. Saha, "Scaling considerations for high performance 25 nm metal-oxide-semiconductor field-effect transistors," *Journal of Vacuum Science & Technology B*, vol. 19, no. 6, pp. 2240–2246, 2001.
18. S. Saha, "Design considerations for 25 nm MOSFET devices," *Solid-State Electronics*, vol. 45, no. 10, pp. 1851–1857, 2001.
19. S. Saha, "Device characteristics of sub-20-nm silicon nanotransistors," in *Proceedings of the SPIE Conference on Design and Process Integration for Microelectronic Manufacturing*, vol. 5042, pp. 172–179, 2003.
20. M. Caymax, G. Eneman, F. Bellenger, C. Merckling, A. Delabie, G. Wang, R. Loo, E. Simoen, J. Mitard, B. De Jaeger, G. Hellings, K. De Meyer, M. Meuris, and M. Heyns, "Germanium for advanced CMOS anno 2009: A SWOT analysis," in *IEEE International Electron Devices Meeting Technical Digest*, pp. 1–4, 2009.
21. M. Radosavljevic, B. Chu-Kung, S. Corcoran, G. Dewey, M.K. Hudait, J.M. Fastenau, J. Kavalieros, W.K. Liu, D. Lubyshev, M. Metz, K. Millard, N. Mukherjee, W. Rachmady, U. Shah, and R. Chau, "Advanced high-K gate dielectric for high-performance short-channel In0.7Ga0.3As quantum well field effect transistors on silicon substrate for low power logic applications," in *IEEE International Electron Devices Meeting Technical Digest*, pp. 1–4, 2009.
22. F. Schwierz, "Graphene transistors," *National Nanotechnology*, vol. 5, pp. 487–496, 2010.
23. B. Radisavljevic, A. Radenovic, J. Brivio, V. Giacometti, and A. Kis, "Single-layer MoS2 transistors," *National Nanotechnology*, vol. 6, no. 3, pp. 147–150, 2011.
24. J. Kanghoon, L. Wei-Yip, P. Patel, K. Chang Yong, O. Jungwoo, A. Bowonder, P. Chanro, C.S. Park, C. Smith, P. Majhi, T. Hsing-Huang, R. Jammy, L. Tsu-Jae King, and H. Chenming, "Si tunnel transistors with a novel silicided source and 46mV/dec swing," in *Symposium on VLSI Technology*, pp. 121–122, 2010.
25. A.I. Khan, D. Bhowmik, P. Yu, S. Joo Kim, X. Pan, R. Ramesh, and S. Salahuddin, "Experimental evidence of ferroelectric negative capacitance in nanoscale heterostructures," *Applied Physics Letters*, vol. 99, no. 11, pp. 113501-1–113501-3, 2011.
26. M. Radosavljevic et al., "Electrostatics improvement in 3-D tri-gate over ultra-thin body planar InGaAs quantum well field effect transistors with high-K gate dielectric and scaled gate-to-drain/gate-to-source separation," in *IEEE International Electron Devices Meeting Technical Digest*, pp. 765–768, 2011.
27. K. Tomioka, M. Yoshimura, and T. Fukui, "Steep-slope tunnel field-effect transistors using III–V nanowire/si-heterojunction," in *Symposium on VLSI Technology*, pp. 47–48, 2012.
28. K.J. Kuhn, "Considerations for ultimate CMOS scaling," *IEEE Transactions on Electron Devices*, vol. 59, no. 7, pp. 1813–1828, 2012.
29. C. Auth et al., "A 22-nm-high performance and low-power CMOS technology featuring fully-depleted tri-gate transistors, self-aligned contacts and high density MIM capacitors," in *Symposium on VLSI Technology*, pp. 131–132, 2012.

30. D. Hisamoto, W.-C. Lee, J. Kedzierski, E. Anderson, H. Takeuchi, K. Asano, K. Tsu-Jae, J. Bokor, and C. Hu, "A folded-channel MOSFET for deep-sub-tenth micron era," in *IEEE International Electron Devices Meeting (IEDM) Technical Digest*, pp. 1032–1034, 1998.

31. X. Huang, W.-C. Lee, C. Kuo, D. Hisamoto, L. Chang, J. Kedzierski, E. Anderson, H. Takeuchi, Y.-K. Choi, K. Asano, V. Subramanian, T.J. King, J. Bokor, and C. Hu, "Sub 50-nm FinFET: PMOS," in *IEEE International Electron Devices Meeting (IEDM) Technical Digest*, pp. 67–70, 1999.

32. B. Yu, L. Chang, S. Ahmed, W. Haihong, S. Bell, C.-Y. Yang, C. Tabery, H. Chau, X. Qi, K. Tsu-Jae, J. Bokor, C. Hu, L. Ming-Ren, and D. Kyser, "FinFET scaling to 10 nm gate length," in *IEEE International Electron Devices Meeting Technical Digest*, pp. 251–254, 2002.

33. C. Jahan, O. Faynot, and M. Cassé, "Omega FETs transistors with TiN metal gate and HfO2 down to 10nm," in *Symposium on VLSI Technology*, pp. 112–113, 2005.

34. F.-L. Yang et al., "5 nm-gate nanowire FinFET," in *Symposium on VLSI Technology*, pp. 196–197, 2004.

35. H. Lee et al., "Sub-5 nm all around gate FinFET for ultimate scaling," in *Symposium on VLSI Technology*, pp. 58–59, 2006.

36. B.S. Doyle, S. Datta, M. Doczy, S. Hareland, B. Jin, J. Kavalieros, T. Linton, A. Murthy, R. Rios, and R. Chau, "High performance full depleted tri-gate CMOS transistors," *IEEE Electron Device Letters*, vol. 24, no. 4, pp. 263–265, 2003.

37. Y.-K. Choi, K. Asano, N. Lindert, V. Subramanian, T.-J. King, J. Bokor, and C. Hu, "Ultrathin-body SOI MOSFET for deep-sub-tenth micron era," *IEEE Electron Device Letters*, vol. 21, no. 5, pp. 254–255, 2000.

38. V. Barral et al., "Strained FDSOI CMOS technology scalability down to 2.5nm _lm thickness and 18nm gate length with a TiN/HfO2 gate stack," in *IEEE International Electron Devices Meeting Technical Digest*, pp. 61–64, 2011.

39. K. Cheng et al., "Fully depleted extremely thin SOI technology fabricated by a novel integration scheme featuring implant-free, zero-silicon-loss, and faceted raised source/drain," in *Symposium on VLSI Technology*, pp. 212–213, 2009.

40. O. Faynot et al., "Planar fully depleted SOI technology: A powerful architecture for the 20nm node and beyond," in *IEEE International Electron Devices Meeting Technical Digest*, pp. 51–53, 2010.

41. C. Auth, "22-nm fully-depleted tri-gate CMOS transistors," in *Proceedings of the IEEE Custom Integrated Circuits Conference*, pp. 1–6, 2012.

42. J. Markoff, *Intel Increases Transistor Speed by Building Upward*, 2011. http://www.nytimes.com/2011/05/05/science/05chip.html, accessed May 4, 2014.

43. R. Merritt, *TSMC taps ARM's V8 on road to 16-nm FinFET*, 2012. http://www.eetimes.com/electronicsnews/4398727/TSMC-taps-ARM-V8-in-road-to-16-nm-FinFET, accessed October 16, 2012.

44. D. McGrath, *Globalfoundries Looks to Leapfrog Fab Rival*, 2012. http://www.eetimes.com/electronicsnews/4396720/Globalfoundries-to-offer-14-nm-process-with-FinFETfinn-2014, accessed September 20, 2012.

45. J.P. Noel, O. Thomas, M. Jaud, O. Weber, T. Poiroux, C. Fenouillet-Beranger, P. Rivallin, P. Scheiblin, F. Andrieu, M. Vinet, O. Rozeau, F. Boeuf, O. Faynot, and A. Amara, "Multi-VT UTBB FDSOI device architectures for low-power CMOS circuit," *IEEE Transactions on Electron Devices*, vol. 58, no. 8, pp. 2473–2482, 2011.

46. K. Fujita, Y. Torii, M. Hori, J. Oh, L. Shifren, P. Ranade, M. Nakagawa, K. Okabe, T. Miyake, K. Ohkoshi, M. Kuramae, T. Mori, T. Tsuruta, S. Thompson, and T. Ema, "Advanced channel engineering achieving aggressive reduction of VT variation for ultra-low-power applications," in *IEEE International Electron Devices Meeting Technical Digest*, pp. 32.3.1–32.3.4, December 2011.
47. S.K. Saha, "Transistors structure and method with an epitaxial layer over multiple halo implants," US Patent Application 14/229,102, 2014.
48. R. Merritt, *Intel's FinFETs Approach Draws Fire from Rivals*, 2012. http://www.eetimes .com/electronicsnews/4403320/Fur-_ies-over-FinFETs-and-future-in-IEDM-panel, accessed December 13, 2012.
49. T. Tanaka, H. Horie, S. Ando, and S. Hijiya, "Analysis of p/sup+ poly Si double-gate thin-film SOI MOSFETs," in *IEEE International Electron Devices Meeting Technical Digest*, pp. 683–686, 1991.
50. N. Paydavosi, S. Venugopalan, Y.S. Chauhan, J.P. Duarte, S. Jandhyala, A. Niknejad, and C.C. Hu, "BSIM—SPICE models enable FinFET and UTB IC design," *IEEE Access*, vol. 1, pp. 201–2016, 2013.
51. C. Maleville, *Wafers for Fully Depleted Soi Devices: Ready for Volume Advanced Subs trate News*, 2010. http://www.advancedsubstratenews.com/2010/12/wafers-for-fully-depleted-soi-devices-ready-for-volume/, accessed December 8, 2010.
52. Y.-K. Choi, D. Ha, T.-J. King, and C. Hu, "Ultra-thin body PMOSFETs with selectively deposited Ge source/drain," in *Symposium on VLSI Technology*, pp. 19–20, 2001.
53. H. Fang, S. Chuang, T. Chia Chang, K. Takei, T. Takahashi, and A. Javey, "High-performance single layered WSe2 p-FETs with chemically doped contacts," *Nano Letters*, vol. 12, no. 7, pp. 3788–3792, 2012.
54. F. Andrieu et al., "Low leakage and low variability ultra-thin body and buried oxide (UT2B) SOI technology for 20 nm low power CMOS and beyond," in *VLSI Symposium on VLSI Technology*, pp. 57–58, 2010.
55. Q. Liu et al., "Ultra-thin-body and BOX (UTBB) fully depleted (FD) device integration for 22nm node and beyond," in *Symposium on VLSI Technology*, pp. 61–62, 2010.
56. K. Cheng et al., "ETSOI CMOS for system-on-chip applications featuring 22 nm gate length, sub-100 nm gate pitch, and 0.08 μm_2 SRAM cell," in *Symposium on VLSI Technology*, pp. 128–129, 2011.
57. H.C. Pao and C.T. Sah, "Effects of diffusion current on characteristics of metal-oxide (insulator)-semiconductor transistors," *Solid-State Electronics*, vol. 9, no. 10, pp. 927–937, 1966.
58. J. Sallese, F. Krummenacher, F. Pregaldiny, C. Lallement, A.S. Roy, and C. Enz, "A design oriented charge-based current model for symmetric DG MOSFET and its correlation with the EKV formalism," *Solid-State Electronics*, vol. 49, no. 3, pp. 485–489, 2005.
59. Y. Taur, X. Liang, W. Wang, and H. Lu, "A continuous, analytic drain current model for DG MOSFETs," *IEEE Electron Device Letters*, vol. 25, no. 2, pp. 107–109, 2004.
60. M.V. Dunga, C.-H. Lin, X. Xi, D.D. Lu, A.M. Niknejad, and C. Hu, "Modeling advanced FET technology in a compact model," *IEEE Transactions on Electron Devices*, vol. 53, no. 9, pp. 1971–1978, 2006.
61. M.V. Dunga, "Nanoscale CMOS modeling," PhD dissertation, Electrical Engineering and Computer Science, University of Berkeley, Berkeley, CA, 2008.

62. Y. Taur, "Analytic solutions of charge and capacitance in symmetric and asymmetric double-gate MOSFETs," *IEEE Transactions on Electron Devices*, vol. 48, no. 12, pp. 2861–2869, 2001.

63. S. Venugopalan, D.D. Lu, Y. Kawakami, P.M. Lee, A.M. Niknejad, and C. Hu, "BSIM-CG: A compact model of cylindrical/surround gate MOSFET for circuit simulations," *Solid-State Electronics*, vol. 67, no. 1, pp. 79–89, 2012.

64. F. Balestra, S. Cristoloveanu, M. Benachir, J. Brini, and T. Elewa, "Double-gate silicon-on-insulator transistor with volume inversion: A new device with greatly enhanced performance," *IEEE Electron Device Letters*, vol. 8, no. 9, pp. 410–412, 1987.

65. M.V. Dunga, C.-H. Lin, D.D. Lu, W. Xiong, C.R. Cleavelin, P. Patruno, J.-R. Hwang, F.-L. Yang, A.M. Niknejad, and C. Hu, "BSIM-MG: A versatile multi-gate FET model for mixed-signal design," in *Symposium on VLSI Technology*, pp. 60–61, 2007.

66. X. Liang and Y. Taur, "A 2-D analytical solution for SCEs in DG MOSFETs," *IEEE Transactions on Electron Devices*, vol. 51, no. 9, pp. 1385–1391, 2004.

67. Q. Chen, E.M. Harrell, and J.D. Meindl, "A physical short-channel threshold voltage model for undoped symmetric double-gate MOSFETs," *IEEE Transactions on Electron Devices*, vol. 50, no. 7, pp. 1631–1637, 2003.

68. K. Suzuki, T. Tanaka, Y. Tosaka, H. Horie, and Y. Arimoto, "Scaling theory for double-gate SOI MOSFETs," *IEEE Transactions on Electron Devices*, vol. 40, no. 12, pp. 2326–2329, 1993.

69. K. Suzuki, Y. Tosaka, and T. Sugii, "Analytical threshold voltage model for short channel n+/p+ double-gate SOI MOSFETs," *IEEE Transactions on Electron Devices*, vol. 43, no. 5, pp. 732–738, 1996.

70. K.K. Young, "Short-channel effect in fully depleted SOI MOSFET's," *IEEE Transactions on Electron Devices*, vol. 36, no. 2, pp. 399–402, 1989.

71. W. Liu, *BSIM4 and MOSFET Modeling for IC Simulation*, World Scientific, Singapore, 2011.

72. M.V. Dunga, C.-H. Lin, A.M. Niknejad, and C. Hu, "BSIM-CMG: A compact model for multi-gate transistors," in FinFETs and Other Multi-Gate Transistors, J.-P. Colinge (ed.). Springer, New York, pp. 113–153, 2008.

73. D. Lu, C.-H. Lin, A.M. Niknejad, and C. Hu, "Multi-gate MOSFET compact model BSIM-MG," in Compact Modeling: Principles, Techniques and Applications, G. Gildenblat (ed.). Springer, New York, pp. 395–429, 2010.

74. S. Khandelwal, J. Duarte, V. Sriramkumar, N. Paydavosi, D. Lu, C.-H. Lin, M. Dunga, S. Yao, T. Morshed, A. Niknejad, and C. Hu, "BSIM-CMG 108.0.0 Multi-Gate MOSFET Compact Model Technical Manual," University of California, Berkeley, CA, 2014.

75. F. Stern, "Electronic properties of two-dimensional systems," *Reviews of Modern Physics*, vol. 54, pp. 437–672, 1982.

76. R. Rios, N.D. Arora, C.-L. Huang, N. Khabil, J. Faricelli, and L. Gruber, "A physical compact MOSFET model, including quantum mechanicals effects, for statistical circuit design applications," in *IEEE International Electron Devices Meeting Technical Digest*, pp. 947–950, 1995.

77. E. Merzbacher, *Quantum Mechanics*, Wiley, New York, 1997.

78. G. Baccarani and S. Reggiani, "A compact double-gate MOSFET model comprising quantum-mechanical and nonstatic effects," *IEEE Transactions on Electron Devices*, vol. 46, no. 8, pp. 1656–1666, 1999.

79. L. Ge and J.G. Fossum, "Analytical modeling of quantization and volume inversion in thin Si-film double gate MOSFETs," *IEEE Transactions on Electron Devices*, vol. 49, no. 2, pp. 287–294, 2002.

80. S. Venugopalan, M.A. Karim, S. Salahuddin, A.M. Niknejad, and C. Hu, "Phenomenological compact model for QM charge centroid in multi gate FETs," *IEEE Transactions on Electron Devices*, vol. 60, no. 4, pp. 480–484, 2013.

81. D. Lu, "Compact models for future generation CMOS," PhD dissertation, Electrical Engineering and Computer Science, University of Berkeley, Berkeley, CA, 2011.

82. S. Khandelwal, Y.S. Chauhan, D.D. Lu, S. Venugopalan, M.A.U. Karim, A.B. Sachid, B.-Y. Nguyen, O. Rozeau, O. Faynot, A.M. Niknejad, and C.C. Hu, "BSIM-IMG: A compact model for ultrathin-body SOI MOSFETs with back-gate control," *IEEE Transactions on Electron Devices*, vol. 59, no. 8, pp. 2019–2026, 2012.

83. Q. Chen, X. Zhong, Y. Wu, N. Zhu, W. Huang, D. Lu, C. Hu, B.Y. Nguyen, and O. Faynot, "An exercise of ET/UTBB SOI CMOS modeling and simulation with BSIM-IMG," in *Proceedings of the IEEE International SOI Conference*, pp. 1–2, 2011.

84. D.E. Ward and R.W. Dutton, "A charge-oriented model for MOS transistor capacitances," *IEEE Journal of Solid-State Circuits*, vol. 13, no. 5, pp. 703–710, 1978.

85. S.Y. Oh, D.E. Ward, and R.W. Dutton, "Transient analysis of MOS transistors," *IEEE Journal of Solid-State Circuits*, vol. 15, no. 4, pp. 636–643, 1980.

Chapter 10

1. P.M. Solomon, "Device Proposals for Beyond Silicon CMOS," *IBM Research Report* RC24963, pp. 1–14. http://domino.watson.ibm.com/library/cyberdig.nsf/papers/C12D0D2E53FE8833852576F90055CF7E/$File/rc24963.pdf, accessed March 19, 2010.

2. K. Galatsis, A. Khitun, R. Ostroumov, K.L. Wang, W.R. Dichtel, E. Plummer, J.F. Stoddart, J.I. Zink, J.Y. Lee, Y.-H. Xie, and K.W. Kim, "Alternate state variables for emerging nanoelectronic devices," *IEEE Transactions on Nanotechnology*, vol. 9, no. 1, pp. 66–75, 2009.

3. A. Chen, A.P. Jacob, C.Y. Sung, K.L. Wang, A. Khitun, and W. Porod, "Collective-effect state variables for post-CMOS logic applications," in *Symposium on VLSI Technology*, pp. 132–133, 2009.

4. P.-F. Wang, K. Hilsenbeck, T. Nirschl, M. Oswald, C. Stepper, M. Weiss, D. Schmitt-Landsiedel, and W. Hansch, "Complementary tunneling transistor for low power application," *Solid-State Electronics*, vol. 48, no. 12, pp. 2281–2286, 2004.

5. K.K. Bhuwalka, J. Schulze, and I. Eisele, "Performance enhancement of vertical tunnel field-effect transistor with SiGe in the $\delta p+$ layer," *Japan Journal of Applied Physics*, vol. 43, no. 7R, pp. 4073–4078, 2004.

6. J. Appenzeller, Y.-M. Lin, J. Knoch, and P. Avouris, "Band-to-band tunneling in carbon nanotube field-effect transistors," *Physical Review Letters*, vol. 93, no. 19, pp. 196805-1–196805-4, 2004.

7. K. Gopalakrishnan, P.B. Griffin, and J.D. Plummer, "I-MOS: A novel semiconductor device with a subthreshold slope lower than kT/q," in *IEEE International Electron Devices Meeting Technical Digest*, pp. 289–292, 2002.

8. W.Y. Choi, J.D. Lee, and B.-G. Park, "Integration process of impact ionization metal-oxide-semiconductor devices with tunneling field-effect transistors and metal-oxide-semiconductor field-effect transistors," *Japan Journal of Applied Physics*, vol. 46, no. 1R, pp. 122–124, 2007.

9. C.-W. Lee, A.N. Nazarov, I. Ferain, N.D. Akhavan, R.Yan, P. Razavi, R.Yu, R.T. Doria, and J.-P. Colinge, "Low subthreshold slope in junctionless multigate transistors," *Applied Physics Letters*, vol. 96, no.10, p. 102106, 2010.

10. E.-H. Toh, G.H. Wang, L. Chan, D. Weeks, M. Bauer, J. Spear, S.G. Thomas, G. Samudra, and Y.-C. Yeo, "Co-integration of in situ doped silicon-carbide source and silicon-carbon i-region in p-channel silicon nanowire impact-ionization transistor," *IEEE Electron Device Letters*, vol. 29, no. 7, pp. 731–733, 2008.

11. S. Salahuddin and S. Datta, "Use of negative capacitance to provide voltage amplification for low power nanoscale devices," *Nano Letters*, vol. 8, no. 2, pp. 405–410, 2008.

12. H. Nathanson, W. Newell, R. Wickstro, and J. Davis, "Resonant gate transistor," *IEEE Transactions on Electron Devices*, vol. ED-14, no. 3, pp. 117–133, 1967.

13. A.M. Ionescu, V. Pott, R. Fritschi, K. Banerjee, M.J. Declercq, P. Renaud, C. Hilbert, P. Fluckiger, and G.A. Racine, "Modeling and design of a low-voltage SOI suspended-gate MOSFET (SG-MOSFET) with a metal-over-gate architecture," in *Proceedings of the International Symposium on Quality Electronic Design*, pp. 496–501, 2002.

14. H. Kam, D.T. Lee, R.T. Howe, and T.-J. King, "A new nano-electro-mechanical field effect transistor (NEMFET) design for low-power electronics," in *IEEE International Electron Devices Meeting Technical Digest*, pp. 463–466, 2005.

15. A.M. Ionescu, K. Boucart, K.E. Moselund, V. Pott, and D. Tsamados, "Small slope micro/nano-electronic switches," in *Proceedings of the International Semiconductor Conference*, pp. 397–402, 2007.

16. A.C. Seabaugh and Q. Zhang, "Low voltage tunnel transistors for beyond CMOS logic," *Proceedings of the IEEE*, vol. 98, no. 12, pp. 2095–2110, 2010.

17. C. Zener, "A theory of the electrical breakdown of solid dielectrics," *Proceedings of the Royal Society of London*, vol. A145, no. 8555, pp. 523–529, 1934.

18. Y. Lu, A. Seabaugh, P. Fay, S.J. Koester, S.E. Laux, W. Haensch, and S.O. Koswatta, "Geometry dependent tunnel FET performance-dilemma of electrostatics vs. quantum confinement," in *Device Research Conference*, pp. 17–18, 2010.

19. C. Hu, "Green transistor as a solution to the IC power crisis," in *9th International Conference on Solid-State and Integrated-Circuit Technology*, pp. 16–20, 2008.

20. O.M. Stuetzer, "Junction fieldistors," *Proceedings of the IRE*, vol. 40, no. 11, pp. 1377–1381, 1952.

21. L. Esaki, "New phenomenon in narrow germanium *p-n* junctions," *Physical Reviews*, vol. 109, no. 2, pp. 603–604, 1958.

22. J. Quinn, G. Kawamoto, and B. McCombe, "Subband spectroscopy by surface channel tunneling," *Surface Science*, vol. 73, no. 1, pp. 190–196, 1978.

23. J.P. Leburton, J. Kolodzey, and S. Biggs, "Bipolar tunneling field-effect transistor: A three-terminal negative differential resistance device for high-speed applications," *Applied Physics Letters*, vol. 52, no. 9, pp. 1608–1620, 1988.

24. T. Baba, "Proposal for surface tunnel transistors," *Japan Journal of Applied Physics*, vol. 31, no. 4B, pp. L455–L457, 1992.
25. T. Uemura and T. Baba, "First observation of negative differential resistance in surface tunnel transistors," *Japan Journal of Applied Physics*, vol. 33, no. 2B, pp. L207–L210, 1994.
26. H. Kawaura, N. Iguchi, and T. Baba, "Fabrication of silicon surface tunnel transistors - Extended Abstract," in *2nd International Workshop on Quantum Functional Devices*, p. 62, May 1995.
27. Y. Omura, "Negative conductance properties in extremely thin silicon-on-insulator (SOI) insulated-gate pn-junction devices (SOI surface tunnel transistors)," *Japan Journal of Applied Physics*, vol. 35, no. 11A, pp. L1401–L1403, 1996.
28. J. Koga and A. Toriumi, "Negative differential conductance at room temperature in three-terminal silicon surface junction tunnel transistor," *Applied Physics Letters*, vol. 70, no. 16, pp. 2138–2140, 1997.
29. J. Koga and A. Toriumi, "Three-terminal silicon surface junction tunnel device for room temperature operation," *IEEE Electron Device Letters*, vol. 20, no. 10, pp. 529–531, 1999.
30. T. Uemura and T. Baba, "First demonstration of a planar-type surface tunnel transistor (STT): Lateral interband tunnel device," *Solid-State Electronics*, vol. 40, no. 1, pp. 519–522, 1996.
31. T. Uemura and T. Baba, "Large enhancement of interband tunneling current densities of over 105 A/cm2 in In0:53Ga0:47As-based surface tunnel transistors," *IEEE Electron Device Letters*, vol. 18, no. 5, pp. 225–227, 1997.
32. Y.J. Chun, T. Uemura, and T. Baba, "Self-aligned surface tunnel transistors fabricated by a regrowth technique," *Japan Journal of Applied Physics*, vol. 38, no. 10B, pt. 2, pp. L1163–L1165, 1999.
33. Y.J. Chun, T. Uemura, and T. Baba, "Fabrication of self-aligned surface tunnel transistors with a 80-nm gate length," *Japan Journal of Applied Physics*, vol. 39, no. 12B, pt. 2, pp. L1273–L1276, 2000.
34. W.M. Reddick and G.A.J. Amartunga, "Silicon surface tunnel transistor," *Applied Physics Letters*, vol. 67, no. 4, pp. 494–496, 1995.
35. W. Hansch, C. Fink, J. Schulze, and I. Eisele, "A vertical MOS-gated Esaki tunneling transistor in silicon," *Thin Solid Films*, vol. 369, vol. 1/2, pp. 387–389, 2000.
36. C. Aydin, A. Zaslevsky, S. Luryi, S. Cristoloveanu, D. Mariolle, D. Fraboulet, and S. Deleonibus, "Lateral interband tunneling transistor in silicon-on-insulator," *Applied Physics Letters*, vol. 84, no. 10, pp. 1780–1782, 2004.
37. Q. Zhang, W. Zhao, and A. Seabaugh, "Analytic expression and approach for low subthreshold-swing tunnel transistors," in *Device Research Conference*, pp. 161–162, 2005.
38. Q. Zhang, W. Zhao, and A. Seabaugh, "Low-subthreshold-swing tunnel transistors," *IEEE Electron Device Letters*, vol. 27, no. 4, 297–300, 2006.
39. Y. Lu, S. Bangsaruntip, X. Wang, L. Zhang, Y. Nishi, and H. Dai, "DNA functionalization of carbon nanotubes for ultrathin atomic layer deposition of high-k dielectrics for nanotube transistors with 60 mV/decade switching," *Journal of American Chemical Soceity*, vol. 128, no. 11, pp. 3518–3519, 2006.
40. W.Y. Choi, B.-G. Park, J.D. Lee, and T.-J.K. Liu, "Tunneling field-effect transistors (TFETs) with subthreshold swing (SS) less than 60 mV/dec," *IEEE Electron Device Letters*, vol. 28, no. 8, pp. 743–745, 2007.

41. F. Mayer, C.L. Royer, J.-F. Damlencourt, K. Romanjek, F. Andrieu, C. Tabone, B. Previtali, and S. Deleonibus, "Impact of SOI, $Si_{1-x}Ge_xOI$ and GeOI substrates on CMOS compatible tunnel FET performance," in *IEEE International Electron Devices Meeting Technical Digest*, pp. 163–166, 2008.

42. K. Jeon, W.-Y. Loh, P. Patel, C.Y. Kang, J. Oh, A. Bowonder, C. Park, C.S. Park, C. Smith, P. Majhi, H.-H. Tseng, R. Jammy, T.-J.K. Liu, and C. Hu, "Si tunnel transistors with a novel silicided source and 46 mV/dec swing," in *Symposium on VLSI Technology*, pp. 121–122, 2010,

43. D. Leonelli, A. Vandooren, R. Rooyackers, A.S. Verhulst, S.D. Gendt, M.M. Heyns, and G. Groeseneken, "Performance enhancement in multi gate tunneling field effect transistors by scaling the fin-width," *Japan Journal of Applied Physics*, vol. 49, no. 4, 04DC10, 2010.

44. T. Krishnamohan, D. Kim, S. Raghunathan, and K. Saraswat, "Double-gate strained-Ge heterostructure tunneling FET (TFET) with record high drive currents and < 60 mV/dec subthreshold slope," in *IEEE International Electron Devices Meeting Technical Digest*, pp. 947–949, 2008.

45. S.H. Kim, H. Kam, C. Hu, and T.-J.K. Liu, "Germanium-source tunnel field effect transistors with record high ION/IOFF," in *Symposium on VLSI Technology*, pp. 178–179, 2009.

46. Y. Khatami and K. Banerjee, "Steep subthreshold slope n- and p-type tunnel-FET devices for low-power and energy-efficient digital circuits," *IEEE Transactions on Electron Devices*, vol. 56, no. 11, pp. 2752–2761, 2009.

47. L.D. Landau and E.M. Lifshitz, *Quantum Mechanics*, Addison-Wesley, Reading, MA, 1990.

48. S.M. Sze, *Physics of Semiconductor Devices*, 3rd edition, John Wiley, New York, 1998.

49. J. Knoch, D. Mantl, and J. Appenzeller, "Impact of the dimensionality on the performance of tunneling FETs: Bulk versus one-dimensional devices," *Solid-State Electronics*, vol. 51, no. 4, pp. 572–578, 2007.

50. M. Luisier and G. Klimeck, "Simulation of nanowire tunneling transistors: From the Wentzel- Kramers–Brillouin approximation to full-band phonon-assisted tunneling" *Journal of Applied Physics*, vol. 107, no. 8, pp. 084507-1–084507-6, 2010.

51. J. Knoch and J. Appenzeller, "A novel concept for field-effect transistors—The tunneling carbon nanotube FET," in *Device Research Conference*, 153–156, 2005.

52. B. Streetman and S. Banerjee, *Solid State Electronic Devices*, 7th edition, Prentice-Hall, Englewood Cliffs, NJ, 2014.

53. J.-P. Colinge, "Multiple-gate SOI MOSFETs," *Solid-State Electronics*, vol. 48, no. 6, pp. 897–905, 2004.

54. K. Bhuwalka, M. Born, M. Schindler, M. Schmidt, T. Sulima, and I. Eisele, "P-channel tunnel field-effect transistors down to sub-50 nm channel lengths," *Japan Journal of Applied Physics*, vol. 45, no. 4B, pp. 3106–3109, 2006.

55. E.-H. Toh, G.H. Wang, L. Chan, D. Sylvester, C.-H. Heng, G. Samudra, and Y.-C. Yeo, "Device design and scalability of a double-gate tunneling field-effect transistor with silicon-germanium source," *Japan Journal of Applied Physics*, vol. 47, no. 4, pp. 2593–2597, 2008.

56. K.K. Bhuwalka, J. Schulze, and I. Eisele, "Scaling the vertical tunnel FET with tunnel bandgap modulation and gate workfunction engineering," *IEEE Transactions on Electron Devices*, vol. 52, no. 5, pp. 909–917, 2005.

57. K. Boucart and A.M. Ionescu, "Double-gate tunnel FET with high-κ gate dielectric," *IEEE Transactions on Electron Devices*, vol. 54, no. 7, 1725–1733, 2007.

58. K. Bhuwalka, J. Schulze, and I. Eisele, "A simulation approach to optimize the electrical parameters of a vertical tunnel FET," *IEEE Transactions on Electron Devices*, vol. 52, no. 7, pp. 1541–1547, 2005.

59. A. Oritiz-Conde, F.J.G. Sanchez, J.J. Liou, A. Cerdeira, M. Estrada, and Y. Yue, "A review of recent MOSFET threshold voltage extraction methods," *Microelectronics Reliability*, vol. 42, no. 4/5, pp. 583–596, 2002.

60. K. Boucart and A.M. Ionescu, "A new definition of threshold voltage in tunnel FETs," *Solid-State Electronics*, vol. 52, no. 9, pp. 1318–1323, 2008.

61. K. Boucart and A.M. Ionescu, "Length scaling of the double gate tunnel FET with a high-κ gate dielectric," *Solid-State Electronics*, vol. 51, no. 11/12, pp. 1500–1507, 2007.

62. C. Sandow, J. Knoch, C. Urban, Q.-T. Zhao, and S. Mantl, "Impact of electrostatics and doping concentration on the performance of silicon tunnel field-effect transistors," *Solid-State Electronics*, vol. 53, no. 10, 1126–1129, 2009.

63. A.M. Ionescu and H. Riel, "Tunnel field-effect transistors as energy-efficient electronic switches," *Nature*, vol. 479, pp. 329–337, 2011.

64. J. Appenzeller, J. Knoch, M.T. Björk, H. Riel, and W. Riess, "Toward nanowire electronics," *IEEE Transactions on Electron Devices*, vol. 55, no. 11, 2827–2845, 2008.

65. A.M. Ionescu, K. Boucart, K.E. Moselund, and V. Pott, *Small Swing Switches*, Cambridge University Press, Cambridge, 2011.

66. D. Leonelli, A. Vandooren, R. Rooyackers, S. De Gendt, M.M. Heyns, and G. Groeseneken, "Optimization of tunnel FETs: impact of gate oxide thickness, implantation and annealing conditions," in *Proceedings of the European Solid-State Device Research Conference*, pp. 170–173, 2010.

67. C. Hu, P. Patel, A. Bowonder, J. Kanghoon, S.-H. Kim, W.Y. Loh, C.Y. Kang, J. Oh, P. Majhi, A. Javey, T.-J.K. Liu, and R. Jammy, "Prospect of tunneling green transistor for 0.1 V CMOS," in *IEEE International Electron Devices Meeting Technical Digest*, pp. 387–390, 2010.

68. R. Asra, M. Shrivastava, K.V.R.M. Murali, R.K. Pandey, H. Gossner, and V.R. Rao, "A tunnel FET for V_{DD} scaling below 0.6V with a CMOS-comparable performance," *IEEE Transactions on Electron Devices*, vol. 58, no. 7, pp. 1855–1863, 2011.

69. L. De Michielis, L. Lattanzio, P. Palestri, L. Selmi, and A.M. Ionescu, "Tunnel-FET architecture with improved performance due to enhanced gate modulation of the tunneling barrier," in *Device Research Conference*, pp. 111–112, 2011.

70. A.S. Verhulst, W.G. Vandenberghe, K. Maex, S. De Gendt, M.M. Heyns, and G. Groeseneken, "Complementary silicon-based heterostructure tunnel-FETs with high tunnel rates," *IEEE Electron Device Letters* vol. 29, no. 12, pp. 1398–1401, 2008.

71. J. Knoch, "Optimizing tunnel FET performance–impact of device structure, transistor dimensions and choice of material," in *Symposium on VLSI Technology*, pp. 45–46, 2009.

72. J. Knoch and J. Appenzeller, "Modeling of high-performance p-type III–V heterojunction tunnel FETs," *IEEE Electron Device Letters*, vol. 31, no. 4, pp. 305–307, 2010.

73. S.O. Koswatta, S.J. Koester, and W. Haensch, "On the possibility of obtaining MOSFET-like performance and sub-60-mV/dec swing in 1-D broken-gap tunnel transistors," *IEEE Transactions on Electron Devices*, vol. 57, no. 12, pp. 3222–3223, 2010.

74. K.E. Moselund, H. Ghoneim, M.T. Bjork, H. Schmid, S. Kang, E. Lortscher, W. Reis, and H. Riel, "Comparison of VLS grown Si NW tunnel FETs with different gate stacks," in *Proceedings of the European Solid-State Device Research Conference*, pp. 448–451, 2009.

75. O.M. Nayfeh, C.N. Chleirigh, J. Hennessy, L. Gomez, J.L. Hoyt, and D.A. Antonidis, "Design of tunneling field-effect transistors using strained-silicon/strained-germanium type-II staggered heterojunctions," *IEEE Electron Device Letters*, vol. 29, no. 9, pp. 1074–1077, 2008.

76. K. Boucart, A.M. Ionescu, and W. Riess, "Asymmetrically strained all-silicon tunnel FETs featuring 1V operation," in *Proceedings of the European Solid-State Device Research Conference*, pp. 452–456, 2009.

77. K. Boucart, W. Riess, and A.M. Ionescu, "Lateral strain profile as key technology booster for all-silicon tunnel FETs," *IEEE Electron Device Letters*, vol. 30, no. 6, pp. 656–658, 2009.

78. C. Le Royer and F. Mayer, "Exhaustive experimental study of tunnel field effect transistors (TFETs): From materials to architecture," in *International Conference on Ultimate Integration of Silicon*, 53–56, 2009.

79. W.-Y. Loh, "Sub-60nm Si tunnel field effect transistors with I_{on} > 100 μA/μm," in *Proceedings of the European Solid-State Device Research Conference*, pp. 162–165, 2010.

80. S. Mookerjea, D. Mohata, R. Krishnan, J. Singh, A. Vallet, A. Ali, T. Mayer, V. Narayanan, D. Schlom, A. Liu, and S. Datta, "Experimental demonstration of 100 nm channel length $In_{0.53}Ga_{0.47}As$-based vertical inter-band tunnel field effect transistors (TFET) for ultra low-power logic and SRAM applications," *IEEE International Electron Devices Meeting Technical Digest*, pp. 137.1–137.4, 2009.

81. H. Zhao, Y. Chen, Y. Wang, F. Zhou, F. Xue, and J. Lee, "InGaAs tunneling field-effect transistors with atomic-layer-deposited gate oxides," *IEEE Transactions on Electron Devices*, vol. 58, no. 9, pp. 2990–2995, 2011.

82. S. Mookerjea, D. Mohata, T. Mayer, V. Narayanan, and S. Datta, "Temperature-dependent characteristics of a vertical tunnel FET," *IEEE Electron Device Letters*, vol. 31, no. 6, pp. 564–566, 2010.

83. L. Wang, E. Yu, Y. Taur Y and P. Asbeck, "Design of tunneling field-effect transistors based on staggered heterojunctions for ultralow-power applications," *IEEE Electron Device Letters*, vol. 31, no. 5, pp. 431–433, 2010.

84. D. Mohata, S. Mookerjea, A. Agrawal, Y. Li, T. Mayer, V. Narayanan, A. Liu, D. Loubychev, J. Fastenau, and S. Datta, "Experimental staggered-source and N+pocket-doped channel III–V tunnel field-effect transistors and their scalabilities," *Applied Physics Express*, vol. 4, no. 2, pp. 024105-1–024105-3, 2011.

85. G. Zhou, Y. Lu, R. Li, Q. Zhang, W. Huang, Q. Liu, T. Vasen, H. Zhu, J. Kuo, S. Koswatta, T. Kosel, M. Wistey, P. Fay, A. Seabaugh, and H. Xing, "Self-aligned $InAs/Al_{0.45}Ga_{0.55}Sb$ vertical tunnel FETs," in *Device Research Conference*, pp. 205–206, 2011.

86. K. Tomioka, J. Motohisa, S. Hara and T. Fukui, "Control of InAs nanowire growth directions on Si," *Nano Letters*, vol. 8, no. 10, pp. 3475–3480, 2008.

87. M.T. Björk, H. Schmid, C.D. Bessire, K.E. Moselund, H. Ghoneim, S. Karg, E. Lortscher, and H. Riel, "Si–InAs heterojunction Esaki tunnel diodes with high current densities," *Applied Physics Letters*, vol. 97, no. 16, pp. 163501-1–163501-3, 2010.

88. C.D. Bessire, M.T. Björk, H. Schmid, A. Schenk, K.B. Reuter, and H. Riel, "Trap-assisted tunneling in Si–InAs nanowire heterojunction tunnel diodes," *Nano Letters*, vol. 11, no. 10, pp. 4195–4199, 2011.

89. Y. Lu, "Geometry dependent tunnel FET performance—Dilemma of electrostatics *vs.* quantum confinement." in *Device Research Conference*, pp. 17–18, 2010.

90. H. Schmid, K.E. Moselund, M.T. Björk, M. Richter, H. Ghoneim, C.D. Bessire, and H. Riel, "Fabrication of vertical InAs–Si heterojunction tunnel field effect transistors," in *Device Research Conference*, pp. 181–182, 2011.

91. S. Poli, S. Reggiani, A. Gnudi, E. Gnani, and G. Baccarani, "Computational study of the ultimate scaling limits of CNT tunneling devices," *IEEE Transactions on Electron Devices*, vol. 55, no. 1, 313–321, 2008.

92. S.O. Koswatta, M.S. Lundstrom, and D.E. Nikonov, "Band-to-band tunneling in a carbon nanotube metal-oxide-semiconductor field-effect transistor is dominated by phonon-assisted tunneling," *Nano Letters*, vol. 7, no. 5, pp. 1160–1164, 2007.

93. J. Appenzeller, Y.-M. Lin, J. Knoch, Z. Chen, and P. Avouris, "Comparing carbon nanotube transistors—The ideal choice: a novel tunneling device design," *IEEE Transactions on Electron Devices*, vol. 52, no. 12, pp. 2568–2576, 2005.

94. M. Luisier and G. Klimeck, "Performance limitations of graphene nano ribbon tunneling FETS due to line edge roughness," in *Device Research Conference*, pp. 201–202, 2009.

95. G. Fiori and G. Iannaccone, "Ultralow-voltage bilayer graphene tunnel FET," *IEEE Electron Device Letters*, vol. 30, no. 10, pp. 1096–1098, 2009.

96. D. Kim, Y. Lee, J. Cai, I. Lauer, L. Chang, S.J. Koster, D. Sylvester, D. Blaauw, "Low power circuit design based on heterojunction tunneling transistors (HETTs)," in *Proceedings of the 2009 ACM/IEEE International Symposium on Low power Electronics and Design*, pp. 219–224, 2009.

97. V. Saripalli, G. Sun, A. Mishra, Y. Xie, S. Datta, and V. Narayanan, "Exploiting heterogeneity for energy efficiency in chip multiprocessor," *IEEE Transactions on Emerging and Selected Topics in Circuits and Systems*, vol. 1, no. 2, pp. 109–119, 2011.

98. Y. Hong, Y. Yang, L. Yang, G. Samudra, C.H. Heng, and Y.C. Yeo, "SPICE behavioral model of the tunneling field-effect transistor for circuit simulation," *IEEE Transactions on Circuits and Systems–II: Express Briefs*, vol. 56, no. 12, pp. 946–950, 2009.

99. M.S. Kim, H. Liu, X. Li, S. Datta, and V. Narayanan, "Design and implementation of low power mixed signal using a steep slope tunnel FET based SAR analog-to-digital converters," *IEEE Transactions on Electron Devices*, vol. 61, no. 11, pp. 3661–3667, 2014.

100. A.S. Verhulst, B. Soree, D. Leonelli, W.G. Vandenberghe, and G. Groeseneken, "Modeling the single-gate, double-gate, and gate-all-around tunnel field-effect-transistor," *Journal of Applied Physics*, vol. 107, no. 2, pp. 024518-1–024518-8, 2010.

101. M.G. Bardon, H.P. Neves, R. Puers, and C.V. Hoof, "Pseudo two-dimensional model for double-gate tunnel FETs considering the junctions depletion regions," *IEEE Transactions on Electron Devices*, vol. 57, no. 4, pp. 827–834, 2010.

102. P. Solomon, D.J. Franke, and S.O. Koswatta, "Compact model and performance estimation for tunneling nanowire FET," in *69th Annual Device Research Conference*, pp. 197–198, 2011.

103. L. Zhang, J. He, and M. Chan, "A compact model for double-gate tunneling field-effect-transistors and its implications on circuit behaviors," in *IEEE International Electron Devices Meeting Technical Digest*, pp. 143–146, 2012.
104. A. Pan and C.O. Chui, "A quasi-analytical model for double-gate tunneling field-effect transistors," *IEEE Electron Device Letters*, vol. 33, no. 10, pp. 1468–1470, 2012.
105. L. Zhang, X. Lin, J. He, and M. Chan, "An analytical charge model for double-gate tunneling FETs," *IEEE Transactions on Electron Devices*, vol. 59, no. 12, pp. 3217–3223, 2012.
106. B. Bhushan, K. Nayak, and V.R. Rao, "DC compact model for SOI tunnel field-effect transistors," *IEEE Transactions on Electron Devices*, vol. 59, no. 10, pp. 2635–2642, 2012.
107. L. Zhang and M. Chan, "SPICE modeling of double-gate tunnel-FETs including channel transports," *IEEE Transactions on Electron Devices*, vol. 61, no. 2, pp. 300–307, 2014.
108. E. Arnold, "Charge-sheet model for silicon carbide inversion layers," *IEEE Transactions on Electron Devices*, vol. 46, no. 3, pp. 497–503, 1999.
109. X. Aymerich-Humet, F. Serra-Mestres, and J. Millan, "A generalized approximation of the Fermi–Dirac integrals," *Journal of Applied Physics*, vol. 54, no. 5, pp. 2850–2851, 1983.
110. Y. Taur, X. Liang, W. Wang, and H. Lu, "A continuous, analytic drain current model for DG MOSFETs," *IEEE Electron Device Letters*, vol. 25, no. 2, pp. 107–109, 2004.
111. P.M. Solomon, J. Jopling, D.J. Frank, C.D'Emic, O. Dokumaci, P. Ronsheim, W. Haensch, "Universal tunneling behavior in technologically relevant P/N junction diodes," *Journal of Applied Physics*, vol. 95, no. 10, pp. 5800–5812, 2004.
112. E.O. Kane, "Theory of tunneling," *Journal of Applied Physics*, vol. 32, no. 1, pp. 83–91, 1961.
113. Q. Huang, R. Huang, Z. Zhan, Y. Qiu, W. Jiang, C. Wu, and Y. Wang, "A novel Si tunnel FET with 36 mV/dec subthreshold slope based on junction depleted-modulation through striped gate configuration," in *IEEE International Electron Devices Meeting Technical Digest*, pp. 187–190, 2012.
114. H. Riel, K.E. Moselund, C. Bessire, M.T. Bjork, A. Schenk, H. Ghoneim, and H. Schmid, "InAs-Si heterojunction nanowire tunnel diodes and tunnel FETs," in *IEEE International Electron Devices Meeting Technical Digest*, pp. 391–394, 2012.

Chapter 11

1. D.J. Roulston, *Bipolar Semiconductor Devices*, McGraw-Hill, New York, 1990.
2. Y. Taur and T.H. Ning, *Fundamentals of Modern VLSI Devices*, Cambridge University Press, Cambridge, 1998.
3. J. Bardeen and W.H. Brattain, "The transistor, a semiconductor triode," *Physical Review*, vol. 74, no. 2, pp. 230–231, 1948.
4. J.J. Ebers and J.L. Moll, "Large-signal behavior of junction transistors," *Proceedings of the IRE*, vol. 42, no. 12, pp. 1761–1772, 1954.
5. I. Getreu, *Modeling the Bipolar Transistor*, Tektronix, Beaverton, OR, 1976.
6. H.K. Gummel and C.C Poon, "An integral charge control model for bipolar transistors," *Bell System Technical Journal*, vol. 49, no. 5, pp. 827–852, 1970.

7. L.W. Nagel, "SPICE2—A computer program to simulate semiconductor circuits," Memo ERL-M250, Electronic Research Laboratory, University of California, Berkeley, CA, May 1975.
8. J.M. Early, "Effects of space charge layer widening in junction transistors," *Proceedings of the IRE*, vol. 40, no. 11, pp. 1401–1406, 1952.
9. C.T. Kirk, Jr., "A theory of transistor cutoff frequency (fT) fall-off at high current densities," *IRE Transactions on Electron Devices*, vol. 9, no. 2, pp. 164–174, 1962.
10. C.T. Sah, R.N. Noyce, and W. Shockley, "Carrier generation and recombination P-N junctions and P-N junction characteristics," *Proceedings of the IRE*, vol. 45, no. 9, pp. 1228–1243, 1957.
11. Si2. n.d. "Si2 (CMC) homepage." https://www.si2.org/cmc_index.php, access date: May 26, 2015.
12. C.C. McAndrew, J.A. Seitchik, D.F. Bowers, M. Dunn, M. Foisy, I. Getreu, M. McSwain, S. Moinian, J. Parker, D.J. Roulston, M. Schroter, P. van Wijnen, and L.F. Wagner, "VBIC95, the vertical bipolar inter-company model," *IEEE Journal of Solid-State Circuits*, vol. 31, no 10, pp. 1476–1483, 1996.
13. R. van der Toorn, T.C.J. Paasschens, and W.J. Kloosterman, "The Mextram bipolar transistor model level 504.11.0," NXP Semiconductors 2006 and Delft University of Technology, the Netherlands, 2012.
14. M. Schroter and A. Chakravorty, *Compact Hierarchical Bipolar Transistor Modeling with HiCUM*, World Scientific, Singapore, 2010.

Chapter 12

1. Synopsys Inc., HSPICE User Guide: Simulation and Analysis, B-2008.09, Synopsys Inc., September 2008.
2. N. Paydavosi, T.H. Morshed, D.D. Lu, W. Yang, M.V. Dunga, X. Xi, J. He, W. Liu, K. cao, X. Jin, J.J. Ou, M. Chan, A.M. Niknejad, and C. Hu, "BSIM4v4.8.0 MOSFET Model User's Manual," University of California, Berkeley, CA, 2013.
3. *BSIMProPlus* Basic Operations Guide, Product Version 2009.4, December 2009, ProPlus Design Solutions, Inc.,
4. Agilent 85190A *IC-CAP User's* Guide, 2008, Agilent Technologies. http://cp.literature.agilent.com/litweb/pdf/iccap2008/pdf/icug.pdf.
5. S.K. Saha, "Managing technology CAD for competitive advantage: An efficient approach for integrated circuit fabrication technology development," *IEEE Transactions on Engineering Management*, vol. 46, no. 2, pp. 221–229, 1999.
6. S.K. Saha, "Modeling process variability in scaled CMOS technology," *IEEE Design & Test of Computers*, vol. 27, no. 2, pp. 8–16, 2010.
7. S.K. Saha, "Compact MOSFET modeling for process variability-aware VLSI circuit design," *IEEE Access*, vol. 2, pp. 104–115, 2014.
8. S.K. Saha, "Process variability modeling for VLSI circuit simulation," in *Proceedings of the WCM*, vol. 2, pp. 751–755, 2011.
9. S.K. Saha, "General methodology to develop statistical compact MOS models for VLSI circuit simulation," http://www.mos-ak.org/india/presentations/Saha_MOS-AK_India12.pdf, accessed March 16, 2012.

Index